Nanofiltration for Sustainability

This book provides a novel exploration of the application of nanofiltration membrane technology for sustainability in various industries, situated in view of recent breakthroughs and the use of reuse, recycle and resource recovery approaches.

Moving from a comprehensive discussion of nanofiltration membrane processes to case studies and real-world applications of nanofiltration technology across society, both successes and potential limitations are considered.

Features:

- Detailed discussion of the fundamentals of nanofiltration technology
- The concepts of reuse, recycle and resource recovery using nanofiltration technology are explored in combination with other technologies to advance circular economy
- Considered across a range of industries, such as textiles, oil, gas, agriculture and pharmaceutics

Written in a thoroughly detailed manner, this book is an essential guide for industry professionals interested in sustainability and working toward a circular economy. Comprehensive discussions of the fundamental processes underpinning nanofiltration technology also make this book particularly appealing to students of industrial chemistry.

Abdul Wahab Mohammad is currently Dean, College of Engineering, University of Sharjah, UAE and Professor in the Chemical and Water Desalination Engineering Programme at the University of Sharjah. He has published more than 350 journal papers with citation exceeding 20,000 and h-index of 64. He is the co-Chief Editor of the *Journal of Water Process Engineering*. Abdul Wahab is a registered Professional Engineer (PEng) in Malaysia and a Chartered Engineer (CEng) in the United Kingdom. He is a Fellow of IChemE and a Fellow of the Academy of Sciences Malaysia.

Teow Yeit Haan is currently an Associate Professor at Universiti Kebangsaan Malaysia and the Coordinator of Water Solutions and Water Technology research area at the Research Centre for Sustainable Process Technology, Universiti Kebangsaan Malaysia. She has published more than 70 journal papers, 4 books and 10 book chapters, and owns 4 patents. She is a registered Professional Engineer (PEng) in Malaysia, an Associate Member of the Institute of Chemical Engineers (IChemE), a member of Young Scientist Network-Academy of Science Malaysia (YSN-ASM), Deputy Chairman of Chemical Engineering Technical Division in The Institution of Engineers Malaysia (IEM) and a Member of Malaysia Membrane Society (MyMembrane).

Nidal Hilal is a Global Network Professor at New York University and the Founding Director of NYUAD Water Research Center. He held professorships at the University of Nottingham and Swansea University in the United Kingdom. He is also an Emeritus Professor of Engineering at Swansea University and the Founding Director of the Centre for Water Advanced Technologies and Environmental Research (CWATER). He is a Chartered Engineer in the UK and an Elected Fellow of both the Institution of Chemical Engineers and the Learned Society of Wales. He was awarded the 2020 Menelaus Medal by the Learned Society of Wales for excellence in engineering and technology. In 2005, he was awarded a DSc from the University of Wales and the Kuwait Prize for Applied Science "Water Resources Development". Hilal has been named in the Highly Cited Researchers 2022 list by Clarivate.

Nanofiltration for Sustainability
Reuse, Recycle and Resource Recovery

Edited by
Abdul Wahab Mohammad,
Teow Yeit Haan and
Nidal Hilal

CRC Press
Taylor & Francis Group
Boca Raton London New York

CRC Press is an imprint of the
Taylor & Francis Group, an **informa** business

Designed cover image: © Canva

First edition published 2024
by CRC Press
6000 Broken Sound Parkway NW, Suite 300, Boca Raton, FL 33487-2742

and by CRC Press
4 Park Square, Milton Park, Abingdon, Oxon, OX14 4RN

CRC Press is an imprint of Taylor & Francis Group, LLC

ISBN: 978-1-032-19949-8 (hbk)
ISBN: 978-1-032-20002-6 (pbk)
ISBN: 978-1-003-26182-7 (ebk)

DOI: 10.1201/9781003261827

Typeset in Times
by codeMantra

Contents

Contributors

Jamaliah Aburabie
NYUAD Water Research Center
New York University Abu Dhabi
Abu Dhabi, United Arab Emirates

Nor Naimah Rosyadah Ahmad
Faculty of Engineering and Built Environment
Centre for Sustainable Process Technology
(CESPRO)
Universiti Kebangsaan Malaysia
Selangor, Malaysia

Neveen AlQasas
Division of Engineering
NYUAD Water Research Center
New York University Abu Dhabi
Abu Dhabi, United Arab Emirates

Míriam Cristina Santos Amaral
Department of Sanitary and Environmental
Engineering
Universidade Federal de Minas Gerais
Belo Horizonte, Brazil

Wei Lun Ang
Department of Chemical and Process
Engineering
Faculty of Engineering and Built Environment
Universiti Kebangsaan Malaysia
Selangor, Malaysia
and
Faculty of Engineering and Built Environment
Centre for Sustainable Process Technology
(CESPRO)
Universiti Kebangsaan Malaysia
Selangor, Malaysia

Alyza Azzura Abd Rahman Azmi
Faculty of Ocean Engineering Technology &
Informatics
Environmental Sustainable Material Research
Interest Group
Universiti Malaysia Terengganu
Terengganu, Malaysia
and
Faculty of Science and Marine Environment
Universiti Malaysia Terengganu
Terengganu, Malaysia

Alper Baba
Civil Engineering Department
Izmir Institute of Technology
Izmir, Turkey

Pui Vun Chai
Department of Chemical & Petroleum
Engineering
Faculty of Engineering, Technology and Built
Environment
UCSI University
Kuala Lumpur, Malaysia

Zhen Hong Chang
Department of Chemical and Petroleum
Engineering
Faculty of Engineering, Technology and Built
Environment
UCSI University
Kuala Lumpur, Malaysia

Aydin Cihanoğlu
Chemical Engineering Department
Ege University
Izmir, Turkey

Flávia Cristina Rodrigues Costa
Department of Sanitary and Environmental
    Engineering
Universidade Federal de Minas Gerais
Belo Horizonte, Brazil

Hazlini Dzinun
Centre for Diploma Studies
Universiti Tun Hussein Onn Malaysia
Johor, Malaysia

Asma Eskhan
Division of Engineering
NYUAD Water Research Center
New York University Abu Dhabi
Abu Dhabi, United Arab Emirates

Nazlee Faisal Ghazali
Faculty of Engineering
School of Chemical and Energy Engineering
Universiti Teknologi Malaysia
Johor, Malaysia

Pei Sean Goh
Advanced Membrane Technology Research
    Centre (AMTEC)
Universiti Teknologi Malaysia
Johor, Malaysia

Enver Güler
Chemical Engineering Department
Atılım University
Ankara, Turkey

Nur Hanis Hayati Hairom
Microelectronics and Nanotechnology –
    Shamsudin Research Center
Institute for Integrated Engineering
Universiti Tun Hussein Onn Malaysia
Johor, Malaysia

Sofiah Hamzah
Faculty of Ocean Engineering Technology &
    Informatics
Environmental Sustainable Material Research
    Interest Group
Universiti Malaysia Terengganu
Terengganu, Malaysia
and
Faculty of Science and Marine Environment
Universiti Malaysia Terengganu
Terengganu, Malaysia

Nick Hankins
Department of Engineering Science
The University of Oxford
Oxford, United Kingdom

Zawati Harun
Faculty of Mechanical and Manufacturing
    Engineering
Universiti Tun Hussein Onn Malaysia
Johor, Malaysia

Raed Hashaikeh
NYUAD Water Research Center
New York University Abu Dhabi
Abu Dhabi, United Arab Emirates

Nidal Hilal
Division of Engineering
NYUAD Water Research Center
New York University Abu Dhabi
Abu Dhabi, United Arab Emirates

Kah Chun Ho
Faculty of Engineering, Built Environment and
    Information Technology
Centre for Water Research
SEGi University
Selangor Darul Ehsan, Malaysia

Ahmad Fauzi Ismail
Advanced Membrane Technology Research
    Centre (AMTEC)
Universiti Teknologi Malaysia
Johor, Malaysia

Yakubu A. Jarma
Chemical Engineering Department
Ege University
Izmir, Turkey

Daniel James Johnson
NYUAD Water Research Center
New York University Abu Dhabi
Abu Dhabi, United Arab Emirates

Nalan Kabay
Chemical Engineering Department
Ege University
Izmir, Turkey

Aleksandra Kasztelewicz
Mineral and Energy Economy Research
    Institute
Polish Academy of Science
Kraków, Poland

Sevde Korkut
Environmental Engineering Department
Istanbul Technical University
Istanbul, Turkey
and
National Research Center on Membrane
    Technologies
Istanbul Technical University
Istanbul, Turkey

Ismail Koyuncu
Environmental Engineering Department
Istanbul Technical University
Istanbul, Turkey
and
National Research Center on Membrane
    Technologies
Istanbul Technical University
Istanbul, Turkey

Li Sze Lai
Department of Chemical & Petroleum
    Engineering
Faculty of Engineering, Technology and Built
    Environment
UCSI University
Kuala Lumpur, Malaysia
and
UCSI-Cheras Low Carbon Innovation Hub
    Research Consortium
Kuala Lumpur, Malaysia

Woei Jye Lau
Advanced Membrane Technology Research
    Centre (AMTEC)
Universiti Teknologi Malaysia
Johor, Malaysia

Leow Hui Ting Lyly
Department of Chemical and Process
    Engineering
Faculty of Engineering and Built Environment
Universiti Kebangsaan Malaysia
Selangor, Malaysia
and
Faculty of Engineering and Built Environment
Centre for Sustainable Process Technology
    (CESPRO)
Universiti Kebangsaan Malaysia
Selangor, Malaysia

Ki Min Lim
Faculty of Engineering
School of Chemical and Energy Engineering
Universiti Teknologi Malaysia
Johor, Malaysia

Aida Isma M. I.
Faculty of Engineering, Built Environment and
    Information Technology
Centre for Water Research
SEGi University
Selangor Darul Ehsan, Malaysia

Rais Hanizam Madon
Faculty of Engineering Technology
Universiti Tun Hussein Onn Malaysia
Johor, Malaysia

Thomas McKean
Ralph E Martin, Department of Chemical
    Engineering
University of Arkansas
Fayetteville, Arkansas

Abdul Wahab Mohammad
Chemical and Water Desalination Engineering
    Program
College of Engineering, University of Sharjah
Sharjah, United Arab Emirates
and
Department of Chemical and Process
    Engineering
Faculty of Engineering and Built Environment
Universiti Kebangsaan Malaysia
Selangor, Malaysia

Dharshini Mohanadas
Department of Chemical and Process
    Engineering
Faculty of Engineering and Built Environment
Universiti Kebangsaan Malaysia
Selangor, Malaysia
and
Faculty of Engineering and Built Environment
Centre for Sustainable Process Technology
    (CESPRO)
Universiti Kebangsaan Malaysia
Selangor, Malaysia

Ummi Kalsum Hasanah Mohd Nadzim
Faculty of Engineering Technology
Universiti Tun Hussein Onn Malaysia
Johor, Malaysia

Haya Nassrullah
NYUAD Water Research Center
New York University Abu Dhabi
Abu Dhabi, United Arab Emirates
and
Chemical and Biomolecular Engineering
    Department
Tandon School of Engineering
New York University
New York, New York

Nadiene Salleha Mohd Nawi
Advanced Membrane Technology Research
    Centre (AMTEC)
Universiti Teknologi Malaysia
Johor, Malaysia

Fozia Parveen
Department of Engineering Science
The University of Oxford
Oxford, United Kingdom

Nagarajan R. Periasamy
Department of Chemical and Process
    Engineering
Faculty of Engineering and Built Environment
Universiti Kebangsaan Malaysia
Selangor, Malaysia
and
Faculty of Engineering and Built Environment
Centre for Sustainable Process Technology
    (CESPRO)
Universiti Kebangsaan Malaysia
Selangor, Malaysia

Rosiah Rohani
Department of Chemical and Process
    Engineering
Faculty of Engineering and Built Environment
Universiti Kebangsaan Malaysia
Selangor, Malaysia
and
Faculty of Engineering and Built Environment
Centre for Sustainable Process Technology
    (CESPRO)
Universiti Kebangsaan Malaysia
Selangor, Malaysia

N. F. M. Roli
Faculty of Chemical and Process Engineering
    Technology
Universiti Malaysia Pahang
Pahang, Malaysia

Carolina Rodrigues dos Santos
Department of Sanitary and Environmental
    Engineering
Universidade Federal de Minas Gerais
Belo Horizonte, Brazil

S. M. Saufi
Faculty of Chemical and Process Engineering
    Technology
Universiti Malaysia Pahang
Pahang, Malaysia

Mei Qun Seah
Advanced Membrane Technology Research
    Centre (AMTEC)
Universiti Teknologi Malaysia
Johor, Malaysia

M. N. Abu Seman
Faculty of Chemical and Process Engineering
    Technology
Universiti Malaysia Pahang
Pahang, Malaysia
and
Earth Resources and Sustainability (ERAS)
    Center
Universiti Malaysia Pahang
Pahang, Malaysia

Dilaeleyana Abu Bakar Sidik
Centre for Diploma Studies
Universiti Tun Hussein Onn Malaysia
Johor, Malaysia

Jing Yao Sum
Department of Chemical & Petroleum
    Engineering
Faculty of Engineering, Technology and Built
    Environment
UCSI University
Kuala Lumpur, Malaysia

Yeit Haan Teow
Department of Chemical and Process
    Engineering
Faculty of Engineering and Built Environment
Universiti Kebangsaan Malaysia
Selangor, Malaysia
and
Faculty of Engineering and Built Environment
Centre for Sustainable Process Technology
    (CESPRO)
Universiti Kebangsaan Malaysia
Selangor, Malaysia

Barbara Tomaszewska
Mineral and Energy Economy Research
    Institute
Polish Academy of Science
Kraków, Poland
and
Faculty of Geology, Geophysics and
    Environmental Protection
AGH-University of Science and Technology
Kraków, Poland

Vahid Vatanpour
Environmental Engineering Department
Istanbul Technical University
Istanbul, Turkey
and
National Research Center on Membrane
    Technologies
Istanbul Technical University
Istanbul, Turkey
and
Department of Applied Chemistry
Faculty of Chemistry
Kharazmi University
Tehran, Iran

Ranil Wickramasinghe
Ralph E Martin, Department of Chemical
    Engineering
University of Arkansas
Fayetteville, Arkansas

Chin Yin Ying
Faculty of Engineering Technology
Universiti Tun Hussein Onn Malaysia
Johor, Malaysia

Ayse Yuksekdag
Environmental Engineering Department
Istanbul Technical University
Istanbul, Turkey
and
National Research Center on Membrane
    Technologies
Istanbul Technical University
Istanbul, Turkey

H. W. Yussof
Faculty of Chemical and Process Engineering
    Technology
Universiti Malaysia Pahang
Pahang, Malaysia

Zhiyuan Zong
Department of Engineering Science
The University of Oxford
Oxford, United Kingdom

1 Role of Nanofiltration Process for Sustainability in Industries
Reuse, Recycle, and Resource Recovery

Wei Lun Ang
Universiti Kebangsaan Malaysia

Abdul Wahab Mohammad
Universiti Kebangsaan Malaysia
University of Sharjah

Nor Naimah Rosyadah Ahmad and Yeit Haan Teow
Universiti Kebangsaan Malaysia

CONTENTS

1.1 INTRODUCTION

According to the United Nations, as much as 2 billion of the world's population are facing with the issues of water security, where dwindling clean water resources and limited water supply could threaten the health and social development of the affected community [1–3]. The drastic increase in demand for clean water associated with the rapid population growth and accelerated industrialization and urbanization, as well as the illegal discharge of pollutants to the waterways and uncontrollable climate change have severely upset the capability of the water utilities to produce clean water that meets the demand of people and economy [4,5]. The water crisis requires collective efforts from all stakeholders to prevent the situation from further deteriorating and harming the ecosystems and human well-being.

Sustainable development could be one competitive strategy to address the issues of water scarcity. Back in 2015, Agenda 2030 for Sustainable Development that includes 17 Sustainable Development Goals (SDGs) has been endorsed and embraced by all United Nations Member States to attain sustainable development [6]. In response to the issues of water scarcity, SDG 6 has been dedicated to water, with the aim to ensure availability and sustainable management of water and sanitation for all. This can be achieved through various approaches, such as to supply safe and affordable drinking water for

DOI: 10.1201/9781003261827-1

1

all, to reduce water pollution through wastewater treatment and minimum release of hazardous chemicals and materials to waterway, to improve water use efficiency across all sectors, and to promote integrated water resource management and reuse technologies [7]. Considering industrial water use composed of a huge portion of total water consumption worldwide, systematic and proper management of industrial effluent could significantly contribute to sustainable development in water industry.

Wastewater management strategies of various hierarchical can be adopted to achieve sustainable development and meet the call for Agenda 2030. For instance, proper treatment of wastewater could prevent the release of harmful pollutants from entering waterways, protecting the ecosystem and water resources [8]. Technological advancement in the past decades not only further improves the wastewater treatment efficiency but also enables the reclamation of treated water for reuse purposes. This could help to cut down the demand on clean water for operation and minimize wastewater discharge from the industries, which is a major step toward sustainable water management in the industry sectors [9]. Recently, a paramount shift has been observed toward the management of wastewater, where wastewater is no longer seen as a waste but could be an alternative source for various resources or valuable compounds. This is particularly interesting as apart from water, wastewater contains certain constituents (e.g., nutrients, organic compounds, and minerals) that could be recovered and reused, provided a proper treatment system is installed to manage the effluent [10]. Such recovery and reuse of resources from wastewater meet the concept of circular economy, align well with the model of production and consumption highly promoted by the government to achieve the aim of sustainable development.

All the aforementioned sustainable wastewater management strategies could contribute to sustainable development in water industry, and nanofiltration (NF) membrane process could directly or indirectly play a key role in attaining the sustainability status in water industry. Past studies have indicated the capability of NF system to function as a treatment process for the removal of pollutants and recovery of water and other valuable resources for reuse purposes or act as an intermediate/enabler process to alter the composition of the effluents into separate streams for subsequent treatment and reclamation purposes [11].

Furthermore, NF process could also be an alternative option over existing conventional processes which are less environmentally friendly. The replacement of conventional processes by NF will enable the industry to operate more sustainably, such as having lower energy consumption, releasing lesser greenhouse gases, and preserving the quality of products. Though not as widely adopted as NF being employed for wastewater treatment, the utilization of NF in industrial processes reflects the potential of NF in promoting sustainable industrialization. Overall, by employing NF in the treatment process or in operation process, it can help various industry sectors such as textile, food, oil and gas, tannery, pharmaceutical, and biorefinery to attain sustainable development.

1.2 SUSTAINABILITY ASPECTS

In this section, the role of NF process in promoting sustainable practices in various industries will be discussed in two categories: NF as enabler for resource recovery and reuse and NF as enabler for alternative sustainable operation and process. The former category is the application of NF to recover valuable resources from various medium, particularly industrial wastewater, for reuse purposes or as a feedstock for other industries. The second category highlights the potential of NF process as an alternative option to the current conventional processes in the industry, in which the NF process could be more sustainable especially in terms of energy consumption and preservation of products' quality. It has to be reminded that this chapter only discusses the contribution of NF process in various industries toward sustainable development. Technical details of the processes would be covered in Chapter 2.

1.2.1 ENABLER FOR RESOURCE RECOVERY AND REUSE

In this section, the role of NF in assisting various industries to achieve resource recovery and reuse from industrial effluents will be discussed. The basic level of wastewater management is the prevention of pollutants from entering waterways by removing it from wastewater. This is to ensure the discharged wastewater complies with the stringent regulation set by the authority and free from pollutants that could harm the ecosystem and living organisms. For instance, NF has been shown to be capable of rejecting various pollutants and preventing those compounds from passing to treated water, such as dyes in textile effluent [12], dissolved solids in dairy effluent [13], phenolic compounds in olive mill wastewater [14], dissolved minerals in produced water [15], metal ions in acid mine drainage [16], tannins in tannery effluent [17], and multivalent ions in pulp and paper industrial effluent [18]. The rejection of pollutants by NF could help to protect the waterways and minimize the impact of industrial wastewater on the environment.

Realizing that wastewater is a source for alternative valuable resources, industry operators have started to seek for the recovery of these resources for reuse purposes. In most cases, NF has been integrated with other treatment technologies to deliver satisfactorily overall removal and treatment efficiencies, which will be further elaborated in Chapter 2. The most widely accepted practices are the recovery of treated water for reuse inside or outside the plants with the aim to cut down the consumption of fresh water. For instance, treated water (through integrated NF process) meeting the desired quality could be reused for housekeeping (equipment washing and floor cleaning) [19–21], irrigation and landscaping [22–24], feedwater for cooling water and steam generation [13,25,26], and processing [15,27–31]. Unlike reverse osmosis membrane that could basically reject all impurities, NF membrane has a constraint in removing pollutants with size smaller than its pores. Nonetheless, the reduced quantity of pollutants in the permeate could be eliminated by integrating post-treatment process after the NF, such as UV irradiation or ozonation which could be installed to further degrade the resilient compounds in NF permeate [32].

For some other cases, the treated water contains constituents that could enable specific reuse purposes. Religa et al. showed that the negatively charged NF membrane could facilitate permeation of chloride ions, and the chloride-rich permeate could be reused as pickling baths in the tannery industry [33]. Agtas et al. demonstrated that the installation of NF after ultrafiltration (UF) treatment has resulted in higher contaminant removal efficiency when treating textile wastewater [34]. The interesting part of their study was that NF, being a membrane that allows the passage of monovalent ions, enabled at least 50% sodium recovered in the permeate. The recovered caustic solution could be reused in the causticization process, which could potentially contribute to caustic recovery of 480 m^3/year, while caustic usage cost could be reduced by 50% if the recovered caustic solution is reused in the processing.

Similar economic benefit has also been reported by Santos et al. where the NF pilot plant was shown to be capable of recovering caustic solution from spent caustic solution in crude oil refinery [35]. The purified permeate caustic solution can be recycled back for further reuse in the refinery process, and economic analysis revealed that such strategy could save about 1.5 M€ per year for the oil refinery industry. The special characteristic of NF membranes that reject larger impurities but allow the passage of monovalent ions and water molecules also benefits the mining industry. Studies have shown that NF allowed high HSO_4^- anion permeation (>82%) when treating mining effluent, producing permeate stream rich in sulfuric acid which was further concentrated by reverse osmosis to 99% [29,36]. The recovered sulfuric acid with high purity could be reused in the mining production process. Hence, it can be shown that the incorporation of NF process enables the recovery of other important compounds from the wastewater, and the reuse of these resources could lead to cost savings and sustainable resource consumption for the industries.

There are other compounds in wastewater that could be recovered as valuable products. Phytotoxic phenolic compounds are normally found in olive mill wastewater. Their antioxidant and anti-inflammatory properties have attracted the attention of food and cosmetic industries. In this context, NF could be employed to separate phenolic compounds from other impurities in olive mill effluent by allowing them to pass to permeate side [37]. Upon further purification, the phenolic compounds could be recovered as value-added feedstock for cosmetic and pharmaceutical industries. However, further study has to be conducted to validate the economic feasibility of recovering phenolic compounds from olive mill effluent. In another study, Seip et al. employed NF as pre-treatment process prior to ion-exchange sorbents for lithium recovery from flowback-produced water [38]. Even though NF membrane did not directly take part in recovering the lithium, its presence as pre-treatment process enabled the removal of small organic molecules that could disrupt the sorbent of lithium in the ion-exchange process. Indirectly, the efficiency of lithium recovery could be enhanced, and this contributes to sustainable resource management since lithium is of high demand with the surge of battery production for electric cars.

A major percentage of phosphate from wastewater is normally transferred into the sewage sludge during the treatment process. This indicates that the phosphorus locked in the sewage sludge could be recovered and reused as fertilizer. Blöcher et al. have developed an integrated phosphorous recovery process consisted of low-pressure wet oxidation, UF, and NF [39]. The dissolved phosphate ($H_2PO_4^-$) ended in NF permeate since the membrane could not remove monovalent ion. However, after passing through the UF and NF processes, the filtered permeate was free from other impurities. This enabled up to 54% phosphorus recovery and the costs of the entire integrated process were as competitive as conventional sewage sludge disposal. Furthermore, additional benefits such as recovery of valuable phosphorus resource and reduced greenhouse emission could also be attained.

Generally, the unwanted impurities in wastewater are rejected by NF membrane and collected as retentate for subsequent treatment and disposal. The impurities concentrated in retentate stream enable the reduction in treatment volume and cost of other technologies. For instance, NF membrane could retain up to 99% of perfluorohexanoic acid (persistent contaminant) in retentate stream, which facilitated the subsequent electrooxidation process to degrade the pollutants into harmless compounds [40]. Without the concentration effect of NF membrane, electrooxidation process will have to deal with a huge volume of wastewater with trace amount of pollutants, which does not appear to be economically feasible. Similar benefit brought by NF process has also been reported in the handling of pharmaceutical pollutants. The presence of NF process helped to concentrate the pharmaceuticals in retentate and subsequently facilitated more efficient degradation of advanced oxidation processes (such as photo-Fenton and ozonation) [41–43]. This reflects that the incorporation of NF in the treatment process could aid in reducing physical footprint and cost of other treatment technologies since the effluent has been conditioned to smaller volume.

On the other hand, the rejected impurities could also be turned into valuable resources for other applications. For instance, NF membrane is known to be a process with great capability in rejecting multivalent ions and retaining the heavy metals in retentate. These heavy metals could be recovered and utilized in the industry. Muller et al. showed that NF membrane treating acid mine drainage could help to recover valuable metals such as copper in the retentate stream [16]. It was estimated that copper loss of around $69,000/year could be saved if the heavy metals were recovered. Chromium ions that have been used in tanning process in tannery industry could also be recovered by NF in retentate stream [33,44]. The chromium-rich retentate could be reused in tanning process after further purification, assisting the tannery industry to cut down the expenses on fresh chromium resource. Chen et al. reported that NF membrane could retain lactose in retentate when filtering dairy wastewater [45]. The lactose-rich retentate was found to be a better source for anaerobic fermentation to produce a higher proportion of volatile fatty acids and biogas for further utilization. Lupanine – a toxic alkaloid found in lupin beans – wastewater could be a valuable resource in pharmaceutical industry since it can be utilized as starting material for the production of other alkaloids (such as sparteine) [31]. The capability of NF membrane to reject as much as 99.5% of lupanine from lupin beans wastewater could produce a feedstock stream of lupanine for pharmaceutical industry once the retentate is further purified and extracted with other technologies.

NF could be used to recover sulfate ions (in retentate) from acid mine drainage and tannery effluent. With its >99% rejection of sulfate ions from acid mine drainage, the tight NF membrane produced retentate rich in sulfate content (~4,000 mg/L) that could be mixed with flowback water to remove barium and strontium ions from flowback water via sulfate precipitation process [25]. Galiana-Aleixandre et al. estimated that with 97% sulfate retention by NF when treating tannery

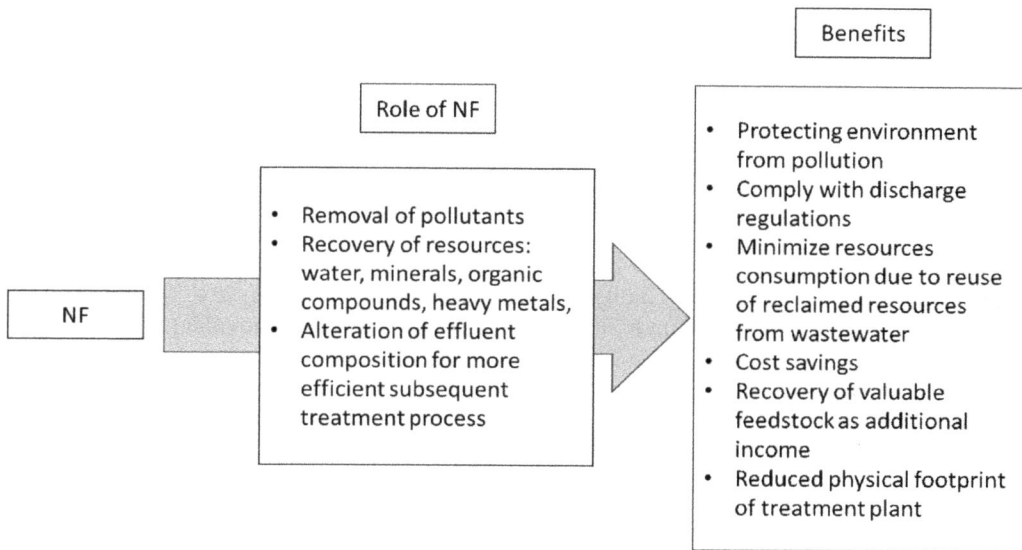

FIGURE 1.1 Role of NF and benefits obtained with the utilization of NF in the processing.

effluent could lead to the recovery of 61.63 t of sulfate per year which can be recycled and reused in tanning drums [46]. This indicates the huge potential to save the operational cost in the processing if NF membrane is installed to enable the recovery and reuse of resources from effluent. In textile industry, the dye effluent consists of high concentration of dyes and salts that would be a waste if treatment process adopted is to degrade and dispose the effluent. However, NF process could be utilized to fractionate the dye and salts, and enable the recovery of water, dye, and salts for reuse purposes. Lin et al. showed that NF process was capable to enrich the dye in retentate from 2.01 to 17.9 g/L while allowing most NaCl to pass through it [47]. The permeate rich in NaCl was then further processed by electrodialysis to separate NaCl from water. Eventually, the integrated process enabled resource extraction and reuse from high-salinity textile wastewater.

Lactic acids and amino acids are two main components that can be found in biorefinery fermentation broth. Both these acids are useful precursors with a wide variety of applications in pharmaceutical, food, and biotechnology products. Separation of these compounds from the fermentation broth is required for subsequent purification and utilization. In this scenario, NF membrane could be used to recover lactic acids and amino acids from fermentation broth, and then separated them into permeate and retentate streams, respectively, for further processing [48]. The NF membrane was also shown to be capable to concentrate the purity of lactic acid up to 85.6% [49]. Figure 1.1 summarizes the role of NF process and the associated benefits with the incorporation of NF in the processing.

1.2.2 ENABLER FOR ALTERNATIVE SUSTAINABLE OPERATION AND PROCESS

As discussed earlier, NF process enables the recovery of resources for reuse or other applications and alters the constituents in permeate or retentate for subsequent treatment or handling of impurities. The contribution of NF process to sustainable development in industry could also be reflected through the replacement of conventional processes by NF process that appears to be more operationally friendlier, for instance, the recovery of xylose from hemicellulose hydrolysate stream. The conventional method of xylose separation is through chromatographic method where the separation process is considered tedious and complex. To address these issues, Sjöman et al. showed that NF could be used to enrich xylose in permeate with xylose as up to 78%–89% of the total dry solids in permeate [50]. The purity of xylose was reported to increase by 1.4–1.7-fold after the NF process, reflecting the separation potential of NF for the xylose recovery from hemicellulose hydrolysate.

Conventional thermal process for the clarification and concentration of juices or liquid products is known to consume a large amount of energy and result in the thermal degradation of beneficial compounds [51,52]. Membrane process (including NF) could be an alternative process that offers several benefits such as less manpower requirement, greater concentration efficiency, shorter processing time, and better quality preservation for compounds susceptible to thermal degradation. For instance, NF has been shown to concentrate various fruit juices such as watermelon, bergamot, orange, pomegranate, and roselle and increase the amount of useful bioactive compounds (e.g., lycopene, flavonoid, ascorbic acid, anthocyanins, and phenolic contents) [53–57]. The elevated concentration of bioactive compounds in the juices with preserved antioxidant activity will be beneficial to the consumers. NF process could also remove haze precursors for juice clarification. Due to its selective property, haze precursors will be retained by NF while allowing the passage of sugars to permeate [58]. This maintains the taste, stability, storage ability, and lightness of the clarified juices.

NF membrane is also found to be an alternative yet competitive process to handle organic solvent separation and purification applications. The membrane process is also known as organic solvent nanofiltration (OSN) and has a great potential to be applied for the management of organic solvents in fine chemical, pharmaceutical, petrochemical, and food industries [59–61]. The advantages of OSN over conventional processes (e.g., preparative chromatography, distillation, extraction, and crystallization) include energy efficiency, cost effectiveness and protection of compounds susceptible to thermal degradation [62]. For instance, conventional extraction and isolation method (such as solid-liquid extraction) of active constituents from the spices and aromatic herbs typically converted the extracts into powder form through evaporation, which could be detrimental to the active constituents as thermal degradation would lead to the loss of desired antioxidant activity [63]. To address this challenge, OSN could be adopted to isolate the active compounds while maintaining the antioxidant property. One example was the extraction of rosmarinic acid (extracts of rosemary) via OSN [64]. The rosmarinic extracts concentrated in NF retentate may be applied directly as preservative and functional ingredient in the foods, cosmetics, nutraceuticals, and medicines since the antioxidant property was preserved after OSN process.

Apart from protecting the valuable compounds, the recovery of solvent via OSN could also be achieved. OSN can be integrated with active pharmaceutical ingredients (API) production process to purify API and recover organic solvent for reuse. Sereewatthanawut et al. demonstrated OSN could remove more than 99% of the impurities from API, producing high-purity API while at the same time reducing the fresh solvent consumption due to the reuse of recovered solvent from OSN [65]. In another industry, ExxonMobil's Beaumont (Texas) refinery process reported promising findings from its large-scale industrial OSN plant for the recovery of dewaxing solvents from lube oil filtrates [66–68]. The OSN process managed to recover solvent mixture (toluene and MEK) with high purity (99%), which could be recycled directly to the chilled feed stream. The installation of OSN also brought many economic benefits, including the reduction of energy consumption per product unit by 20%, increment of the average base oil production by 25 vol.%, reduction of the volatile organic compounds' emissions, and cutdown in the use of cooling water. This proved the potential of NF process in helping the industry to progress toward sustainable development by improving the efficiency of resource consumption (solvent, cooling water, and energy), reducing greenhouse gases emission and saving cost.

The introduction of OSN enables the adoption of more sustainable practices in the industry. Bio-derived solvents such as terpenes are good replacement over conventional solvent (n-hexane) for vegetable oil extraction industry. However, the conventional solvent recovery technology would be quite costly to recover the bio-solvent since high operation temperature is required due to the high boiling points and heat of vaporization of terpenes. Operating the solvent recovery process at high temperature also deteriorates the quality (antioxidant property) of the oil [69,70]. In this context,

```
┌─────────────────────────┐
│     Benefits of NF as   │
│   alternative process   │
└─────────────────────────┘
```

```
┌──────────────────────────────────────────────┐
│  •  Ease of operation                          │
│  •  Efficient separation, filtration, and recovery │
│     processes                                  │
│  •  Lower energy consumption                   │
│  •  Protect valuable compounds susceptible to  │
│     thermal degradation                        │
│  •  Cost savings                               │
│  •  Recovery of solvents                       │
│  •  Promote sustainable practices in industry  │
└──────────────────────────────────────────────┘
```

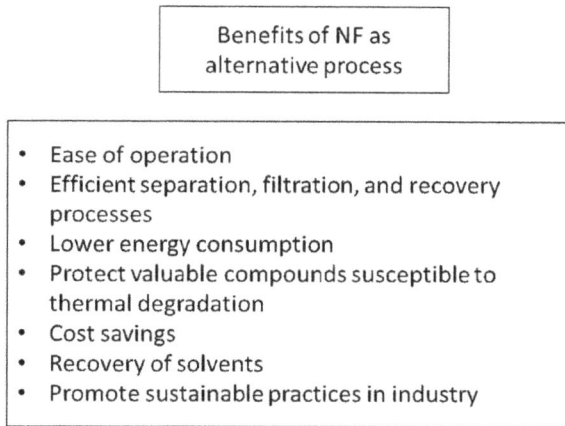

FIGURE 1.2 Benefits of NF as an alternative process.

OSN process that operates at low temperature could be an alternative process for the recovery of terpenes solvent, as shown by Abdellah et al. in canola oil extraction [71]. OSN also offers a more cost-effective process for the recovery of catalyst from organic solvents in hydroformylation and hydrogenation reactions. Conventional distillation method is costly and could degrade the catalytic activity during the recovery stage [72]. The adoption of OSN enabled the recovery of catalysts in retentate, and technological evaluation simulation revealed that recovery using OSN could potentially cut down the energy and costs by 85% and 75%, respectively, as compared to conventional distillation method [73].

NF process has also been used as an alternative technology to produce clean drinking water. The drinking water treatment plant with NF installed as tertiary process could enhance the overall treatment efficiency by removing organic matter and pesticides that conventional treatment processes failed to remove [74]. It was reported that the operational cost of having NF as tertiary treatment process was lower than a traditional plant with refining using ozone and carbon. Similar observation has also been reported by García-Vaquero et al. where compared to conventional treatment system (coagulation, flocculation, sedimentation, filtration, and filtration), NF process possessed greater treatment capability with better removal efficiency toward micropollutants in water [75]. Additional benefits such as the removal of hardness and color could also be acquired with the use of NF process for groundwater treatment as compared to conventional aeration and filtration processes [76,77]. The NF process could replace the softening and carbon filtration and deliver similar treatment efficiency in a single process. Hence, adoption of NF over conventional processes could lead to a more sustainable operation in water industry.

Figure 1.2 depicts the benefits of NF as alternative process. Despite the technical feasibility of NF process in resource recovery and reuse and as an alternative process for more sustainable operation in various industries, more study is required to understand its performance in real field and the associated overall cost. This information will further validate the feasibility of NF or integrated NF process in assisting the industry to attain sustainable development.

1.3 SUSTAINABILITY ASSESSMENT

Knowing the extent of NF process in improving the sustainability of industries, be it in resource recovery and reuse or operation process, would definitely help to convince the decision makers or stakeholders to adopt the NF in their operation for attaining better sustainability status. However, such information is scarce, and the sustainability claims of NF process are normally made solely based on technical or economic feasibility without support by quantitative data. It is to be clarified that sustainability is a complex concept that does not only focus on technical feasibility but also include other issues associated with the process itself, such as greenhouse gas emissions, energy consumption, land use, costs, removal of pollutants, public acceptance, reliability, and many more factors [78]. Past study has demonstrated that the typical cost-oriented selection approach did not necessarily give the best choice of wastewater treatment processes, as one has to balance between the treatment efficiency while meeting the sustainable criteria (e.g., reduce greenhouse gas emissions and energy consumption) [79]. Typically, the treatment process that enables resource recovery (such as biogas valorization and water recovery) would achieve higher ranking in the selection process due to the additional boost to sustainability from the perspective of sustainable resource recovery and consumption [80]. Other factors such as chemical consumption, energy consumption, and environmental emissions associated with the particular process could also affect the environmental and economic impacts caused by the adopted process. Therefore, the sustainability of NF process in various industries is also a complicated case, and the sustainable development brought by the adoption of NF process has to be properly evaluated before conclusion can be drawn on the sustainability status of the whole industry. Systematic assessment criteria will have to be developed and a comprehensive evaluation should be conducted to address this challenge.

The development of sustainability criteria is an arduous task as the criteria might vary between different processes or industries, small and large-scale processes, well-developed and developing or underdeveloped countries, and established or emerging new technologies [79,81,82]. Thus, the sustainability criteria have to be contextualized according to different scenarios since the NF can be adopted in the industries in different ways, either as enabler for resource recovery and reuse, enabler for the adoption of more sustainable practices in the operation, or as alternative technology over existing conventional processes, as discussed in the previous section. A few sustainability categories and indicators used in the sustainability assessment of wastewater treatment plants can be used as a guide to determine the sustainability of NF process in various industries [81–84]. These include technical category (treatment efficiency, compatibility with other processes, ease of implementation, process stability, reliability, complexity of construction and operation and maintenance, health and safety risks), environmental category (waste generation, energy use, global warming potential, land area, quality of effluent, potential recycling, release of chemical substances, CO_2 emission, water reuse potential, resource recovery, climate change, and human toxicity), economic category (investments, operating costs, capital costs, affordability, and cost effectiveness), and social category (odor impact, noise and visual impacts, public acceptance, employee satisfaction, expertise, local employment, safe and healthy conditions, and training). Life cycle assessment, coupled with the inputs from stakeholders, could provide the required information for the sustainability assessment. Hence, the contribution of NF process toward sustainable development in the industries could be measured by first developing the sustainability criteria for different scenarios followed by assessment through data collection. The finding would then be valuable to promote and convince the industry operators to adopt NF process in their operation for attaining sustainable development.

1.4   CONCLUSION

As demonstrated in the past studies, NF process could play a significant role in promoting resource recovery and reuse practices in various industries. These include the recovery of water and other resources such as minerals and acids that could be reused in the processing stages. The adoption of NF process also enables the recovery of valuable compounds as high-value raw materials in other industries. Furthermore, NF process could be an alternative option for existing conventional processes which are less environmentally friendly. All these indicate that the adoption of NF process could promote sustainable development in the industries. However, it is challenging to verify the extent of sustainability contribution from NF process. Sustainability criteria and assessment related to NF process in the industries should be developed such that solid evidence could be obtained and used to convince the industry operators to adopt NF process for a more sustainable operation.

REFERENCES

1. United Nations, Water, 2018. http://www.un.org/en/sections/issues-depth/water/.
2. United Nations, *The United Nations World Water Development Report 2017. Wastewater: The Untapped Resource*, Paris, UNESCO, 2017. http://unesdoc.unesco.org/images/0024/002471/247153e.pdf.
3. United Nations, *The United Nations World Water Development Report 2016: Water and Jobs*, Paris, UNESCO, 2016. ISBN 978–92–3–100146–8.
4. W.L. Ang, A.W. Mohammad, *Integrated and Hybrid Process Technology*, Elsevier Inc., 2019. https://doi.org/10.1016/b978-0-12-816170-8.00009–0.
5. W.L. Ang, A.W. Mohammad, Integrated and hybrid process technology, in: A.W. Mohammad, W.L. Ang (Eds.), *Integr. Hybrid Process Technol. Water Wastewater Treat.*, Elsevier, 2021: pp. 1–15. https://doi.org/10.1016/B978–0–12–823031–2.00027–6.
6. United Nations Development Programme, Sustainable Development Goals, 2021. https://www.undp.org/sustainable-development-goals.
7. United Nations Development Programme, Goal 6: Clean water and sanitation, 2019. https://www.undp.org/content/undp/en/home/sustainable-development-goals/goal-6-clean-water-and-sanitation.html.
8. W.L. Ang, W.J. Lau, Y.H. Tan, E. Mahmoudi, A.W. Mohammad, An overview on development of membranes incorporating branched macromolecules for water treatment, *Sep. Purif. Rev.* (2022) 1–23. https://doi.org/10.1080/15422119.2021.2008434.
9. G. Pérez, P. Gómez, I. Ortiz, A. Urtiaga, Techno-economic assessment of a membrane-based wastewater reclamation process, *Desalination*. 522 (2022) 115409. https://doi.org/10.1016/j.desal.2021.115409.
10. W.L. Ang, A.W. Mohammad, Recent development in nanofiltration process applications, in: N. Hilal, A.F. Ismail, M. Khayet, D.B.T.-O.E. Johnson (Eds.), *Osmosis Eng.*, Elsevier, 2021: pp. 97–129. https://doi.org/10.1016/B978–0–12–821016–1.00010–3.
11. N.N.R. Ahmad, W.L. Ang, Y.H. Teow, A.W. Mohammad, N. Hilal, Nanofiltration membrane processes for water recycling, reuse and product recovery within various industries: A review, *J. Water Process Eng.* 45 (2022) 102478. https://doi.org/10.1016/j.jwpe.2021.102478.
12. Y.K. Ong, F.Y. Li, S.-P. Sun, B.-W. Zhao, C.-Z. Liang, T.-S. Chung, Nanofiltration hollow fiber membranes for textile wastewater treatment: Lab-scale and pilot-scale studies, *Chem. Eng. Sci.* 114 (2014) 51–57. https://doi.org/10.1016/j.ces.2014.04.007.
13. L.H. Andrade, F.D.S. Mendes, J.C. Espindola, M.C.S. Amaral, Nanofiltration as tertiary treatment for the reuse of dairy wastewater treated by membrane bioreactor, *Sep. Purif. Technol.* 126 (2014) 21–29. https://doi.org/10.1016/j.seppur.2014.01.056.
14. J.M. Ochando-Pulido, J.R. Corpas-Martínez, A. Martinez-Ferez, About two-phase olive oil washing wastewater simultaneous phenols recovery and treatment by nanofiltration, *Process Saf. Environ. Prot.* 114 (2018) 159–168. https://doi.org/10.1016/j.psep.2017.12.005.
15. M. Gamal Khedr, Nanofiltration of oil field-produced water for reinjection and optimum protection of oil formation, *Desalin. Water Treat.* 55 (2015) 3460–3468. https://doi.org/10.1080/19443994.2014.939497.
16. M. Mullett, R. Fornarelli, D. Ralph, Nanofiltration of mine water: Impact of feed pH and membrane charge on resource recovery and water discharge, *Membrane*. 4 (2014). https://doi.org/10.3390/membranes4020163.
17. E.M. Romero-Dondiz, J.E. Almazán, V.B. Rajal, E.F. Castro-Vidaurre, Comparison of the performance of ultrafiltration and nanofiltration membranes for recovery and recycle of tannins in the leather industry, *J. Clean. Prod.* 135 (2016) 71–79. https://doi.org/10.1016/j.jclepro.2016.06.096.
18. M. Kamali, Z. Khodaparast, Review on recent developments on pulp and paper mill wastewater treatment, *Ecotoxicol. Environ. Saf.* 114 (2015) 326–342. https://doi.org/10.1016/j.ecoenv.2014.05.005.
19. C.F. Couto, L.S. Marques, M.C.S. Amaral, W.G. Moravia, Coupling of nanofiltration with microfiltration and membrane bioreactor for textile effluent reclamation, *Sep. Sci. Technol.* 52 (2017) 2150–2160. https://doi.org/10.1080/01496395.2017.1321670.
20. M.S.H. Ghani, T.Y. Haan, A.W. Lun, A.W. Mohammad, R. Ngteni, K.M.M. Yusof, Fouling assessment of tertiary palm oil mill effluent (POME) membrane treatment for water reclamation, *J. Water Reuse Desalin.* 8 (2017) 412–423. https://doi.org/10.2166/wrd.2017.198.
21. Y. Kaya, Z.B. Gönder, I. Vergili, H. Barlas, The effect of transmembrane pressure and pH on treatment of paper machine process waters by using a two-step nanofiltration process: Flux decline analysis, *Desalination*. 250 (2010) 150–157. https://doi.org/10.1016/j.desal.2009.06.034.
22. K. Li, Q. Liu, F. Fang, X. Wu, J. Xin, S. Sun, Y. Wei, R. Ruan, P. Chen, Y. Wang, M. Addy, Influence of nanofiltration concentrate recirculation on performance and economic feasibility of a pilot-scale membrane bioreactor-nanofiltration hybrid process for textile wastewater treatment with high water recovery, *J. Clean. Prod.* 261 (2020) 121067. https://doi.org/10.1016/j.jclepro.2020.121067.

23. A. Azaïs, J. Mendret, S. Gassara, E. Petit, A. Deratani, S. Brosillon, Nanofiltration for wastewater reuse: Counteractive effects of fouling and matrice on the rejection of pharmaceutical active compounds, *Sep. Purif. Technol.* 133 (2014) 313–327. https://doi.org/10.1016/j.seppur.2014.07.007.

24. J.M. Ochando-Pulido, M.D. Victor-Ortega, G. Hodaifa, A. Martinez-Ferez, Physicochemical analysis and adequation of olive oil mill wastewater after advanced oxidation process for reclamation by pressure-driven membrane technology, *Sci. Total Environ.* 503–504 (2015) 113–121. https://doi.org/10.1016/j.scitotenv.2014.06.109.

25. S.S. Wadekar, T. Hayes, O.R. Lokare, D. Mittal, R.D. Vidic, Laboratory and pilot-scale nanofiltration treatment of abandoned mine drainage for the recovery of products suitable for industrial reuse, *Ind. Eng. Chem. Res.* 56 (2017) 7355–7364. https://doi.org/10.1021/acs.iecr.7b01329.

26. P.B. Moser, B.C. Ricci, B.G. Reis, L.S.F. Neta, A.C. Cerqueira, M.C.S. Amaral, Effect of MBR-H_2O_2/UV Hybrid pre-treatment on nanofiltration performance for the treatment of petroleum refinery wastewater, *Sep. Purif. Technol.* 192 (2018) 176–184. https://doi.org/10.1016/j.seppur.2017.09.070.

27. J. Wang, K. Li, Y. Wei, Y. Cheng, D. Wei, M. Li, Performance and fate of organics in a pilot MBR–NF for treating antibiotic production wastewater with recycling NF concentrate, *Chemosphere.* 121 (2015) 92–100. https://doi.org/10.1016/j.chemosphere.2014.11.034.

28. M. Khosravi, G. Badalians Gholikandi, A. Soltanzadeh Bali, R. Riahi, H.R. Tashaouei, Membrane process design for the reduction of wastewater color of the mazandaran pulp-paper industry, Iran, *Water Resour. Manag.* 25 (2011) 2989–3004. https://doi.org/10.1007/s11269-011-9794-1.

29. B.C. Ricci, C.D. Ferreira, A.O. Aguiar, M.C.S. Amaral, Integration of nanofiltration and reverse osmosis for metal separation and sulfuric acid recovery from gold mining effluent, *Sep. Purif. Technol.* 154 (2015) 11–21. https://doi.org/10.1016/j.seppur.2015.08.040.

30. H. Chang, B. Liu, B. Yang, X. Yang, C. Guo, Q. He, S. Liang, S. Chen, P. Yang, An integrated coagulation-ultrafiltration-nanofiltration process for internal reuse of shale gas flowback and produced water, *Sep. Purif. Technol.* 211 (2019) 310–321. https://doi.org/10.1016/j.seppur.2018.09.081.

31. T. Esteves, A.T. Mota, C. Barbeitos, K. Andrade, C.A.M. Afonso, F.C. Ferreira, A study on lupin beans process wastewater nanofiltration treatment and lupanine recovery, *J. Clean. Prod.* 277 (2020) 123349. https://doi.org/10.1016/j.jclepro.2020.123349.

32. M.P. Lopes, C.T. Matos, V.J. Pereira, M.J. Benoliel, M.E. Valério, L.B. Bucha, A. Rodrigues, A.I. Penetra, E. Ferreira, V.V. Cardoso, M.A.M. Reis, J.G. Crespo, Production of drinking water using a multi-barrier approach integrating nanofiltration: A pilot scale study, *Sep. Purif. Technol.* 119 (2013) 112–122. https://doi.org/10.1016/j.seppur.2013.09.002.

33. P. Religa, A. Kowalik, P. Gierycz, Effect of membrane properties on chromium(III) recirculation from concentrate salt mixture solution by nanofiltration, *Desalination.* 274 (2011) 164–170. https://doi.org/10.1016/j.desal.2011.02.006.

34. M. Ağtaş, Ö. Yılmaz, M. Dilaver, K. Alp, İ. Koyuncu, Pilot-scale ceramic ultrafiltration/nanofiltration membrane system application for caustic recovery and reuse in textile sector, *Environ. Sci. Pollut. Res.* 28 (2021) 41029–41038. https://doi.org/10.1007/s11356-021-13588-0.

35. B. Santos, J.G. Crespo, M.A. Santos, S. Velizarov, Oil refinery hazardous effluents minimization by membrane filtration: An on-site pilot plant study, *J. Environ. Manage.* 181 (2016) 762–769. https://doi.org/10.1016/j.jenvman.2016.07.027.

36. R.L. Ramos, L.B. Grossi, B.C. Ricci, M.C.S. Amaral, Membrane selection for the Gold mining pressure-oxidation process (POX) effluent reclamation using integrated UF-NF-RO processes, *J. Environ. Chem. Eng.* 8 (2020) 104056. https://doi.org/10.1016/j.jece.2020.104056.

37. C.M. Sánchez-Arévalo, Á. Jimeno-Jiménez, C. Carbonell-Alcaina, M.C. Vincent-Vela, S. Álvarez-Blanco, Effect of the operating conditions on a nanofiltration process to separate low-molecular-weight phenolic compounds from the sugars present in olive mill wastewaters, *Process Saf. Environ. Prot.* 148 (2021) 428–436. https://doi.org/10.1016/j.psep.2020.10.002.

38. A. Seip, S. Safari, D.M. Pickup, A. V Chadwick, S. Ramos, C.A. Velasco, J.M. Cerrato, D.S. Alessi, Lithium recovery from hydraulic fracturing flowback and produced water using a selective ion exchange sorbent, *Chem. Eng. J.* 426 (2021) 130713. https://doi.org/10.1016/j.cej.2021.130713.

39. C. Blöcher, C. Niewersch, T. Melin, Phosphorus recovery from sewage sludge with a hybrid process of low pressure wet oxidation and nanofiltration, *Water Res.* 46 (2012) 2009–2019. https://doi.org/10.1016/j.watres.2012.01.022.

40. Á. Soriano, D. Gorri, A. Urtiaga, Efficient treatment of perfluorohexanoic acid by nanofiltration followed by electrochemical degradation of the NF concentrate, *Water Res.* 112 (2017) 147–156. https://doi.org/10.1016/j.watres.2017.01.043.

41. S. Miralles-Cuevas, I. Oller, J.A.S. Pérez, S. Malato, Removal of pharmaceuticals from MWTP effluent by nanofiltration and solar photo-Fenton using two different iron complexes at neutral pH, *Water Res.* 64 (2014) 23–31. https://doi.org/10.1016/j.watres.2014.06.032.

42. N. Klamerth, S. Malato, A. Agüera, A. Fernández-Alba, Photo-Fenton and modified photo-Fenton at neutral pH for the treatment of emerging contaminants in wastewater treatment plant effluents: A comparison, *Water Res.* 47 (2013) 833–840. https://doi.org/10.1016/j.watres.2012.11.008.

43. S. Miralles-Cuevas, F. Audino, I. Oller, R. Sánchez-Moreno, J.A. Sánchez Pérez, S. Malato, Pharmaceuticals removal from natural water by nanofiltration combined with advanced tertiary treatments (solar photo-Fenton, photo-Fenton-like Fe(III)–EDDS complex and ozonation), *Sep. Purif. Technol.* 122 (2014) 515–522. https://doi.org/10.1016/j.seppur.2013.12.006.

44. L.M. Gando-Ferreira, J.C. Marques, M.J. Quina, Integration of ion-exchange and nanofiltration processes for recovering Cr(III) salts from synthetic tannery wastewater, *Environ. Technol.* 36 (2015) 2340–2348. https://doi.org/10.1080/09593330.2015.1027284.

45. Z. Chen, J. Luo, X. Chen, X. Hang, F. Shen, Y. Wan, Fully recycling dairy wastewater by an integrated isoelectric precipitation–nanofiltration–anaerobic fermentation process, *Chem. Eng. J.* 283 (2016) 476–485. https://doi.org/10.1016/j.cej.2015.07.086.

46. M.-V. Galiana-Aleixandre, J.-A. Mendoza-Roca, A. Bes-Piá, Reducing sulfates concentration in the tannery effluent by applying pollution prevention techniques and nanofiltration, *J. Clean. Prod.* 19 (2011) 91–98. https://doi.org/10.1016/j.jclepro.2010.09.006.

47. J. Lin, Q. Chen, X. Huang, Z. Yan, X. Lin, W. Ye, S. Arcadio, P. Luis, J. Bi, B. Van der Bruggen, S. Zhao, Integrated loose nanofiltration-electrodialysis process for sustainable resource extraction from high-salinity textile wastewater, *J. Hazard. Mater.* 419 (2021) 126505. https://doi.org/10.1016/j.jhazmat.2021.126505.

48. J. Ecker, T. Raab, M. Harasek, Nanofiltration as key technology for the separation of LA and AA, *J. Memb. Sci.* 389 (2012) 389–398. https://doi.org/10.1016/j.memsci.2011.11.004.

49. J. Sikder, S. Chakraborty, P. Pal, E. Drioli, C. Bhattacharjee, Purification of lactic acid from microfiltrate fermentation broth by cross-flow nanofiltration, *Biochem. Eng. J.* 69 (2012) 130–137. https://doi.org/10.1016/j.bej.2012.09.003.

50. E. Sjöman, M. Mänttäri, M. Nyström, H. Koivikko, H. Heikkilä, Xylose recovery by nanofiltration from different hemicellulose hydrolyzate feeds, *J. Memb. Sci.* 310 (2008) 268–277. https://doi.org/10.1016/j.memsci.2007.11.001.

51. K. Nath, H.K. Dave, T.M. Patel, Revisiting the recent applications of nanofiltration in food processing industries: Progress and prognosis, *Trends Food Sci. Technol.* 73 (2018) 12–24. https://doi.org/10.1016/j.tifs.2018.01.001.

52. C. Bhattacharjee, V.K. Saxena, S. Dutta, Fruit juice processing using membrane technology: A review, *Innov. Food Sci. Emerg. Technol.* 43 (2017) 136–153. https://doi.org/10.1016/j.ifset.2017.08.002.

53. N.P. Kelly, A.L. Kelly, J.A. O'Mahony, Strategies for enrichment and purification of polyphenols from fruit-based materials, *Trends Food Sci. Technol.* 83 (2019) 248–258. https://doi.org/10.1016/j.tifs.2018.11.010.

54. C. Conidi, A. Cassano, F. Caiazzo, E. Drioli, Separation and purification of phenolic compounds from pomegranate juice by ultrafiltration and nanofiltration membranes, *J. Food Eng.* 195 (2017) 1–13. https://doi.org/10.1016/j.jfoodeng.2016.09.017.

55. M. Cissé, F. Vaillant, D. Pallet, M. Dornier, Selecting ultrafiltration and nanofiltration membranes to concentrate anthocyanins from roselle extract (Hibiscus sabdariffa L.), *Food Res. Int.* 44 (2011) 2607–2614. https://doi.org/10.1016/j.foodres.2011.04.046.

56. J. Warczok, M. Ferrando, F. López, C. Güell, Concentration of apple and pear juices by nanofiltration at low pressures, *J. Food Eng.* 63 (2004) 63–70. https://doi.org/10.1016/S0260-8774(03)00283-8.

57. C. Conidi, A. Cassano, E. Drioli, A membrane-based study for the recovery of polyphenols from bergamot juice, *J. Memb. Sci.* 375 (2011) 182–190. https://doi.org/10.1016/j.memsci.2011.03.035.

58. V. Vivekanand, M. Iyer, S. Ajlouni, Clarification and stability enhancement of pear juice using loose nanofiltration, *J. Food Process. Technol.* 3 (2012) 1–6.

59. M. Amirilargani, M. Sadrzadeh, E.J.R. Sudhölter, L.C.P.M. de Smet, Surface modification methods of organic solvent nanofiltration membranes, *Chem. Eng. J.* 289 (2016) 562–582. https://doi.org/10.1016/j.cej.2015.12.062.

60. S. Hermans, H. Mariën, C. Van Goethem, I.F.J. Vankelecom, Recent developments in thin film (nano) composite membranes for solvent resistant nanofiltration, *Curr. Opin. Chem. Eng.* 8 (2015) 45–54. https://doi.org/10.1016/j.coche.2015.01.009.

61. Y. Ji, W. Qian, Y. Yu, Q. An, L. Liu, Y. Zhou, C. Gao, Recent developments in nanofiltration membranes based on nanomaterials, *Chinese J. Chem. Eng.* 25 (2017) 1639–1652. https://doi.org/10.1016/j.cjche.2017.04.014.

62. G. Szekely, M.F. Jimenez-Solomon, P. Marchetti, J.F. Kim, A.G. Livingston, Sustainability assessment of organic solvent nanofiltration: From fabrication to application, *Green Chem.* 16 (2014) 4440–4473. https://doi.org/10.1039/C4GC00701H.

63. S. Başkan, N. Öztekin, F.B. Erim, Determination of carnosic acid and rosmarinic acid in sage by capillary electrophoresis, *Food Chem.* 101 (2007) 1748–1752. https://doi.org/10.1016/j.foodchem.2006.01.033.

64. D. Peshev, L.G. Peeva, G. Peev, I.I.R. Baptista, A.T. Boam, Application of organic solvent nanofiltration for concentration of antioxidant extracts of rosemary (Rosmarinus officiallis L.), *Chem. Eng. Res. Des.* 89 (2011) 318–327. https://doi.org/10.1016/j.cherd.2010.07.002.

65. I. Sereewatthanawut, F.W. Lim, Y.S. Bhole, D. Ormerod, A. Horvath, A.T. Boam, A.G. Livingston, Demonstration of molecular purification in polar aprotic solvents by organic solvent nanofiltration, *Org. Process Res. Dev.* 14 (2010) 600–611.

66. L.S. White, A.R. Nitsch, Solvent recovery from lube oil filtrates with a polyimide membrane, *J. Memb. Sci.* 179 (2000) 267–274. https://doi.org/10.1016/S0376–7388(00)00517–2.

67. R.M. Gould, L.S. White, C.R. Wildemuth, Membrane separation in solvent lube dewaxing, *Environ. Prog.* 20 (2001) 12–16. https://doi.org/10.1002/ep.670200110.

68. L.S. White, Development of large-scale applications in organic solvent nanofiltration and pervaporation for chemical and refining processes, *J. Memb. Sci.* 286 (2006) 26–35. https://doi.org/10.1016/j.memsci.2006.09.006.

69. R. Subramanian, M. Nakajima, K.S.M.S. Raghavarao, T. Kimura, Processing vegetable oils using nonporous denser polymeric composite membranes, *J. Am. Oil Chem. Soc.* 81 (2004) 313. https://doi.org/10.1007/s11746-004-0901-z.

70. S. Arora, S. Manjula, A.G. Gopala Krishna, R. Subramanian, Membrane processing of crude palm oil, *Desalination.* 191 (2006) 454–466. https://doi.org/10.1016/j.desal.2005.04.129.

71. M.H. Abdellah, L. Liu, C.A. Scholes, B.D. Freeman, S.E. Kentish, Organic solvent nanofiltration of binary vegetable oil/terpene mixtures: Experiments and modelling, *J. Memb. Sci.* 573 (2019) 694–703. https://doi.org/10.1016/j.memsci.2018.12.026.

72. A. Cano-Odena, P. Vandezande, D. Fournier, W. Van Camp, F.E. Du Prez, I.F.J. Vankelecom, Solvent-resistant nanofiltration for product purification and catalyst recovery in click chemistry reactions, *Chem. – A Eur. J.* 16 (2010) 1061–1067. https://doi.org/10.1002/chem.200901659.

73. W.L. Peddie, J.N. van Rensburg, H.C.M. Vosloo, P. van der Gryp, Technological evaluation of organic solvent nanofiltration for the recovery of homogeneous hydroformylation catalysts, *Chem. Eng. Res. Des.* 121 (2017) 219–232. https://doi.org/10.1016/j.cherd.2017.03.015.

74. B. Cyna, G. Chagneau, G. Bablon, N. Tanghe, Two years of nanofiltration at the Méry-sur-Oise plant, France, *Desalination.* 147 (2002) 69–75. https://doi.org/10.1016/S0011–9164(02)00578–7.

75. N. García-Vaquero, E. Lee, R. Jiménez Castañeda, J. Cho, J.A. López-Ramírez, Comparison of drinking water pollutant removal using a nanofiltration pilot plant powered by renewable energy and a conventional treatment facility, *Desalination.* 347 (2014) 94–102. https://doi.org/10.1016/j.desal.2014.05.036.

76. W.G.J. Van der Meer, J.C. Van Winkelen, Method for purifying water, in particular groundwater, under anaerobic conditions, using a membrane filtration unit, a device for purifying water, as well as drinking water obtained by such a method, 2002.

77. P. Hiemstra, J. van Paassen, B. Rietman, J. Verdouw, Aerobic versus anaerobic nanofiltration: Fouling of membranes, in: *Proc. AWWA Membr. Conf.*, Long Beach, CA, 1999.

78. W.L. Ang, A.W. Mohammad, Design approach and sustainability of advanced integrated treatment, in: *Integr. Hybrid Process Technol. Water Wastewater Treat.*, Elsevier, 2021. https://doi.org/10.1016/B978-0-12-823031-2.00001-X.

79. A. Gherghel, C. Teodosiu, M. Notarnicola, S. De Gisi, Sustainable design of large wastewater treatment plants considering multi-criteria decision analysis and stakeholders' involvement, *J. Environ. Manage.* 261 (2020) 110158. https://doi.org/10.1016/j.jenvman.2020.110158.

80. A. Arias, C.R. Behera, G. Feijoo, G. Sin, M.T. Moreira, Unravelling the environmental and economic impacts of innovative technologies for the enhancement of biogas production and sludge management in wastewater systems, *J. Environ. Manage.* 270 (2020) 110965. https://doi.org/10.1016/j.jenvman.2020.110965.

81. C. Cossio, J. Norrman, J. McConville, A. Mercado, S. Rauch, Indicators for sustainability assessment of small-scale wastewater treatment plants in low and lower-middle income countries, *Environ. Sustain. Indic.* 6 (2020) 100028. https://doi.org/10.1016/j.indic.2020.100028.

82. C. Seifert, T. Krannich, E. Guenther, Gearing up sustainability thinking and reducing the bystander effect – A case study of wastewater treatment plants, *J. Environ. Manage.* 231 (2019) 155–165. https://doi.org/10.1016/j.jenvman.2018.09.087.

83. M. Kamali, K.M. Persson, M.E. Costa, I. Capela, Sustainability criteria for assessing nanotechnology applicability in industrial wastewater treatment: Current status and future outlook, *Environ. Int.* 125 (2019) 261–276. https://doi.org/10.1016/j.envint.2019.01.055.

84. A. Padilla-Rivera, L.P. Güereca, A proposal metric for sustainability evaluations of wastewater treatment systems (SEWATS), *Ecol. Indic.* 103 (2019) 22–33. https://doi.org/10.1016/j.ecolind.2019.03.049.

2 Applying Nanofiltration for Sustainability
Case Studies in Various Industries

Nor Naimah Rosyadah Ahmad
Universiti Kebangsaan Malaysia
Universiti Tenaga Nasional

Abdul Wahab Mohammad
Universiti Kebangsaan Malaysia
University of Sharjah

Wei Lun Ang and Yeit Haan Teow
Universiti Kebangsaan Malaysia

CONTENTS

2.1   INTRODUCTION

Water is an essential resource for civilization and its function is more important nowadays due to the climate change situation. Water consumption is predicted to rise annually, and it is anticipated that by 2030, the world will need about 6,900 billion m^3 of clean water. This amount of water is about 64% more than the amount of water that most countries have access to [1]. Concerning this issue, the desalination process and water recycling and reuse practice can be applied to increase the clean water supply in this world. It has been reported that water recycling and reuse operations have expanded significantly during the previous decade [2]. Several initiatives have been launched for treating various industrial wastewaters, including municipal sources, for either non-potable or potable applications. Currently, adsorption, membrane separation, and advanced oxidation technologies have all been utilized in the tertiary treatment of wastewater. Even so, more research exploration is

DOI: 10.1201/9781003261827-2

required to promote the water recycling and reuse practice. This is because a study by Global Water Market 2017 showed that only 1.2 billion m^3 of water per year have been recycled and reused as of 2017, corresponding to 4% of the total estimated wastewater [3].

Over the past years, membrane-based processes including microfiltration (MF), ultrafiltration (UF), nanofiltration (NF), reverse osmosis (RO), and membrane bioreactor (MBR) have been frequently applied for water recycling and reuse. Numerous sectors have embraced these membrane technologies due to their compact footprint, high efficiency, high productivity, modular design, and clean process. NF is a unique process compared to UF since it exhibits superior retention of small-size compounds (molecular weight of 200–2,000 Da) such as amino acids, peptides, and sugars [4]. Unlike RO, NF exhibits greater fluxes and selectivity toward divalent/polyvalent ions, while still enabling the passage of monovalent ions and smaller molecules less than 100 Da. Based on these properties, NF is a promising option for water recycling, reuse, and product recovery. Hence, this chapter is intended to discuss several case studies on the application of NF for water recycling, reuse, and product recovery from various types of wastewater including those from agriculture activities, and domestic and industrial sources.

Generally, NF is commonly integrated with other technologies for the purpose of water recycling, reuse, and product recovery in numerous types of wastewater treatment (Figure 2.1). Centrifugation, sedimentation, flocculation, electrocoagulation (EC), activated sludge, aeration, floatation, isoelectric precipitation, activated sludge process, UF, MF, MBR, and advanced oxidation process (AOP) are examples of water treatment techniques that can be used to pre-treat the wastewater before entering the NF unit. Installation of pre-treatment units is a common practice to prolong the membrane lifetime, minimize the fouling issue, and enhance the separation performance. Most of the reported works have demonstrated that using NF at the end of a treatment train is sufficient to generate permeate up to reusable water quality, depending on the reuse application. In some cases, such as in the textile industry, the NF permeate is further treated using ultraviolet (UV) disinfection before reuse in textile production to ensure safety. Meanwhile, the NF retentate can be post-treated using technologies such as resin adsorption, solvent extraction, and electrochemical treatment to recover valuable products from the concentrate stream, depending on the properties of the targeted compounds. The implementation

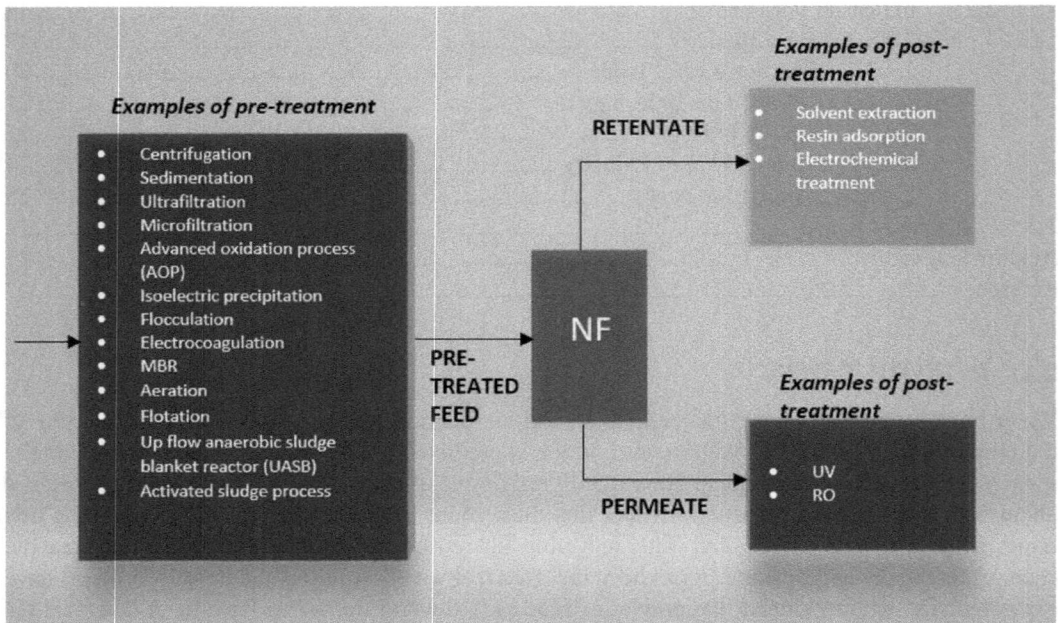

FIGURE 2.1   Overview of the integrated process comprising NF technology for water recycling, reuse, and product recovery in wastewater treatment.

of water recycling, reuse, and product recovery will help toward achieving sustainable development in various sectors such as agriculture and industrial activities. The details of the case studies regarding NF application in each sector are discussed in the following section.

2.2 APPLICATION OF NANOFILTRATION FOR REUSE, RECYCLING, AND RESOURCE RECOVERY (3R)

2.2.1 AGRICULTURAL WASTEWATER

The agricultural sector consumes considerable water for crop production and livestock watering. Crops such as rice, wheat, and maize are produced in large amounts throughout the world, resulting in increased water demand relative to overall crop production. Meanwhile, it has been reported that livestock consumes more water since they require a large amount of feed (crops), drinking water, and service water. Consequently, water shortage will affect irrigation for human food production and livestock production (in which water is required for both feeding and watering purposes) [5]. To minimize the depletion of natural water resources and enhance agricultural output, water reuse and recycling in agricultural activities should be promoted. Some regions of the world already practice water treatment and reuse for agricultural purposes. This may be seen in Spain where 71% of the treated wastewaters are used in agriculture [6]. Nevertheless, the high cost of water production is a major concern when using treated water. Thus, it is important to recover the wastewater at a low cost to promote wastewater reuse and recycling in agricultural activities.

Membrane technology is a promising option for wastewater reclamation. Previously, NF has been combined with other technologies for wastewater reclamation and reuse in agricultural activities. For instance, NF can be applied to recover water from slurry produced in high-density livestock farming. Basically, a slurry is a mixture of animal excrement, urea, and technological water that is used in agriculture. In addition to its high nutrient content (i.e., phosphorus and nitrogen), slurry also contains high water content (>90%) which can be extracted using pressure-driven membrane technology. Konieczny et al. [7] developed an integrated system comprised of centrifugation-UF-NF to recover water from slurry for reuse purposes. The application of NF after two steps of UF treatment was effective in obtaining final permeate with 33% of initial crude slurry volume where its permeate quality is suitable for reusing as sanitary safe industrial water such as for cleaning farmhouses and animals. In addition to waste volume reduction, the integrated process is able to maintain the nutrients properties of slurry in the form of the retentate which can be further applied as fertilizer on the farm or to generate biogas.

Besides nutrients, carboxylic acid is another potential resource that can be recovered from agricultural wastewater. Due to the negative charge and low molecular weight of carboxylic acid, NF is the most suitable process for its recovery, as NF leverages a variety of mechanisms, including steric exclusions (namely, size or molecular weight), shape, and charge. For instance, Zacharof et al. [8] compared the performance of five commercial NF membranes, namely HL, DL, DK, NF270 and LF10 for recovering carboxylic acid from pre-treated agricultural wastewater. In terms of retention and permeate flux, the best candidates for carboxylic acid separation and concentration from these complex effluents were NF270, DK, and DL membranes. These membranes can attain about 75% retention ratio with retentate concentrations up to 53.94 mM for acetate and 28.38 mM for butyrate. This finding is promising to encourage the shifting into chemicals recovery from the natural resource which can reduce the carbon footprint in the chemical production.

2.2.2 DOMESTIC WASTEWATER

To address the growing global dilemma of water scarcity, domestic wastewater (DWW) must be viewed as a potentially valuable water resource rather than as waste. Greywater is DWW produced from households, including kitchen sinks, bathrooms, washing machines, and laundry. The common pollutants in greywater include surfactants, dissolved organic/inorganic matter, particulate, and micropollutants [9]. At the moment, enormous amounts of wastewater are generated by households as a result of increased urbanization and population growth. It has been estimated that greywater make up about 40%–80% of all municipal wastewater [10,11]. Greywater reclamation has received a lot of attention in the previous two decades since no higher degree of treatment is required to treat this type of wastewater and it may have more public acceptance compared to municipal wastewater reuse [9].

Recently, membrane-based separation appears as a promising approach for greywater reclamation. In comparison to MF/UF, NF and RO processes can generate high-quality treated greywater due to their membrane properties. Several studies have shown the potential of NF in treating the laundry greywater from various sources such as industrial laundry and cruise ship [12,13]. For instance, a study conducted by Guilbaud et al. [12] applied direct NF process based on tubular membrane to reclaim water from simulated laundry greywater on board. The NF treatment at 35 bar and 25°C using AFC80 membrane was capable to generate a permeate with a quality that allows for the recycling of 80% greywater which can be reused as inlets to the washing machine. However, moderate fouling was reported and the techno-economic analysis indicated that water production is quite costly (5–6 €/m³ of treated water).

It has been proposed that the permeate quality from MBR is sufficient for processing water, but it does not fulfill the recycled water criteria (for laundry application) which require more stringent water quality [14]. Mozia et al. [13] showed that the treatment of laundry wastewater using a hybrid moving bed biofilm reactor (HMBBR) has reduced the total organic carbon (TOC) content by about 91%, but its final concentration was still higher than that of TOC concentration in freshwater for laundry activity (1.6 mg/L). Thus, an integrated process comprised of AOP-UF-NF was proposed to polish the HMBBR effluent quality [13]. The AOP process was used to mitigate the organic fouling issue in the pressure-driven membrane since this process can eliminate the organic compounds. Meanwhile, the NF treatment was to enhance the quality of UF permeate as it still has a high content of inorganic pollutants (conductivity higher than 2.3 mS/cm). The combination of all these technologies has successfully generated permeate with quality higher than the conventionally treated groundwater, allowing its reuse at any stage of the laundry process such as for washing or rinse cycle step.

Kitchen wastewater (KWW) is another type of greywater that has received much interest in wastewater treatment due to its oil and grease as well as organic pollutant contents. The release of untreated KWW causes detrimental consequences on the ecology, including eutrophication of water bodies and clogging of sewer lines. Meanwhile, the stagnant KWW promotes the spread of deadly diseases such as malaria, filariasis, and dengue [15]. Due to the current water crisis, the reuse of KWW has also been promoted. Conventional methods for KWW treatment such as activated sludge, coagulation, and chemical treatments require high energy and high cost, and could not produce high-quality reclaimed water [16,17]. Thus, Chandrasekhar et al. [18] proposed the integration of NF membrane into a side-stream MBR for treating the KWW up to reusable criteria (Figure 2.2). To improve the antifouling properties of NF, the polyamide membrane hydrophilicity was enhanced by coating with polyvinyl alcohol layer. It was found that such an integrated system has completely removed odor and color from the treated water while chemical oxygen demand (COD), total dissolved solids (TDS), total suspended solids, and turbidity were removed by 97.6%, 93.7%, 94.6%, and 98.5%, respectively. During continuous operation of the MBR for 12 days, permeate flux of 79 L/m²h and maximum water recovery of 82% were recorded. The system was successfully built on a pilot scale, producing highly cleaned water that can be reused for a variety of purposes, including landscaping irrigation, gardening, groundwater replenishment, and water body recovery. The MBR's aerobic configuration versatility allows it to operate in a variety of situations, which is a significant benefit for application in remote locations.

2.2.3 INDUSTRIAL WASTEWATER

2.2.3.1 Food Industry

The food industry including the dairy manufacturing industry consumes a huge amount of water in the production process, general cleaning, and housekeeping and sanitation activities. Annually, about 10 billion liters of dairy wastewater are emitted in Australia's Victoria province [19]. It has been reported that dairy wastewater exhibits a TDS range of 1,800–2,700 mg/L since it contains

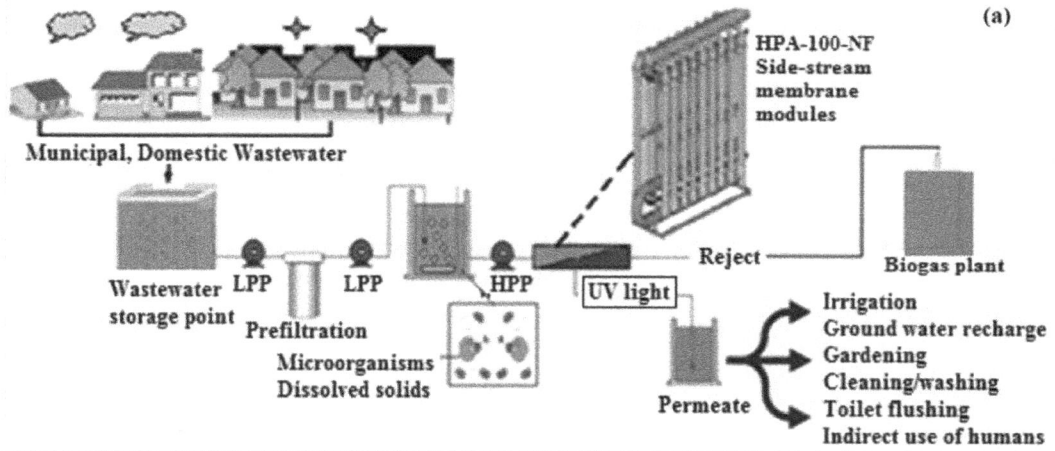

FIGURE 2.2 Integrated process for water recovery from kitchen wastewater [18]. (Copyright 2022. Reproduced with permission from Elsevier Science Ltd.)

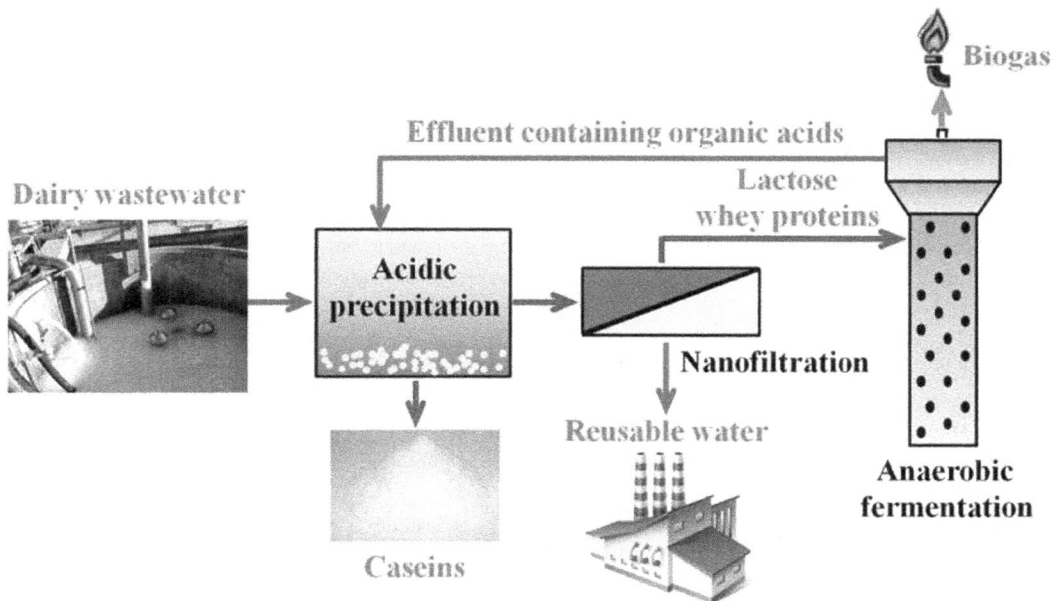

FIGURE 2.3 An integrated isoelectric precipitation-NF-anaerobic fermentation process for fully recycling dairy wastewater [21]. (Copyright 2016. Reproduced with permission from Elsevier Science Ltd.)

salts that originate from cheese-salting and whey-processing [19]. Instead of being used for treatment, the NF membrane might be used to recover water for reuse in the food-processing industry and can extract value-added compounds contained in the effluents. Several studies have integrated NF with other technologies for the water reuse target in the dairy industry [20,21]. In an integrated system, the implementation of the pre-treatment process prior to NF is essential to address the fouling issue. Caseins, the main protein in dairy effluent, can foul membrane operation. Thus, a study conducted by Chen et al. [21] applied the isoelectric precipitation as pre-treatment to remove the caseins from the dairy wastewater (Figure 2.3). Implementation of this pre-treatment step successfully minimized the fouling in NF which was indicated by the slight increment of transmembrane pressure (TMP) (from 2 to 3 bar) as compared to without pre-treatment (TMP rose from 2 to 34 bar). The retentate without caseins can be a resource for anaerobic fermentation to generate volatile fatty acids or biogas while the permeate from this integrated process could be reused in the plant. The NF process also has been used to further polish the permeate quality of MBR which treats the real dairy effluent. Even though the treated water from MBRs might be applicable for reuse in unrestricted irrigation or for recreational purposes after color removal, additional treatment such as NF is required after MBR process if the water reuse is targeted for industrial application [22]. It has been demonstrated in Andrade et al.'s study [20] that the integration of MBR with the NF process can remove 99% COD and 93% total solids (TS) from the dairy effluent where the NF permeate fulfilled the criteria for water standards in cooling and low-pressure steam generation.

Apart from the dairy industry, the olive mill industry also discharges a huge amount of wastewater. On a daily basis, a medium-sized olive oil factory produces tens of cubic meters of olive mill wastewater and $1\,m^3/t$ of washed olives (olive washing wastewater) [23]. This culminates in a yearly outflow of millions of cubic meters and the consumption of the same quantity of potable water. Conventional physicochemical treatments including natural precipitation, flocculation-sedimentation, Fenton-like reaction, and olive stone filtration in series were incapable of removing dissolved ions from olive mill effluent. This restricted the reuse or discharge of treated water into the environment. As an alternative, the installation of NF after these treatment steps can help to polish the permeate quality of the olive mill effluent up to reusable criteria as demonstrated by Ochando-Pulido, Victor-Ortega [24]. In their case, the final electrical conductivity value (1.5 mS/cm) for the generated permeate has fulfilled the standard water criteria for irrigation purposes according to the Food and Agriculture Organization.

Meanwhile, olive mill wastewater also contains phenolic compounds that are detrimental to the environment. Phenolic compounds are phytotoxic and antibacterial, making their decomposition under normal circumstances challenging and negatively affecting the survival of microorganisms and plants. Even so, the antioxidant and anti-inflammatory features of phenolic compounds are promising for application in the pharmaceutical, food, and cosmetics industries. Thus, olive mill effluent appears to be a viable source for the recovery of phenolic compounds that can be reused for various purposes [25,26]. For instance, a study conducted by Ochando-Pulido et al. [23] showed that the application of NF is capable to recover the phenols up to 75.7% (1,315.7 mg/L) in the retentate line, which could be a potential source to satisfy the demand in other industries. Since the profit from phenols recovery can potentially offset a portion of the NF process' capital and operating expenses, it could help encourage stakeholders to adopt the NF treatment in the food industry wastewater treatment. To confirm this benefit, however, pilot research and economic review should be done.

2.2.3.2 Pharmaceutical Industry

Pharmaceutical manufacturing encompasses the production, extraction, processing, purification, and packaging of chemical and biological ingredients as solids and liquids for use in human and animal medicine [27]. The pharmaceutical industry is expected to develop at a compound annual growth rate of 11.34% from 2021 to 2028, making it one of the fastest-growing industries [28]. Since water is primarily employed in pharmaceutical production processes such as chemical synthesis and fermentation stages, it is projected that more pharmaceutical wastewater will be generated in the future. Organic and inorganic pollutants, pharmaceutically active chemicals (PhACs) (e.g., tranquilizers, antibiotics, diuretics, and psychiatric drugs), and endocrine-disrupting compounds (EDCs) can all be found in pharmaceutical effluent [29]. Due to the detrimental effects of PhACs and EDCs to human health and the environment, removing these compounds from pharmaceutical wastewater has become one of the major concerns. Furthermore, the pharmaceutical industry has placed a greater emphasis on water reuse in recent years [30]. As a result, the development of effective technologies for treating pharmaceutical wastewater is crucial.

As most PhACs have a molecular weight larger than 250 Da [27], efficient removal of PhACs can be accomplished using membrane technologies such as NF. In this circumstance, tight NF with a low molecular weight cut-off (MWCO) (<500 Da) is a promising strategy that has been used to remove PhACs. Several studies have demonstrated the potential of tight NF membrane for pharmaceutical wastewater reclamation [31–33]. For instance, a study conducted by Azaïs et al. [32] showed that the commercial NF90 membrane was capable to remove more than 90% PhACs from the pretreated real effluent spiked with PhACs such as acetaminophen (ACT), atenolol, carbamazepine (CBZ), and diatrizoic acid. It was reported that the recovered water employing the NF90 membrane met the irrigation water requirement, indicating that a tight NF membrane such as the NF90 is a feasible option for wastewater reuse. The performance of NF90 membrane was also compared with NF200 and RO membrane during the treatment of synthetic wastewater containing numerous PhACs and EDCs [33]. Unlike NF200, NF90 demonstrated higher rejection due to a combination of electrostatic repulsion and size exclusion. Since the performance of NF90 in terms of removal of ionic and neutral compounds was comparable to RO membrane, it was suggested that the tight NF membranes can be a promising alternative to RO since they can also effectively remove PhACs or EDCs at a reduced operating cost.

Despite the good PhACs removal efficiency of tight NF membrane, the irreversible fouling remains a challenge in the operation of tight NF with low MWCO. To address this issue, researchers have proposed the use of loose NF with higher MWCO for treating pharmaceutical wastewater. For instance, NF50 membrane (HYDRACoRe) with MWCO of 1,000 Da is promising to recover water with minimal fouling since its negatively charged surface could repel the negatively charged drugs molecules. Maryam et al. [34] who assessed the performance of NF50 for the removal of PhACs from synthetic wastewater found that this loose NF membrane can remove 99.7% of Diclofenac at pH 3 and 80.5% of Ibuprofen at neutral pH. However, low rejection (36.2%) for Paracetamol was reported.

To enhance the quality of recovered water up to reusable water standard, NF can be integrated as a polishing unit after the MBR treatment of pharmaceutical wastewater [35,36]. Installation of a pilot-scale MBR-NF system at a pharmaceutical company in Wuxi, China have successfully produced permeate with turbidity of 0.15 NTU, conductivity of 2.5 mS/cm, TOC of 5.52 mg/L, and total phosphorus (TP) of 0.34 mg/L, complying with China's water quality standard for industrial usage [35]. High water recovery (92%) was attained by recycling part of the NF retentate to the MBR. Another work [36] used an ozone-based AOP process in between MBR and NF treatment to mitigate the NF fouling. The experiment was conducted by using the real MBR-effluent, taken from a wastewater treatment plant spiked with four pharmaceuticals (i.e., ACT, CBZ, sulfamethoxazole, and tetracyclin) and herbicide. The NF fouling resistance was reduced by about 40% since preozonation step has turned degraded organic matters into more hydrophilic components with less tendency for irreversible fouling.

Besides water recovery, NF has also been used to recover organic solvent and PhACs from waste stream generated by the pharmaceutical production process. For instance, it has been demonstrated that commercial membrane such as NF90 and Duramem 150 can retain more than 60% of 1-(5-bromo-fur-2-il)-2-bromo-2-nitroethane (G-1) from residual ethanol stream generated by G-1 purification, suggesting the NF potential for PhACs recovery [37]. Although the reuse of byproducts in the pharmaceutical industry is often limited by strict quality control rules, the waste exchange can be used as an alternative to encourage recycling and reuse. This means that the recovered products can be sent to another industry or company.

2.2.3.3 Oil and Gas Industry

In the oil and gas industry, the hydraulic fracturing activities for extracting oil and gas produce a large volume of waste, known as produced water (PW). The annual production of PW is estimated to exceed 70 billion barrels [38,39]. PW typically contains hydrocarbons, corrosion inhibitors, salts, dissolved organic carbon, heavy metals, suspended particles, and dissolved gases (e.g., H_2S and CO_2) [40]. Meanwhile, the oil refinery process also required a large amount of water for cracking, reforming, and topping activities. Approximately, 246–340 L of water is needed for every barrel of crude oil produced, emitting the effluent which 0.4–1.6 times the volume of oil processed [41]. To help the oil and gas industry minimize its reliance on freshwater, the reuse of treated water has been encouraged, which necessitates the development of effective systems for reclaiming water from PW and oil refinery effluent.

Numerous researches have been undertaken to determine the efficacy of NF in treating oil and gas industry wastewater for subsequent reuse [42–44]. In the case of PW treatment, most of the studies focused on the water recovery for internal reuse of hydraulic fracturing and reinjection purpose to enhance oil production. It has been reported that injecting improperly treated PW with a high SO_4^{2-} level would ruin the oil formation's porous structure. Therefore, Gamal Khedr [42] applied the coagulation/filtration-NF process for treating the PW taken from the Suez Gulf region. The role of NF treatment was to enhance the removal efficiency of SO_4^{2-}, uranium, and other hardness cations from PW. It was found that the use of an NF membrane as a polishing step was able to improve the quality of treated PW for reinjection, which is critical for preventing scale and corrosion in pipes. The divalent cations in the PW also need to be removed prior to internal reuse for hydraulic fracturing since their presence could affect the shale gas production due to the formation of stable carbonate and sulfate precipitation [45,46]. For this purpose, the application of the NF membrane is advantageous for treating PW because it is capable of removing divalent ions. Several studies have applied NF as a polishing step during PW recovery for hydraulic fracturing reuse [43,46]. For example, a study conducted by Chang et al. [46] demonstrated that the installation of NF after coagulation-UF treatment has attained 72.8%–91.7% removal efficiency for SO_4^{2-}, Mg^{2+}, Ca^{2+}, Sr^{2+}, and Ba^{2+} cations where the final permeate complied with the flowback PW reuse standard in Marcellus shale play.

In the context of resource recovery, it has been reported that PW is a potential source for lithium recovery [47]. Interest in lithium sources has grown in recent years because of its numerous applications, which include lubricant production, lithium-based air purification, ceramic and glass production, metallurgy, polymer products, and pharmaceuticals, to name a few [47]. A recent work by Seip et al. [48] showed that NF can be integrated with manganese-based ion-exchange sorbents for lithium recovery from flowback PW. In that study, the role of NF was to remove the small organic molecules (<250 Da) before the lithium recovery step by the manganese-based ion-exchange sorbents. The NF pre-treatment is important because the presence of these small organic molecules can reduce the manganese and lead to sorbent loss via reductive dissolution.

The potential of NF for water and sodium recovery in refinery industry effluent has been reported in several works [44]. In the Portuguese oil refinery, it has been identified that the spent caustic emitted from the kerosene caustic washing unit has contributed to the polar oil and grease contamination of wastewater. Since the spent caustic solution is toxic to bacteria used in wastewater treatment, it is essential to treat this type of effluent at its original source to enable recycling and reuse within refineries. For this purpose, Santos et al. [49] investigated the performance of composite polymeric NF (SeIRO® MPS-34) for treating the spent caustic solution originating from a kerosene caustic washing unit. It has been reported that the NF membrane operation at 15 bar was capable to remove polar oil and grease and COD by about 99.9% and 97.7%, respectively. Moreover, the NF treatment has successfully recovered the sodium on permeate side, enabling the reuse of purified caustic solution in the refinery process.

2.2.3.4 Textile Industry

The textile industry utilizes vast amounts of water throughout the various phases of manufacturing and generates up to 200–350 m^3 of wastewater per tonne of completed products [50]. A significant amount of color (dyes), COD, inorganic salts, suspended particles, and trace heavy metals are typically present in the wastewater created [51,52]. The persistence, carcinogenicity, and low biodegradability of dyes make them the most troublesome among these contaminants. Without proper treatment, the release of poorly treated textile wastewater will have a negative impact on the flora and fauna present in water bodies and consequently reduce the clean water resources. Innovative treatment processes are needed to deal with the textile effluent that contains recalcitrant dyes and inorganic salts, which could both be recovered and reused in the textile industry.

In general, size exclusion and Donnan exclusion effects could be used to explain the separation mechanism of NF for treating textile wastewater. It is easier for dyes with bigger molecular weights to be rejected by the NF membrane compared to inorganic salt ions. This can be reflected in Ji et al.'s study [53], where dyes with high molecular weight such as Congo Red and Direct Yellow were rejected by more than 99% and 97%, respectively, after passing through NF, while low-molecular-weight dye (Acid Orange 10) was removed about 78.5%. In another study [54], the anionic dye with a high molecular weight (RR195, 1,136.3 Da) was removed by about 97.2% compared to cationic dye with a lower molecular weight (MB, 319.8 Da, 90.2% removal) after passing through a negatively charged NF membrane. Besides size exclusion, the stronger electrostatic repulsion (Donnan exclusion) which repels anionic dyes from the negatively charged NF explains why the RR195 dye removal was higher.

Even though NF is promising for dye/salt separation, the viability of its process is hampered by fouling, which is a common issue in the membrane system. One of the feasible fouling mitigation options is the integrated process, which integrate the membrane unit with other processes in a treatment train [55]. The goal of an integrated process is to produce greater separation performance than any of the component processes by minimizing or eliminating the weaknesses of the stand-alone process [56]. Several works have integrated NF with other processes to reduce the fouling effect while producing reusable water in the textile industry [57–59]. For example, the feasibility of an MF-NF-AOP integrated system for water recovery from textile wastewater was studied by Lebron et al. [57]. The systems which comprise either MF-NF-AOP(c) (i.e., NF concentrate treated by AOP) or MF-AOP-NF were tested and compared to seek a suitable configuration for the textile wastewater treatment. Since the first configuration demonstrated a higher NF permeate flux (38 LMH) and requires a lower operating cost (0.421 US$/m^3) compared to the second configuration (flux of 19 LMH, 0.736 US$/m^3), it was proposed that it is more appropriate for textile effluent treatment due to its cost-effectiveness. The NF permeate from the system fulfilled the water criteria for equipment washing down and yarn washing-off, enabling its reuse. Besides MF, another process such as EC has been proposed as the pre-treatment before entering the NF unit. EC is a well-known technique for lowering COD to around 800 mg/L and removing a wide spectrum of dyes, and thus, can act as a primary treatment step to reduce the fouling propensity in the membrane process. In a study conducted by Tavangar et al. [60], the installation of EC has successfully reduced the foulant layer thickness on NF membrane since COD and suspended solids were removed by EC. As a result, the EC-NF integrated process has achieved a higher flux recovery ratio (67.99%) and steady membrane flux of almost 15 LMH as compared to the stand-alone NF process (11.68%, ~2 LMH). In the case of water reuse application in fabric dyeing activity, some studies have investigated the effect of recycled water quality on the quality of the fabric product to promote water reuse in the textile industry. For instance, a study by Cinperi et al. [61] reported that the reclaimed water from an integrated system comprised of MBR-NF-UV (Figure 2.4) showed no adverse effects on the product quality in the dyeing process, highlighting the potential of this process for water recycling and reuse.

In the context of product recovery, it has been reported that the NaOH can be extracted from highly alkaline effluent produced by caustic main bath discharges, which can be further reused in the causticization process [62]. For that purpose, a ceramic membrane is preferable over a polymeric membrane owing to its chemical stability when dealing with extreme pH conditions. Furthermore, ceramic membrane demonstrates high thermal stability, good mechanical strength, ease of cleaning procedure, and long membrane lifetime, making it as an interesting alternative to the polymeric membrane [63]. A recent study by Ağtaş et al. [62] showed that the application of commercial ceramic NF such as ATECH can be used to recover caustic chemicals from caustic-containing textile effluent. The combination of the UF and NF process successfully recovered at least 50% sodium while reducing the TOC, COD, color, and total hardness by about 67%, 71%, 92%, and 42%, respectively. Besides the sodium recovery, it was proposed that the reclaimed water could be reused in appropriate processes within the facility.

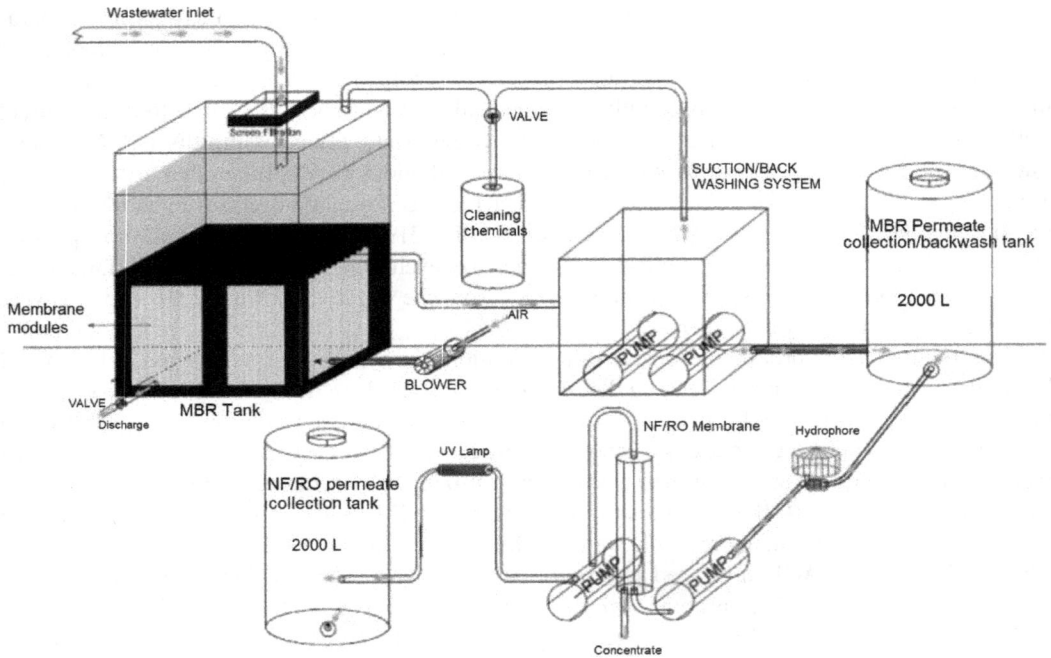

FIGURE 2.4 Integrated system comprised of MBR-NF-UV for water recycling and reuse in textile industry [61]. (Copyright 2019. Reproduced with permission from Elsevier Science Ltd.)

2.2.3.5 Pulp and Paper Industry

In the pulp and paper industry, the production of paper and pulp requires a substantial amount of clean water (273–455 m^3 per tonne of paper) and results in the discharge of a sizeable quantity of polluted wastewater (220–380 m^3 per tonne of paper) [64,65]. The types of contaminants and the composition of the effluent depend on the manufacturing process (i.e., pulping, bleaching, and papermaking) and the type of substrate (e.g., recycled paper, softwood, and hardwood) [66]. As an example, the papermaking process generates effluent with high COD, phosphate, TDS, sulfate, and chloride while the effluent from bleaching activity contains toxic contaminants including chlorinated organic compounds, phenols, and organic halogen which are harmful to live organisms [66]. Meanwhile, the effluent from pulping process possesses a high lignin level (11,000–25,000 mg/L), pH of 6.3–6.8, and COD of 500–115,000 mg/L.

Efforts have been made to limit the amount of freshwater consumption in the pulp and paper industry by developing innovative methods of water treatment, such as membrane technology, which can reclaim water for recycling and reuse in the production process. In this case, NF is one of the effective membrane treatments since it that may reject multivalent ions from pulp and paper manufacturing effluent. Conventional biological treatment such as the activated sludge process that has been used in most paper mills is not sufficient for generating water that can be reused for producing paper grades [67]. Therefore, a desalination process employing membrane technologies such as NF is necessary to refine the water quality for reuse in industrial processes. Several studies have reported the potential of NF to recover water from pulp and paper industry wastewater [67–71]. For instance, NF was integrated as a post-treatment in a treatment train consisting of flotation, up-flow anaerobic sludge blanket bioreactor, and an activated sludge process to reclaim water from simulated thermo-mechanical pulp mill effluent [68]. It was observed that the Cu^{2+}, Mn^{2+}, Fe^{2+}, Ca^{2+}, and Mg^{2+} ions were removed by about 89%, 99%, 82%, 74%, and 57%, respectively, after passing through the NF unit, generating treated water that can be reused in industrial processes. Moreover, it has been demonstrated that the pulp quality which is based on brightness and brightness reversion criteria was not significantly affected, even though 100% of the treated water was reused in the bleaching step.

Membrane fouling, which causes a rapid flow drop, has emerged as the most significant constraint in pulp and paper wastewater treatment using membrane methods. Operational conditions and membrane selection are crucial factors in preventing membrane fouling. Thus, a study was conducted by Gönder et al. [69] to investigate the effect of operating conditions such as volume reduction factor (VRF), pH, temperature, and TMP on the membrane fouling during the two-step NF process for recovering water from real pulp and paper wastewater. Utilizing FM NP010 membrane (loose NF) in the initial stage of the NF process, 92% total hardness, 91% COD, and 98% sulfate were rejected under optimal conditions. The use of FM NP030 membrane (tight NF) in the second step of the NF treatment system consequently created permeate of the same quality as the real process water. Increasing the feed pH up to pH10 could reduce the solutes' adsorption onto the membrane due to the increase in electrostatic repulsion between the negatively charged membrane and negatively charged pollutants. Meanwhile, a high-temperature condition is not preferable since it can cause membrane pores expansion, inducing pores plugging and fouling. Thus, the optimum performance with minimal fouling was achieved under optimal conditions, i.e., pH 10, 25°C, VRF: 4, and TMP of 12 bar. The application of two-step NF process which consists of loose and tight NF membranes for the water recovery from paper machine whitewater was also investigated by others [70]. It has been demonstrated that the use of loose NF followed by tight NF such as FM NP030 membrane successfully reduced the UV absorption, organic carbon, and color of the permeate but is less effective for retaining monovalent ions (e.g., chloride and nitrate ions). Even so, the two-step NF treatment is sufficient for producing permeate with quality that can be reused as shower water for paper machines. Additional treatment such as RO can be installed after the NF to further polish the permeate quality if chloride removal is needed. It can be suggested that NF is still an interesting way to recover clean water for reuse applications (e.g., papermaking process) as long as the paper industry effluent does not contain a high level of monovalent ions [67].

In the context of resource recovery, it has been reported that the brown effluents from bleached sulfite pulp mills can be a potential resource for lignin recovery. Lignin is one of major components in lignocellulosic biomass, which can be applied in the manufacturing of fine chemicals, biofuels, and resins. Previously, a study was conducted by Ebrahimi et al. [71] to assess the potential of ceramic tubular membranes for recovering lignin from pulp and paper industry wastewater. Utilizing ceramic tubular membranes for the treatment of alkaline bleaching effluent has significant economic and environmental benefits. The test conducted using the UF-NF process has removed about 35%–40% and 45%–66% of COD and lignin, respectively. However, a performance comparison between different pairs of membranes showed that the MF-UF configuration showed better treatment efficiency than the two-stage UF-NF.

2.2.3.6 Mining Industry

Acid mine drainage (AMD) is one of the major effluent produced by the mining industry. Commonly, AMD is characterized by a low pH (2–4), a high sulfate concentration (1–20 g/L), a high concentration of heavy metals, and the presence of hazardous components [63,72]. This type of effluent could be formed during exposure of sulfur-containing rocks to the air and water [63,73]. Meanwhile, the pressure-oxidation process in gold mining activity emits a huge amount of acidic wastewater having a substantial metal concentration. Previously, the lime neutralization method has been used to treat the mining effluent since this technique can precipitate metals and sulfate [74,75]. Unfortunately, this process generates an enormous volume of sludge that poses an environmental risk and must be disposed of properly [72, 76]. The implementation of membrane technologies such as NF is a viable option that can provide permeate for reuse and permit the metals and acid recovery from mining industrial effluent, as described in a number of studies [72,76–82].

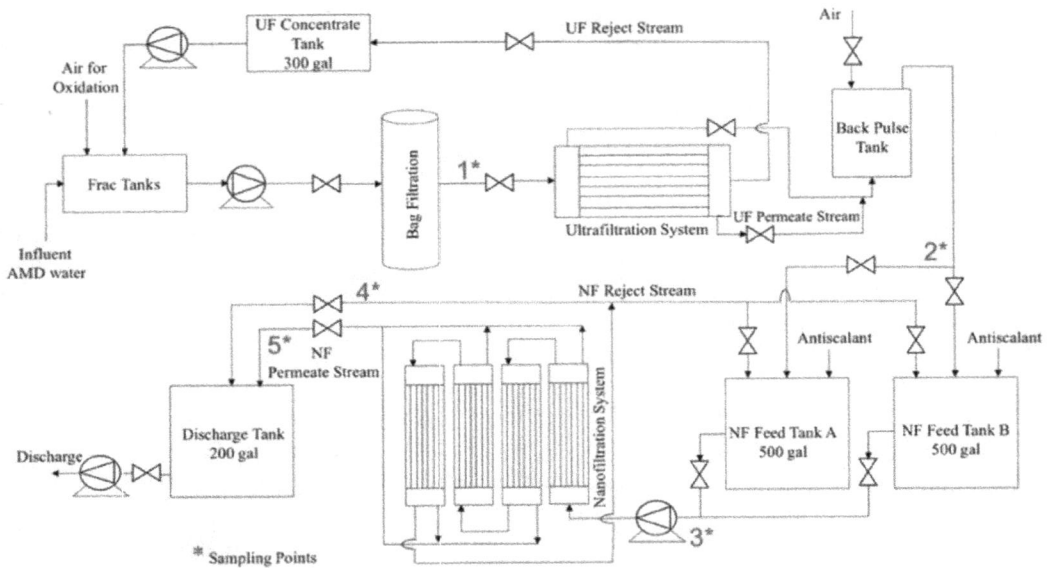

FIGURE 2.5 Pilot-scale system for water recovery from AMD water [72]. (Copyright 2017. Reproduced with permission from American Chemical Society.)

Several commercial polymeric NF membranes such as NF90 and NF270 have been used in the treatment of AMD for water reuse. In terms of filtration performance, the polyamide NF90 membrane with a smaller pore size is preferable compared to NF270 when dealing with real AMD since it can reject most of the ions including sulfate (removal >97%) [72]. The pre-treatment steps are essential to prevent inorganic fouling and ensure the long-term stability of NF operation. Thus, the aeration, sedimentation, bag filtration, and UF process were used to pre-treat the AMD prior to the NF process in pilot-scale AMD treatment (Figure 2.5) [72]. It was observed that the integrated system utilizing the NF90 membrane can reject more than 98% of TDS and attain a stable water recovery (57%) under long-term operation (208 hours) with minimal scaling occurrence. This process successfully recovered clean water from AMD that can be reused for various applications such as irrigation and cooling system in power plant. Besides water reusability, the NF concentrate with high sulfate concentration (~4,000 mg/L) could be applied in the treatment of flowback water since it has been reported that the mixing of some sulfate-rich concentrate and flowback water can remove the barium and strontium ions via sulfate precipitation step [83].

Aside from AMD, some studies have investigated the performance of commercial polymeric NF membranes for water recovery from gold mining effluent [77–79]. For instance, it has been reported that the NF90 also demonstrated promising performance in terms of the permeate quality during the treatment of gas scrubber effluent which is taken from a gold mining plant in Brazil [79]. However, the TDS removal (86%) from gold mining effluent is quite low compared to that of AMD (>98%) since gold mining effluent possesses a level of a much higher contaminant. The performance of the same NF membrane can vary depending on the type of effluent and its contaminants level. Despite the fact that the NF permeate did not meet the water quality criterion for reuse in boilers or cooling systems, it was suggested that it can be reused in other mining processes that utilize water with a low pH value, hence reducing the cost of pH adjustment.

Since the gold mining effluent generated by the pressure-oxidation process is highly acidic, it is important for the NF membrane to possess good chemical stability. The membrane selection is a crucial aspect of constructing processes to recover water and precious resources from challenging effluent. Previous work by Ramos et al. [80] demonstrated that the DK membrane, a commercial NF membrane produced by GE Osmonic, is promising for treating the gold mining effluent compared to the other commercial NF membranes, namely Duracid, NF90, NF270, and MPF-34. Besides high permeate flux and low fouling propensity, the DK membrane also exhibited higher chemical stability under long-term operation (180 days) during the treatment of acidic gold mining effluent. Ramos et al. [80] also found that the integrated system consisting of UF-NF-RO has successfully produced permeate water with a low acid concentration (pH 2.5) that could be reused in the gold mining process.

Mining effluent such as AMD contains valuable metals such as copper that can be recovered for further reuse. Thus, AMD shall be considered a potential resource for metal recovery rather than being disposed of. It is known that the metals rejection performance of RO and NF is comparable [84,85]. Nonetheless, NF has been recommended as the preferred treatment due to its higher fluxes at lower pressure, which results in less capital investment and cheaper operating and maintenance costs [85]. The impact of feed pH on the performance of commercial NF membranes (Dow NF270 and Trisep TS 80) for metal recovery from AMD has been assessed in the previous study [76]. Compared to NF270, the feed pH has imparted a more negligible effect on the TS80 membrane performance, where higher multivalent ions retention can be achieved across the studied pH range. Mullet et al. [76] also revealed that elevating the feed pH to a higher level than the membrane's isoelectric point will result in a loss of 2.4 kg/h of copper in the permeate. Their results indicated the need to comprehend the link between solution chemistry and membrane properties in order to achieve optimal recovery and avoid considerable capital and operating cost losses.

2.3 CONCLUSIONS

The ability of NF to separate divalent/polyvalent ions from monovalent ions and small molecules has attracted a great deal of interest in wastewater reclamation, recycling, reuse, and resource recovery applications. Most of the studies applied the NF at the end of an integrated process to improve the permeate quality up to the reusable water criteria and recover other valuable components. Moreover, the integration of NF with other technologies can reduce the fouling propensity, increase the membrane lifetime, and enhance the membrane performance upon exposure to actual wastewater. To facilitate the implementation of NF technology for water recycling, reuse, and resource recovery, issues regarding membrane materials, NF concentrate treatment, long-term membrane stability, fouling, cost assessment, design, process configuration, process scaling up, and reuse application can be further investigated.

ACKNOWLEDGMENT

The authors gladly acknowledge the financial support from the MRUN research grant (grant number: KK-2019-001) and UKM Postdoctoral Scheme (Modal Insan).

REFERENCES

1. Maryam, B. and H. Büyükgüngör, Wastewater reclamation and reuse trends in Turkey: Opportunities and challenges. *Journal of Water Process Engineering*, 2019. **30**: p. 100501.
2. Yang, J., et al., Membrane-based processes used in municipal wastewater treatment for water reuse: State-of-the-art and performance analysis. *Membranes*, 2020. **10**(6): p. 131.
3. Tejero, I.F.G. and V.H.D. Zuazo, *Water Scarcity and Sustainable Agriculture in Semiarid Environment: Tools, Strategies, and Challenges for Woody Crops.* 2018: Academic Press.
4. Luo, J. and Y. Wan, Effects of pH and salt on nanofiltration – A critical review. *Journal of Membrane Science*, 2013. **438**: p. 18–28.
5. Quist-Jensen, C.A., F. Macedonio, and E. Drioli, Membrane technology for water production in agriculture: Desalination and wastewater reuse. *Desalination*, 2015. **364**: p. 17–32.
6. Iglesias, R., et al., Water reuse in Spain: Data overview and costs estimation of suitable treatment trains. *Desalination*, 2010. **263**(1–3): p. 1–10.

7. Konieczny, K., A. Kwiecińska, and B. Gworek, The recovery of water from slurry produced in high density livestock farming with the use of membrane processes. *Separation and Purification Technology*, 2011. **80**(3): p. 490–498.

8. Zacharof, M.-P., et al., Nanofiltration of treated digested agricultural wastewater for recovery of carboxylic acids. *Journal of Cleaner Production*, 2016. **112**: p. 4749–4761.

9. Wu, B., Membrane-based technology in greywater reclamation: A review. *Science of the Total Environment*, 2019. **656**: p. 184–200.

10. Oh, K.S., et al., A review of greywater recycling related issues: Challenges and future prospects in Malaysia. *Journal of Cleaner Production*, 2018. **171**: p. 17–29.

11. Vuppaladadiyam, A.K., et al., Simulation study on comparison of algal treatment to conventional biological processes for greywater treatment. *Algal Research*, 2018. **35**: p. 106–114.

12. Guilbaud, J., et al., Laundry water recycling in ship by direct nanofiltration with tubular membranes. *Resources, Conservation and Recycling*, 2010. **55**(2): p. 148–154.

13. Mozia, S., et al., A system coupling hybrid biological method with UV/O_3 oxidation and membrane separation for treatment and reuse of industrial laundry wastewater. *Environmental Science and Pollution Research*, 2016. **23**(19): p. 19145–19155.

14. Hoinkis, J. and V. Panten, Wastewater recycling in laundries – From pilot to large-scale plant. *Chemical Engineering and Processing: Process Intensification*, 2008. **47**(7): p. 1159–1164.

15. Parwin, R. and K. Karar Paul, Overview of applications of kitchen wastewater and its treatment. *Journal of Hazardous, Toxic, and Radioactive Waste*, 2020. **24**(2): p. 04019041.

16. Burton, F.L., et al., *Wastewater Engineering: Treatment and Reuse*. 2003: McGraw-Hill, New York.

17. Zoungrana, A., et al., Treatability of municipal wastewater with direct contact membrane distillation. *Sigma*, 2017. **8**(3): p. 245–254.

18. Chandrasekhar, S.S., et al., Performance assessment of a side-stream membrane bioreactor for the treatment of kitchen wastewater. *Biochemical Engineering Journal*, 2022. **180**: p. 108366.

19. Chen, G., et al., A review of salty waste stream management in the Australian dairy industry. *Journal of Environmental Management*, 2018. **224**: p. 406–413.

20. Andrade, L., et al., Nanofiltration as tertiary treatment for the reuse of dairy wastewater treated by membrane bioreactor. *Separation and Purification Technology*, 2014. **126**: p. 21–29.

21. Chen, Z., et al., Fully recycling dairy wastewater by an integrated isoelectric precipitation–nanofiltration–anaerobic fermentation process. *Chemical Engineering Journal*, 2016. **283**: p. 476–485.

22. Andrade, L., et al., Reuse of dairy wastewater treated by membrane bioreactor and nanofiltration: Technical and economic feasibility. *Brazilian Journal of Chemical Engineering*, 2015. **32**: p. 735–747.

23. Ochando-Pulido, J., J. Corpas-Martínez, and A. Martinez-Ferez, About two-phase olive oil washing wastewater simultaneous phenols recovery and treatment by nanofiltration. *Process Safety and Environmental Protection*, 2018. **114**: p. 159–168.

24. Ochando-Pulido, J.M., et al., Physicochemical analysis and adequation of olive oil mill wastewater after advanced oxidation process for reclamation by pressure-driven membrane technology. *Science of the Total Environment*, 2015. **503**: p. 113–121.

25. Ochando-Pulido, J., et al., Optimization of polymeric nanofiltration performance for olive-oil-washing wastewater phenols recovery and reclamation. *Separation and Purification Technology*, 2020. **236**: p. 116261.

26. Sánchez-Arévalo, C.M., et al., Effect of the operating conditions on a nanofiltration process to separate low-molecular-weight phenolic compounds from the sugars present in olive mill wastewaters. *Process Safety and Environmental Protection*, 2021. **148**: p. 428–436.

27. Gadipelly, C., et al., Pharmaceutical industry wastewater: Review of the technologies for water treatment and reuse. *Industrial & Engineering Chemistry Research*, 2014. **53**(29): p. 11571–11592.

28. Research, G.V. *Pharmaceutical manufacturing market size, share & trends analysis report by molecule type, by drug development type, by formulation, by routes of administration, by sales channel, by age group, and segment forecasts, 2021–2028*. 2022; Available from: https://www.grandviewresearch.com/industry-analysis/pharmaceutical-manufacturing-market.

29. Mandal, M.K., et al., Membrane technologies for the treatment of pharmaceutical industry wastewater, in *Water and Wastewater Treatment Technologies*. 2019: Springer. p. 103–116.

30. Strade, E., D. Kalnina, and J. Kulczycka, Water efficiency and safe re-use of different grades of water – Topical issues for the pharmaceutical industry. *Water Resources and Industry*, 2020. **24**: p. 100132.

31. de Souza, D.I., et al., Nanofiltration for the removal of norfloxacin from pharmaceutical effluent. *Journal of Environmental Chemical Engineering*, 2018. **6**(5): p. 6147–6153.

32. Azaïs, A., et al., Nanofiltration for wastewater reuse: Counteractive effects of fouling and matrice on the rejection of pharmaceutical active compounds. *Separation and Purification Technology*, 2014. **133**: p. 313–327.
33. Yangali-Quintanilla, V., et al., Proposing nanofiltration as acceptable barrier for organic contaminants in water reuse. *Journal of Membrane Science*, 2010. **362**(1–2): p. 334–345.
34. Maryam, B., et al., A study on behavior, interaction and rejection of Paracetamol, Diclofenac and Ibuprofen (PhACs) from wastewater by nanofiltration membranes. *Environmental Technology & Innovation*, 2020. **18**: p. 100641.
35. Wang, J., et al., Performance and fate of organics in a pilot MBR–NF for treating antibiotic production wastewater with recycling NF concentrate. *Chemosphere*, 2015. **121**: p. 92–100.
36. Yacouba, Z.A., et al., Removal of organic micropollutants from domestic wastewater: The effect of ozone-based advanced oxidation process on nanofiltration. *Journal of Water Process Engineering*, 2021. **39**: p. 101869.
37. Martínez, M.B., et al., Effect of impurities in the recovery of 1-(5-bromo-fur-2-il)-2-bromo-2-nitro-ethane using nanofiltration. *Chemical Engineering and Processing: Process Intensification*, 2013. **70**: p. 241–249.
38. Al-Ghouti, M.A., et al., Produced water characteristics, treatment and reuse: A review. *Journal of Water Process Engineering*, 2019. **28**: p. 222–239.
39. Alzahrani, S., et al., Identification of foulants, fouling mechanisms and cleaning efficiency for NF and RO treatment of produced water. *Separation and Purification Technology*, 2013. **118**: p. 324–341.
40. Zolghadr, E., et al., The role of membrane-based technologies in environmental treatment and reuse of produced water. *Frontiers in Environmental Science*, 2021. **9**: p. 71.
41. Alva-Argáez, A., A.C. Kokossis, and R. Smith, The design of water-using systems in petroleum refining using a water-pinch decomposition. *Chemical Engineering Journal*, 2007. **128**(1): p. 33–46.
42. Gamal Khedr, M., Nanofiltration of oil field-produced water for reinjection and optimum protection of oil formation. *Desalination and Water Treatment*, 2015. **55**(12): p. 3460–3468.
43. Riley, S.M., et al., Hybrid membrane bio-systems for sustainable treatment of oil and gas produced water and fracturing flowback water. *Separation and Purification Technology*, 2016. **171**: p. 297–311.
44. Moser, P.B., et al., Effect of MBR-H$_2$O$_2$/UV hybrid pre-treatment on nanofiltration performance for the treatment of petroleum refinery wastewater. *Separation and Purification Technology*, 2018. **192**: p. 176–184.
45. Gregory, K.B., R.D. Vidic, and D.A. Dzombak, Water management challenges associated with the production of shale gas by hydraulic fracturing. *Elements*, 2011. **7**(3): p. 181–186.
46. Chang, H., et al., An integrated coagulation-ultrafiltration-nanofiltration process for internal reuse of shale gas flowback and produced water. *Separation and Purification Technology*, 2019. **211**: p. 310–321.
47. Jang, E., Y. Jang, and E. Chung, Lithium recovery from shale gas produced water using solvent extraction. *Applied Geochemistry*, 2017. **78**: p. 343–350.
48. Seip, A., et al., Lithium recovery from hydraulic fracturing flowback and produced water using a selective ion exchange sorbent. *Chemical Engineering Journal*, 2021. **426**: p. 130713.
49. Santos, B., et al., Oil refinery hazardous effluents minimization by membrane filtration: An on-site pilot plant study. *Journal of Environmental Management*, 2016. **181**: p. 762–769.
50. Ranganathan, K., K. Karunagaran, and D. Sharma, Recycling of wastewaters of textile dyeing industries using advanced treatment technology and cost analysis – Case studies. *Resources, Conservation and Recycling*, 2007. **50**(3): p. 306–318.
51. Samsami, S., et al., Recent advances in the treatment of dye-containing wastewater from textile industries: Overview and perspectives. *Process Safety and Environmental Protection*, 2020. **143**: p. 138–163.
52. Thamaraiselvan, C. and M. Noel, Membrane processes for dye wastewater treatment: Recent progress in fouling control. *Critical Reviews in Environmental Science and Technology*, 2015. **45**(10): p. 1007–1040.
53. Ji, D., et al., Preparation of high-flux PSF/GO loose nanofiltration hollow fiber membranes with dense-loose structure for treating textile wastewater. *Chemical Engineering Journal*, 2019. **363**: p. 33–42.
54. Salahshoor, Z., A. Shahbazi, and S. Maddah, Magnetic field–influenced nanofiltration membrane blended by CS–EDTA–mGO as multi–functionality green modifier to enhance nanofiltration performance, efficient removal of Na$_2$SO$_4$/Pb^{2+}/RR195 and cyclic wastewater treatment. *Chemosphere*, 2021. **278**: p. 130379.
55. Ang, W.L. and A.W. Mohammad, Integrated and hybrid process technology, in *Sustainable Water and Wastewater Processing*. 2019: Elsevier. p. 279–328.

56. Ang, W.L., et al., A review on the applicability of integrated/hybrid membrane processes in water treatment and desalination plants. *Desalination*, 2015. **363**: p. 2–18.

57. Lebron, Y.A.R., et al., Integrated photo-Fenton and membrane-based techniques for textile effluent reclamation. *Separation and Purification Technology*, 2021. **272**: p. 118932.

58. Couto, C.F., et al., Coupling of nanofiltration with microfiltration and membrane bioreactor for textile effluent reclamation. *Separation Science and Technology*, 2017. **52**(13): p. 2150–2160.

59. Li, K., et al., Influence of nanofiltration concentrate recirculation on performance and economic feasibility of a pilot-scale membrane bioreactor-nanofiltration hybrid process for textile wastewater treatment with high water recovery. *Journal of Cleaner Production*, 2020. **261**: p. 121067.

60. Tavangar, T., et al., Toward real textile wastewater treatment: Membrane fouling control and effective fractionation of dyes/inorganic salts using a hybrid electrocoagulation–nanofiltration process. *Separation and Purification Technology*, 2019. **216**: p. 115–125.

61. Cinperi, N.C., et al., Treatment of woolen textile wastewater using membrane bioreactor, nanofiltration and reverse osmosis for reuse in production processes. *Journal of Cleaner Production*, 2019. **223**: p. 837–848.

62. Ağtaş, M., et al., Pilot-scale ceramic ultrafiltration/nanofiltration membrane system application for caustic recovery and reuse in textile sector. *Environmental Science and Pollution Research*, 2021. **28**(30): p. 41029–41038.

63. Samaei, S.M., S. Gato-Trinidad, and A. Altaee, The application of pressure-driven ceramic membrane technology for the treatment of industrial wastewaters – A review. *Separation and Purification Technology*, 2018. **200**: p. 198–220.

64. Badar, S. and I.H. Farooqi, Pulp and paper industry – Manufacturing process, wastewater generation and treatment, in *Environmental Protection Strategies for Sustainable Development*. 2012: Springer. p. 397–436.

65. Kamali, M. and Z. Khodaparast, Review on recent developments on pulp and paper mill wastewater treatment. *Ecotoxicology and Environmental Safety*, 2015. **114**: p. 326–342.

66. Toczyłowska-Mamińska, R., Limits and perspectives of pulp and paper industry wastewater treatment– A review. *Renewable and Sustainable Energy Reviews*, 2017. **78**: p. 764–772.

67. Khosravi, M., et al., Membrane process design for the reduction of wastewater color of the Mazandaran pulp-paper industry, Iran. *Water Resources Management*, 2011. **25**(12): p. 2989–3004.

68. Caldeira, D.C.D., et al., A case study on the treatment and recycling of the effluent generated from a thermo-mechanical pulp mill in Brazil after the installation of a new bleaching process. *Science of the Total Environment*, 2021. **763**: p. 142996.

69. Gönder, Z.B., S. Arayici, and H. Barlas, Advanced treatment of pulp and paper mill wastewater by nanofiltration process: Effects of operating conditions on membrane fouling. *Separation and Purification Technology*, 2011. **76**(3): p. 292–302.

70. Kaya, Y., et al., The effect of transmembrane pressure and pH on treatment of paper machine process waters by using a two-step nanofiltration process: Flux decline analysis. *Desalination*, 2010. **250**(1): p. 150–157.

71. Ebrahimi, M., et al., Treatment of the bleaching effluent from sulfite pulp production by ceramic membrane filtration. *Membranes*, 2015. **6**(1): p. 7.

72. Wadekar, S.S., et al., Laboratory and pilot-scale nanofiltration treatment of abandoned mine drainage for the recovery of products suitable for industrial reuse. *Industrial & Engineering Chemistry Research*, 2017. **56**(25): p. 7355–7364.

73. Akcil, A. and S. Koldas, Acid mine drainage (AMD): Causes, treatment and case studies. *Journal of Cleaner Production*, 2006. **14**(12–13): p. 1139–1145.

74. Simate, G.S. and S. Ndlovu, Acid mine drainage: Challenges and opportunities. *Journal of Environmental Chemical Engineering*, 2014. **2**(3): p. 1785–1803.

75. Santomartino, S. and J.A. Webb, Estimating the longevity of limestone drains in treating acid mine drainage containing high concentrations of iron. *Applied Geochemistry*, 2007. **22**(11): p. 2344–2361.

76. Mullett, M., R. Fornarelli, and D. Ralph, Nanofiltration of mine water: Impact of feed pH and membrane charge on resource recovery and water discharge. *Membranes*, 2014. **4**(2): p. 163–180.

77. Ricci, B.C., et al., Integration of nanofiltration and reverse osmosis for metal separation and sulfuric acid recovery from gold mining effluent. *Separation and Purification Technology*, 2015. **154**: p. 11–21.

78. Ricci, B.C., et al., Assessment of the chemical stability of nanofiltration and reverse osmosis membranes employed in treatment of acid gold mining effluent. *Separation and Purification Technology*, 2017. **174**: p. 301–311.

79. Reis, B.G., et al., Comparison of nanofiltration and direct contact membrane distillation as an alternative for gold mining effluent reclamation. *Chemical Engineering and Processing-Process Intensification*, 2018. **133**: p. 24–33.
80. Ramos, R.L., et al., Membrane selection for the Gold mining pressure-oxidation process (POX) effluent reclamation using integrated UF-NF-RO processes. *Journal of Environmental Chemical Engineering*, 2020. **8**(5): p. 104056.
81. Amaral, M.C., et al., Integrated UF–NF–RO route for gold mining effluent treatment: From bench-scale to pilot-scale. *Desalination*, 2018. **440**: p. 111–121.
82. Andrade, L., et al., Nanofiltration and reverse osmosis applied to gold mining effluent treatment and reuse. *Brazilian Journal of Chemical Engineering*, 2017. **34**: p. 93–107.
83. He, C., T. Zhang, and R.D. Vidic, Co-treatment of abandoned mine drainage and Marcellus Shale flow-back water for use in hydraulic fracturing. *Water Research*, 2016. **104**: p. 425–431.
84. Zhong, C.-M., et al., Treatment of acid mine drainage (AMD) by ultra-low-pressure reverse osmosis and nanofiltration. *Environmental Engineering Science*, 2007. **24**(9): p. 1297–1306.
85. Carvalho, A., et al., Separation of potassium clavulanate and potassium chloride by nanofiltration: Transport and evaluation of membranes. *Separation and Purification Technology*, 2011. **83**: p. 23–30.

3 Development of Nanofiltration Membranes

From Early Membranes to Nanocomposites

Daniel James Johnson, Neveen AlQasas,
Asma Eskhan, and Nidal Hilal
New York University Abu Dhabi

CONTENTS

3.1 INTRODUCTION

Nanofiltration (NF), though long established, is in some ways still an immature and developing technology, with much applications and research in the areas of pre-treatment of seawater prior to desalination and in the treatment, reuse and recycling of waste water sources [1–4]. NF membranes are capable of reducing monovalent salts and efficiently removing almost all multivalent ions [5,6], reducing hardness and successfully rejecting microorganisms [7,8] and the majority of organic molecules [9,10]. In comparison with the much tighter, and hence more resistive, reverse osmosis (RO) membranes, specific flux is higher allowing lower pumping pressures, which means for applications where

DOI: 10.1201/9781003261827-3

high monovalent salt removal is not the aim it is often preferable to RO, such as for polishing of treated effluent streams [11] and direct treatment of freshwater for potable water applications. In addition, the high organic and hardness removal rates make it a candidate for RO pre-treatment tasks [12–14].

Research has been focussed on the development of NF membranes tasked predominantly with either seawater desalination or wastewater treatment applications [15,16]. This has been aimed at delivering membranes with high water fluxes, high rejection rates and good fouling resistance. Much of the recent research has involved the use of nanomaterials as fillers to create more open structures, to improve membrane transport (often at the sacrifice of some selectivity) or by introducing antifouling functionalities [17–21].

3.1.1 Nanofiltration Membranes: Categories and Definitions

Membrane pore size or size of excluded particles (typically particle or molecule size at which a rejection rate greater than 90% is observed) forms the basis of the categorisation of membranes by the applications they are suitable for [3,22]. Here, NF membranes are positioned in the range which operates between tight ultrafiltration (UF) membranes and relatively loose RO membranes, the latter being the original definition of an NF membrane. NF membranes have average critical pore diameters of no more than 2 nm, with removal of particles as small as 0.9–10 nm being generally observed for different membranes. This allows NF membranes to be capable of removing colloids, nanoparticles, most organics and some ions.

NF membranes were first derived from RO membrane research and development due to the need for looser membranes with low flow resistance and hence higher flux rates for applications where this was more important than monovalent salt rejection, such as removal of hardness, product recovery and recycling of process waste solutions [2,23,24]. Whilst much of the reduced operating pressures observed with NF membranes is due to the more open structure with lower flow resistance compared with tighter RO membranes, it must be remembered that the lower osmotic pressure difference between the feed and a permeate containing monovalent salts will also reduce the operating pressure needed to establish a particular membrane flux.

The term of nanofiltration first entered the academic literature in the mid-1980s, following commercial development of membrane modules marketed using that term [25]. This has expanded to include other applications, including as a desalination pre-treatment step, concentration or removal of dissolved small organic molecules, fruit juice concentration and further treatment of conventionally treated wastewater effluent for reuse where high water quality is required [23].

3.1.2 Nanofiltration Transport Mechanisms

Once the presence of pores in NF membranes was demonstrated, appropriate models for mass transport through NF membranes had to be developed [26]. Separation of solvent from the solute or suspended colloids by the selective layer is due to a combination of size, electrostatic, dielectric, transport and adsorption mechanisms [27–29]. Selection is dominated by steric interactions in the absence of charge. When the solute consists of charged species, the Gibbs–Donnan effect, due to charge interactions, becomes significant in determining rejection of ions, as well as other effects such as concentration polarisation. For charged solute and colloids, dielectric effects also play a role in specificity. Here the passage of ions through regions of solvent with different relative permittivity values, due to influence of the surrounding membrane, can result in the formation of an energy barrier which must be overcome before an ion may complete its journey through the membrane [30–32].

The Nernst–Planck equation can be used to model the flow of charged solutes due to both thermally driven diffusion and electrostatic interactions and was modified by Schlögl by the addition of an advective flow term, which was applied to the passage of charged solutes through an ion exchange membrane [33]. For a single dissolved constituent passing through a membrane, this can be approximated as [29]:

$$J_i = -D_i \, \nabla c_i + c_i \, v - z_i \, v_i \, c_i \, \nabla \Phi \, F, \tag{3.1}$$

where J_i is the flux of dissolved species i, D_i is the diffusivity, c_i is the concentration of i, v is the solvent velocity, z_i is the charge number of i, v_i is the mobility of i, Φ is the electrical potential and F is the Faraday constant. This extended Nernst–Planck equation was adjusted over a number of years by a number of researchers to account for advection-only transport [34], the restriction of liquid transport through void spaces of a similar size to solute molecules [35] and several other adaptations for specific circumstances [36–40].

The Donnan-Steric Pore Model (DSPM) was developed by Bowen et al. [41] in the 1990s. Here, the extended Nernst–Planck equation used contained Donnan and steric terms to describe partitioning due to electrical and size exclusion effects. This model was successful in predicting the performance of membranes for filtration of a solution containing dyes and salts [42]. This model assumed the, not yet verified, existence of nanometre-scale pores in the NF membrane selective layer, leading to much improved predictions of experimental filtration data. This model was later extended to include dielectric exclusion (often termed the DSPM-DE model) [26,43]. This latter model has not been substantively bettered for prediction of transport of ionic solutes through an NF membrane [44]. Soon afterwards atomic force microscopy (AFM) scans of NF membrane surfaces provided the first direct experimental evidence that small holes are indeed present on the membrane selective layer, which were presumed to be the mouths of selective layer crossing pores.

In the DSPM-DE model, pores are assumed to be cylindrical (i.e. straight and of uniform circular cross section), allowing the flow of solvent to be modelled according to a Hagen–Poiseuille-based relationship [42]:

$$J_v = \frac{r_P^2 \, (\Delta P - \Delta \pi)}{8\eta \left(\dfrac{x}{\varphi}\right)}, \tag{3.2}$$

where J_v is the volumetric flow rate, r_P is the mean pore radius, ΔP is the hydraulic transmembrane pressure, $\Delta \pi$ is the osmotic pressure difference across the membrane active layer, η is the effective fluid viscosity within the membrane void space, x is the effective membrane thickness and φ is the effective membrane porosity. The solute flux though the membrane pores is modelled using the extended Nernst–Planck equation [44].

For uncharged solutes, Donnan and dielectric effects can be ignored, with only steric mechanisms driving solute rejection. This allows the experimental determination of pore size to be made from flow measurements at a wide range of operating pressures fitted to the following relationship:

$$R = 1 - \frac{c_P}{c_F} = 1 - \frac{(K_c - Y)\Phi}{1 - \left[1 - (K_C - Y)\Phi\right]\exp(-Pe')}, \tag{3.3}$$

where R is the solute rejection, c_P and c_F are the permeated and feed solute concentrations, respectively, K_c is the uncharged solute hindrance factor for advection, Φ is the uncharged steric partition coefficient of the ion and Pe' is the modified Peclet number. Y is a dimensionless solute function independent of the solution concentration:

$$Y = \frac{D_P \, V_S \, 8 \, \eta}{RT \, r_P^2}, \tag{3.4}$$

where D_P is the solute pore diffusion coefficient, V_S is the partial molar volume of the solute, R is the gas constant and T is the absolute temperature. From this, the structural properties of the membrane can be determined. Further rejection experiments with charged solvents can then be used to determine the pore dielectric constant and membrane volumetric charge density.

Since the DSPM-DE model was first proposed, several other NF transport models have been developed. These include the coupled-series parallel resistance model, which is specific for infiltration of inorganic membranes by organic solvents [45]; the solution-diffusion electro-migration model [46]; and the pore-blockage cake filtration model [47].

3.2 NANOFILTRATION MEMBRANE CATEGORIES

NF membranes can be manufactured from polymers or ceramic materials or mixtures consisting of a dispersal of one type within another (often termed mixed matrix membranes). Polymeric NF membranes are categorised based on the chemical make-up of the selective layer or by the overall structure of the membrane (symmetric, asymmetric, thin-film composite [TFC], etc.). Choosing the correct membrane for a particular application is based on both physical and chemical parameters. In terms of physical parameters, pore size, porosity and permeability are important as they determine the flow rates for a given transmembrane pressure as well as selectivity due to size exclusion mechanisms. Chemical factors, such as the types of polymers used in the selective-layer construction, are important as they determine the chemical stability of the membrane to attack by chlorine compounds during disinfection or cleaning or from extremes of pH which may be encountered when using some feed waters. In addition, the chemical functionalities at the surface of the membrane and around the membrane pores affect selectivity, particularly of ions, and resistance to deposition of fouling compounds carried by the feed water. However, both the physical and chemical parameters of the membrane are functions of the chosen polymers, techniques used to form the selective layer and composition and temperature of the casting or polymerisation solution.

There are a number of commercially available polymeric NF membranes and membrane modules available, and a few of them are listed, along with a few relevant parameters, in Table 3.1. The vast majority of membranes consist of TFC membranes with polyamide (PA) selective layers, although a few other membrane types are available. More detailed descriptions of the exact composition of the membranes are typically proprietary information, so whether and how the membranes may be chemically modified is not readily available. So far as the authors are aware, nanomaterial-modified membranes are yet to feature at any large scale in terms of commercially available membranes, even though this has featured prominently in the academic research into NF membrane materials and fabrication published in recent years [21,48]. Industry is naturally conservative with regards to large-scale adoption of new technologies, especially when few of the membranes produced and tested at the lab scale have not been demonstrated successfully at scales relevant to commercial operations or over very long filtration time scales. It must also be remembered that what works well on a bench scale may be difficult and/or expensive to manufacture at commercial scale. Nevertheless, there is much room for investigation into improved NF membranes, often tailored to specific applications.

3.2.1 CELLULOSIC NANOFILTRATION MEMBRANES

Not surprisingly, as a technology that developed from RO membranes, and has often been seen as loose RO membranes, many of the same materials have been used. Initial NF membranes were made of materials derived from cellulose, with the first example being cellulose acetate (CA) fabricated using phase inversion techniques [49], with other examples being cellulose diacetate [50] and cellulose triacetate (CTA) [51]. Such membranes are typically asymmetric, with a cellulosic selective layer mated to a non-woven backing to provide mechanical support. Cellulosic membranes generally exhibit high resistance to attack by chlorine, an important property where bleaching agents are routinely used for cleaning or pre-treatment. However, these membranes often have lower permeability coefficients than optimal for the applications NF membranes are often used for. In order to make them suitable for NF applications, a number of strategies can be adopted to increase their water permeability. This can involve adjusting the degree of acetylation of the cellulose units or by the addition of other polymers or chemical agents to the casting or polymerisation mix [52]. Nanoparticles, when added to the selective layer, have been demonstrated to adjust the polymeric

TABLE 3.1

Commercially Available Polymeric Nanofiltration Membranes and Membrane Modules

Manufacturer	Membrane	Active Layer Material	Operating pH	Max Pressure (bar)	Salt Rejection
CSM	NE	Polyamide TFC	1–13	41	40%–70% NaCl; 45%–70% CaCl$_2$; 97% MgSO$_4$
Dow	NF90	Polyamide TFC	2–11	41	>97% Salt
	NF200	Polyamide TFC	3–10	41	40%–60% CaCl$_2$; 97% MgSO$_4$
	NF270	Polyamide TFC	3–10	41	>97% MgSO$_4$
Hydronautics	ESNA	Composite polyamide	2–10	41	92% CaCl$_2$
	Nano-BW	Composite polyamide	3–9	41	99.7% MgSO$_4$
	Nano-SW	Composite polyamide	3–9	41	99.8% MgSO$_4$
Koch	SelRO	Proprietary TFC	1–11	35	10% NaCl
	DairyPro NF	Proprietary TFC	4–10	55	–
Microdyne	Nadir NP010	PES	0–14	–	35%–75% NaSO$_4$
	Nadir NP030	PES	0–14	–	80%–95% NaSO$_4$
nx	dNF40	Modified PES	2–12	6	91% MgSO$_4$
	dNF80	Modified PES	2–12	6	76% MgSO$_4$
Snowpure	ExcellNano NF-2	Polyamide	4–9	27	98%–99% MgSO$_4$; 10%–70% NaCl
Suez	DK	Proprietary TFC	3–9	41	98% MgSO$_4$
	NF1 4040	Proprietary TFC	3–9	40	98 MgSO$_4$
Synder	NFS	Proprietary polyamide	3–10.5	41	99.5% MgSO$_4$; 50% NaCl
	NFX	Proprietary polyamide	3–10.5	41	99.0% MgSO$_4$; 40% NaCl
	NFW	Proprietary polyamide	4–10	41	97% MgSO$_4$; 20% NaCl
	NFG	Proprietary polyamide	4–10	41	50% MgSO$_4$; 10% NaCl
Toray	SU-620	Cross-linked polyamide	3–8	10	55% NaCl; 99.0% MgSO$_4$
	DL	Proprietary TFC	3–9	41	96% MgSO$_4$
	Duraslick	Proprietary TFC	3–9	41	98% MgSO$_4$
	HL	Proprietary TFC	3–9	31	98% MgSO$_4$
	C	Cellulose acetate	5–6.5	31	97% MgSO$_4$
	CK	Cellulose acetate	5–6.5	14	97% MgSO$_4$
	Muni NF	TFC	3–9	41	98.8% MgSO$_4$

structure of the layer by reducing the amount of polymer cross-links and increasing the porosity and size of channels. Nanoparticles may be designed and incorporated, which increase other beneficial properties, such as providing an adsorptive moiety to improve collection of specific dissolved chemicals [51], adding a biocidal functionality [53] or some other desirable quality [54,55].

3.2.2 THIN-FILM COMPOSITE NANOFILTRATION MEMBRANES

TFC membranes have an asymmetric structure composing a relatively tight thin selective (or sometimes termed active layer), cast onto a more porous support layer. This structure is then typically wedded to a non-woven support formed from a mechanically stable polymer, such as poly(ethyl teraphthalate). The selective layer faces the feed and acts as the primary barrier to solute and colloid penetration of the membrane. This tight layer is kept as thin as possible to keep hydraulic resistance low, which leads to poor strength and hence the need for a support layer. For most membranes, the selective layer is manufactured using interfacial polymerisation onto the ready cast support, preformed using phase inversion techniques, which results in relatively large water channels and a more open porous network.

Adjusting a number of fabrication parameters including composition of the monomers used, polymerisation bath temperature and the use of additives can make profound changes in the chemical and physical properties, allowing experienced membrane technologists to adjust the membrane properties to the intended application [56,57]. PA, as is apparent from Table 3.2, is the polymer of

TABLE 3.2
Highlights of Nanocomposite NF Membranes Discussed in This Work

Nanomaterial		Active Layer Material	Pure Water Permeability (L/m²h bar)	Rejection Rate	Ref.
Nanodiamond	0D	Polyamide	15.0	~98% NaH$_2$SO$_4$	[106]
Tannic acid-coated TiO$_2$	0D	Polyester	5.6	58% NaCl, 95% Na$_2$SO$_4$	[107]
Cellulose nanocrystals/ silver	0D	Polyamide	25.4	99% Na$_2$SO$_4$	[166]
Dopamine/melanin nanosphere	0D	Cellulose triacetate	2.2	97% Na$_2$SO$_4$, 92% MgCl$_2$, 76% NaCl	[131]
Hollow poly-pyrrole nanospheres	0D	Poly(piperazine amide)	18.0	99% Na$_2$SO$_4$, 99% MgSO$_4$, 69% MgCl$_2$, 49% NaCl	[130]
Modified bentonite nanoclay	0D	Polyamide	>10	~10% MgSO$_4$	[135]
Halloysite nanotubes	1D	Polyamide	20.5	94.9% Na$_2$SO$_4$, 91.5% MgSO$_4$	[136]
Hydrous manganese oxide	1D	Polyamide	4.25	30.3% NaCl, 97.4% Na$_2$SO$_4$, 96.7% MgSO$_4$	[137]
PHFBA = poly(2,2,3,4,4,4)-hexafluorobutyl acrylate modified hydrous manganese oxide	1D	Polyamide	5.18	43.2% NaCl, 98.6% Na$_2$SO$_4$, 97.6% MgSO$_4$	[137]
Modified silica-coated Fe$_3$O$_4$	1D	Poly(ether sulfone)	13.0–15.5	~25% NaCl, ~80% Na$_2$SO$_4$, ~58% MgSO$_4$	[138]
MIL-101(Cr) MOF	MOF	Polyamide-PVA	~5.1–~8.9	32% NaCl, 97% MgSO$_4$, 88% Na$_2$SO$_4$, 36% MgCl	[210]
Carboxy-functionalised CNT	1D	Polyamide	~6–22	<10% NaCl, 98% methyl blue	[152]
β-CD-functionalised MWCNT = multiwalled carbon nanotube	1D	Poly(ether sulfone)	5.25	98% Direct Red 16	[155]
Boron nitride nanotubes	1D	Polyamide	4.5	90% MgSO$_4$, 80% CaCl$_2$	[158]
Cellulose nanofibres	1D	Poly ethylene imine	32.7	90% MgCl$_2$, 65% MgSO$_4$, 44% NaCl, 39% Na$_2$SO$_4$	[165]
Graphene oxide	2D	Electro-sprayed GO film	11.1–22.3	61.1% Na$_2$SO$_4$, 27.9% NaCl, 41.8 MgSO$_4$, 15.0% MgCl$_2$	[186]
HBE = hyperbranched epoxy	2D	Poly(piperazine amide)	6.7–8	~43% NaCl, ~98% Na$_2$SO$_4$, ~96% MgSO$_4$, ~85% MgCl$_2$	[187]
GO/Co(OH)$_2$ nanosheets	2D	Hybrid 2D nanomaterial	3.8–4.2	48.7% NaCl, 85.4% Na$_2$SO$_4$	[193]
GO nanoplatelets	2D	GO nanoplatelet assemblage	1,171.6	–	[194]
g-C$_3$N$_4$ nanoplatelets	2D	PDA/PEI	28.4	2.9% NaCl, 7.6% N$_2$SO$_4$	[197]
g-C$_3$N$_4$ nanoplatelets	2D	Polyamide	~20–23 (MgCl$_2$ sol.)	>90% MgCl$_2$, >90% MgSO$_4$, ~55%–65% NaCl, ~50%–55% LiCl, ~45%–55% Na$_2$SO$_4$	[211]

choice for most commercially available NF membranes. There are a wide number of selective-layer polymers which have been studied at the research level for various NF applications, including trimesoyl chloride, diethylenetriamine, piperazine, bisphenol A, isophthaloyl chloride, tannic acid and diethylene triamine [58–63].

3.2.3 MIXED MATRIX NANOFILTRATION MEMBRANES

Mixed matrix membranes consist of similar material types blended together to make a membrane, typically inorganic or metallic fillers (including nanomaterials) blended into a polymer matrix. Here, the presence of the filler may greatly alter the properties of membranes from that seen with a pure polymer. Much of the research activity here has been the introduction of a wide range of inorganic nanomaterials into polymer films to enhance water transport and improve selectivity [64–68], as well as to introduce biocidal functionality to reduce formation and development of biofilms which may hinder water transport [69,70].

3.3 NANOCOMPOSITE MEMBRANES

The term nanocomposite membrane can be used as a general categorisation for polymeric membranes incorporating any of a variety of nanomaterials. These can then be sub-categorised depending on the way the nanomaterials are located within the membrane (see Figure 3.1 for a schematic of each type). Conventional nanocomposites are symmetrical membranes with the nanomaterials deposited evenly throughout. Surface-located nanocomposites have nanomaterials attached to the feed-facing surface of the membrane rather than placed within the polymer structure. Thin-film nanocomposites (TFNs), possibly the most seen in the research literature, are asymmetric TFC membranes with nanomaterials placed throughout the selective layer. Finally, the nanomaterials may be placed in the porous support of a TFC membrane, rather than the selective layer. When discussing nanomaterials, general definitions agree that they are materials which contain at least one

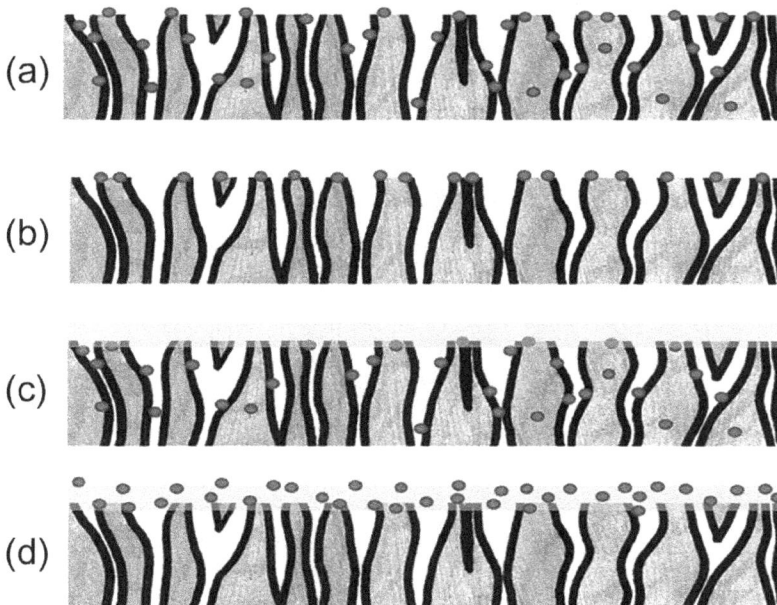

FIGURE 3.1 The different types of nanocomposite membranes, according to placement of nanoparticles within the structure: (a) conventional nanocomposite (CN); (b) surface-located nanocomposite; (c) thin-film nanocomposite (TFN); and (d) TFC membrane with nanocomposite support. Nanoparticles are represented as dots, with overlaid transparent selective layer and grey support. (Reproduced with permission from Ref. [21].)

dimension where the size is in the nanometre range or has an internal or surface structure on the nanoscale [71–73], and by extension any composite material which contains a substantial proportion of nanoparticulates, although there is still much debate on the precise definition [74].

Nanomaterials can be created from a top-down or bottom-up approach. The former involves taking bulk material and breaking it down into smaller and smaller fragments, whether by mechanical methods such as by milling, laser ablation or sputtering [75–78] or exfoliation, which has been used for graphene [79]. Conversely, bottom-up approaches involve the creation of nanomaterials from assembling single molecules or atoms using any of a wide number of approaches [80].

3.3.1 Modification of NF Membrane Properties by Nanomaterials

Membrane technologists may decide to incorporate nanomaterials into polymer membranes to modify a number of different relevant parameters. The most common is to improve membrane transport or alter permselectivity by adjusting the pore structure and porosity of the selective layer. This is dealt with in more detail in the next section. In this section, we will consider some of the other properties which are of interest for certain applications, including allowing adsorption of dissolved substances in the feed water, reducing biofouling by use of nanomaterials which have some antimicrobial action and by the addition of photocatalytic materials to membranes to allow the active disruption of fouling layers.

3.3.1.1 Adsorption by Membrane Bound Nanomaterials

Removal, or rejection, of some dissolved species is difficult for NF membranes to achieve by size exclusion alone, due to the size of the pores present. In some instances, it may be desirable to add an adsorptive capacity to the membrane to remove any problematic small molecules or ions, preventing them from contaminating the permeate water and thus obviating the need for further downstream treatment steps. One obvious area where this may be needed is in the treatment of industrial wastewaters contaminated by heavy metals or groundwater extracted from certain sources (naturally contaminated or contaminated from mine water leaching). Typically, metal ions will pass easily through a conventional NF membrane. This also applies to other feed waters which may be metal-contaminated, such as some groundwater supplies. Nanoparticles of suitable adsorbent material may be in these cases included in a NF membrane to remove low concentrations of these contaminants [81,82]. Nanomaterials are of interest as absorbents in these instances due to their small size and very high specific surface areas allowing them to be well dispersed in the membrane structures and to allow a very high adsorptive area with relatively little material. Similarly, NF membranes have been reported electrospun from nanofibrous absorbent materials for the same applications [83].

One group utilised poly(ether sulfone) (PES) NF membranes which were fabricated by incorporating graphene oxide (GO)-modified magnetic Fe_2O_3 nanoparticles for copper and dye containing wastewater treatment [84]. Further, functionalisation of the nanoparticles with surface amine groups enhanced the metal and dye adsorption significantly. In addition, the nanoparticle incorporation resulted in a more hydrophilic membrane, which was used to explain the higher specific water fluxes observed with the modified membrane. However, with prolonged filtration times, concentration polarisation effects were observed at the membrane surface, due to ion collection, which lead to reduced flux rates.

Nanoparticles of the aluminium hydroxide oxide mineral boehmite were reported to have been functionalised with curcumin and incorporated into PES NF membranes for the removal of a variety of problematic metal ions, including iron, copper, manganese, lead, zinc and nickel [85]. Nanoparticle–metal complexes were found to form, leading to reduced concentration of those metals in the permeate.

PES NF membranes were also used as the basis for a chitosan nanoparticle-containing membrane for removal of nitrite from wastewater [86]. Despite an increase in water permeability, nitrite rejection also increased, suggesting an increase in permselectivity. Protonation of the nanoparticle surfaces at low feedwater pH values was reported to enhance nitrite removal.

One issue with the use of adsorbent materials is that sooner or later the adsorbent will become saturated with the adsorbed solute, in a concentration-dependent manner, leading to a drop in collection efficiency. When this happens, the adsorbent typically needs to be regenerated or replaced before removal efficiency can be regained. When the adsorbent is a component which is difficult to produce and expensive nanocomposite membrane, regular replacement is likely to be prohibitively expensive in terms of commercial considerations. Regeneration also leads to downtime and does not necessarily imply a full recovery of adsorptive capacity. As such, use of adsorbent membranes may have a future role for removal of low-concentration problematic solutes or ions which are hard to remove by simple membranes alone, perhaps as a final polishing step. Adsorptive nanocomposite membranes may, however, not be suitable for long-term filtration of concentrated feeds, where the membrane may become quickly saturated.

3.3.1.2 Antimicrobial Nanomaterials

When membrane elements encounter feedwaters containing microorganisms, development of biofilms coating any surface exposed to those feedwaters will inevitably occur. When that biofilm is formed over the selective layer or within the membrane void spaces, it will inevitably lead to a greater transport resistance for that membrane. Flux rates will either decline, or higher pumping pressures will need to be applied on the feed side to maintain a desired water flux. This may also lead to chemical agents used for either pre-treatment of feedwater or post-fouling cleaning of membrane units, potential downtime whilst units are cleaned. Chemical treatments, such as with hypochlorite, may damage the membrane surface, leading to reduced lifetime. Overall these effects of biofouling will lead to increased specific economic and environmental costs to produce the treated water. As such, prevention or mitigation of biofouling by incorporating some form of antimicrobial moiety to NF membranes is of considerable research interest. Nanoparticles have extremely high specific surface area, which allows rapid dissolution rate of the functional ions or increased contact surface, depending on the particular mechanism of antimicrobial action.

Various metals have long been known to have inherently antimicrobial properties, with surfaces constructed of or incorporating such metals as reducing or preventing bacterial growth, due to disruption of bacterial cell walls caused by accumulation of metal ions [87], leading to inhibition of biofilm growth [88]. Studies of effectiveness of metal action on microbial growth have shown copper, closely followed by silver, as being the most effective, with metals such as titanium and tin as being less effective [89]. As a result, these metals have been much studied in antimicrobial polymer membranes. For instance, Zhu et al. used nanocomposite copper-reduced GO as filler material when fabricating the selective layer of NF membranes [90]. This membrane was reported to significantly restrict the viability and activity of *Escherichia Coli*. Akar et al. [91] incorporated a combination of copper and selenium nanoparticles into a PES membrane. Lower fouling rates were achieved by an oxidative mechanism, although only proteins were examined as the foulants, rather than live microbial cells. Silver nanoparticles were examined by Liu et al. [92] who attached silver nanoparticles (AgNP) onto the feed-facing surface of TFC NF membranes, leading to retarded growth rates of the bacteria studied.

As the main mechanism for antimicrobial activity involves the dissolution of the material itself into a dissolved form which can enter living cells, it would be expected that their action would be time-limited. This is likely to be enhanced by the high specific area of the nanoparticles, which, ironically, is a plus point when it comes to their effectiveness. However, the use of metal nanomaterials in antibacterial surfaces elsewhere has had successful commercial application [93,94], suggesting that incorporation of antimicrobial nanoparticles into membranes do indeed have a place.

3.3.1.3 Photocatalytic Nanomaterials

Addition of photocatalytic nanoparticles to the membrane surface is capable of adding an easily controllable self-cleaning function. For instance, TiO_2 has long been known to have photosensitisation effects [95,96], including the generation of reactive oxygen species when exposed to short wavelength (<400 nm) light, and has seen general application for disruption of biofilms and self-cleaning of surfaces [97]. TiO_2 nanoparticles doped with potassium, boron and nitrogen were incorporated into a PES membrane by Zangeneh et al. [98]. Intermittent exposure to light allowed substantial recovery of flux rates when the membranes had been previously fouled with milk powder. Membranes fouled from filtration tests with wastewater from palm oil processing were visibly cleaner. Later, the same researchers investigated the use of boron doped-TiO_2-SiO_2/$CoFe_2O_4$ nanoparticles [99] and found these nanoparticles capable of reversing fouling within the membrane voids, in addition to those on the surface.

As well as foulant removal, the use of photocatalytic nanomaterials in a NF membrane can simultaneously allow degradation of unwanted solutes which may otherwise pass through a conventional NF membrane, as an alternative approach to the use of adsorbent nanomaterials. This was demonstrated by Soria et al. [100] who applied TiO_2 and ZnO nanoparticles mixed with polydopamine as a surface treatment to a commercially available membrane. As well as preventing organic and biological fouling, the membrane was capable of photocatalytic degradation of dissolved dye molecules.

3.4 NANOMATERIAL-MODIFIED NANOFILTRATION MEMBRANES

3.4.1 MODIFICATION OF MEMBRANE STRUCTURE USING NANOMATERIALS

As noted above, as well as altering the functionality of membrane materials, addition of nanoparticles alter physical properties such as pore size, porosity and permselectivity [101,102] and chemical properties such as surface energy and hydrophilicity [103–105]. The presence of and cross-linking to nanoparticles reduces the number of cross-links between polymers and reduces the density of the polymer matrix, contributing to a more open and porous structure with typically greater critical pore sizes.

As well as perturbing the structure of a polymer membrane merely by their presence, researchers have investigated the use of nanomaterials as initiation sites for polymerisation during membrane fabrication. In one instance researchers investigated the use of 'nanodiamonds' when forming PA selective layers [106] (see Figure 3.2). By adjusting the size and shape of the nanodiamonds in the reaction mixture, their degree of aggregation and hence the size of the interface with the surrounding mixture, the polymerisation of the PA precursors was affected and hence, the membrane structure was altered. Differences in morphology of the nanodiamonds affected their aggregation behaviour, which in turn affected the polymerisation of the surrounding polymer and the resultant selective-layer structure. Nanocomposite membranes were more hydrophilic, had increased water permeability and were still able to reject multivalent ions.

Li et al. investigated functionalisation of TiO_2 nanoparticles with tannic acid in the formation of polyester TFN NF membranes [107] in a one-step procedure in which nanoparticles became functionalised simultaneously with active layer fabrication by interfacial polymerisation. The use of the

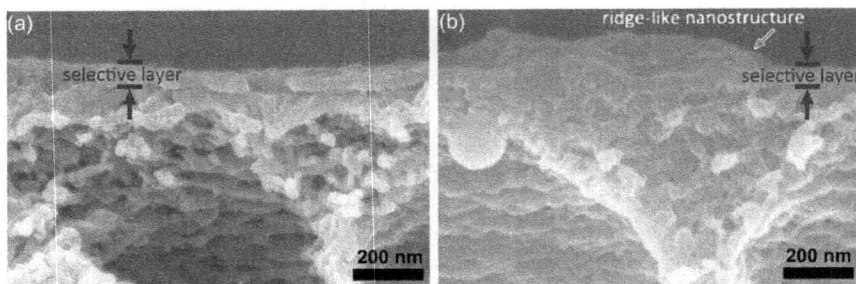

FIGURE 3.2 Comparison of (a) conventional and (b) nanodiamond-modified polyamide membranes. (Reproduced with permission from Ref. [106].)

tannic acid was to reduce the interaction between nanoparticles, allowing reduced aggregation and better dispersal within the polymer network.

3.4.2 NANOCOMPOSITE NF MEMBRANES

Nanomaterials incorporated into polymer membranes can be categorised in several ways, most notable in terms of their physical form, or according to their chemistry. They are often described as 0-, 1- or 2-dimensional nanomaterials, depending on the number of physical dimensions which are not sized in the nanoscale range. This size range is often given as between 1 and 100 nm, which excludes both larger colloidal particles and dissolved ions and small organics [108]. For instance, 0D nanomaterials are the classic nanoparticles, with all dimensions between about 1 and 100 nm [109]. Whilst these nanoparticles are not necessarily regular spheres, they are often approximated as spheres to allow easier size comparison. This category also includes particles with more complex architecture, such as nanospheres [110,111], which consist of concentric shells of different materials, nanocapsules [112], which consist of a fluid core encapsulated by a polymeric shell, and nanodroplets, which are nanoparticles formed from liquids suspended in a medium which acts as a poor solvent [113]. In addition, there are a number of types of supramolecular assembly which may have sizes in the nanoscale range, including metal organic frameworks (MOFs) or covalent organic frameworks. These may be of interest in incorporation into polymer membranes due to their open porous structure and their great potential as adsorbent materials.

Nanomaterials with sizes in a single dimension larger than the nanoscale are often termed 1D nanomaterials and include those with filamentous shapes, including carbon nanotubes (CNTs), nanofilaments and nanowires. Those materials with only one dimension in the nanoscale range are 2D nanomaterials, which include graphene and GO sheets and nanoplatelets, as well as similarly arranged structures formed from other materials.

3.4.2.1 NF Membranes Incorporating Nanoparticles

One bottom-up method which has been applied to direct fabrication of polymer membranes is by the self-assembly of amphiphilic block co-polymers [114,115]. Ma et al. [116] used such a technique to create a polymer membrane. Here, di-block copolymer nanoparticles were spray-coated into a suitable support material, with the spray alternating between nanoparticles of opposite charge to create an UF membrane, although the authors suggested that by adjusting nanoparticle size, pore sizes could be reduced to within the NF pore size range. Stability tests demonstrated that the membrane was robust enough to withstand pressures of at least 3.5 bar. pH-responsive membranes were created using a similar approach by another group which achieved typical NF pore sizes, using alternate layers of iron oxide and copolymer nanoparticles [117]. The membrane fabrication scheme for the different types of nanomaterials investigated is reproduced in Figure 3.3. This created a membrane with pore sizes as small as 2 nm, well within the typical NF membrane range. It was found that flux values varied according to the pH of the feed solution, due to changing charges of the membrane polymers, with flow being severely reduced at pH values below the pKa of the membrane polymers.

Nanoparticles are not necessarily homogenous pieces of materials, albeit at small scale, and can contain internal structure. One such type of nanoparticle is those which consist of an external, more or less spherical shell, generally porous, encapsulating another solid material, termed nanospheres [118,119]. The majority of past research has involved use in targeted drug delivery [120–123], as part of the manufacture of advanced energy storage devices [124], to function as catalysts [125] or in lithography applications [126,127]. More recently, they have been explored as potential nanofillers for polymer separation membranes [128,129], including for NF applications.

For instance, Ding et al. incorporated polypyrrole hollow nanospheres into poly(piperazine amine) to produce nanocomposite NF membranes [130]. Membranes containing the hollow nanospheres had over double the water flux values than membranes without. Divalent ion rejection rates were not adversely affected by the membrane modification. The membranes also had decreased

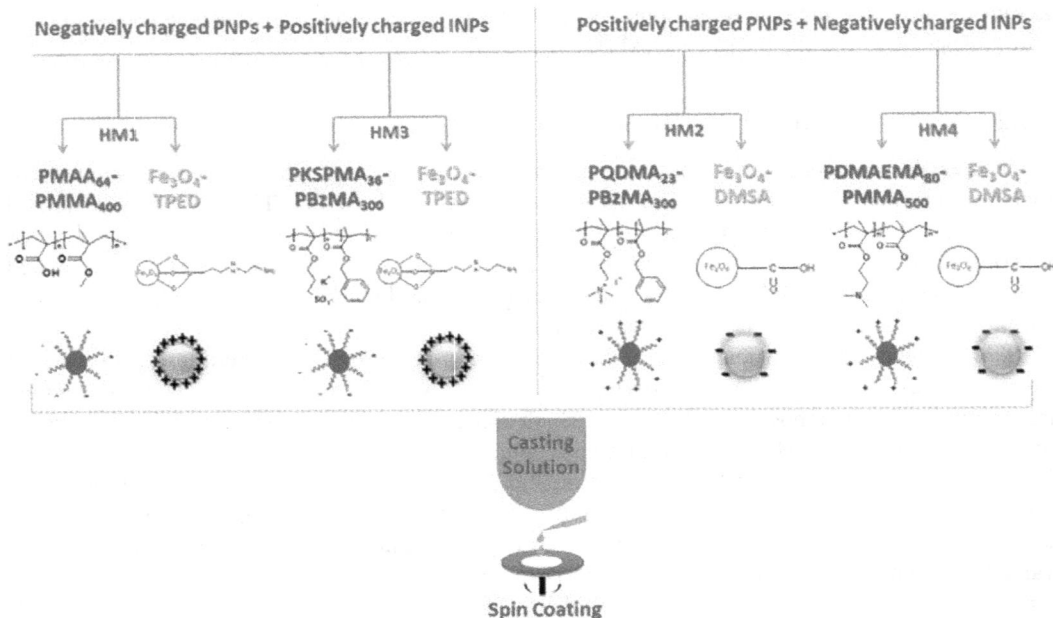

FIGURE 3.3 Fabrication scheme for nanofiltration membranes fabricated by spin coating alternate layers of inorganic and polymeric nanoparticles. (Reproduced with permission from Ref. [117].)

fouling by humic acid solutions, which was explained by increased hydrophilicity of the selective layers. In another work, Song et al. [131] added polymeric dopamine-melanin nanospheres to CTA membranes. The presence of amine and carboxyl radicals on the surface of the nanospheres had a number of positive effects, including increasing compatibility with the CTA polymer, increasing hydrophilicity of the membrane and increasing selectivity for mono and multivalent salts due to charge interactions with dissolved ions.

Clays are naturally occurring mechanically plastic materials when wet, which are composed of small particles of hydrated phyllosilicates or similar materials of particles sizes less than 4 µm [132]. When the clay particles are separated on the basis of size to allow particles in the nanosized range to be selected, they are often termed nanoclays, and exist in a number of forms based on the chemistry of the particles [133,134]. These materials are an attractive material for use as nanofillers in membranes due to their easy availability and easy handling with resultant low cost, their low toxicity and high hydrophilicity and porosity.

Bentonite nanoclay was incorporated by Maalige et al. [135] into the PA selective layer of an NF membrane. To further increase the hydrophilicity of the membrane, the bentonite was silanised to populate the surface with sulfonate groups. Selective layers containing the clay nanofiller had smoother surfaces which promoted fouling reduction. This was demonstrated through long-duration filtration experiments (50 hours) using humic acid and green dye. After cleaning, flux recovery was 3–4 times greater, suggesting a reduction in irreversible fouling. Liu et al. combined halloysite nanotubes with graphitic carbon nitride into the PA selective layer of a TFN membrane [136] by mixing with piperazine solution sprayed onto the support before adding trimesoyl chloride, which initiated polymerisation. It was found that the nanofiller material increased the surface roughness, giving a 'crumpled' effect.

Much of the research on nanoparticle use in NF membranes consists of metal, or metal oxide, nanoparticles. These nanoparticles are then usually modified afterwards by addition of surface functional groups to alter their properties, such as to improve cross-linking with the surrounding material. Lau et al. [137] used chemical vapour deposition to modify manganese oxide nanoparticles with poly(hexafluorobutyl acrylate). This increased the dispersion of the nanoparticles within the polymerisation solution. This increased dispersion increased water permeability by 22% compared with membranes incorporating unmodified nanoparticles, and by as much as 67% when no nanoparticles were present at all, whilst maintaining the rejection rate of multivalent ions. The increased surface charge of the membranes also led to reduction in fouling by bovine serum albumin (BSA), due to electrostatic repulsion effects. Kamari and Shahbazi [138] used amorphous silica obtained from sustainable sources to coat Fe_3O_4 nanoparticles, before functionalising with 3-aminopropyl(trimethoxy silane). PES membranes containing these nanoparticles were used for removal of salts, metals and textile dye components. A positive correlation was found between the proportion of nanoparticles added to the phase inversion bath and the removal efficiency of both cadmium and dye, due to adsorption of these dissolved components at the nanoparticle surface. It was found that addition of nanoparticles to the PES also increased the hydrophilicity, which was used to explain the observed increase in pure water flux.

3.4.2.2 NF Membranes Incorporating 1D Nanomaterials

The next category of nanomaterials we will look at are often termed 1D nanomaterials. That is, those who have one dimension larger than the nanosize range – i.e. those termed nanofibres, nanofilaments, nanotubes, etc. These have been made from a variety of materials, including metals or metal oxides, carbonaceous materials, for instance.

The materials in this category have attracted a great amount of interest in membrane development because, as well as having some of the advantages of nanoparticles, such as producing more open membrane structures and high specific surface areas, the networks they form with each other and the polymer substrate can potentially lead to far more mechanically strong and durable membranes [139–141].

With regard to 1D nanomaterials, the one which has possibly been utilised the most for polymer membrane fabrication are CNTs. They consist essentially of layers of graphene rolled into a closed cylindrical form and come in both single-walled and multi-walled varieties, depending on the number of graphene layers present [142]. They have high mechanical strength and are relatively easily modified [143,144], as well as being able to easily cross-link with membrane polymers [145,146]. Simple CNTs are non-polar and hence exhibit relatively high water contact angles, but this can be easily reduced by appropriate chemical modification of the surface, by addition of new functional groups or simply by acid washing with concentrated HNO_3, a treatment often used during purification to make the nanotubes more easily dispersed in water [141,147,148]. The reduction in polymer–polymer cross-linking by the disruptive presence of the nanotubes, such as for nanoparticles, leads to a more porous membrane and increased water transport, as well as potential water transport through the core of the nanotube itself [149–151].

Ma et al. [152] used the simultaneous electro-spraying of multi-walled carboxy-CNTs (Figure 3.4) with monomer droplets onto a support layer to form TFN PA membranes using interfacial polymerisation. This allowed the depth of the selective layer to be precisely controlled, which in turn affects the flux rates, with thicker membranes exhibiting greater flow resistance. Beyond the effects of precisely controlled film deposition, presence of the carboxy-CNTs increased water flux due to effects on both porosity and hydrophilicity. However, salt selectivity was little affected, meaning favourable changes in the permselectivity.

The cyclic oligosaccharide β-cyclodextrin (β-CD) has been used as an antifouling agent in membranes through its ability to increase their surface hydrophilicity and charge density [153]. Whilst they have a surface containing largely hydrophilic groups, its hydrophobic core makes it also compatible with hydrophobic molecules [154]. Rahimi et al. used β-CD to functionalise multi-walled CNTs which were incorporated into the PES selective layer of TFN membranes [155]. The β-CD-decorated CNTs had improved dispersion through the selective layer, improving membrane flux, through increased pore size, porosity and hydrophilicity, as well as increasing dye rejection and lowering organic fouling compared with plain PES surfaces. A similar approach has recently been

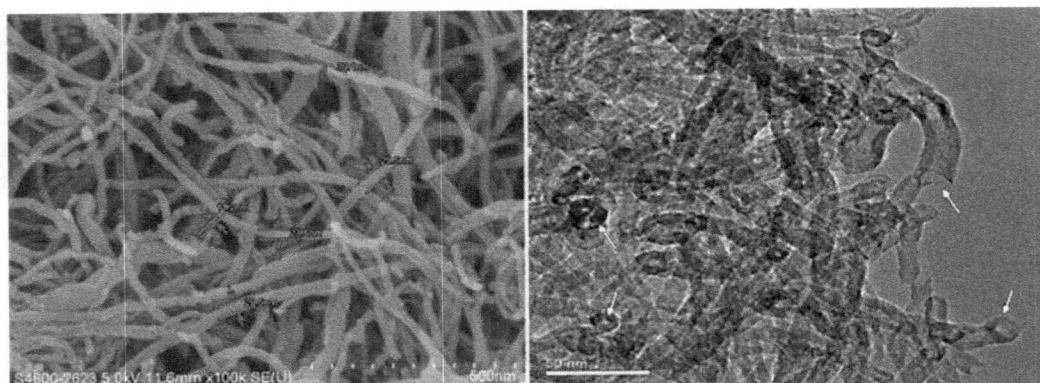

FIGURE 3.4 Carboxy-functionalised multi-walled carbon nanotubes: (a) scanning electron microscopy (SEM) image and (b) transmission electron microscopy (TEM) image. (Reproduced with permission from Ref. [152].)

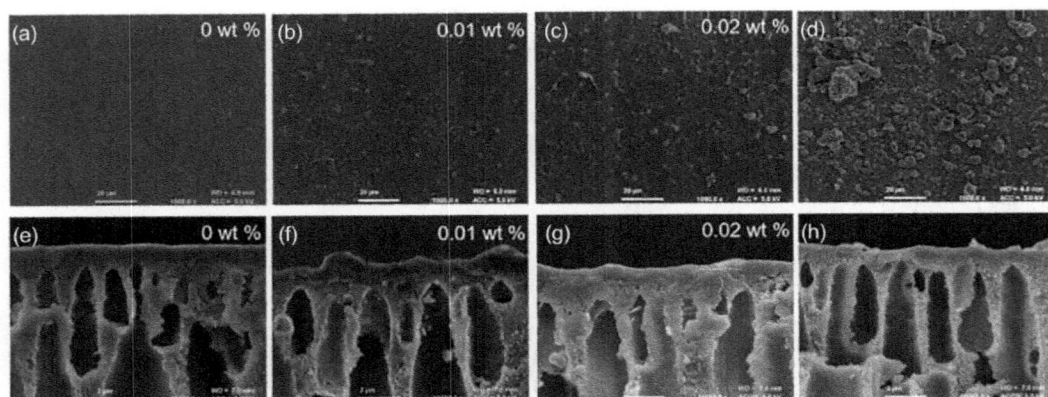

FIGURE 3.5 Surface and cross-section images of polyamide membranes embedded with increasing loading of boron nitride nanotubes: (a, e) pristine membrane; (b, f) 0.01 wt%; (c, g) 0.02 wt% (d, h) 0.03 wt%. (Reproduced with permission from Ref. [158].)

reported using GO platelets functionalised with β-CD and embedded into polymeric membranes for the treatment and reuse of oil-produced water [156].

Inkjet printing is another method which has been used to build up films of polymers and CNTs. Park et al. [157] used this technology to alternately add a layer of single-walled CNTs followed by a selective PA skin onto a PES support material. Repeat applications allowed layer thickness to be tightly controlled. Addition of CNTs reduced PES pore size and roughness. This in turn reduced the thickness of the PA film needed. As the membrane resistance is related to the selective-layer thickness, this allowed increased flow rates without compromising selectivity.

Nanotubes based on metallic materials have also been a subject of study when developing new NF membrane materials and design. Casanova et al. fabricated PA NF membranes incorporating boron nitride nanotubes (BNNTs) [158] (Figure 3.5). High performance improvements were found with low quantities of BNNTs applied to the selective layer, with flux losses after filtration of salt, humic acid and dye solutions higher than observed for PA membranes with no BNNTs, along with higher flux recovery after simple cleaning. However, the performance was heavily dependent on the loading of the nanomaterials, with peak-performing membranes having no more than 0.02% BNNT by weight. Above this value performance parameters decreased.

As well as nanotubes, solid nanofibres are of interest in membrane fabrication. For instance, cellulose nanofibres are fabricated by a number of methods from plant derived cellulose, a polymer naturally occurring in plant material [159,160]. Due to the cheap and readily available source material, low toxicity and amenability to modification, they have attracted interest as potential membrane materials [161–163]. Cellulose nanomaterials may be characterised as nanofibres or nanocrystals, depending on how the materials are structured [159]. For instance, Zhang et al. [164] used cellulose nanofibres to fabricate nanoporous separation membranes by filtration of cellulose nanofibres dispersed in aqueous solution through an aluminium support containing 200 nm diameter pores. The resultant cellulose nanofibre mats had pore sizes of between 2 and 12 nm. Specific water flux rates of 3.7–23 L/m^{2}h bar were obtained for the membranes as filtration layer thickness decreased from 47 to 23 nm. More recently, Soyekwo et al. [165] used a layer of cellulose nanofibres placed on a support material, into which a polyethylene imine selective layer was formed through interfacial polymerisation. This led to the creation of a nanostructured selective layer with pore size of 0.45 nm and high water permeability (33 L/m^{2}h bar).

Xu et al. investigated the use of nanocomposite materials formed from cellulose nanocrystals and silver embedded into PA layers to form NF membranes [166]. Cellulose nanocrystals were selected due to expected low toxicity and it being a cheap material. The cellulose nanocrystals increased water flux and rejection of multivalent ions, leading to more favourable permselectivity. The presence of the silver reduced the viability of bacteria in growth tests with *E. coli*. In addition, after filtration of organic fouling solution containing humic acid, flux recovery ratios as high as 93% were obtained after cleaning, suggesting that the majority of the fouling was loosely adhered reversible fouling.

3.4.2.3 NF Membranes Incorporating 2D Nanomaterials

Nanomaterials in this category are those for which only one dimension is in the nanosized range. As such, they consist of flat sheets of variable size, with nanosized thickness.

3.4.2.3.1 Organic 2D Nanomaterials

The most well-known family of organic materials in this category belongs to graphene and those materials with structures based on graphene, such as GO and reduced graphene oxide (rGO) [167]. Graphene consists of single or small numbers of layers of graphite, and hence graphene layers consist of carbon atoms each connected to three other carbon atoms to make a hexagonal repeating pattern, similar in appearance to a honeycomb. Due to the small size of the repeating structure, graphene sheets are impermeable to passage of atoms and molecules [168,169]. As such, to allow permeability for applications in membranes, this needs to be overcome, either by deliberately producing defects in the graphene surface [170–172], at least where large continuous sheets of graphene are to be used, or by the use of smaller pieces of graphene immersed in a polymer matrix. In this latter case, water channels of small enough dimension to retain selectivity based on size exclusion may be arrived at by using closely spaced stacks of graphene particles with fluid flow happening around the edges and between graphene layers [173–175].

The first described production and isolation of graphene involved the delamination of a graphite source using application of adhesive tape [79]. Whilst other lab-scale production methods have been developed since then, production at the large scale needed for commercial operations directly from graphite remains problematic [176,177]. Alternatively, production of GO from suitable treated graphite, is much easier, largely due to its hydrophilicity and subsequent swelling in aqueous environments, meaning graphite oxide is more easily delaminated [178–181]. When a more hydrophobic material closer to graphene is required, GO can be converted into rGO by suitable reduction reactions [182,183].

The simplest method for preparation of GO membranes is the dead-end filtration of GO suspensions through a suitable support, to produce membranes with low flow resistance [184,185] or through deposition of GO material using electro-spraying of a porous surface [186]. However, the durability of such membranes is limited by the strength of the interactions between the particles, which may be limited or dependent on the particular environment. For instance, electrostatic

FIGURE 3.6 Graphene oxide platelets imaged with SEM: (a) GO; (b) amine-GO; (c) acyl chloride-GO. (Reproduced with permission from Ref. [188].)

interactions are likely to diminish significantly in the presence of electrolyte solutions. To improve membrane strength, more durable bonds are needed, such as by cross-linking with a polymer or embedding within a polymer matrix. Careful functionalisation of the nanomaterial can be used to engineer improved compatibility and promote cross-linking with associated polymers. For instance, Xie et al. [187] created a NF membrane which had a poly(piperazine amide) (PPA) selective layer formed through interfacial polymerisation. Prior to polymerisation, particles of GO, which had been functionalised with a polyester-based material, to improve cross-linking with the PPA and hence the strength of the selective layer, were added to the polymerisation mixture. This allowed the fabrication of membranes which had the same selectivity as PPA, but could be made thinner, improving water transport properties. More recently, García-Picazo [188] fabricated PA TFC NF membranes incorporating GO particles functionalised with PA monomers (amine and acyl chloride – see Figure 3.6), leading to the GO being directly cross-linked with the polymer membrane during the interfacial polymerisation process. Compared to membranes fabricated from PA alone, improved flux rates and selectivity were reported.

The increased hydrophilicity of GO is valued by those interested in water separation membranes due to its positive effects on water transport and fouling [189,190]. However, this increased hydrophilicity is not without balancing unwanted effects. For instance, the high compatibility with water means that GO will swell due to hydration, which in turn decreases interactions between neighbouring platelets and increases separation. This increased separation makes for larger water channels [185], limiting the ability to reject smaller ions and molecules through size exclusion mechanisms, increased molecular weight cut-off and decreased mechanical stability. This has somewhat led to a focus on adjusting water channel size by GO functionalisation [162,191,192].

Dong et al. [193] formed hybrid nanosheets from the growth of $Co(OH)_2$ thin films on the surface of GO platelets. Vacuum filtration of suspensions of this nanomaterial onto a PES support layer was used to create tight membranes with a selective layer consisting of stacked nanosheets. A more gentle approach to GO nanoparticle assembly was taken by Rode et al. [194] who used diffusion-based self-assembly of GO nanoplatelets. Thin selective layers, containing as few as three GO sheets deep, allowed high water flux to be achieved in haemodialysis application. Another approach to the fabrication of membranes formed from stacked GO platelets was reported by Wang et al. [195] who used GO functionalised by polydopamine to form a selective layer using inkjet printing technology. This allowed fine-tuning of the selective-layer thickness by adjusting the number of applications.

Another carbon-based 2D nanomaterial which has been used for membrane manufacture and filtration is graphene-like carbon nitride (g-C_3N_4) [196]. For instance, Ye et al. [197] mixed g-C_3N_4 with the polymers poly ethylene imine (PEI) and polydiacetylene (PDA), with nanochannels formed around the g-C_3N_4 platelets (Figure 3.7). Flux rates were improved, but whilst good dye rejection was achieved during filtration experiments, salt rejection was poor. Nadig et al. [198] introduced

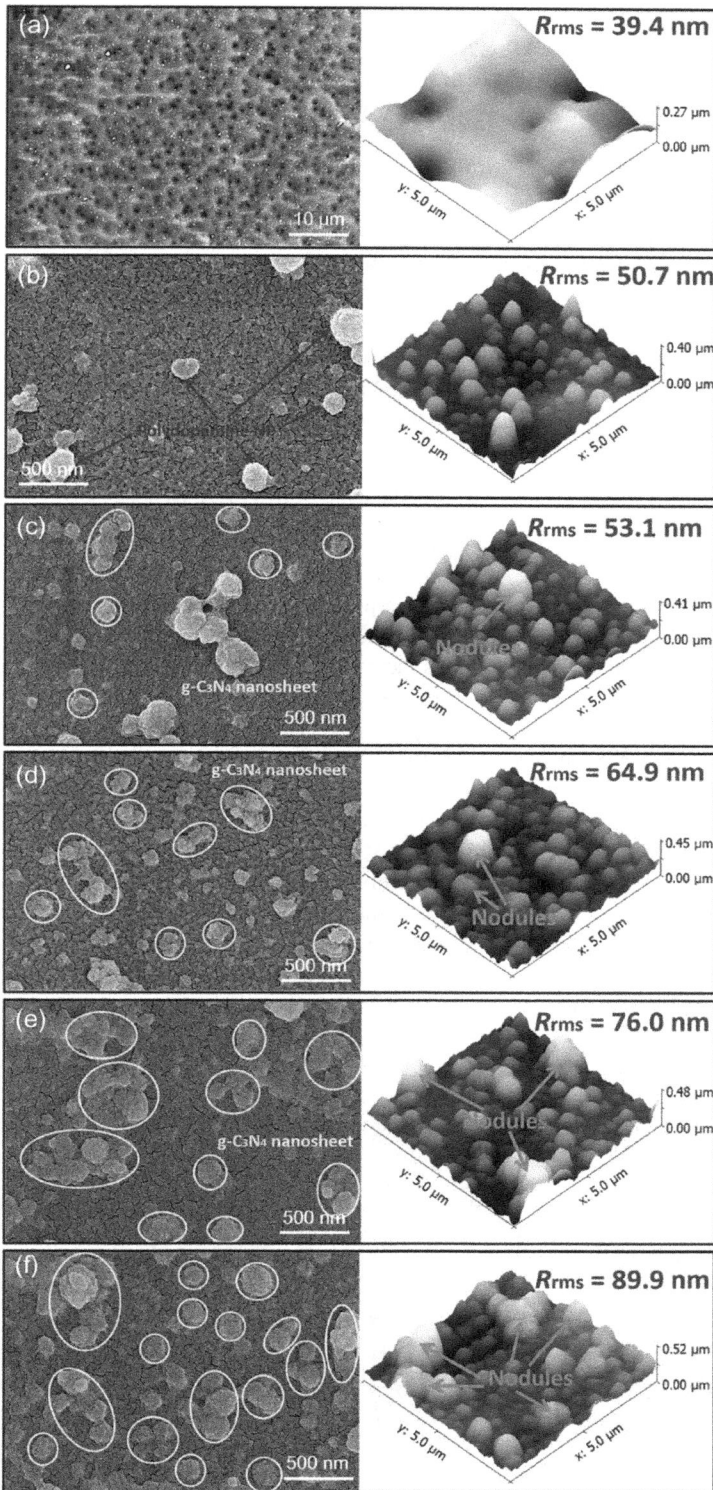

FIGURE 3.7 SEM and AFM images of NF membranes incorporating successively greater loadings of g-C$_3$N$_4$ nanoplatelets. (Reproduced with permission from Ref. [197].)

g-C_3N_4 platelets into a polysulfone matrix and investigated the effects on heavy metal removal from wastewater. The addition of the g-C_3N_4 increased the hydrophilicity of the membranes and achieved high rejection rates for lead (>95%), cadmium (>80%) and arsenic (>70%). In addition, fouling recovery rates after filtration of BSA solutions were increased by the nanofiller addition.

3.4.2.3.2 Inorganic 2D Nanomaterials

Boron nitride nanosheets functionalised with amine (BN(NH)$_2$) were used by Abdikheibari et al. [199] to form nanocomposite polymer membranes. It was found that incorporating the negatively charged edges of the amine-functionalised BN(NH)$_2$ nanocomposite membranes into the feed-facing selective layer reduced the severity of organic fouling as well as increasing water flux.

Pandey et al. [200] recently fabricated a mixed matrix NF membrane using MXene (a class of transition metal-based 2D nanomaterials with the general formula $Ti_3C_2T_x$) embedded into a CA selective layer. Compared to pure CA selective layer, water flux was much improved and bacterial growth tests on the membrane surfaces showed decreased cell viability.

3.4.3 Membranes Fabricated Using Metal Organic Frameworks

MOFs consist of metal ions complexed with organic ligands, with a 1, 2 or 3-dimensional form. They often have highly porous structures and very high specific surface areas [201], which makes them an attractive candidate for a number of adsorption applications, which include incorporation into polymer membranes [202–205]. These properties make them especially attractive for use in NF membranes where their high adsorption efficiency makes them useful for collection of low concentrations of small molecules and ions, such as pollutants, which may otherwise present in the permeate water [206,207].

Wang et al. [208] used nanoparticles of the MOF ZIF-8 as templates to adjust the physical formation of a selective layer fabricated by interfacial polymerisation (Figure 3.8). The ZIF-8 was placed across the support layer prior to selective-layer formation. This resulted in a crumpled, rougher selective-layer surface, allowing a greater interaction surface with the feedwater. This resulted in increased specific fluxes, with minimal effects on selectivity. The MOF nanoparticles played no role in the function of the membranes, as they were removed after membrane fabrication prior to filtration tests.

A similar effect on the membrane surface morphology was observed by another group when using the zirconium-based MOF UiO-66-NH$_2$ modified with palmitoyl chloride to reduce formation of aggregates [209]. Inclusion of the MOF in the selective layer led to the formation of repeated ridge-valley structures, which greatly enhanced the surface area of the membranes, with a concomitant improvement in water flux, although at the loss of some salt selectivity.

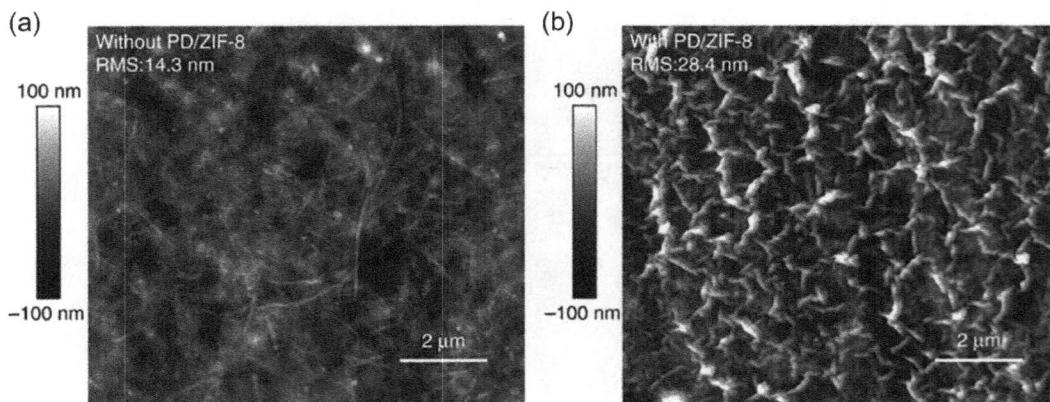

FIGURE 3.8 AFM images of (a) pristine PA membrane and (b) PA membrane modified with ZIF-8 MOF, exhibiting ridge and valley 'Turing' structures. (Reproduced with permission from Ref. [208].)

Song et al. [210] also reported similar ridge-valley structures when using a PA selective layer containing poly(vinyl acetate) (PVA). When the authors added the chromium-based MOF MIL101(Cr) to the selective layer, more complex surface structures were seen due to the nature of the solvent flow through porous structure of the MOFs, further enhancing the surface area. However, it was noted that the increased roughness did increase the fouling propensity of the membranes.

3.5 CONCLUSIONS

In recent years, there has been much interest in investigating the use of novel materials for the fabrication of NF membranes, particularly where the goal of improving water flux for low-pressure filtration applications. One focus of this research has been the investigation of a range of nanomaterials as fillers in polymer membranes. This has been with the aim of improving water permeation, improving selectivity of salt and smaller ions (although for NF membranes this is often of less interest than water flux), decreasing fouling propensity, increasing membrane strength and durability.

The introduction of nanomaterials to polymer membranes are capable of opening up the polymer structure through disrupting cross-linking, enhancing as a result the width of water channels and porosity; introducing bactericidal properties; and improving adsorption rates of undesirable small molecules and ions which might otherwise enter the permeate. This work has been carried out with nanomaterials in various physical forms, such as 0D nanoparticles and nanospheres, 2D nanotubes and nanofilaments, as well as 2D nanosheets and platelets. Each of these forms have high specific surface areas and hence high levels of interaction with the surrounding polymer and fluid materials. However, filamentous 1D nanomaterials have the advantage of increasing membrane strength, whilst 2D platelets, such as GO particles, can form stacked structures between which the water channels form, capable of being fine-tuned by appropriate functionalisation. This is a rich and ever-evolving field, as evidenced by the large and growing number of publications devoted to this field.

ACKNOWLEDGEMENTS

The authors would like to thank Tamkeen for funding the NYUAD Water Research Center under the NYUAD Research Institute Award (project CG007). This work was also supported by New York University Abu Dhabi (NYUAD) Faculty research funds (AD330).

REFERENCES

1. A.W. Mohammad, Y.H. Teow, W.L. Ang, Y.T. Chung, D.L. Oatley-Radcliffe, N. Hilal, Nanofiltration membranes review: Recent advances and future prospects, *Desalination*, 356 (2015) 226–254.
2. J. Cadotte, R. Forester, M. Kim, R. Petersen, T. Stocker, Nanofiltration membranes broaden the use of membrane separation technology, *Desalination*, 70 (1988) 77–88.
3. N. Hilal, H. Al-Zoubi, N. Darwish, A. Mohammad, M.A. Arabi, A comprehensive review of nanofiltration membranes: Treatment, pretreatment, modelling, and atomic force microscopy, *Desalination*, 170 (2004) 281–308.
4. L.W. Jye, A.F. Ismail, *Nanofiltration Membranes: Synthesis, Characterization, and Applications*, CRC Press, 2016.
5. N. Hilal, H. Al-Zoubi, A. Mohammad, N. Darwish, Nanofiltration of highly concentrated salt solutions up to seawater salinity, *Desalination*, 184 (2005) 315–326.
6. N. Hilal, H. Al-Zoubi, N.A. Darwish, A.W. Mohammad, Performance of nanofiltration membranes in the treatment of synthetic and real seawater, *Separation Science Technology*, 42 (2007) 493–515.
7. A. Lesimple, S.Y. Jasim, D.J. Johnson, N. Hilal, The role of wastewater treatment plants as tools for SARS-CoV-2 early detection and removal, *Journal of Water Process Engineering*, 38 (2020) 101544.
8. R. Singh, R. Bhadouria, P. Singh, A. Kumar, S. Pandey, V.K. Singh, Nanofiltration technology for removal of pathogens present in drinking water, in: *Waterborne Pathogens*, Elsevier, 2020, pp. 463–489.
9. I. Koyuncu, O.A. Arikan, M.R. Wiesner, C. Rice, Removal of hormones and antibiotics by nanofiltration membranes, *Journal of Membrane Science*, 309 (2008) 94–101.

10. S. Wang, L. Li, S. Yu, B. Dong, N. Gao, X. Wang, A review of advances in EDCs and PhACs removal by nanofiltration: Mechanisms, impact factors and the influence of organic matter, *Chemical Engineering Journal*, 406 (2021) 126722.

11. H.K. Shon, S. Phuntsho, D.S. Chaudhary, S. Vigneswaran, J. Cho, Nanofiltration for water and wastewater treatment – A mini review, *Drinking Water Engineering and Science*, 6 (2013) 47–53.

12. B. Cyna, G. Chagneau, G. Bablon, N. Tanghe, Two years of nanofiltration at the Méry-sur-Oise plant, France, *Desalination*, 147 (2002) 69–75.

13. A. Houari, D. Seyer, K. Kecili, V. Heim, P.D. Martino, Kinetic development of biofilm on NF membranes at the Méry-sur-Oise plant, France, *Biofouling*, 29 (2013) 109–118.

14. A. Khalik, V.S. Praptowidodo, Nanofiltration for drinking water production from deep well water, *Desalination*, 132 (2000) 287–292.

15. Y. Teow, K. Ho, J. Sum, A.W. Mohammad, Principles of nanofiltration membrane processes, in: N. Hilal, A. Ismail, M. Khayet, D. Johnson (Eds.) *Osmosis Engineering*, Elsevier, 2021.

16. A. Wahab Mohammad, N. Hilal, M. Nizam Abu Seman, A study on producing composite nanofiltration membranes with optimized properties, *Desalination*, 158 (2003) 73–78.

17. Y.H. Teow, A.W. Mohammad, New generation nanomaterials for water desalination: A review, *Desalination*, 451 (2019) 2–17.

18. D. Johnson, F. Galiano, S.A. Deowan, J. Hoinkis, A. Figoli, N. Hilal, Adhesion forces between humic acid functionalized colloidal probes and polymer membranes to assess fouling potential, *Journal of Membrane Science*, 484 (2015) 35–46.

19. F. Galiano, I. Friha, S.A. Deowan, J. Hoinkis, Y. Xiaoyun, D. Johnson, R. Mancuso, N. Hilal, B. Gabriele, S. Sayadi, A. Figoli, Novel low-fouling membranes from lab to pilot application in textile wastewater treatment, *Journal of Colloid and Interface Science*, 515 (2018) 208–220.

20. M.A. Seman, M. Khayet, N. Hilal, Comparison of two different UV-grafted nanofiltration membranes prepared for reduction of humic acid fouling using acrylic acid and N-vinylpyrrolidone, *Desalination*, 287 (2012) 19–29.

21. D.J. Johnson, N. Hilal, Nanocomposite nanofiltration membranes: State of play and recent advances, *Desalination*, 524 (2022) 115480.

22. T. Peters, Membrane technology for water treatment, *Chemical Engineering Technology*, 33 (2010) 1233–1240.

23. Y.H. Teow, J.Y. Sum, K.C. Ho, A.W. Mohammad, 3 – Principles of nanofiltration membrane processes, in: N. Hilal, A.F. Ismail, M. Khayet, D. Johnson (Eds.) *Osmosis Engineering*, Elsevier, 2021, pp. 53–95.

24. C. Linder, O. Kedem, History of Nanofiltration Membranes from 1960 to 1990, in: *Nanofiltration*, 2021, pp. 1–34.

25. W.J. Conlon, Pilot field test data for prototype ultra low pressure reverse osmosis elements, *Desalination*, 56 (1985) 203–226.

26. W.R. Bowen, J.S. Welfoot, Modelling the performance of membrane nanofiltration – Critical assessment and model development, *Chemical Engineering Science*, 57 (2002) 1121–1137.

27. Y.H. Teow, J.Y. Sum, K.C. Ho, A.W. Mohammad, Principles of nanofiltration membrane processes, in: *Osmosis Engineering*, Elsevier, 2021, pp. 53–95.

28. O. Labban, C. Liu, T.H. Chong, J.H. Lienhard, Relating transport modeling to nanofiltration membrane fabrication: Navigating the permeability-selectivity trade-off in desalination pretreatment, *Journal of Membrane Science*, 554 (2018) 26–38.

29. V. Burganos, 1.2 Modeling and simulation of membrane structure and transport properties, *Comprehensive Membrane Science and Engineering*, (2017) 17.

30. Y. Zhu, H. Zhu, G. Li, Z. Mai, Y. Gu, The effect of dielectric exclusion on the rejection performance of inhomogeneously charged polyamide nanofiltration membranes, *Journal of Nanoparticle Research*, 21 (2019) 1–13.

31. N.S. Suhalim, N. Kasim, E. Mahmoudi, I.J. Shamsudin, A.W. Mohammad, F. Mohamed Zuki, N.L.-A. Jamari, Rejection mechanism of ionic solute removal by nanofiltration membranes: An overview, 12 (2022) 437.

32. H. Zhu, B. Hu, Dielectric properties of aqueous electrolyte solutions confined in silica nanopore: Molecular simulation vs. continuum-based models, 12 (2022) 220.

33. R. Schlögl, Membrane permeation in systems far from equilibrium, *Berichte der Bunsengesellschaft für physikalische Chemie*, 70 (1966) 400–414.

34. L. Dresner, Some remarks on the integration of the extended Nernst-Planck equations in the hyperfiltration of multicomponent solutions, *Desalination*, 10 (1972) 27–46.

35. W. Deen, Hindered transport of large molecules in liquid-filled pores, *AIChE Journal*, 33 (1987) 1409–1425.
36. T. Tsuru, S.-i. Nakao, S. Kimura, Calculation of ion rejection by extended Nernst–Planck equation with charged reverse osmosis membranes for single and mixed electrolyte solutions, *Journal of Chemical Engineering of Japan*, 24 (1991) 511–517.
37. F. Morrison Jr, J. Osterle, Electrokinetic energy conversion in ultrafine capillaries, *The Journal of Chemical Physics*, 43 (1965) 2111–2115.
38. G. Rios, R. Joulie, S. Sarrade, M. Carles, Investigation of ion separation by microporous nanofiltration membranes, *AIChE Journal*, 42 (1996) 2521–2528.
39. X.-L. Wang, T. Tsuru, S.-i. Nakao, S. Kimura, The electrostatic and steric-hindrance model for the transport of charged solutes through nanofiltration membranes, *Journal of Membrane Science*, 135 (1997) 19–32.
40. C. Combe, C. Guizard, P. Aimar, V. Sanchez, Experimental determination of four characteristics used to predict the retention of a ceramic nanofiltration membrane, *Journal of Membrane Science*, 129 (1997) 147–160.
41. W.R. Bowen, A.W. Mohammad, N. Hilal, Characterisation of nanofiltration membranes for predictive purposes – Use of salts, uncharged solutes and atomic force microscopy, *Journal of Membrane Science*, 126 (1997) 91–105.
42. W. Richard Bowen, A. Wahab Mohammad, Diafiltration by nanofiltration: Prediction and optimization, *AIChE Journal*, 44 (1998) 1799–1812.
43. S. Bandini, D. Vezzani, Nanofiltration modeling: The role of dielectric exclusion in membrane characterization, *Chemical Engineering Science*, 58 (2003) 3303–3326.
44. R. Wang, S. Lin, Pore model for nanofiltration: History, theoretical framework, key predictions, limitations, and prospects, *Journal of Membrane Science*, (2020) 118809.
45. S. Darvishmanesh, A. Buekenhoudt, J. Degrève, B. Van der Bruggen, Coupled series–parallel resistance model for transport of solvent through inorganic nanofiltration membranes, *Separation and Purification Technology*, 70 (2009) 46–52.
46. A. Yaroshchuk, M.L. Bruening, E. Zholkovskiy, Modelling nanofiltration of electrolyte solutions, *Advances in Colloid and Interface Science*, 268 (2019) 39–63.
47. J. Park, K. Jeong, S. Baek, S. Park, M. Ligaray, T.H. Chong, K.H. Cho, Modeling of NF/RO membrane fouling and flux decline using real-time observations, *Journal of Membrane Science*, 576 (2019) 66–77.
48. S. Bandehali, F. Parvizian, H. Ruan, A. Moghadassi, J. Shen, A. Figoli, A.S. Adeleye, N. Hilal, T. Matsuura, E. Drioli, S.M. Hosseini, A planned review on designing of high-performance nanocomposite nanofiltration membranes for pollutants removal from water, *Journal of Industrial and Engineering Chemistry*, 101 (2021) 78–125.
49. H.K. Lonsdale, U. Merten, R.L. Riley, Transport properties of cellulose acetate osmotic membranes, *Journal of Applied Polymer Science*, 9 (1965) 1341–1362.
50. E. Ferjani, R.H. Lajimi, A. Deratani, M.S. Roudesli, Bulk and surface modification of cellulose diacetate based RO/NF membranes by polymethylhydrosiloxane preparation and characterization, *Desalination*, 146 (2002) 325–330.
51. S. Yu, Q. Cheng, C. Huang, J. Liu, X. Peng, M. Liu, C. Gao, Cellulose acetate hollow fiber nanofiltration membrane with improved permselectivity prepared through hydrolysis followed by carboxymethylation, *Journal of Membrane Science*, 434 (2013) 44–54.
52. G. Arthanareeswaran, S.A. Kumar, Effect of additives concentration on performance of cellulose acetate and polyethersulfone blend membranes, *Journal of Porous Materials*, 17 (2010) 515–522.
53. W.L. Chou, D.G. Yu, M.C. Yang, The preparation and characterization of silver-loading cellulose acetate hollow fiber membrane for water treatment, *Polymers for Advanced Technologies*, 16 (2005) 600–607.
54. S.F. Anis, B.S. Lalia, A. Lesimple, R. Hashaikeh, N. Hilal, Electrically conductive membranes for contemporaneous dye rejection and degradation, *Chemical Engineering Journal*, 428 (2022) 131184.
55. V. Vatanpour, S.S. Mousavi Khadem, M. Masteri-Farahani, N. Mosleh, M.R. Ganjali, A. Badiei, E. Pourbashir, A.H. Mashhadzadeh, M. Tajammal Munir, G. Mahmodi, P. Zarrintaj, J.D. Ramsey, S.-J. Kim, M.R. Saeb, Anti-fouling and permeable polyvinyl chloride nanofiltration membranes embedded by hydrophilic graphene quantum dots for dye wastewater treatment, *Journal of Water Process Engineering*, 38 (2020) 101652.
56. L.Y. Ng, A.W. Mohammad, C.Y. Ng, A review on nanofiltration membrane fabrication and modification using polyelectrolytes: Effective ways to develop membrane selective barriers and rejection capability, *Advances in Colloid and Interface Science*, 197 (2013) 85–107.

57. N. Hilal, M. Khayet, C.J. Wright, *Membrane Modification: Technology and Applications*, CRC Press, 2012.
58. Y. Zhang, Y. Su, J. Peng, X. Zhao, J. Liu, J. Zhao, Z. Jiang, Composite nanofiltration membranes prepared by interfacial polymerization with natural material tannic acid and trimesoyl chloride, *Journal of Membrane Science*, 429 (2013) 235–242.
59. Y. Li, Y. Su, Y. Dong, X. Zhao, Z. Jiang, R. Zhang, J. Zhao, Separation performance of thin-film composite nanofiltration membrane through interfacial polymerization using different amine monomers, *Desalination*, 333 (2014) 59–65.
60. S. Li, S. Liu, F. Huang, S. Lin, H. Zhang, S. Cao, L. Chen, Z. He, R. Lutes, J. Yang, Engineering, preparation and characterization of cellulose-based nanofiltration membranes by interfacial polymerization with piperazine and trimesoyl chloride, *ACS Sustainable Chemistry & Engineering*, 6 (2018) 13168–13176.
61. M.A. Seman, M. Khayet, N. Hilal, Nanofiltration thin-film composite polyester polyethersulfone-based membranes prepared by interfacial polymerization, 348 (2010) 109–116.
62. S. Yu, M. Ma, J. Liu, J. Tao, M. Liu, C. Gao, Study on polyamide thin-film composite nanofiltration membrane by interfacial polymerization of polyvinylamine (PVAm) and isophthaloyl chloride (IPC), *Journal of Membrane Science*, 379 (2011) 164–173.
63. S.H. Chen, D.J. Chang, R.M. Liou, C.S. Hsu, S.S. Lin, Preparation and separation properties of polyamide nanofiltration membrane, *Journal of Applied Polymer Science*, 83 (2002) 1112–1118.
64. Q. Liu, L. Li, Z. Pan, Q. Dong, N. Xu, T. Wang, Inorganic nanoparticles incorporated in polyacrylonitrile-based mixed matrix membranes for hydrophilic, ultrafast, and fouling-resistant ultrafiltration, *Journal of Applied Polymer Science*, 136 (2019) 47902.
65. A. Abdel-Karim, M.E. El-Naggar, E. Radwan, I.M. Mohamed, M. Azaam, E.-R. Kenawy, High-performance mixed-matrix membranes enabled by organically/inorganic modified montmorillonite for the treatment of hazardous textile wastewater, *Chemical Engineering Journal*, 405 (2021) 126964.
66. M. Delavar, G. Bakeri, M. Hosseini, N. Nabian, Fabrication and characterization of polyvinyl chloride mixed matrix membranes containing high aspect ratio anatase titania and hydrous manganese oxide nanoparticle for efficient removal of heavy metal ions: Competitive removal study, *The Canadian Journal of Chemical Engineering*, 98 (2020) 1558–1579.
67. M.H.D.A. Farahani, V. Vatanpour, A comprehensive study on the performance and antifouling enhancement of the PVDF mixed matrix membranes by embedding different nanoparticulates: Clay, functionalized carbon nanotube, SiO2 and TiO2, *Separation and Purification Technology*, 197 (2018) 372–381.
68. J. García-Ivars, M.-J. Corbatón-Báguena, M.-I. Iborra-Clar, Chapter 6: Development of mixed matrix membranes: Incorporation of metal nanoparticles in polymeric membranes, in: S. Thomas, D. Pasquini, S.-Y. Leu, D.A. Gopakumar (Eds.) *Nanoscale Materials in Water Purification*, Elsevier, 2019, pp. 153–178.
69. L. Upadhyaya, B. Oliveira, V.J. Pereira, M.T.B. Crespo, J.G. Crespo, D. Quemener, M. Semsarilar, Nanocomposite membranes from nano-particles prepared by polymerization induced self-assembly and their biocidal activity, *Separation and Purification Technology*, 251 (2020) 117375.
70. S. Al Aani, V. Gomez, C.J. Wright, N. Hilal, Fabrication of antibacterial mixed matrix nanocomposite membranes using hybrid nanostructure of silver coated multi-walled carbon nanotubes, *Chemical Engineering Journal*, 326 (2017) 721–736.
71. ISO/TR 18401:2017(en) Nanotechnologies – Plain language explanation of selected terms from the ISO/IEC 80004 series, https://www.iso.org/obp/ui/#iso:std:iso:tr:18401:ed-1:v1:en.
72. E.P.C.o.t.E. Union, Regulation (EU) 2017/745 of the European Parliament and of the Council of 5 April 2017 on medical devices, amending Directive 2001/83/EC, Regulation (EC) No 178/2002 and Regulation (EC) No 1223/2009 and repealing Council Directives 90/385/EEC and 93/42/EEC, *Official Journal of the EU*, L117 (2017) 1–175.
73. EC, Commission Regulation (EU) 2018/1881 of 3 December 2018 amending Regulation (EC) No 1907/2006 of the European Parliament and of the Council on the Registration, Evaluation, Authorisation and Restriction of Chemicals (REACH) as regards Annexes I, III, VI, VII, VIII, IX, X, XI, and XII to address nanoforms of substances, *Official Journal of the EU*, L308 (2018).
74. M. Miernicki, T. Hofmann, I. Eisenberger, F. von der Kammer, A. Praetorius, Legal and practical challenges in classifying nanomaterials according to regulatory definitions, *Nature Nanotechnology*, 14 (2019) 208–216.
75. H. Lyu, B. Gao, F. He, C. Ding, J. Tang, J.C. Crittenden, Ball-milled carbon nanomaterials for energy and environmental applications, *ACS Sustainable Chemistry & Engineering*, 5 (2017) 9568–9585.
76. M. Kumar, X. Xiong, Z. Wan, Y. Sun, D.C. Tsang, J. Gupta, B. Gao, X. Cao, J. Tang, Y.S. Ok, Ball milling as a mechanochemical technology for fabrication of novel biochar nanomaterials, *Bioresource Technology*, 312 (2020) 123613.

77. H. Zeng, X.W. Du, S.C. Singh, S.A. Kulinich, S. Yang, J. He, W. Cai, Nanomaterials via laser ablation/irradiation in liquid: A review, *Advanced Functional Materials*, 22 (2012) 1333–1353.

78. P. Ayyub, R. Chandra, P. Taneja, A. Sharma, R. Pinto, Synthesis of nanocrystalline material by sputtering and laser ablation at low temperatures, *Applied Physics A*, 73 (2001) 67–73.

79] K.S. Novoselov, A.K. Geim, S.V. Morozov, D.-e. Jiang, Y. Zhang, S.V. Dubonos, I.V. Grigorieva, A.A. Firsov, Electric field effect in atomically thin carbon films, *Science*, 306 (2004) 666–669.

80. N. Baig, I. Kammakakam, W. Falath, Nanomaterials: A review of synthesis methods, properties, recent progress, and challenges, *Materials Advances*, 2 (2021) 1821–1871.

81. B.A.M. Al-Rashdi, D.J. Johnson, N. Hilal, Removal of heavy metal ions by nanofiltration, *Desalination*, 315 (2013) 2–17.

82. Q. Zia, M. Tabassum, J. Meng, Z. Xin, H. Gong, J. Li, Polydopamine-assisted grafting of chitosan on porous poly (L-lactic acid) electrospun membranes for adsorption of heavy metal ions, *International Journal of Biological Macromolecules*, 167 (2021) 1479–1490.

83. F. Zhu, Y.-M. Zheng, B.-G. Zhang, Y.-R. Dai, A critical review on the electrospun nanofibrous membranes for the adsorption of heavy metals in water treatment, *Journal of Hazardous Materials*, 401 (2021) 123608.

84. G. Abdi, A. Alizadeh, S. Zinadini, G. Moradi, Removal of dye and heavy metal ion using a novel synthetic polyethersulfone nanofiltration membrane modified by magnetic graphene oxide/metformin hybrid, *Journal of Membrane Science*, 552 (2018) 326–335.

85. G. Moradi, S. Zinadini, L. Rajabi, A. Ashraf Derakhshan, Removal of heavy metal ions using a new high performance nanofiltration membrane modified with curcumin boehmite nanoparticles, *Chemical Engineering Journal*, 390 (2020) 124546.

86. N. Ghaemi, P. Daraei, F.S. Akhlaghi, Polyethersulfone nanofiltration membrane embedded by chitosan nanoparticles: Fabrication, characterization and performance in nitrate removal from water, *Carbohydrate Polymers*, 191 (2018) 142–151.

87. M. Yasuyuki, K. Kunihiro, S. Kurissery, N. Kanavillil, Y. Sato, Y. Kikuchi, Antibacterial properties of nine pure metals: A laboratory study using *Staphylococcus aureus* and *Escherichia coli*, *Biofouling*, 26 (2010) 851–858.

88. J.J. Harrison, H. Ceri, C.A. Stremick, R.J. Turner, Biofilm susceptibility to metal toxicity, *Environmental Microbiology*, 6 (2004) 1220–1227.

89. I.D. Akhidime, F. Saubade, P.S. Benson, J.A. Butler, S. Olivier, P. Kelly, J. Verran, K.A. Whitehead, The antimicrobial effect of metal substrates on food pathogens, *Food and Bioproducts Processing*, 113 (2019) 68–76.

90. J. Zhu, J. Wang, A.A. Uliana, M. Tian, Y. Zhang, Y. Zhang, A. Volodin, K. Simoens, S. Yuan, J. Li, J. Lin, K. Bernaerts, B. Van der Bruggen, Mussel-inspired architecture of high-flux loose nanofiltration membrane functionalized with antibacterial reduced graphene oxide–copper nanocomposites, *ACS Applied Materials & Interfaces*, 9 (2017) 28990–29001.

91. N. Akar, B. Asar, N. Dizge, I. Koyuncu, Investigation of characterization and biofouling properties of PES membrane containing selenium and copper nanoparticles, *Journal of Membrane Science*, 437 (2013) 216–226.

92. S. Liu, F. Fang, J. Wu, K. Zhang, The anti-biofouling properties of thin-film composite nanofiltration membranes grafted with biogenic silver nanoparticles, *Desalination*, 375 (2015) 121–128.

93. J. Pulit-Prociak, M. Banach, Silver nanoparticles – A material of the future…? *Journal of Open Chemistry*, 14 (2016) 76–91.

94. B. Nowack, H.F. Krug, M. Height, 120 years of nanosilver history: Implications for policy makers, *Environmental Science & Technology*, 45 (2011) 1177–1183.

95. C.F. Goodeve, J.A. Kitchener, Photosensitisation by titanium dioxide, *Transactions of the Faraday Society*, 34 (1938) 570–579.

96. C.F. Goodeve, J.A. Kitchener, The mechanism of photosensitisation by solids, *Transactions of the Faraday Society*, 34 (1938) 902–908.

97. B. Jalvo, M. Faraldos, A. Bahamonde, R. Rosal, Antimicrobial and antibiofilm efficacy of self-cleaning surfaces functionalized by TiO_2 photocatalytic nanoparticles against *Staphylococcus aureus* and *Pseudomonas putida*, *Journal of Hazardous Materials*, 340 (2017) 160–170.

98. H. Zangeneh, A.A. Zinatizadeh, S. Zinadini, M. Feyzi, D.W. Bahnemann, A novel photocatalytic self-cleaning PES nanofiltration membrane incorporating triple metal-nonmetal doped TiO_2 (K-B-N-TiO2) for post treatment of biologically treated palm oil mill effluent, *Reactive and Functional Polymers*, 127 (2018) 139–152.

99. H. Zangeneh, A.A. Zinatizadeh, S. Zinadini, M. Feyzi, D.W. Bahnemann, Preparation and characterization of a novel photocatalytic self-cleaning PES nanofiltration membrane by embedding a visible-driven photocatalyst boron doped-TiO$_2$SiO$_2$/CoFe$_2$O$_4$ nanoparticles, *Separation and Purification Technology*, 209 (2019) 764–775.

100. R. Bahamonde Soria, J. Zhu, I. Gonza, B. Van der Bruggen, P. Luis, Effect of (TiO$_2$: ZnO) ratio on the anti-fouling properties of bio-inspired nanofiltration membranes, *Separation and Purification Technology*, 251 (2020) 117280.

101. I. Soroko, A. Livingston, Impact of TiO2 nanoparticles on morphology and performance of crosslinked polyimide organic solvent nanofiltration (OSN) membranes, *Journal of Membrane Science*, 343 (2009) 189–198.

102. E.P. Chan, W.D. Mulhearn, Y.-R. Huang, J.-H. Lee, D. Lee, C.M. Stafford, Tailoring the permselectivity of water desalination membranes via nanoparticle assembly, *Langmuir*, 30 (2014) 611–616.

103. S. Balta, A. Sotto, P. Luis, L. Benea, B. Van der Bruggen, J. Kim, A new outlook on membrane enhancement with nanoparticles: The alternative of ZnO, *Journal of Membrane Science*, 389 (2012) 155–161.

104. S.M. Hosseini, F. Karami, S.K. Farahani, S. Bandehali, J. Shen, E. Bagheripour, A. Seidypoor, Tailoring the separation performance and antifouling property of polyethersulfone based NF membrane by incorporating hydrophilic CuO nanoparticles, *Korean Journal of Chemical Engineering*, 37 (2020) 866–874.

105. B. Van der Bruggen, Chemical modification of polyethersulfone nanofiltration membranes: A review, *Journal of Applied Polymer Science*, 114 (2009) 630–642.

106. D. Qin, G. Huang, D. Terada, H. Jiang, M.M. Ito, A. H. Gibbons, R. Igarashi, D. Yamaguchi, M. Shirakawa, E. Sivaniah, B. Ghalei, Nanodiamond mediated interfacial polymerization for high performance nanofiltration membrane, *Journal of Membrane Science*, 603 (2020).

107. T. Skorjanc, D. Shetty, F. Gándara, L. Ali, J. Raya, G. Das, M.A. Olson, A. Trabolsi, Remarkably efficient removal of toxic bromate from drinking water with a porphyrin–viologen covalent organic framework, *Chemical Science*, 11 (2020) 845–850.

108. C. Jiang, L. Tian, Z. Zhai, Y. Shen, W. Dong, M. He, Y. Hou, Q.J. Niu, Thin-film composite membranes with aqueous template-induced surface nanostructures for enhanced nanofiltration, *Journal of Membrane Science*, 589 (2019) 117244.

109. M.M. Modena, B. Rühle, T.P. Burg, S. Wuttke, Nanoparticle characterization: What to measure? 31 (2019) 1901556.

110. R. Bardhan, S. Mukherjee, N.A. Mirin, S.D. Levit, P. Nordlander, N.J. Halas, Nanosphere-in-a-nanoshell: A simple nanomatryushka, *The Journal of Physical Chemistry C*, 114 (2010) 7378–7383.

111. M.S. Fleming, T.K. Mandal, D.R. Walt, Nanosphere–microsphere assembly: Methods for core–shell materials preparation, *Chemistry of Materials*, 13 (2001) 2210–2216.

112. P.N. Ezhilarasi, P. Karthik, N. Chhanwal, C. Anandharamakrishnan, Nanoencapsulation techniques for food bioactive components: A review, *Food and Bioprocess Technology*, 6 (2013) 628–647.

113. E. Zdrali, Y. Chen, H.I. Okur, D.M. Wilkins, S. Roke, The molecular mechanism of nanodroplet stability, *ACS Nano*, 11 (2017) 12111–12120.

114. S. Förster, M. Antonietti, Amphiphilic block copolymers in structure-controlled nanomaterial hybrids, *Advanced Materials*, 10 (1998) 195–217.

115. S.S. Su, I. Chang, Review of production routes of nanomaterials, in: D. Brabazon, E. Pellicer, F. Zivic, J. Sort, M. Dolors Baró, N. Grujovic, K.-L. Choy (Eds.) *Commercialization of Nanotechnologies–A Case Study Approach*, Springer International Publishing, Cham, 2018, pp. 15–29.

116. J. Ma, H.M. Andriambololona, D. Quemener, M. Semsarilar, Membrane preparation by sequential spray deposition of polymer PISA nanoparticles, *Journal of Membrane Science*, 548 (2018) 42–49.

117. U. Farooq, L. Upadhyaya, A. Shakeel, G. Martinez, M. Semsarilar, pH-responsive nano-structured membranes prepared from oppositely charged block copolymer nanoparticles and iron oxide nanoparticles, *Journal of Membrane Science*, 611 (2020) 118181.

118. K. Letchford, H. Burt, A review of the formation and classification of amphiphilic block copolymer nanoparticulate structures: Micelles, nanospheres, nanocapsules and polymersomes, *European Journal of Pharmaceutics and Biopharmaceutics*, 65 (2007) 259–269.

119. B. Zuo, W. Li, X. Wu, S. Wang, Q. Deng, M. Huang, Recent advances in the synthesis, surface modifications and applications of core-shell magnetic mesoporous silica nanospheres, 15 (2020) 1248–1265.

120. B. Sui, X. Liu, J. Sun, Dual-functional dendritic mesoporous bioactive glass nanospheres for calcium influx-mediated specific tumor suppression and controlled drug delivery in vivo, *ACS Applied Materials & Interfaces*, 10 (2018) 23548–23559.

121. K. AbouAitah, W. Lojkowski, Delivery of natural agents by means of mesoporous silica nanospheres as a promising anticancer strategy, 13 (2021) 143.

122. R. Gref, A. Domb, P. Quellec, T. Blunk, R.H. Müller, J.M. Verbavatz, R. Langer, The controlled intravenous delivery of drugs using PEG-coated sterically stabilized nanospheres, *Advanced Drug Delivery Reviews*, 16 (1995) 215–233.
123. Y. Tai, L. Wang, G. Yan, J.-m. Gao, H. Yu, L. Zhang, Recent research progress on the preparation and application of magnetic nanospheres, 60 (2011) 976–994.
124. M. Wang, Y. Huang, X. Chen, K. Wang, H. Wu, N. Zhang, H. Fu, Synthesis of nitrogen and sulfur co-doped graphene supported hollow ZnFe2O4 nanosphere composites for application in lithium-ion batteries, *Journal of Alloys and Compounds*, 691 (2017) 407–415.
125. Q. Chen, Q. Zhang, H. Liu, J. Liang, W. Peng, Y. Li, F. Zhang, X. Fan, Preparation of hollow cobalt–iron phosphides nanospheres by controllable atom migration for enhanced water oxidation and splitting, 17 (2021) 2007858.
126. X. Xu, Q. Yang, N. Wattanatorn, C. Zhao, N. Chiang, S.J. Jonas, P.S. Weiss, Multiple-patterning nanosphere lithography for fabricating periodic three-dimensional hierarchical nanostructures, *ACS Nano*, 11 (2017) 10384–10391.
127. J. Li, Y. Hu, L. Yu, L. Li, D. Ji, L. Li, W. Hu, H. Fuchs, Recent advances of nanospheres lithography in organic electronics, *Small*, 17 (2021) 2100724.
128. J. Wu, Y. Ding, J. Wang, T. Li, H. Lin, J. Wang, F. Liu, Facile fabrication of nanofiber-and micro/nanosphere-coordinated PVDF membrane with ultrahigh permeability of viscous water-in-oil emulsions, *Journal of Materials Chemistry A*, 6 (2018) 7014–7020.
129. J. Zhang, J.A. Schott, Y. Li, W. Zhan, S.M. Mahurin, K. Nelson, X.G. Sun, M.P. Paranthaman, S. Dai, Membrane-based gas separation accelerated by hollow nanosphere architectures, *Advanced Materials*, 29 (2017) 1603797.
130. X. Ding, X. Li, H. Zhao, J. Yao, Y. Zhang, Improvement of poly(Piperazine-amide) composite nanofiltration membranes by incorporating of hollow polypyrrole nanospheres with mesoporous shells, *Desalination and Water Treatment*, 208 (2020) 32–42.
131. X. Song, Y. Zhang, Y. Wang, M. Huang, S. Gul, H. Jiang, Nanocomposite membranes embedded with dopamine-melanin nanospheres for enhanced interfacial compatibility and nanofiltration performance, *Separation and Purification Technology*, 242 (2020).
132. F. Bergaya, G. Lagaly, *Handbook of Clay Science*, Newnes, 2013.
133. O. Gupta, S. Roy, Chapter 2: Recent progress in the development of nanocomposite membranes, in: M. Sadrzadeh, T. Mohammadi (Eds.) *Nanocomposite Membranes for Water and Gas Separation*, Elsevier, 2020, pp. 29–67.
134. F. Constantinescu, O.A. Boiu Sicuia, Chapter 13: Phytonanotechnology and plant protection, in: N. Thajuddin, S. Mathew (Eds.) *Phytonanotechnology*, Elsevier, 2020, pp. 245–287.
135. R. Nidhi Maalige, K. Aruchamy, A. Mahto, V. Sharma, D. Deepika, D. Mondal, S.K. Nataraj, Low operating pressure nanofiltration membrane with functionalized natural nanoclay as antifouling and flux promoting agent, *Chemical Engineering Journal*, 358 (2019) 821–830.
136. Y. Liu, X. Wang, X. Gao, J. Zheng, J. Wang, A. Volodin, Y.F. Xie, X. Huang, B. Van der Bruggen, J. Zhu, High-performance thin film nanocomposite membranes enabled by nanomaterials with different dimensions for nanofiltration, *Journal of Membrane Science*, 596 (2020).
137. G.S. Lai, W.J. Lau, P.S. Goh, M. Karaman, M. Gürsoy, A.F. Ismail, Development of thin film nanocomposite membrane incorporated with plasma enhanced chemical vapor deposition-modified hydrous manganese oxide for nanofiltration process, *Composites Part B: Engineering*, 176 (2019) 107328.
138. S. Kamari, A. Shahbazi, Biocompatible Fe_3O_4@ SiO_2-NH_2 nanocomposite as a green nanofiller embedded in PES–nanofiltration membrane matrix for salts, heavy metal ion and dye removal: Long–term operation and reusability tests, *Chemosphere*, 243 (2020) 125282.
139. P. Zhang, L. Yang, L. Li, M. Ding, Y. Wu, R. Holze, Enhanced electrochemical and mechanical properties of P (VDF-HFP)-based composite polymer electrolytes with SiO2 nanowires, *Journal of Membrane Science*, 379 (2011) 80–85.
140. Y. Feng, K. Wang, C.H. Davies, H. Wang, Carbon nanotube/alumina/polyethersulfone hybrid hollow fiber membranes with enhanced mechanical and anti-fouling properties, *Nanomaterials*, 5 (2015) 1366–1378.
141. Y.M. Manawi, K. Wang, V. Kochkodan, D.J. Johnson, M.A. Atieh, M.K. Khraisheh, Engineering the surface and mechanical properties of water desalination membranes using ultralong carbon nanotubes, *Membranes*, 8 (2018) 106.
142. M.S. Dresselhaus, G. Dresselhaus, P. Eklund, A. Rao, Carbon nanotubes, in: *The Physics of Fullerene-Based and Fullerene-Related Materials*, Springer, 2000, pp. 331–379.

143. Z. Abousalman-Rezvani, P. Eskandari, H. Roghani-Mamaqani, M. Salami-Kalajahi, Functionalization of carbon nanotubes by combination of controlled radical polymerization and "grafting to" method, *Advances in Colloid and Interface Science*, 278 (2020) 102126.

144. M. Sianipar, S.H. Kim, F. Iskandar, I.G. Wenten, Functionalized carbon nanotube (CNT) membrane: Progress and challenges, *RSC Advances*, 7 (2017) 51175–51198.

145. J. Cha, J. Kim, S. Ryu, S.H. Hong, Comparison to mechanical properties of epoxy nanocomposites reinforced by functionalized carbon nanotubes and graphene nanoplatelets, *Composites Part B: Engineering*, 162 (2019) 283–288.

146. Q. Duan, S. Wang, Q. Wang, T. Li, S. Chen, M. Miao, D. Zhang, Simultaneous improvement on strength, modulus, and elongation of carbon nanotube films functionalized by hyperbranched polymers, *ACS Applied Materials & Interfaces*, 11 (2019) 36278–36285.

147. M. Pumera, B. Šmíd, K. Veltruská, Influence of nitric acid treatment of carbon nanotubes on their physico-chemical properties, *Journal of Nanoscience and Nanotechnology*, 9 (2009) 2671–2676.

148. Ü. Anik, S. Cevik, M. Pumera, Effect of nitric acid "washing" procedure on electrochemical behavior of carbon nanotubes and glassy carbon μ-particles, *Nanoscale Research Letters*, 5 (2010) 846–852.

149. H.D. Lee, H.W. Kim, Y.H. Cho, H.B. Park, Experimental evidence of rapid water transport through carbon nanotubes embedded in polymeric desalination membranes, *Small*, 10 (2014) 2653–2660.

150. Y. Li, Z. Li, F. Aydin, J. Quan, X. Chen, Y.-C. Yao, C. Zhan, Y. Chen, T.A. Pham, A. Noy, Water-ion permselectivity of narrow-diameter carbon nanotubes, *Science Advances*, 6 (2020) eaba9966.

151. A. Berezhkovskii, G. Hummer, Single-file transport of water molecules through a carbon nanotube, *Physical Review Letters*, 89 (2002) 064503.

152. X.-H. Ma, H. Guo, Z. Yang, Z.-K. Yao, W.-H. Qing, Y.-L. Chen, Z.-L. Xu, C.Y. Tang, Carbon nanotubes enhance permeability of ultrathin polyamide rejection layers, *Journal of Membrane Science*, 570–571 (2019) 139–145.

153. H. Wu, B. Tang, P. Wu, Preparation and characterization of anti-fouling β-cyclodextrin/polyester thin film nanofiltration composite membrane, *Journal of Membrane Science*, 428 (2013) 301–308.

154. T. Wimmer, Cyclodextrins, in: *Ullmann's Encyclopedia of Industrial Chemistry*.

155. Z. Rahimi, A.A. Zinatizadeh, S. Zinadini, M. van Loosdrecht, β-cyclodextrin functionalized MWCNTs as a promising antifouling agent in fabrication of composite nanofiltration membranes, *Separation and Purification Technology*, 247 (2020) 116979.

156. A.Q. Al-Gamal, T.A. Saleh, F.I. Alghunaimi, Nanofiltration membrane with high flux and oil rejection using graphene oxide/β-cyclodextrin for produced water reuse, *Materials Today Communications*, 31 (2022) 103438.

157. M.J. Park, C. Wang, D.H. Seo, R.R. Gonzales, H. Matsuyama, H.K. Shon, Inkjet printed single walled carbon nanotube as an interlayer for high performance thin film composite nanofiltration membrane, *Journal of Membrane Science*, 620 (2021) 118901.

158. S. Casanova, T.-Y. Liu, Y.-M.J. Chew, A. Livingston, D. Mattia, High flux thin-film nanocomposites with embedded boron nitride nanotubes for nanofiltration, *Journal of Membrane Science*, 597 (2020) 117749.

159. M. Muqeet, R.B. Mahar, T.A. Gadhi, N. Ben Halima, Insight into cellulose-based-nanomaterials – A pursuit of environmental remedies, *International Journal of Biological Macromolecules*, 163 (2020) 1480–1486.

160. J. Moohan, S.A. Stewart, E. Espinosa, A. Rosal, A. Rodríguez, E. Larrañeta, R.F. Donnelly, J. Domínguez-Robles, Cellulose nanofibers and other biopolymers for biomedical applications. A review, 10 (2020) 65.

161. N. Dizge, E. Shaulsky, V. Karanikola, Electrospun cellulose nanofibers for superhydrophobic and oleophobic membranes, *Journal of Membrane Science*, 590 (2019) 117271.

162. S. Kim, R. Ou, Y. Hu, X. Li, H. Zhang, G.P. Simon, H. Wang, Non-swelling graphene oxide-polymer nanocomposite membrane for reverse osmosis desalination, *Journal of Membrane Science*, 562 (2018) 47–55.

163. P.R. Sharma, S.K. Sharma, T. Lindström, B.S. Hsiao, Nanocellulose-enabled membranes for water purification: Perspectives, 4 (2020) 1900114.

164. Q.G. Zhang, C. Deng, F. Soyekwo, Q.L. Liu, A.M. Zhu, Sub-10 nm wide cellulose nanofibers for ultra-thin nanoporous membranes with high organic permeation, *Advanced Functional Materials*, 26 (2016) 792–800.

165. F. Soyekwo, Q. Zhang, R. Gao, Y. Qu, C. Lin, X. Huang, A. Zhu, Q. Liu, Cellulose nanofiber intermediary to fabricate highly-permeable ultrathin nanofiltration membranes for fast water purification, *Journal of Membrane Science*, 524 (2017) 174–185.

166. C. Xu, W. Chen, H. Gao, X. Xie, Y. Chen, Cellulose nanocrystal/silver (CNC/Ag) thin-film nanocomposite nanofiltration membranes with multifunctional properties, *Environmental Science: Nano*, 7 (2020) 803–816.

167. D.J. Johnson, N. Hilal, Can graphene and graphene oxide materials revolutionise desalination processes? *Desalination*, 500 (2021) 114852.
168. O. Leenaerts, B. Partoens, F. Peeters, Graphene: A perfect nanoballoon, *Applied Physics Letters*, 93 (2008) 193107.
169. V. Berry, Impermeability of graphene and its applications, *Carbon*, 62 (2013) 1–10.
170. M.D. Fischbein, M. Drndić, Electron beam nanosculpting of suspended graphene sheets, *Applied Physics Letters*, 93 (2008) 113107.
171. C.J. Russo, J.A. Golovchenko, Atom-by-atom nucleation and growth of graphene nanopores, *Proceedings of the National Academy of Sciences*, 109 (2012) 5953–5957.
172. L. Liu, S. Ryu, M.R. Tomasik, E. Stolyarova, N. Jung, M.S. Hybertsen, M.L. Steigerwald, L.E. Brus, G.W. Flynn, Graphene oxidation: Thickness-dependent etching and strong chemical doping, *Nano Letters*, 8 (2008) 1965–1970.
173. H. Abadikhah, E. Naderi Kalali, S. Khodi, X. Xu, S. Agathopoulos, Multifunctional thin-film nanofiltration membrane incorporated with reduced graphene oxide@ TiO_2@ Ag nanocomposites for high desalination performance, dye retention, and antibacterial properties, *ACS Applied Materials & Interfaces*, 11 (2019) 23535–23545.
174. Y.T. Chung, E. Mahmoudi, A.W. Mohammad, A. Benamor, D. Johnson, N. Hilal, Development of polysulfone-nanohybrid membranes using ZnO-GO composite for enhanced antifouling and antibacterial control, *Desalination*, 402 (2017) 123–132.
175. M. Abbaszadeh, D. Krizak, S. Kundu, Layer-by-layer assembly of graphene oxide nanoplatelets embedded desalination membranes with improved chlorine resistance, *Desalination*, 470 (2019) 114116.
176. C. Berger, Z. Song, X. Li, X. Wu, N. Brown, C. Naud, D. Mayou, T. Li, J. Hass, A.N. Marchenkov, Electronic confinement and coherence in patterned epitaxial graphene, *Science*, 312 (2006) 1191–1196.
177. J. Wintterlin, M.-L. Bocquet, Graphene on metal surfaces, *Surface Science*, 603 (2009) 1841–1852.
178. R. Muzyka, M. Kwoka, Ł. Smędowski, N. Díez, G. Gryglewicz, Oxidation of graphite by different modified Hummers methods, *New Carbon Materials*, 32 (2017) 15–20.
179. W.S. Hummers Jr, R.E. Offeman, Preparation of graphitic oxide, *Journal of the American Chemical Society*, 80 (1958) 1339–1339.
180. M. Acik, Y.J. Chabal, A review on thermal exfoliation of graphene oxide, *Journal of Materials Science Research*, 2 (2013) 101.
181. D.C. Marcano, D.V. Kosynkin, J.M. Berlin, A. Sinitskii, Z. Sun, A. Slesarev, L.B. Alemany, W. Lu, J.M. Tour, Improved synthesis of graphene oxide, *ACS Nano*, 4 (2010) 4806–4814.
182. A.N. Banerjee, Graphene and its derivatives as biomedical materials: Future prospects and challenges, *Interface Focus*, 8 (2018) 20170056.
183. A.T. Smith, A.M. LaChance, S. Zeng, B. Liu, L. Sun, Synthesis, properties, and applications of graphene oxide/reduced graphene oxide and their nanocomposites, *Nano Materials Science*, 1 (2019) 31–47.
184. B. Mi, Graphene oxide membranes for ionic and molecular sieving, *Science*, 343 (2014) 740–742.
185. R. Joshi, P. Carbone, F.-C. Wang, V.G. Kravets, Y. Su, I.V. Grigorieva, H. Wu, A.K. Geim, R.R. Nair, Precise and ultrafast molecular sieving through graphene oxide membranes, *Science*, 343 (2014) 752–754.
186. L. Chen, J.-H. Moon, X. Ma, L. Zhang, Q. Chen, L. Chen, R. Peng, P. Si, J. Feng, Y. Li, J. Lou, L. Ci, High performance graphene oxide nanofiltration membrane prepared by electrospraying for wastewater purification, *Carbon*, 130 (2018) 487–494.
187. Q. Xie, S. Zhang, H. Ma, W. Shao, X. Gong, Z. Hong, A novel thin-film nanocomposite nanofiltration membrane by incorporating 3D hyperbranched polymer functionalized 2D graphene oxide, *Polymers*, 10 (2018) 1253.
188. F.J. García-Picazo, S. Pérez-Sicairos, G.A. Fimbres-Weihs, S.W. Lin, M.I. Salazar-Gastélum, B. Trujillo-Navarrete, Preparation of thin-film composite nanofiltration membranes doped with N-and Cl-functionalized graphene oxide for water desalination, *Polymers*, 13 (2021) 1637.
189. B. Van der Bruggen, Chemical modification of polyethersulfone nanofiltration membranes: A review, 114 (2009) 630–642.
190. A.L. Ahmad, B.S. Ooi, A. Wahab Mohammad, J.P. Choudhury, Development of a highly hydrophilic nanofiltration membrane for desalination and water treatment, *Desalination*, 168 (2004) 215–221.
191. J. Abraham, K.S. Vasu, C.D. Williams, K. Gopinadhan, Y. Su, C.T. Cherian, J. Dix, E. Prestat, S.J. Haigh, I.V. Grigorieva, Tunable sieving of ions using graphene oxide membranes, *Nature Nanotechnology*, 12 (2017) 546–550.
192. D. Xu, H. Liang, X. Zhu, L. Yang, X. Luo, Y. Guo, Y. Liu, L. Bai, G. Li, X. Tang, Metal-polyphenol dual crosslinked graphene oxide membrane for desalination of textile wastewater, *Desalination*, 487 (2020) 114503.

193. Y. Dong, C. Lin, S. Gao, N. Manoranjan, W. Li, W. Fang, J. Jin, Single-layered GO/LDH hybrid nanoporous membranes with improved stability for salt and organic molecules rejection, *Journal of Membrane Science*, 607 (2020) 118184.

194. R.P. Rode, H.H. Chung, H.N. Miller, T.R. Gaborski, S. Moghaddam, Trilayer interlinked graphene oxide membrane for wearable hemodialyzer, *Advanced Materials Interfaces*, 8 (2021) 2001985.

195. C. Wang, M.J. Park, D.H. Seo, H.K. Shon, Inkjet printing of graphene oxide and dopamine on nanofiltration membranes for improved anti-fouling properties and chlorine resistance, *Separation and Purification Technology*, 254 (2021) 117604.

196. G. Akonkwa Mulungulungu, T. Mao, K. Han, Two-dimensional graphitic carbon nitride-based membranes for filtration process: Progresses and challenges, *Chemical Engineering Journal*, 427 (2022) 130955.

197. W. Ye, H. Liu, F. Lin, J. Lin, S. Zhao, S. Yang, J. Hou, S. Zhou, B. Van der Bruggen, High-flux nanofiltration membranes tailored by bio-inspired co-deposition of hydrophilic gC₃N₄ nanosheets for enhanced selectivity towards organics and salts, *Environmental Science: Nano*, 6 (2019) 2958–2967.

198. A.R. Nadig, N.S. Naik, M. Padaki, R.K. Pai, S. Déon, Impact of graphitic carbon nitride nanosheets in mixed-matrix membranes for removal of heavy metals from water, *Journal of Water Process Engineering*, 41 (2021) 102026.

199. S. Abdikheibari, W. Lei, L.F. Dumée, A.J. Barlow, K. Baskaran, Novel thin film nanocomposite membranes decorated with few-layered boron nitride nanosheets for simultaneously enhanced water flux and organic fouling resistance, *Applied Surface Science*, 488 (2019) 565–577.

200. R.P. Pandey, P.A. Rasheed, T. Gomez, R.S. Azam, K.A. Mahmoud, A fouling-resistant mixed-matrix nanofiltration membrane based on covalently cross-linked Ti3C2TX (MXene)/cellulose acetate, *Journal of Membrane Science*, 607 (2020) 118139.

201. H.-L. Jiang, Q. Xu, Porous metal–organic frameworks as platforms for functional applications, *Chemical Communications*, 47 (2011) 3351–3370.

202. X. Li, Y. Liu, J. Wang, J. Gascon, J. Li, B. Van der Bruggen, Metal–organic frameworks based membranes for liquid separation, *Chemical Society Reviews*, 46 (2017) 7124–7144.

203. M.S. Denny, J.C. Moreton, L. Benz, S.M. Cohen, Metal–organic frameworks for membrane-based separations, *Nature Reviews Materials*, 1 (2016) 1–17.

204. Y. Shi, B. Liang, R.-B. Lin, C. Zhang, B. Chen, Gas separation via hybrid metal–organic framework/polymer membranes, *Trends in Chemistry*, 2 (2020) 254–269.

205. Q. Xin, M. Zhao, J. Guo, D. Huang, Y. Zeng, Y. Zhao, T. Zhang, L. Zhang, S. Wang, Y. Zhang, Light-responsive metal–organic framework sheets constructed smart membranes with tunable transport channels for efficient gas separation, *RSC Advances*, 12 (2022) 517–527.

206. J.E. Efome, D. Rana, T. Matsuura, C.Q. Lan, Insight studies on metal-organic framework nanofibrous membrane adsorption and activation for heavy metal ions removal from aqueous solution, *ACS Applied Materials & Interfaces*, 10 (2018) 18619–18629.

207. D. Li, X. Tian, Z. Wang, Z. Guan, X. Li, H. Qiao, H. Ke, L. Luo, Q. Wei, Multifunctional adsorbent based on metal-organic framework modified bacterial cellulose/chitosan composite aerogel for high efficient removal of heavy metal ion and organic pollutant, *Chemical Engineering Journal*, 383 (2020) 123127.

208. Z. Wang, Z. Wang, S. Lin, H. Jin, S. Gao, Y. Zhu, J. Jin, Nanoparticle-templated nanofiltration membranes for ultrahigh performance desalination, *Nature Communications*, 9 (2018) 1–9.

209. H. Liu, M. Zhang, H. Zhao, Y. Jiang, G. Liu, J. Gao, Enhanced dispersibility of metal–organic frameworks (MOFs) in the organic phase via surface modification for TFN nanofiltration membrane preparation, *RSC Advances*, 10 (2020) 4045–4057.

210. N. Song, X. Xie, D. Chen, G. Li, H. Dong, L. Yu, L. Dong, Tailoring nanofiltration membrane with three-dimensional turing flower protuberances for water purification, *Journal of Membrane Science*, 621 (2021).

211. Q. Bi, C. Zhang, J. Liu, X. Liu, S. Xu, Positively charged zwitterion-carbon nitride functionalized nanofiltration membranes with excellent separation performance of Mg2+/Li+ and good antifouling properties, *Separation and Purification Technology*, 257 (2021) 117959.

4 Nanofiltration in Desalination, Separation Mechanism, and Performance

Jamaliah Aburabie
New York University Abu Dhabi

Haya Nassrullah
New York University Abu Dhabi
New York University

Raed Hashaikeh
New York University Abu Dhabi

CONTENTS

4.1 RESEARCH TREND AND GLOBAL MARKET OF NANOFILTRATION MEMBRANES IN DESALINATION

Desalination records registered nearly 16,000 operational desalination plants in 177 countries across all major world regions. The production capacity reached around 95 million m^3/day of desalinated water for human use, of which 48% is produced in the Middle East and North Africa regions [1]. The reverse osmosis (RO) membrane process dominates the desalination market, where it produces

DOI: 10.1201/9781003261827-4

65% of the total global desalination capacity. RO membranes can reject more than 90% of monovalent salts such as sodium chloride. Despite the high performance, RO experiences some drawbacks, such as high operating pressure, low water permeability, and membrane fouling [2]. Concentrated brine is another serious issue; it heavily impacts the environment and is costly to handle [3–5].

Porous membranes (1–100 nm) that can reject sucrose and raffinose yet pass all micro-ions are classified as ultrafiltration (UF) membranes, whereas nonporous, monovalent-rejecting membranes are distinctly marked as RO. The transition area between porous and nonporous regions is anticipated to cover membranes with very small pores (0.5–2 nm). Presumably, in this range, membranes have good rejections of divalent ions, most organic solutes, and low rejection of monovalent ions. Such regions also recruit membranes that operate at lower pressures than RO; these membranes are called loose RO or nanofiltration (NF) membranes. NF is the latest of the four types of pressure-driven membrane technologies. Since its development in the late 1970s, NF has been fruitful in many fields, such as wastewater treatment, resource recovery, and solution purification and/or concentration [6]. As the name suggests, NF has lower monovalent salt rejections than RO but higher water fluxes with molecular weight cut-offs (MWCO) for neutral organic solutes of 200–2,000 Da [6,7]. NF membranes mostly carry sufficient surface charge, which contributes to the rejection of charged solutes [8]. Typically, NF membranes have low energy consumption (operating pressure of 3–15 bar), high water flux [9,10], high removal efficiency of multivalent salt ions, and low rejection of monovalent salts (sodium chloride). The low rejection rate of monovalent salts disqualifies NF as a standalone process to match the requirement of desalination despite their high permeations and low energy consumption. NF has potential for brackish and seawater desalination, especially if the monovalent ion rejection performance is improved while maintaining high permeability. Accordingly, reducing membrane pore size or enhancing electrostatic interaction may be the effective and available strategies to improve the rejection performance of the NF membrane. There is an increase number of publications [11] related to the use of NF in the desalination applications whether as a replacement for or in combination with traditional technologies or in some cases a replacement of RO, Figure 4.1 shows the research trends for desalination using NF membranes.

As shown in Figure 4.1, the demand for employing NF in desalination has increased over the past 15 years. According to the following research results, the energy–water–environment nexus is, in fact, behind the necessity to explore NF in desalination:

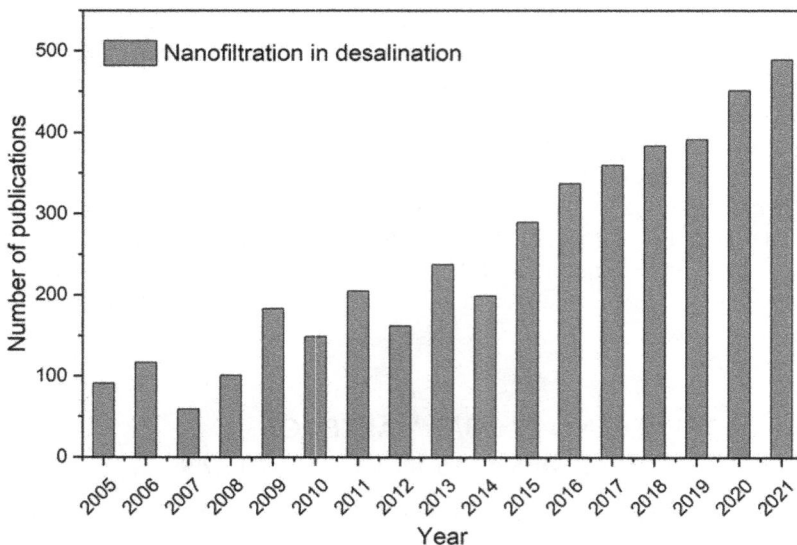

FIGURE 4.1   Number of publications on desalination using nanofiltration membranes; web of science database was used to generate the trend using "nanofiltration desalination" as a keyword search.

- 2%–3% annual increase in extraction rate of the groundwater supplies causes such resources to diminish [12–14].
- Around 1.2 billion people live in water scarcity regions, where water availability is less than 1,000 m^3 per capita per year [15].
- An increase of more than 55% of global water demand is expected by 2050, mainly due to the high gross domestic product growth rate, meaning increased water demands for manufacturing by 400%, power generation by 140%, and domestic sector use by 130% [16].
- Presently, there are >20,000 desalination plants worldwide, with total global installed desalination capacity stands at 36 billion m^3 per year [17].
- Desalination is the most energy-demanding water treatment process that drains 75.2 TWh of electricity annually, about 0.4% of global electricity [18,19].
- Desalination processes powered by fossil fuels are a major source of CO_2 emissions. Desalination processes currently contribute to 76 million tons of CO_2 annually and are expected to grow to 218 million tons of CO_2 annually by 2040 [20]. Figure 4.2 summarizes the increased percentage of CO_2 emissions, population, and world water withdrawals & consumption between the years of 1900 and 2040. It can be seen that the CO_2 emission is over 1,500%, and it is expected to grow to 2,200% by 2040 [16]. Expectedly, according to the above facts, water withdrawals and consumption increased to over 1,000%. It can be seen that the growth rate of CO_2 emission is the highest, and it is speculated to continue growing.
- The current worldwide market share of membrane-based desalination processes is about 67%, of which RO membrane technology accounts for 62.4% [21]. Figure 4.3 demonstrates that 59% of desalination processes depend on seawater as desalination feed, 23% on brackish water, and 5% on wastewater [21,22].
- **Energy consumption perspective**: Two-stage NF desalination systems successfully eliminated ions from seawater feed while consuming 20%–30% less energy than the conventional one-stage RO system [10]. Harrison et al. [23] reported that the permeate total dissolved solids (TDS) was <400 mg/L when the high-rejection NF-90 membrane was used in two-stages NF systems. Altaee et al. [24] evaluated the performance of two-stage desalination systems using ROSA simulation software. Parameters were set with feed salinity at 35,000 mg/L and the recovery of the first and second stages at 59% and 67%, respectively. RO system was first evaluated as a standalone system; permeate TDS of

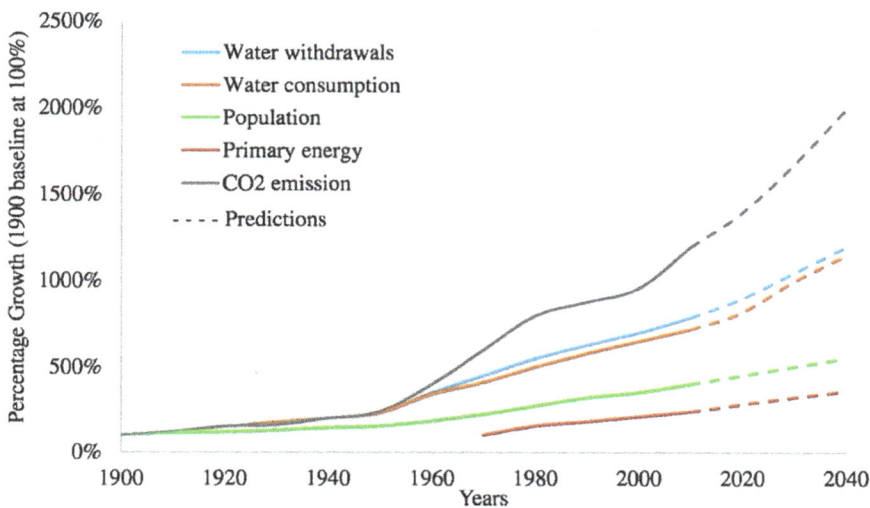

FIGURE 4.2 CO_2 emission, population, water withdrawals, and consumption growth rate percentages from 1900 to 2040 [16].

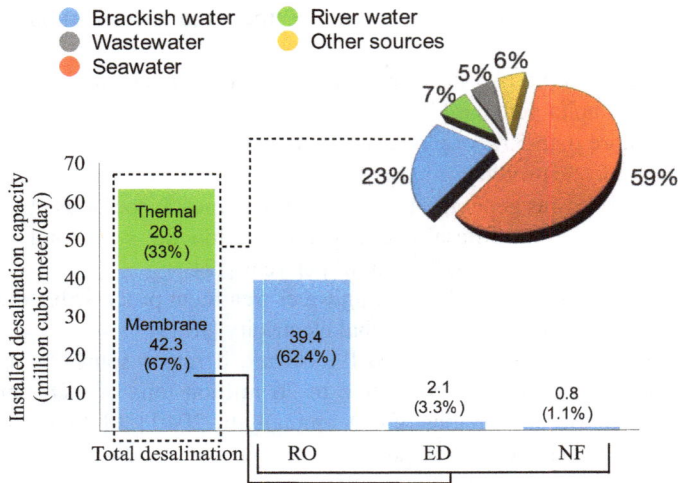

FIGURE 4.3 Thermal and membrane-desalination market share and desalination feed resources.

<100 mg/L was achieved with energy consumption of 3.7 kWh/m³. On the other hand, the permeate TDS of the two-stage NF90 system was 254 mg/L and the energy consumption was 3.35 kWh/m³ while the two-stage NF90-RO system scored a TDS of <100 mg/L with energy consumption of 3.25 kWh/m³. Erikssona et al. reported that using NF as pretreatment for RO, the hardness content is dropped to 220 ppm [25], water recovery was elevated from 28% to 56%, while the power consumption was decreased from 9.596 to 5.858 kWh/m³ [26]. NF pretreatment had shown in multiple studies the efficiency in removing divalent ions and lowering RO feed water osmotic pressure. In viewing of water quality and energy consumption, the dual-stage NF seawater desalination process is an encouraging seawater desalination alternative.

4.2 TRANSPORT OF WATER AND IONIC SOLUTES IN NANOFILTRATION MEMBRANES

4.2.1 MECHANISMS OF ION TRANSPORT

Understanding the ion rejection mechanism of NF membranes is essential to designing membranes tailored for specific applications and achieving the desired performance. The mechanisms discussed in this section are related only to the interactions between the ions and the wet membrane. The rejection performance of NF membranes is also affected by operating conditions and solution composition, which are discussed in Section 4.2.2. There are many discrepancies in salt rejection patterns in NF membranes reported in the literature, which suggest that different NF membranes exhibit a combination of varying rejection mechanisms. Therefore, this section will discuss the general rejection mechanisms of salts by NF membranes without referring to a specific type of membrane. The mechanism of ion rejection is complex as it depends on micro-hydrodynamics and interfacial interactions occurring at the membrane surface and inside the pores [27]. Typically, ion rejection is attributed to a combination of sieving (steric) and non-sieving (Donnan and dielectric) effects, as shown in Figure 4.4.

4.2.1.1 Steric Exclusion

Steric exclusion is the simplest and most basic mechanism affecting the rejection performance of NF membranes. Uncharged molecules which are larger than the membrane pore size are rejected. Steric exclusion might be considered a rejection mechanism for uncharged solutes only; however, charged

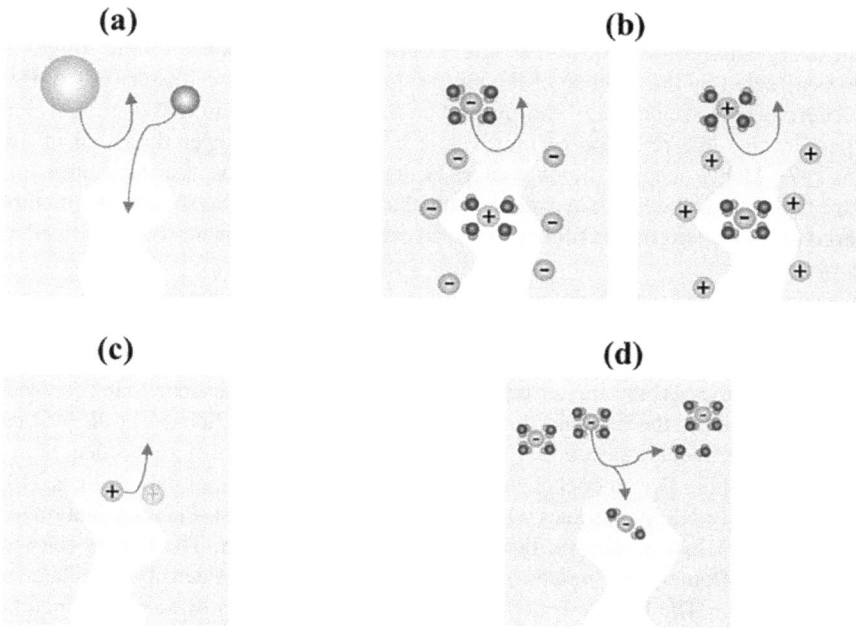

FIGURE 4.4 Mechanism of ion transport: (a) steric, (b) Donnan, (c) image force, and (d) Born effects.

solutes also undergo more significant steric hindrance with smaller membrane pores [28]. The pore size of NF membranes is often evaluated using MWCO, which represents the molecular weight of a reference compound that is rejected at 90%. The MWCO of an NF membrane is determined by filtering several reference uncharged molecules through the membrane. Then, a plot of measured rejection percentage versus the molecular weight of the reference molecules is constructed. The MWCO of NF membranes is typically 200–1,000 g/mol [29]. The MWCO is taken as an indication of the size of the molecule; however, this is not entirely accurate as it does not provide information about the geometry of the molecule. In some cases, molecules with low molecular weight which have a branched structure are larger than molecules with higher molecular weight and a compact structure. Therefore, defining the rejection capability of NF membranes by MWCO is misleading because MWCO only provides a rough estimate of size exclusion [30,31]. Alternatively, models such as Spiegler–Kedem [32] and steric hindrance pore model [33] are used to evaluate the pore size of the membrane. These models provide a more accurate estimation of the pore size of NF membranes compared to MWCO. In an NF process, charged solutes smaller than the estimated pore size of the membrane are partially rejected, indicating that these solutes undergo several rejection mechanisms other than size exclusion.

4.2.1.2 Donnan Exclusion

Donnan exclusion is one of the main non-sieving ion rejection mechanisms in NF membranes [34]. The Donnan effects result from the electrostatic interactions between ionic solutes in the feed solution and the membrane surface charge [31,35,36]. The ion selectivity is controlled by the value and sign of membrane surface charge, which rises from the acid/base dissociation of membrane functional groups, coordination between ionic solutes in the feed and membrane functional groups, and accumulation or adsorption of charged ions from the feed solution onto the membrane surface [31,36,37]. Electrochemical characterizations of NF membranes have shown that the membrane surface and pore charge are highly dependent on the feed pH, as well as the type and concentration of ionic solutes [37]; more details are provided in Section 4.2.2. The Donnan theory states that membranes with a particular charge attract ions with opposite charges (counter-ions) and repel ions with similar charges (co-ions) [38,39]. The strength of interaction between ionic solutes and membrane pores is affected by the valence of the ionic solute; higher valence results in stronger interactions [35]. Therefore, the rejection of multivalent co-ions is higher than that of monovalent co-ions [40]. On the contrary, the rejection of multivalent counter-ions is lower than that of monovalent counter-ions [27]. Debye length, the characteristic length for electrostatic interactions, is another critical factor that affects the Donnan exclusion mechanism [41,42]. The smaller the membrane pore size compared to Debye length, the higher the electrostatic interactions between ions and membrane surface charge.

Initially, it was believed that Donnan exclusion was the primary ion rejection mechanism of NF membranes. This conclusion was made, as it was observed that NF membranes have high rejection for multivalent co-ions and low rejection for multivalent counter-ions [43]. However, it was also observed that some NF membranes exhibit higher rejection of multivalent counter-ions than monovalent counter-ions, which contradicts the Donnan theory [44,45]. Furthermore, Schaep et al. [46] carried out several filtration experiments on commercial NF membranes using different salts. They used the Donnan-steric partitioning pore model (DSPM) to analyze the rejection of ionic solutes. DSPM is the most common model for NF membranes which uses an extended Nernst–Planck equation to represent ion transport due to diffusion, electric field gradient, and convection. The results showed that the membranes yielded extremely high values of effective membrane charge density, calculated as a fitting parameter in the model. The high rejection of multivalent counter-ions by some NF membranes and the physically unrealistic values of membrane charge density obtained by the DSPM model indicated that there was an additional non-sieving mechanism other than the Donnan exclusion [37].

4.2.1.3 Dielectric Exclusion

Dielectric exclusion is the most complicated phenomenon for ion rejection, as the exact nature of the mechanism is still a subject of debate in the NF community. Yaroshchuk [31,43] was among the first researchers to suggest that dielectric exclusion may be occurring. He proposed that the ion polarizes the membrane matrix and solvent according to their relative dielectric constant. The polarization charge builds at the interface between the membrane and solvent, creating a repulsive image force (Figure 4.4c) for anions and cations if the dielectric constant of the membrane is lower than that of the solvent [47]. Ionic solutes interact with the image force induced by them and by other ions present in the solution [31]. The magnitude of the image force depends on three factors: the ion valence, membrane pore size, and relative dielectric constant between the membrane and the solvent [27,43]; a detailed description of this mechanism can be found elsewhere [43]. Bowen and Welfoot [48] proposed another theory for dielectric exclusion, known as the "Born effect" (Figure 4.4d). This theory states that the ion solvation ability of water in confined pores is lower than that in the bulk solution. This is because water has a lower dielectric constant in the pores than its dielectric constant in the bulk solution. This creates a positive solvation energy barrier which affects the rejection of all ionic solutes irrespective of the sign of membrane or ion charge. This mechanism justifies the high rejection of multivalent counter-ions by some NF membranes. It is important to note that the Born effect is influenced by the valence of ionic solutes; higher valence results in better rejection.

The question is whether image force, Born effect, or both contribute to dielectric exclusion. Saliha et al. [49] used a steric, electric, and dielectric exclusion model to describe the rejection of multi-ionic solutions. They performed a series of filtration tests to verify the results obtained from the model. Variations between the experimental and modeling results were found when both dielectric exclusion mechanisms were considered in the model. Conversely, the agreement between the model and experimental results was satisfactory when only the Born effect was used for dielectric exclusion. In another study by Oatley et al. [27], the dielectric constant of water inside the pores of commercial NF membranes was calculated using the results obtained from a series of filtration tests of salt solutions. Based on their results, they concluded that the Born theory provides a more appropriate representation for the dielectric exclusion mechanism of NF membranes.

Most NF membranes exhibit lower selectivity toward monovalent ions compared to multivalent ions. NF membranes remove monovalent ions to different extents. In some cases, the difference in the hydrated radii between monovalent ions is insignificant, yet the membrane exhibits a noticeable difference in the rejection of these ions. For example, the rejection of chloride (Cl^-) is higher than nitrate (NO_3^-) even though they have the same net charge and similar hydrated radii. The different selectivity toward monovalent ions by NF membranes has been recently justified in several studies by the dehydration phenomenon, which results from the Born dielectric effects [50,51]. For a hydrated ion to pass through the membrane pores, it should strip its hydration layer temporarily by overcoming an energy barrier known as hydration energy [28]. The hydration energy of the ion controls the level of dehydration; higher hydration energy corresponds to a more significant barrier for dehydration, which translates into fewer chances for the ion to infiltrate through the membrane [50,52]. The hydration energy depends on the ionic size and charge; ions with a smaller ionic radius and higher charge density have higher hydration energy [53–55]. Studying the rejection performance of NF membranes experimentally has increased our knowledge of the dehydration phenomenon. However, these experiments are not sufficient to evaluate the theoretical energy barriers. Recently, it has been proposed that ions need to overcome an activation energy barrier to pass through the membrane. Chu et al. [55] obtained a higher rejection of acetate (Ac^-) than chloride (Cl^-). The results were explained by the dehydration phenomenon and apparent activation energy. The ionic radius of Ac^- is smaller than Cl^- and thus, the former has higher ionic charge density and higher hydration energy; hence, it is more difficult for Ac^- to pass through the membrane pores than Cl^-. Furthermore, Ac^- requires more apparent activation energy for ion transport.

4.2.2 Overview of the Effect of Solution Properties and Process Parameters on Nanofiltration Performance

4.2.2.1 Feed pH

The selectivity of an NF membrane can be controlled by changing the pH of the feed solution. The effect of pH on the membrane surface charge is dependent on the materials used for the fabrication of an NF membrane; acidic, basic, or both functional groups may be present [27]. The surface charge of NF membranes changes at different pH values based on the dissociation constants (pKa) of the functional groups present on the membrane. Most NF membranes are amphoteric; they exhibit an isoelectric point (IEP) at a certain pH [56]. As discussed in Section 4.2.1.2, the membrane surface charge is the main membrane property that affects the Donnan exclusion mechanism. The electrostatic interactions between the ions and the membrane surface can be improved with increased membrane charge and reduced around the IEP of the membrane. Labbez et al. [57] investigated the rejection of a range of symmetric and asymmetric single inorganic salt solutions of KCl, LiCl, $MgSO_4$, K_2SO_4, and $MgCl_2$ at different pH values (Figure 4.5). Two retention behaviors were observed; the V-shaped retention curves that describe the symmetric salts and the S-shaped retention curves that describe the asymmetric salts. At any pH, the rejection of symmetric salts was governed by the electrostatic interactions between the membrane and the co-ions. For example, at a pH of 4, the membrane was positively charged; hence, the rejection was controlled by the interaction with the cations (K^+, Li^+, and Mg^{+2}). Similarly, at pH values higher than the IEP, the membrane was negatively charged, and thus the rejection was controlled by the interactions with the anions (Cl^- and SO_4^{-2}). The rejection of symmetric salts was lowest at the IEP, where the membrane was not charged, and electrostatic interactions no longer contributed to ion rejection. A different trend was observed with asymmetric salt, as the salt rejection was influenced by the ions with the higher valence regardless of the membrane charge, hence the S-shaped retention.

Filtration tests using single-salt solutions provide sufficient data to understand the general effect of pH on ion rejection by NF membranes. However, improving our knowledge of the effect of pH on rejection mechanisms of NF requires studying the desalination performance using more complex

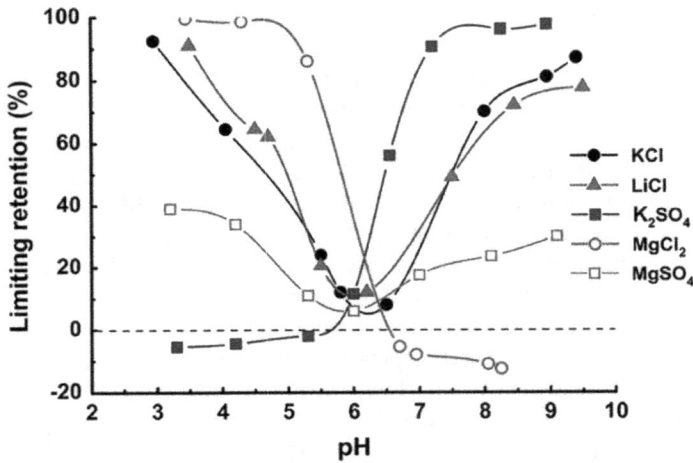

FIGURE 4.5 Retention of single-salt solutions as a function of pH at a fixed salt concentration of 1 mM [58].

(a)   **(b)**

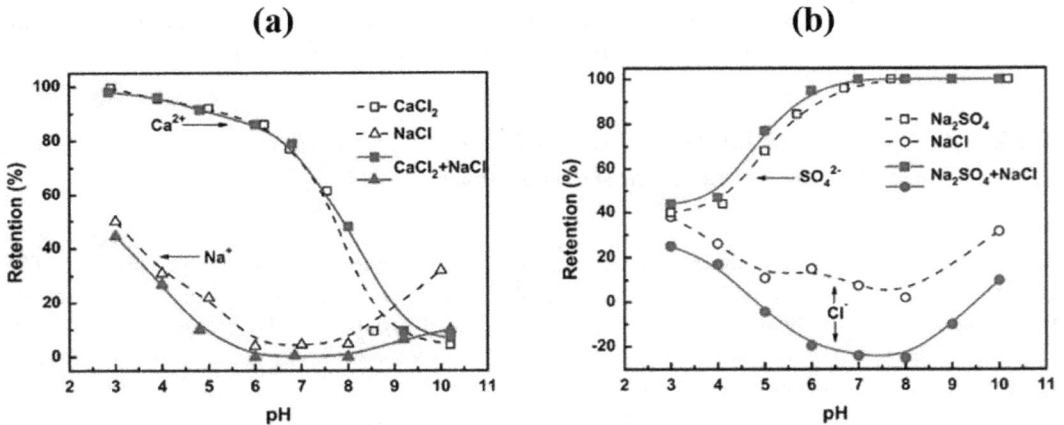

FIGURE 4.6 Effect of mixing electrolyte on retention at different pH: (a) 500 ppm NaCl + 620 ppm $CaCl_2$ and (b) 500 ppm NaCl + 500 ppm Na_2SO_4 [58].

feed solutions with a mixture of different salts. The effect of pH on solute rejection for multi-electrolyte systems was experimentally investigated by Szoke et al. [59]. As shown in Figure 4.6, the rejection of the asymmetric salts seemed to be the same whether they were in a mixture with symmetric salts or separate. On the contrary, the rejection of symmetric salts dropped when asymmetric salts were present in the solution; when NaCl and $CaCl_2$ were mixed, the retention of Na^+ decreased at higher pH levels (Figure 4.6a). The addition of Ca^{+2} created the co-ions competition effect at high pH, which accelerated the passage of Na^+. Above IEP, the rejection is governed by Cl^-; however, with Ca^{+2} around, it dominated, replaced, and overpowered the monovalent ion. A similar trend was observed for Na_2SO_4 and NaCl mixture (Figure 4.6b).

Besides altering the membrane surface charge, the pH of the feed solution may influence the membrane separation performance by changing the pore size of the selective membrane layer. For instance, Freger et al. [60] observed that the permeate flux decreased steadily as the feed pH increased. They suspected that this reduction in permeate flux was due to the shrinkage in the separation layer caused by the hydration differences between membrane functional groups and counter-ions at different pH values. The theories and studies presented in this section provide some understanding of the effect of solution pH on solute selectivity and water permeation. In reality, identifying this effect might not be accurate. This is because the effect of pH on membrane performance is influenced by other parameters such as the type and concentration of solutes in the feed solution, as well as concentration polarization. These are key parameters that also affect the permeability and selectivity of the membrane. Thus, several conflicting results have been presented in the literature [58,61,62].

4.2.2.2 Concentration of Ionic Solutes

It is common to observe a reduction in permeate flux, as salt concentration increases at constant pH. High salt content causes a variation in water viscosity in membrane pores [63]. As the salt content in the feed increases, more counter-ions adsorb and accumulate onto the membrane pores forming an electric double layer which increases the electroviscous force; thus, flux is reduced [58]. Alternatively, the increase in salt concentration leads to a higher concentration polarization effect as the back-diffusion of solutes away from the membrane decreases due to increased bulk viscosity [64].

The effect of salt concentration on rejection is highly dependent on the types of salt in the solution. With single symmetric salts, such as NaCl or KCl, increasing the concentration of salts is often associated with a decrease in rejection. Several theories have been proposed to explain this trend. First, increasing the amount of salt in the feed solution leads to weaker electrostatic interactions due to screening by counter-ions. Second, the adsorption or accumulation of ions on the membrane leads to concentration polarization which reduces the rejection performance of the membrane. Different rejection trends are observed with asymmetric salts such as $CaCl_2$; rejection increases as the concentration of salt increases [37]. This may be related to the physical–chemical interactions between the membrane and the divalent salts. At fixed pH values, increasing the concentration of $CaCl_2$ results in more coordination between Ca^{+2} ions and the membrane functional groups, thus switching the charge of the membrane from negative to positive [65–67]. On the contrary, Teixeira et al. [68] showed that the interaction between Ca^{+2} ions and the membrane caused a shift in the zeta potential of the membrane. The membrane became less negatively charged; therefore, electrostatic interactions were weakened, and the salt rejection was reduced. These results suggest that the effect of the concentration of asymmetric salts on the desalination performance of NF membranes is case-specific.

4.2.2.3 Pressure

Transmembrane pressure is one of the key parameters which affect the desalination performance of NF membranes in terms of solute rejection and permeate flux. The effect of pressure varies depending on the pore size of the NF membrane and the type of solutes in the feed solution [69]. As shown in Figure 4.7, Loose NF membranes show a retention performance similar to that of UF membranes,

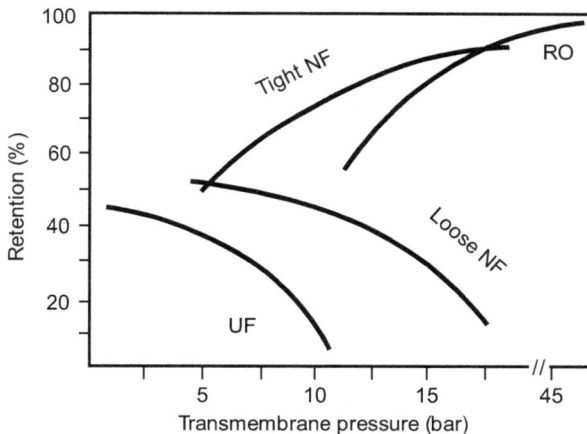

FIGURE 4.7 Effect of transmembrane pressure on the retention of UF, NF, and RO membranes (lines represent general trends only) [71].

where retention may decrease as pressure increases. In a UF process, this behavior is attributed to the rise in concentration polarization which raises the transport of solutes across the membrane. On the other hand, the performance of tight NF membranes is typically similar to that of RO membranes, where the rejection increases toward an asymptote as the pressure increases (Figure 4.7). In a RO process, it is widely accepted that the primary mechanism of rejection is solution diffusion, in which the transport of water and solutes is uncoupled. Therefore, applying higher pressure facilitates water flow through the membrane, diluting the solute in the permeate. Generally, in NF processes, increasing the applied pressure leads to increase in the permeate flux, since the pressure drop across the membrane, the driving force for water transport, increases [29,70].

4.2.2.4 Temperature

The temperature-induced changes in water and ion mobility across NF membranes are relevant to numerous industrial applications, including hybrid NF with thermal desalination units. Understanding the effect of temperature on NF performance is critical for system design and efficient separation. Roy et al. [72] carried out a detailed theoretical study at different temperatures to capture the variation in water permeation and ion rejection. The results showed that increasing the temperature resulted in enhanced water flux and decreased rejection of NaCl. These results were attributed mainly to changes in membrane structural parameters, which were accounted for in the model. They relied on previous studies which showed that increasing the temperature led to an increase in the pore size of the selective membrane layer and a decrease in the membrane thickness [73]. According to Amar et al. [73], the increase in the membrane pore radius with increasing temperature was due to thermal pore dilation and expansion of the polymer structure. Other parameters which affected the membrane performance were solvent viscosity and ion diffusivity. As the temperature increased, water kinematic viscosity decreased [74], and ion diffusivity increased [75]. Reduction in water viscosity led to enhanced mass transfer coefficient, hence higher water permeation and lower solute rejection. Moreover, it was found that the change in membrane properties had a more prominent effect than the change in solute diffusivity and solvent viscosity. Therefore, membranes with low structural changes with temperature are desirable for high-temperature applications. Similarly, Rui et al. [76] showed that increasing the feed temperature enhanced the pure water flux due to the increase in the pore size of the membrane selective layer.

4.3 NANOFILTRATION MEMBRANE FABRICATION AND PROPERTIES

As theories predict and studies confirm, membrane performance is highly directed by its structural and physicochemical properties. Properties such as thickness, porosity, pore size, roughness, hydrophilicity, and surface charge density play a crucial role. Membrane properties are affected heavily by the fabrication methods and conditions. Understanding the membrane fabrication–property–performance relationships will surely assist in optimizing the design of NF membranes [77]. Membranes are categorized into isotropic membranes and anisotropic membranes according to membrane morphology. Isotropic membranes have unvaried material composition and structure throughout; they can be either porous or dense. Both types of isotropic membranes are not explored for NF; the dense, relatively thick membranes have too low transmembrane flux for practical separation processes. The porous isotropic membranes are widely used as microfiltration membranes. Anisotropic membranes consist of layered structures where the pore size and composition change from top to bottom. The anisotropic membrane has a relatively thin, dense skin layer mounted on thicker, microporous support [7]. The skin layer imparts selectivity and is the main barrier to flow through the membrane. The porous support layer establishes mechanical integrity. If both layers are made from the same material, the membranes are referred to as asymmetric membranes; if the layers are made from different materials, the membranes are called thin-film composite (TFC). Methods of fabrication are summarized in the flowchart presented in Figure 4.8a. The importance of anisotropic membranes was recognized by Loeb and Sourirajan. They prepared the first

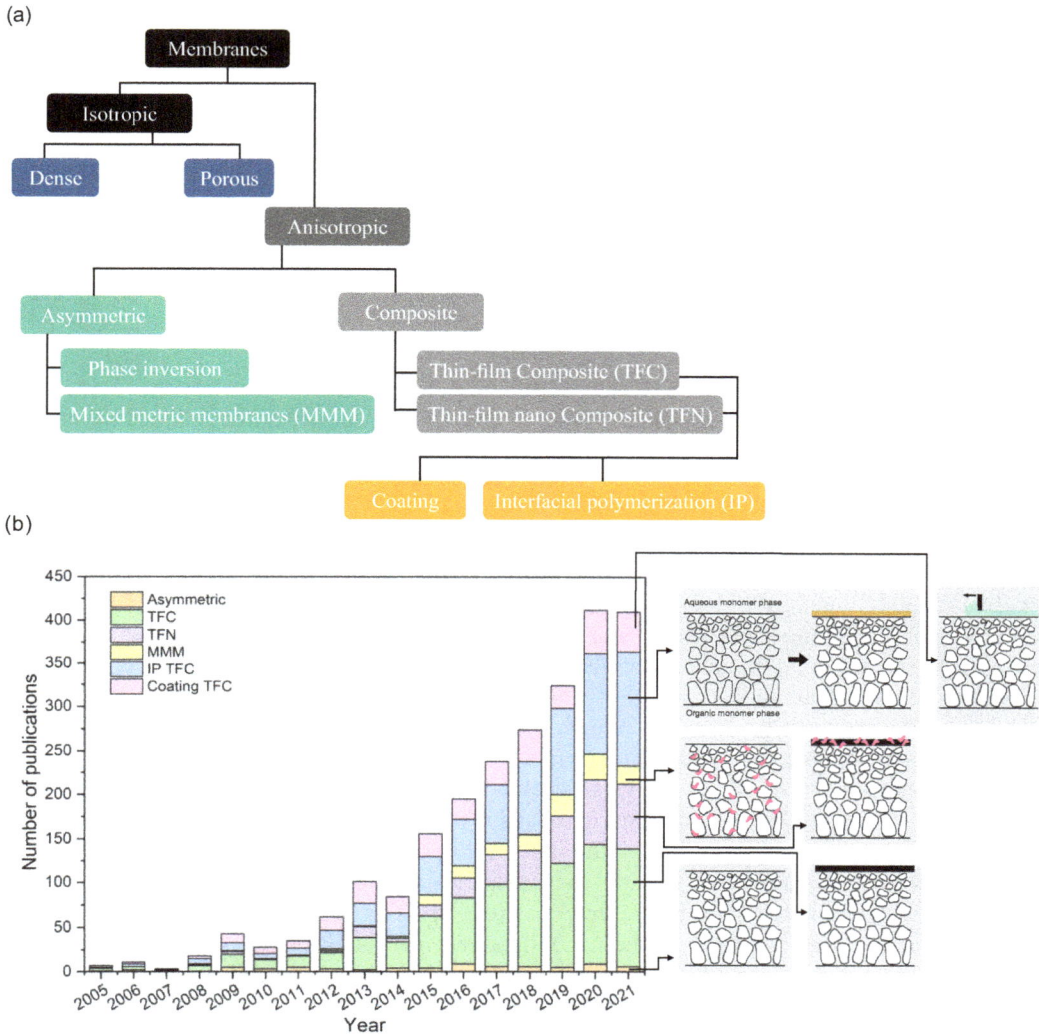

FIGURE 4.8 (a) Types of membranes and their fabrication methods. (b) Number of publications on different types of nanofiltration membranes used for desalination. Web of science database was used to generate the trend using "nanofiltration desalination" and "asymmetric"/"thin-film composite (TFC)"/"Thin-film nanocomposite (TFN)"/"mixed matrix membranes (MMM)"/"interfacial polymerization (IP)"/"coating" as a keyword search.

high-flux, anisotropic RO membranes by what is now known as the Loeb–Sourirajan technique. The Loeb–Sourirajan technique involves first, casting of a polymer solution and second, precipitation (solidification) by immersion in a nonsolvent bath. Loeb and Sourirajan's discovery was a critical breakthrough in membrane technology. The advantageous features of the anisotropic structure encouraged the development of other methods and other modifications. The following are the main NF membrane fabrication techniques for desalination:

- **Phase-inversion asymmetric membranes**: This method depends on the following concept; a thermodynamically stable polymer solution is transformed from liquid to solid phase in a controlled manner under the influence of a certain factor. This solidification is preceded by a liquid–liquid demixing. The demixing, provoked by certain factor, separates into a polymer-rich and a polymer-lean phase. The polymer-rich phase solidifies either through gelation, vitrification, or crystallization forming the solid membrane matrix. The polymer-lean phase will convert into the pores of the membrane. The demixing can be induced by: (i) immersion precipitation (immersion in a nonsolvent bath), (ii) thermal precipitation (lowering temperature), (iii) controlled evaporation (evaporation of volatile solvent from the cast solution), or (iv) precipitation from the vapor phase (placing the cast film in a nonsolvent vapor phase) [78].

- **Interfacial polymerization (IP) composite membranes**: IP membranes consist of a polymerized thin layer of polymer (A) at the surface of a different porous polymer (B) layer (support) [79]. The possibilities to optimize each layer independently to achieve maximum performance and the facile scalable production portray IP membranes as the most explored membrane fabrication for desalination. To date, IP remains the most established fabrication method.
- **Solution-coating composite membranes**: Polymer solution is deposited on top of a porous support; deposition can be via casting, dip, or spin coating [7].

4.3.1 Nanofillers in Nanofiltration Membranes (Nanocomposite Membranes)

Polymeric NF membranes entrapping a range of nanomaterials (organic or inorganic) can be commonly categorized under the term nanocomposite membranes. Several nanomaterials are used as nanofillers in membranes, including metal oxides (inorganic) (such as titania, zeolite, silica, alumina, etc.) and carbon-based nanomaterials (organic) (such as carbon nanotubes (CNTs), graphene oxide (GO), carbon quantum dots, metal-organic frameworks (MOFs)) [80,81]. Combining polymers with nanomaterials seems to be the suitable fabrication method to achieve the qualities of the two moieties in one outcome. Nanocomposite membranes aims to merge the flexibility, low manufacturing cost, outstanding selectivity, and high packing density of the polymeric component with the long-term stabilities, high mechanical strength, and regeneration capability of inorganic materials. In addition, the presence of nanofillers in the skin layer bestows additional pathways to impart high permeation while maintaining the original molecular sieving capacity, leading to higher water flux membranes without alteration in their selectivities [82–84]. The improved mass transfer is a result of both intrinsic nanosized pores of the nanofillers and interface voids between polymer chains and fillers. Depending on the position of the nanoparticles within the membrane structure, nanocomposite membranes can be subcategorized. These sub-categories are conventional nanocomposite membranes or mixed matrix membranes (MMM), surface-located nanocomposite membranes, thin-film nanocomposite (TFN) membranes, or TFC with nanocomposite substrate [85]. Among the four, MMM and TFN are the most explored for NF in desalination. MMM membranes are traditional asymmetric membranes with the nanomaterial incorporated throughout the membrane cross-section. TFN membranes are TFC membranes with the nanomaterial deposited only and specifically at the selective layer. Figure 4.8b represents the number of NF publications for desalination topic sectioned for each type of membrane. It can be seen that the two most explored fabrication techniques are TFC and TFN membranes, especially the interracially polymerized membranes. Table 4.1 summarizes the recent reports on nanocomposite NF membranes for desalination.

4.3.2 Permeability/Salt Rejection Trade-Off

One critical issue researchers face in developing NF membranes for desalination is the trade-off between water permeability and selectivity. The trade-off manifest that higher water permeability mostly effectuates decreased salt rejection and vice versa. How to achieve a balance in the membrane's performance is the question. The trade-off between membrane permeability and selectivity was first recognized by Robeson for gas separation membranes. At first, the majority of the membranes did not surpass the "Robeson upper bound". Nonetheless, with the advances in membrane

TABLE 4.1

Summary of Recent Nanocomposite NF Membranes for Desalination

Type	Nanomaterial	Active Layer Material	Water Permeability (L/m²h bar)	Salt Rejection	Reference
Inorganic	Nanodiamond	Polyamide	15	~98% NaH_2SO_4	[86]
	TiO_2 coated with Tannic acid	Polyester	5.6	58% NaCl, 95% Na_2SO_4	[87]
	Manganese oxide	Polyamide	4.25	30.3% NaCl, 97.4% Na_2SO_4, 96.7% $MgSO_4$	[88]
	Halloysite nanotubes	Polyamide	20.5	94.9% Na_2SO_4, 91.5% $MgSO_4$	[89]
	Poly(hexafluorobutyl acrylate (PHFBA)-modified hydrous manganese oxide	Polyamide	5.18	98.6% Na_2SO_4, 97.6% $MgSO_4$, 43.2% NaCl	[88]
	Silica-coated Fe_3O_4	Polyether sulfone (PES)	13.0–15.5	~25% NaCl, ~80% Na_2SO_4, ~58% $MgSO_4$	[90]
	Born nitride nanotubes	Polyamide	4.5	90% $MgSO_4$, 80% $CaCl_2$	[91]
	Beta (β) zeolite	Polyamide	40.6	17% NaCl, 94% Na_2SO_4, 81% $MgSO_4$	[92]
	ZIF-8	Polyamide	9.2	95% Na_2SO_4	[93]
	Aluminum hydroxide nanospheres	Polyamide	6.1	95% $MgSO_4$, 31% NaCl	[94]
Carbon-based	Cellulose nanocrystals	Polyamide	25.4	99% Na_2SO_4	[95]
	MOF	Polyamide-PVA	~5.1-~8.9	32% NaCl, 97% $MgSO_4$, 88% Na_2SO_4, 36% MgCl	[96]
	Cellulose nanofibers	Polyethyleneimine	32.7	90% MgCl2, 65% $MgSO_4$, 44% NaCl, 39% Na_2SO_4	[97]
	Graphene oxide (GO)	GO film	11.1–22.3	61.1% Na_2SO_4, 27.9% NaCl, 41.8% $MgSO_4$, 15.0% $MgCl_2$	[98]
	Hyperbranched polyester (HBE)-functionalized GO	Poly (piperazine amide)	6.7–8	~43% NaCl, ~98% Na_2SO_4, ~96% $MgSO_4$, ~85% $MgCl_2$	[99]
	g-C_3N_4 nanoplatelets	PDA/PEI	28.4	2.9% NaCl, 7.6% Na_2SO_4	[100]
	g-C_3N_4 nanoplatelets	Polyamide	~20–23	>90% $MgCl_2$, >90% $MgSO_4$, 65% NaCl, 55% Na_2SO_4	[101]

fabrication, novel membranes progressively exceeding the upper bound. Subsequently, Robeson published a revised version for his original correlation, which insinuates that the trade-off between membrane permeability and selectivity can be coordinated through smart scenarios for membrane fabrication [102]. Freshly, the same concept was quoted for TFC membranes works in the context of RO and NF desalination [103]; the article provided a critical tool for the evaluation and benchmarking of future desalination and water reuse membrane development. In the interest of achieving the balance between high rejection and high permeation, the Van der Bruggen group has discussed dividing the active functional layer in NF membranes into four different mass transfer regions, including (i) solution–diffusion region (nonporous, water, and salt ions permeabilities depend heavily on solubility coefficients), (ii) water molecule passage region (nonporous or porous, water molecule channels), (iii) partial salt passage region (porous, steric hindrance and electrostatic repulsion), and (iv) absolute salt passage region (large pores, maybe defects) (Figure 4.9a). Based on the mass transfer properties of these different regions, the relationship between water permeability and salt rejection was explained and the trade-off could be coordinated [104].

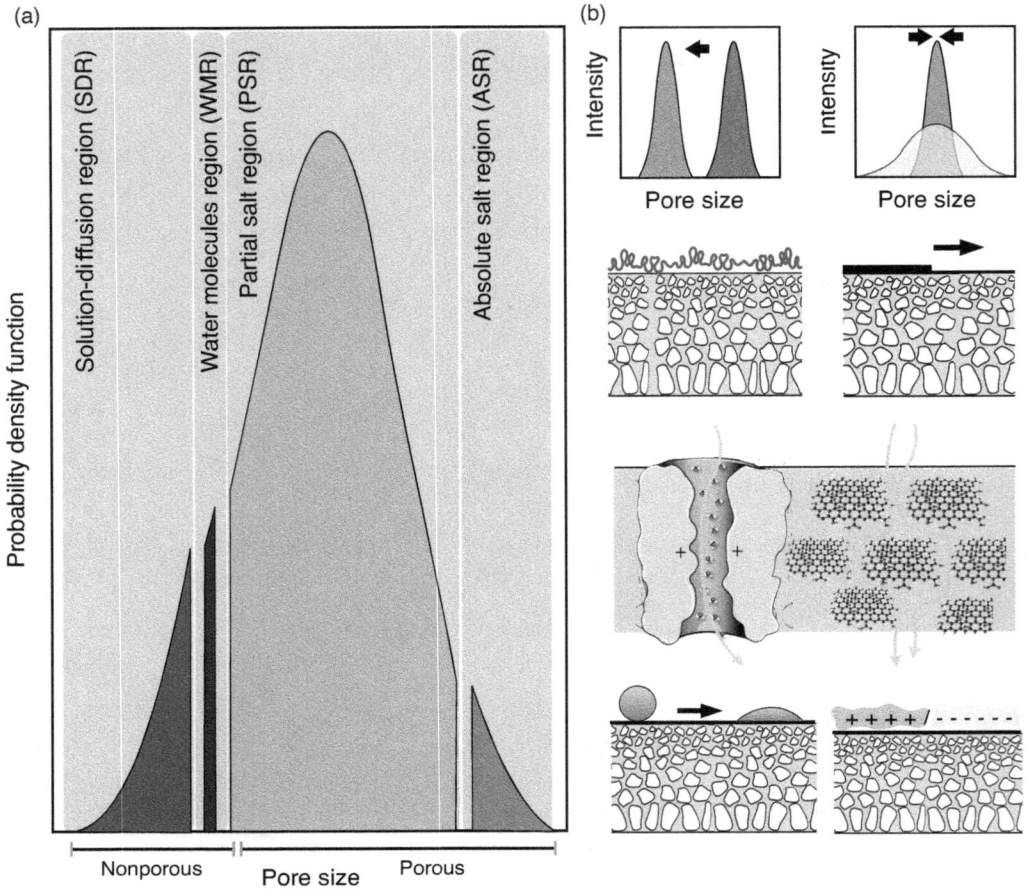

FIGURE 4.9 (a) Typical pore size distribution of an NF membrane divided into four mass transfer regions. (b) Membrane structure and physiochemical properties manipulation for better NF performance; water permeability can be enhanced by reducing thickness, increasing the effective surface area (roughness), increasing water transport channels, and boosting the surface hydrophilicity. The selectivity can be regulated by reducing pore size, adjusting surface charge, and narrowing pore size distribution.

Throughout the past decade, improving the membrane separation performance has been the focus of many researchers who have proposed various solutions from different perspectives. Technical approaches to coordinate the trade-off between water permeability and salt rejection are summarized and discussed below and also depicted in Figure 4.9b.

- **Promoting water transfer** by surface/membrane modification, for example, (I) increasing membrane surface hydrophilicity with surface grafting [105] or surface irradiation [106]; (II) establishing water molecules channels within the functional layer such as aquaporins [107], graphene sheets [108], or CNTs [109]. A key point to be considered in this approach is the defects in the active layer which might form by such fillers.
- **Increase effective membrane area** by assembling a nano-wrinkled active layer. Although such an approach improves permeability and salt rejection significantly, it increases the membrane surface roughness and hence fouling tendency.
- **Decreasing the thickness of the functional layer** in NF membranes, more precisely, the thickness, which is pointless in salt rejection but has a negative impact on water permeation.

For example, sub-10nm polyamide (PA) film was prepared on UF support using a special substrate of deposited cadmium hydroxide Nanostrands. The ultrathin layer encouraged ultrafast solvent permeation through NF membranes [110].

• **Narrowing the pore size distribution** of NF membranes, that is, producing isoporous membranes. Post-treatments such as acid or oxidation treatment are feasible for adjusting membrane structure and narrowing the pore size distribution. One promising strategy to fabricate isoporous membranes is the self-assembly of block copolymers. Peinemann et al. reported an NF membrane with a pore size of 1.5nm, MWCO of 600 g/mol, and flux of one order magnitude higher than commercial NF membranes. The membranes were prepared via the self-assembly of mixtures of two chemically interacting copolymers [111]. A novel strategy was proposed to narrow down the pore size distribution of ready-made NF membranes by selectively anchoring small molecules on the pore wall that is larger than the radius of the anchored molecules. Du et al. have demonstrated using module membrane of polydopamine (PDA)/polyethylenimine (PEI) deposited on the surface of polyacrylonitrile UF support. Filtration of the reactive solution of MPSI was done; the larger pores of the membrane were decorated with 3-mercapto-1-propanesulfonate ions (MPSI), reducing the pore size. This approach increased the selectivity of the membranes for divalent ions with increased reaction time; however, the NaCl rejection was not improved [112].

• **Solvent activation** via immersing the active PA layer in good solvent (solubility coefficient similar to the membrane active layer) is a promising method. It has a dual effect of healing existing defects and unblocking the permeation pathways by dissolving unwanted fragments and residual reactants [113]. Applying such a technique is promising; however, the challenge is that the support of the membrane should be resistant to harsh solvents.

4.3.3 Interfacial Polymerization Nanofiltration Membranes

IP-based NF membranes have dominated the water industry market, with rising research to improve separation performance [77,114,115]. The preparation procedure used to form these membranes is IP of a thin PA layer by the reaction of amine and acyl chloride monomers on top of a porous phase-inversion prepared support. The formation of this interfacial layer results in acid groups attached to the polymeric backbone, which causes the membrane to acquire a charge. The IP process is governed by many parameters, which each has an impact on the outcome membrane. Parameters such as the structure and morphology of the substrate; the type, concentration, and time of the monomers reaction; and the physicochemical properties of the reaction solution all impact the final selective layer of the membrane, hence the separation performance [7]. Notably, with more than 40 years of development, the IP process still has room and possibilities for exploration [116]. Additionally, investigating the membrane fabrication–property–performance nexus is essential, especially since the IP technique now involves nanomaterials and interlayers. Commonly, boosting the performance of NF membranes is fulfilled through one of two routes. The first is to exploit and regulate the IP process and push the boundaries to accomplish the controllable design of the active layer characteristics [79,117]. The recurring improvements in water permeability of commercial NF membranes during the past decades while keeping a similar MWCO are a prominent proof. For example, decreasing the concentration of the aqueous monomer during IP may considerably increase the density of negative charges on the membrane surface [118]. The second route is to introduce other controlled reaction processes, innovative methods, and new materials for finer fabrication and better tuning of the NF membrane properties [84]. Indeed, the latter has the potential to showcase excellent performance in comparison with the IP-fabricated membranes. Commercialization, however, is acutely hindered because of difficulties linked to defects, complexity, scaling-up, cost, and longevity. Besides, there is a performance improvement ceiling by virtue of the trade-off between water permeability and selectivity for NF membranes.

4.3.4 ADVANCES IN IP NANOFILTRATION MEMBRANES FOR DESALINATION

Achieving high selectivity for monovalent ions remains one of the challenges for NF membranes. Numerous advances in NF reflected high efficiency for the removal of monovalent ions. In addition, NF membranes with superior flux and excellent salt rejection are attractive to the current market. Listed below are some of the advances for improved IP membranes performance:

- **IP on hydrogels**: Bart Van der Bruggen et al. have developed an interesting and promising approach. A successful development of a novel m-phenylenediamine-based TFC NF membrane on a nanofibrous hydrogel substrate using Kevlar aramid nanofiber (ANF) followed by dimethylformamide (DMF) solvent activation step was investigated. The DMF solvent activation tuned the membrane surface morphology by eliminating the top "leaf-like" PA layers. In addition, the surface hydrophilicity enhanced significantly, resulting in higher water permeability. It was found that the desalination performance of the ANF TFC membrane was shifted from the RO zone to the NF zone. The ANF TFC membrane has a pure water permeability of 14.4 L/m^{2}h bar and excellent rejections of divalent ions (Na$_2$SO$_4$=100%, MgSO$_4$=99.4%, MgCl$_2$=92.7%) and a high rejection of monovalent ions (NaCl=80.3%) which outperformed reported piperazine (PIP)-based NF membranes [119].
- **TFN membranes**: Assembling nanomaterials within membranes has resolved major challenges, such as biofouling, scaling, low permeation rate, selectivity, and degradation. To mention a few, nanoporous single-layer graphene and stacked GO sheets are promising materials for NF membranes. Graphene is considered impermeable, and if used as a membrane, therefore, porosity must be imposed. Monolayer porous graphene was used as a desalination membrane, and nanometer-sized pores were fabricated using an oxygen plasma etching process. The membrane portrayed a salt rejection rate of 100% and rapid water transport. Water fluxes of up to 103 L/m^{2}s were measured using pressure difference as a driving force [120]. This work represents a proof of concept of the effectiveness and potential of nanoporous graphene for desalination applications; nevertheless, scaling up these membranes for commercial processes remains a significant challenge. GO sheets have antifouling properties that are highly advantageous for improving membrane properties. In addition, there is a channeling created between the sheets that facilitates the water permeability. Ali et al. prepared TFC membranes via the IP technique, and GO was implemented as a modifier to improve the membrane properties. Incorporating GO into the membrane improved the hydrophilicity, water permeability, chlorine resistance, and antifouling properties of the membrane. Good compatibility is guaranteed since the sheets bonded with the PA layer through hydrogen bonding. A 39% increase in water flux was scored when the amine phase contained 100 ppm of GO during IP compared to a pristine TFC membrane. At 15 bar, a water flux of 29.6 L/m^{2}h and a salt rejection of ≥97% were obtained for a saline solution (2,000 ppm of NaCl) [121]. Another interesting technique reported by Wang et al. is the fabrication of crumpled PA TFC NF membrane on a CNTs/polyethersulfone composite support impregnated with MOF nanoparticles. The nanoparticles acted as a template for forming a crumpled nanostructured PA active layer. The obtained NF membrane demonstrated high permeances of 53.5 L/m^{2}h bar with Na$_2$SO$_4$ rejection above 95%. This work exceeded other reported literature regarding water permeability; however, the monovalent ion rejection was still low [122].
- **Turing structure**: A Turing structure arises when imbalances between diffusion rates make a stable steady-state system sensitive to heterogeneous disruptions. This structure consists of bumps, voids, and islands, which is desirable for water desalination. An innovative approach to form thin-film nanofibrous composite NF membranes with Turing structures was reported. Tailored Turing structure was obtained for various NaCl content in an aqueous-phase solution during IP for effective desalination. The addition of salt in the

aqueous phase would increase the interfacial tension with the organic phase, hence hindering the mass transfer of PIP monomer, constructing an ultrathin and loose PA layer. Meanwhile, the continuous complexation of Na^+ with the PA carbonyl groups may motivate the uneven distribution of the IP reaction, leading to a change in the morphology from the crisscrossed ridge networks to nodular arrays. Besides, a hydrophilic PA layer was formed due to the Na^+/carbonyl groups reaction with the of trimesoyl chloride (TMC), which accelerated the hydrolysis of TMC. With 20.0 wt% NaCl content, the pore size in the PA skin layer shifted from 0.670 to 0.868 nm, and the thickness of the skin layer reduced from around 85 to 55 nm. Overall, PA20/PAN composite membranes showed impressive NF performance with an ultra-high water flux of 129.0 L/m^2h and Na_2SO_4 rejection of 99.1%. This strategy could be easily extended to commercial applications, seeing that the salt-tailored IP technique is cost-effective, easily handled, and significantly productive [123]. To form Turing structures, the diffusion coefficient of the inhibitor must be larger than that of the activator. Accordingly, PIP (the activator) was encouraged to interact via hydrogen bonding with polyvinyl alcohol (PVA). The interaction is expected to increase the viscosity of the aqueous phase and hence, to reduce the diffusion of PIP. Due to the diffusion-driven instability, a spotted and tube-like Turing morphology was fabricated successfully [117].

- **Double IP fabrication**: Acyl chloride-terminated hyperbranched polyester (HBPAC) was prepared and crosslinked into the PA layer of TFC membranes using the double IP technique. HBPAC with a highly branched structure increased the hydrophilic sites between PA chains, which benefitted the water permeation. Residual carboxyl groups of HBPAC act as hydrophilic sites, which promoted fast water permeation. The HBPAC compatibility and stability with the PA layer were imposed via possible crosslinking with the *m*-phenylene diamine (MPD) monomer. The prepared HBPAC-based TFC membranes demonstrated a smoother surface, enhanced surface hydrophilicity, and exhibited an increase in water flux from 30 to 45 L/m^2h without sacrificing salt rejection. This work shows great potential for flux enhancement by integrating a new acyl chloride-terminated hyperbranched polymer. Although the prepared membranes as RO membranes, it is more accurate to consider this membrane an NF membrane, since the TFC membrane prepared without HBPAC showed low NaCl rejections [124].

- **Surfactant-mediated IP fabrication**: An interesting research has been published recently aiming at the preparation of active PA layer with more uniform sub-nanometer pores via surfactant assembly-regulated interfacial polymerization (SARIP). The technique with the self-assembled surfactants promotes homogeneous diffusion of amine monomers across the water/hexane interface during IP. Fascinating results of step-wise transition from low rejection to near-perfect rejection over a solute size range smaller than half Angstrom were achieved [125]. PA TFC NF membranes were prepared via IP using PIP and TMC monomers with an alteration of adding the surfactant sodium dodecyl sulfate. The PA membrane formed via conventional IP had a very wide range of rejection for solutes with Stokes radius (r_s) between 2.5 and 5.0 Å. The wide distribution of rejection for similar size solutes only suggests that the mechanism governing the separation is a mix of electrostatic, ion dehydration, and size-based sieving. The use of the surfactant in the SARIP method dramatically changed the solute separation behavior. Remarkably, the rejection curve demonstrates a sharp step-wise transition at r_s ~2.7 Å; the separation of solutes here is dependent on the ion size, where monovalent and divalent cations were distinguished with miraculous precision.

- **Substrate modification/manipulation**: The physical/chemical properties of membranes substrate, such as roughness, wettability, and surface charge, affect the distribution/diffusion of aqueous monomers within the substrate. The uniform distribution of aqueous monomers is a precondition for forming defect-free ultrathin PA. Livingston et al. [110] prepared a defect-free, high flux, and ultrathin (thickness of 8.4 nm) PA membrane with the aid of interlayer. The cadmium hydroxide nanowires sacrificial layer was impregnated with an aqueous monomer to eliminate the influence of the substrate. The interlayer provided a smooth surface, high porosity, and good hydrophilicity, which encouraged even distribution of the aqueous monomer phase. MPD-TMC interfacially polymerized PA layer was synthesized on top of a packed, vertically aligned CNT substrate. Owing to the exceptional porosity and hydrophobic nature of the CNTs, the composite PA/CNTs membrane showed an impressive desalination performance with a flux of 8.3 L/m^2h bar and NaCl rejection of 98.3%. The water permeation reported was much higher than reported for the conventional PA membranes on polysulfone/CNT support [126]. For the sake of ultra-porosity, the water molecules path through the PA layer into the water channel of the PA/CNTs support is relatively shorter than traditional PA/PS membranes due to challenges accessing the substrate pores.

4.4 CONCLUSIONS

NF has paved its way in desalination applications due to its unique performance, including higher water flux and lower operating pressure than RO membranes, as well as high rejection of divalent ions and some neutral organics. However, NF suffers from low selectivity toward monovalent ion and thus, cannot be used as a standalone process for desalination. Alternatively, NF is hybridized with other desalination technologies to achieve the desired permeate quality.

Understanding the ion transport mechanisms of NF is key to designing membranes tailored for specific applications and achieving the desired performance. This chapter provided an overview of the ion rejection mechanisms of NF membranes and the effect of several solution and process parameters on membrane performance. The main mechanisms were discussed separately, although, in reality, ion transport is governed by combinations of these mechanisms. Moreover, these mechanisms are not entirely independent and could affect each other. For instance, dielectric exclusion can be enhanced with steric effect as the dielectric properties of water change depending on whether it is in membrane pores or the bulk solution. Image force induced by ions (dielectric effect) may be shielded by the charge of the membrane (Donnan effect). Meanwhile, the dielectric effect reduces ions' concentration in the membrane pores, leading to a lower screening effect, thus enhancing the Donnan effect. Knowledge of the true nature of ion rejection mechanisms by NF membranes is still limited due to the lack of proper techniques to study the physical structure and measure the electrical properties of NF membranes. Therefore, future studies should be focused on finding simple techniques to investigate the membrane properties.

Studying the performance of NF as a function of one specific testing condition is not accurate. This is because the performance of NF is a result of the interplay between all the conditions, including feed pH, solute concentration, pressure, and temperature. For instance, the effect of pH might vary depending on the concentration of solutes present in the solution. Therefore, discrepancies in literature are acceptable as each NF process is unique with its membrane material and testing conditions. Moreover, investigating the relationship between all these parameters is complicated in an experimental setting, and thus producing comprehensive mathematical models is essential.

The structural and physiochemical properties of NF membranes play an important role in determining their performance. These properties can be manipulated through innovations in materials and fabrication techniques. Therefore, understanding the link between fabrication and performance is essential for optimized design. This chapter discussed the main membrane fabrication methods with the light focused on IP. IP has a great potential for improvement since it dominates the market for NF membranes. Optimization of the IP reaction process and incorporation of new materials, fillers, and/or reactions are basically the two prime strategies to grasp tuned design of NF membranes. Although many of the novel NF membranes based on new materials have shown excellent performance, they have not overshadowed IP-based NF membranes in terms of commercialization. The ever-present trade-off effect between water permeability and salt rejection is one of the main obstacles faced in further elevating membrane performance. The fabrication conditions of IP have a direct impact on the final membrane properties; hence, the fabrication–property relationships are important to understand. Once the IP fabrication–property relationships are understood, maximizing the performance accordingly is a straightforward process. In particular, the selective layer thickness can be decreased to obtain high permeation by (i) reducing the concentration of diamine monomers, (ii) employing a hydrophilic and smooth substrate, and (iii) slowing down the diffusion rate of amine phase. The effective membrane surface area for water transport can be increased by creating imbalances between diffusion rates of monomers. Additionally, integrating porous nanomaterials within the selective layer is a successful method to fashion additional water transport channels for continuous water permeation paths. As for managing the selectivity of NF membranes, the membrane pore size can be regulated based on the monomers used for IP. Moreover, the membrane selectivity can be improved by enhancing the pore uniformity by post-treatment of the defective areas. The fabrication–property–performance relationships of IP-based NF membranes discussed here are derived from thorough analysis of separation performance of many reported NF membranes. These complex relationships can educate readers about the membrane fabrication procedures and methods for high-performance membranes and advocate their applications in water treatment to satisfy requirements for sustainable development.

REFERENCES

1. E. Jones, M. Qadir, M.T.H. van Vliet, V. Smakhtin, S.-m. Kang, The state of desalination and brine production: A global outlook, *Science of the Total Environment*, 657 (2019) 1343–1356.
2. S.S. Shenvi, A.M. Isloor, A.F. Ismail, A review on RO membrane technology: Developments and challenges, *Desalination*, 368 (2015) 10–26.
3. C.D. Peters, D. Li, Z. Mo, N.P. Hankins, Q. She, Exploring the limitations of osmotically assisted reverse osmosis: Membrane fouling and the limiting flux, *Environmental Science & Technology*, 56 (2022) 6678–6688.
4. T.V. Bartholomew, L. Mey, J.T. Arena, N.S. Siefert, M.S. Mauter, Osmotically assisted reverse osmosis for high salinity brine treatment, *Desalination*, 421 (2017) 3–11.
5. X. Chen, N.Y. Yip, Unlocking high-salinity desalination with cascading osmotically mediated reverse osmosis: Energy and operating pressure analysis, *Environmental Science & Technology*, 52 (2018) 2242–2250.
6. S. Guo, Y. Wan, X. Chen, J. Luo, Loose nanofiltration membrane custom-tailored for resource recovery, *Chemical Engineering Journal*, 409 (2021) 127376.
7. R.W. Baker, *Membrane Technology and Applications*, John Wiley & Sons, 2004.
8. Y. Zhao, T. Tong, X. Wang, S. Lin, E.M. Reid, Y. Chen, Differentiating solutes with precise nanofiltration for next generation environmental separations: A review, *Environmental Science & Technology*, 55 (2021) 1359–1376.
9. A.W. Mohammad, Y.H. Teow, W.L. Ang, Y.T. Chung, D.L. Oatley-Radcliffe, N. Hilal, Nanofiltration membranes review: Recent advances and future prospects, *Desalination*, 356 (2015) 226–254.
10. D. Zhou, L. Zhu, Y. Fu, M. Zhu, L. Xue, Development of lower cost seawater desalination processes using nanofiltration technologies – A review, *Desalination*, 376 (2015) 109–116.
11. D.L. Oatley-Radcliffe, M. Walters, T.J. Ainscough, P.M. Williams, A.W. Mohammad, N. Hilal, Nanofiltration membranes and processes: A review of research trends over the past decade, *Journal of Water Process Engineering*, 19 (2017) 164–171.
12. B.L. Morris, A.R. Lawrence, P. Chilton, B. Adams, R.C. Calow, B.A. Klinck, Groundwater and its susceptibility to degradation: A global assessment of the problem and options for management (2003).
13. S.S.D. Foster, P.J. Chilton, Groundwater: The processes and global significance of aquifer degradation, *Philosophical Transactions of the Royal Society B*, 358 (2003) 1957–1972.
14. L. Aquilina, V. Vergnaud-Ayraud, A.A. Les Landes, H. Pauwels, P. Davy, E. Pételet-Giraud, T. Labasque, C. Roques, E. Chatton, O. Bour, S. Ben Maamar, A. Dufresne, M. Khaska, C.L.G. La Salle, F. Barbecot, Impact of climate changes during the last 5 million years on groundwater in basement aquifers, *Scientific Reports*, 5 (2015) 14132.
15. S.M. Salman, United Nations general assembly resolution: International decade for action, water for life, 2005–2015: A water forum contribution, *Water International*, 30 (2005) 415–418.

16. M.W. Shahzad, M. Burhan, L. Ang, K.C. Ng, Energy-water-environment nexus underpinning future desalination sustainability, *Desalination*, 413 (2017) 52–64.

17. M. Sanz, Dynamic growth for desalination and water reuse in 2019, in: *World Water*, 2019.

18. G.F. Houngbo, The role of UN-water as an inter-agency coordination mechanism for water and sanitation, in: *UN Chronicle*, 2018, pp. 19–23.

19. H. Nassrullah, S.F. Anis, R. Hashaikeh, N. Hilal, Energy for desalination: A state-of-the-art review, *Desalination*, 491 (2020) 114569.

20. IEA, Redrawing the energy-climate map, in: *World Energy Outlook Special Report*, 2013, pp. 134.

21. A.N. Mabrouk, H.E.S. Fath, Technoeconomic study of a novel integrated thermal MSF–MED desalination technology, *Desalination*, 371 (2015) 115–125.

22. M.A. Eltawil, Z. Zhengming, L. Yuan, Renewable energy powered desalination systems: Technologies and economics-state of the art, in: *Twelfth International Water Technology Conference, IWTC12*, 2008, pp. 1–38.

23. C.J. Harrison, Y.A. Le Gouellec, R.C. Cheng, A.E. Childress, Bench-scale testing of nanofiltration for seawater desalination, *Journal of Environmental Engineering*, 133 (2007) 1004–1014.

24. A. AlTaee, A.O. Sharif, Alternative design to dual stage NF seawater desalination using high rejection brackish water membranes, *Desalination*, 273 (2011) 391–397.

25. P. Eriksson, M. Kyburz, W. Pergande, NF membrane characteristics and evaluation for sea water processing applications, *Desalination*, 184 (2005) 281–294.

26. A. Hassan, A. Farooque, A. Jamaluddin, M. Al-Sofi, A. Al-Ajlan, O. Hamed, A. Al-Amoudi, A. Al-Rubaian, N. Kither, A. Al-Azzaz, Conversion and operation of the commercial umm Lujj SWRO plant from a single SWRO desalination process to the new dual NF-SWRO desalination process, in: *IDA Conference*, Manama Bahrain, 2002.

27. D.L. Oatley, L. Llenas, R. Pérez, P.M. Williams, X. Martínez-Lladó, M. Rovira, Review of the dielectric properties of nanofiltration membranes and verification of the single oriented layer approximation, *Advances in Colloid and Interface Science*, 173 (2012) 1–11.

28. R. Epsztein, E. Shaulsky, N. Dizge, D.M. Warsinger, M. Elimelech, Role of ionic charge density in donnan exclusion of monovalent anions by nanofiltration, *Environmental Science & Technology*, 52 (2018) 4108–4116.

29. R.W. Baker, *Membrane Technology and Applications*, Third edition, John Wiley & Sons, United Kingdom, 2012.

30. B. Van der Bruggen, J. Schaep, D. Wilms, C. Vandecasteele, Influence of molecular size, polarity and charge on the retention of organic molecules by nanofiltration, *Journal of Membrane Science*, 156 (1999) 29–41.

31. A.E. Yaroshchuk, Rejection mechanisms of NF membranes, *Membrane Technology*, 1998 (1998) 9–12.

32. K.S. Spiegler, O. Kedem, Thermodynamics of hyperfiltration (reverse osmosis): Criteria for efficient membranes, *Desalination*, 1 (1966) 311–326.

33. S.-I. Nakao, S. Kimura, Models of membrane transport phenomena and their applications for ultrafiltration data, *Journal of Chemical Engineering of Japan*, 15 (1982) 200–205.

34. N.S. Suhalim, N. Kasim, E. Mahmoudi, I.J. Shamsudin, A.W. Mohammad, F. Mohamed Zuki, N.L.-A. Jamari, Rejection mechanism of ionic solute removal by nanofiltration membranes: An overview, *Nanomaterials*, 12 (2022) 437.

35. H. Zhang, Q. He, J. Luo, Y. Wan, S.B. Darling, Sharpening nanofiltration: Strategies for enhanced membrane selectivity, *ACS Applied Materials & Interfaces*, 12 (2020) 39948–39966.

36. I.I. Ryzhkov, A.V. Minakov, Theoretical study of electrolyte transport in nanofiltration membranes with constant surface potential/charge density, *Journal of Membrane Science*, 520 (2016) 515–528.

37. S. Bandini, L. Bruni, Transport phenomena in nanofiltration membranes, in: E. Drioli, L. Giorno (Eds.) *Comprehensive Membrane Science and Engineering*, Elsevier, Oxford, 2010, pp. 67–89.

38. A.E. Yaroshchuk, Non-steric mechanisms of nanofiltration: Superposition of Donnan and dielectric exclusion, *Separation and Purification Technology*, 22–23 (2001) 143–158.

39. A.E. Childress, M. Elimelech, Relating nanofiltration membrane performance to membrane charge (electrokinetic) characteristics, *Environmental Science & Technology*, 34 (2000) 3710–3716.

40. O. Labban, C. Liu, T.H. Chong, J.H. Lienhard V, Fundamentals of low-pressure nanofiltration: Membrane characterization, modeling, and understanding the multi-ionic interactions in water softening, *Journal of Membrane Science*, 521 (2017) 18–32.

41. L.D. Nghiem, A.I. Schäfer, M. Elimelech, Role of electrostatic interactions in the retention of pharmaceutically active contaminants by a loose nanofiltration membrane, *Journal of Membrane Science*, 286 (2006) 52–59.

42. A.M. Smith, A.A. Lee, S. Perkin, The electrostatic screening length in concentrated electrolytes increases with concentration, *The Journal of Physical Chemistry Letters*, 7 (2016) 2157–2163.

43. A.E. Yaroshchuk, Dielectric exclusion of ions from membranes, *Advances in Colloid and Interface Science*, 85 (2000) 193–230.

44. Y. Zhu, H. Zhu, G. Li, Z. Mai, Y. Gu, The effect of dielectric exclusion on the rejection performance of inhomogeneously charged polyamide nanofiltration membranes, *Journal of Nanoparticle Research*, 21 (2019) 217.

45. S. Bandini, D. Vezzani, Nanofiltration modeling: The role of dielectric exclusion in membrane characterization, *Chemical Engineering Science*, 58 (2003) 3303–3326.

46. J. Schaep, C. Vandecasteele, A. Wahab Mohammad, W. Richard Bowen, Modelling the retention of ionic components for different nanofiltration membranes, *Separation and Purification Technology*, 22–23 (2001) 169–179.

47. J.N. Israelachvili, *Intermolecular and Surface Forces*, Third Edition, Academic Press, San Diego, 2011.

48. W.R. Bowen, J.S. Welfoot, Modelling the performance of membrane nanofiltration – Critical assessment and model development, *Chemical Engineering Science*, 57 (2002) 1121–1137.

49. B. Saliha, F. Patrick, S. Anthony, Investigating nanofiltration of multi-ionic solutions using the steric, electric and dielectric exclusion model, *Chemical Engineering Science*, 64 (2009) 3789–3798.

50. B. Tansel, J. Sager, T. Rector, J. Garland, R.F. Strayer, L. Levine, M. Roberts, M. Hummerick, J. Bauer, Significance of hydrated radius and hydration shells on ionic permeability during nanofiltration in dead end and cross flow modes, *Separation and Purification Technology*, 51 (2006) 40–47.

51. L.A. Richards, A.I. Schäfer, B.S. Richards, B. Corry, The importance of dehydration in determining ion transport in narrow pores, *Small*, 8 (2012) 1701–1709.

52. L.A. Richards, B.S. Richards, B. Corry, A.I. Schäfer, Experimental energy barriers to anions transporting through nanofiltration membranes, *Environmental Science & Technology*, 47 (2013) 1968–1976.

53. B. Tansel, Significance of thermodynamic and physical characteristics on permeation of ions during membrane separation: Hydrated radius, hydration free energy and viscous effects, *Separation and Purification Technology*, 86 (2012) 119–126.

54. M. Misin, M.V. Fedorov, D.S. Palmer, Hydration free energies of molecular ions from theory and simulation, *The Journal of Physical Chemistry B*, 120 (2016) 975–983.

55. C.-H. Chu, C. Wang, H.-F. Xiao, Q. Wang, W.-J. Yang, N. Liu, X. Ju, J.-X. Xie, S.-P. Sun, Separation of ions with equivalent and similar molecular weights by nanofiltration: Sodium chloride and sodium acetate as an example, *Separation and Purification Technology*, 250 (2020) 117199.

56. A.E. Childress, M. Elimelech, Effect of solution chemistry on the surface charge of polymeric reverse osmosis and nanofiltration membranes, *Journal of Membrane Science*, 119 (1996) 253–268.

57. C. Labbez, P. Fievet, A. Szymczyk, A. Vidonne, A. Foissy, J. Pagetti, Analysis of the salt retention of a titania membrane using the "DSPM" model: Effect of pH, salt concentration and nature, *Journal of Membrane Science*, 208 (2002) 315–329.

58. J. Luo, Y. Wan, Effects of pH and salt on nanofiltration – A critical review, *Journal of Membrane Science*, 438 (2013) 18–28.

59. S. Szoke, G. Patzay, L. Weiser, Characteristics of thin-film nanofiltration membranes at various pH-values, *Desalination*, 151 (2003) 123–129.

60. V. Freger, T.C. Arnot, J.A. Howell, Separation of concentrated organic/inorganic salt mixtures by nanofiltration, *Journal of Membrane Science*, 178 (2000) 185–193.

61. M. Dalwani, N.E. Benes, G. Bargeman, D. Stamatialis, M. Wessling, A method for characterizing membranes during nanofiltration at extreme pH, *Journal of Membrane Science*, 363 (2010) 188–194.

62. M. Dalwani, N.E. Benes, G. Bargeman, D. Stamatialis, M. Wessling, Effect of pH on the performance of polyamide/polyacrylonitrile based thin film composite membranes, *Journal of Membrane Science*, 372 (2011) 228–238.

63. W.R. Bowen, H.N.S. Yousef, Effect of salts on water viscosity in narrow membrane pores, *Journal of Colloid and Interface Science*, 264 (2003) 452–457.

64. J. Luo, L. Ding, Y. Su, ShaopingWei, Y. Wan, Concentration polarization in concentrated saline solution during desalination of iron dextran by nanofiltration, *Journal of Membrane Science*, 363 (2010) 170–179.

65. S. Bason, V. Freger, Phenomenological analysis of transport of mono- and divalent ions in nanofiltration, *Journal of Membrane Science*, 360 (2010) 389–396.

66. S. Déon, A. Escoda, P. Fievet, A transport model considering charge adsorption inside pores to describe salts rejection by nanofiltration membranes, *Chemical Engineering Science*, 66 (2011) 2823–2832.

67. L. Bruni, S. Bandini, The role of the electrolyte on the mechanism of charge formation in polyamide nanofiltration membranes, *Journal of Membrane Science*, 308 (2008) 136–151.

68. M.R. Teixeira, M.J. Rosa, M. Nyström, The role of membrane charge on nanofiltration performance, *Journal of Membrane Science*, 265 (2005) 160–166.
69. A.I. Schäefer, A.G. Fane, *Nanofiltration: Principles, Applications, and New Materials*, John Wiley & Sons, 2021.
70. J. Wang, D.S. Dlamini, A.K. Mishra, M.T.M. Pendergast, M.C.Y. Wong, B.B. Mamba, V. Freger, A.R.D. Verliefde, E.M.V. Hoek, A critical review of transport through osmotic membranes, *Journal of Membrane Science*, 454 (2014) 516–537.
71. A.I. Schäfer, A.G. Fane, *Nanofiltration: Principles, Applications, and New Materials*, 2nd Edition, John Wiley & Sons, 2021.
72. Y. Roy, D.M. Warsinger, J.H. Lienhard, Effect of temperature on ion transport in nanofiltration membranes: Diffusion, convection and electromigration, *Desalination*, 420 (2017) 241–257.
73. N. Ben Amar, H. Saidani, A. Deratani, J. Palmeri, Effect of temperature on the transport of water and neutral solutes across nanofiltration membranes, *Langmuir*, 23 (2007) 2937–2952.
74. M.L. Huber, R.A. Perkins, A. Laesecke, D.G. Friend, J.V. Sengers, M.J. Assael, I.N. Metaxa, E. Vogel, R. Mareš, K. Miyagawa, New international formulation for the viscosity of H_2O, *Journal of Physical Chemical Reference Data*, 38 (2009) 101–125.
75. B. Tomaszewska, M. Rajca, E. Kmiecik, M. Bodzek, W. Bujakowski, K. Wątor, M. Tyszer, The influence of selected factors on the effectiveness of pre-treatment of geothermal water during the nanofiltration process, *Desalination*, 406 (2017) 74–82.
76. R. Xu, M. Zhou, H. Wang, X. Wang, X. Wen, Influences of temperature on the retention of PPCPs by nanofiltration membranes: Experiments and modeling assessment, *Journal of Membrane Science*, 599 (2020) 117817.
77. K. Wang, X. Wang, B. Januszewski, Y. Liu, D. Li, R. Fu, M. Elimelech, X. Huang, Tailored design of nanofiltration membranes for water treatment based on synthesis–property–performance relationships, *Chemical Society Reviews*, 51 (2022) 672–719.
78. M. Mulder, J. Mulder, *Basic Principles of Membrane Technology*, Springer Science & Business Media, 1996.
79. Y. Song, J.-B. Fan, S. Wang, Recent progress in interfacial polymerization, *Materials Chemistry Frontiers*, 1 (2017) 1028–1040.
80. M. Kumar, M.A. Khan, H.A. Arafat, Recent developments in the rational fabrication of thin film nanocomposite membranes for water purification and desalination, *ACS Omega*, 5 (2020) 3792–3800.
81. Y. Ji, W. Qian, Y. Yu, Q. An, L. Liu, Y. Zhou, C. Gao, Recent developments in nanofiltration membranes based on nanomaterials, *Chinese Journal of Chemical Engineering*, 25 (2017) 1639–1652.
82. M.L. Lind, A.K. Ghosh, A. Jawor, X. Huang, W. Hou, Y. Yang, E.M.V. Hoek, Influence of Zeolite Crystal Size on Zeolite-Polyamide Thin Film Nanocomposite Membranes, *Langmuir*, 25 (2009) 10139–10145.
83. W.J. Lau, S. Gray, T. Matsuura, D. Emadzadeh, J. Paul Chen, A.F. Ismail, A review on polyamide thin film nanocomposite (TFN) membranes: History, applications, challenges and approaches, *Water Research*, 80 (2015) 306–324.
84. J. Wang, J. Zhu, Y. Zhang, J. Liu, B. Van der Bruggen, Nanoscale tailor-made membranes for precise and rapid molecular sieve separation, *Nanoscale*, 9 (2017) 2942–2957.
85. D.J. Johnson, N. Hilal, Nanocomposite nanofiltration membranes: State of play and recent advances, *Desalination*, 524 (2022) 115480.
86. D. Qin, G. Huang, D. Terada, H. Jiang, M.M. Ito, A. H. Gibbons, R. Igarashi, D. Yamaguchi, M. Shirakawa, E. Sivaniah, B. Ghalei, Nanodiamond mediated interfacial polymerization for high performance nanofiltration membrane, *Journal of Membrane Science*, 603 (2020) 118003.
87. T. Li, Y. Xiao, D. Guo, L. Shen, R. Li, Y. Jiao, Y. Xu, H. Lin, In-situ coating TiO2 surface by plant-inspired tannic acid for fabrication of thin film nanocomposite nanofiltration membranes toward enhanced separation and antibacterial performance, *Journal of Colloid and Interface Science*, 572 (2020) 114–121.
88. G.S. Lai, W.J. Lau, P.S. Goh, M. Karaman, M. Gürsoy, A.F. Ismail, Development of thin film nanocomposite membrane incorporated with plasma enhanced chemical vapor deposition-modified hydrous manganese oxide for nanofiltration process, *Composites Part B: Engineering*, 176 (2019) 107328.
89. Y. Liu, X. Wang, X. Gao, J. Zheng, J. Wang, A. Volodin, Y.F. Xie, X. Huang, B. Van der Bruggen, J. Zhu, High-performance thin film nanocomposite membranes enabled by nanomaterials with different dimensions for nanofiltration, *Journal of Membrane Science*, 596 (2020) 117717.
90. S. Kamari, A. Shahbazi, Biocompatible $Fe_3O_4@SiO_2-NH_2$ nanocomposite as a green nanofiller embedded in PES–nanofiltration membrane matrix for salts, heavy metal ion and dye removal: Long–term operation and reusability tests, *Chemosphere*, 243 (2020) 125282.

91. S. Casanova, T.-Y. Liu, Y.-M.J. Chew, A. Livingston, D. Mattia, High flux thin-film nanocomposites with embedded boron nitride nanotubes for nanofiltration, *Journal of Membrane Science*, 597 (2020) 117749.

92. B. Borjigin, l. Liu, L. Yu, L. Xu, C. Zhao, J. Wang, Influence of incorporating beta zeolite nanoparticles on water permeability and ion selectivity of polyamide nanofiltration membranes, *Journal of Environmental Sciences*, 98 (2020) 77–84.

93. F. Xiao, B. Wang, X. Hu, S. Nair, Y. Chen, Thin film nanocomposite membrane containing zeolitic imidazolate framework-8 via interfacial polymerization for highly permeable nanofiltration, *Journal of the Taiwan Institute of Chemical Engineers*, 83 (2018) 159–167.

94. H. Li, W. Shi, Y. Zhang, Q. Du, X. Qin, Y. Su, Improved performance of poly(piperazine amide) composite nanofiltration membranes by adding aluminum hydroxide nanospheres, *Separation and Purification Technology*, 166 (2016) 240–251.

95. C. Xu, W. Chen, H. Gao, X. Xie, Y. Chen, Cellulose nanocrystal/silver (CNC/Ag) thin-film nanocomposite nanofiltration membranes with multifunctional properties, *Environmental Science: Nano*, 7 (2020) 803–816.

96. N. Song, X. Xie, D. Chen, G. Li, H. Dong, L. Yu, L. Dong, Tailoring nanofiltration membrane with three-dimensional turing flower protuberances for water purification, *Journal of Membrane Science*, 621 (2021) 118985.

97. F. Soyekwo, Q. Zhang, R. Gao, Y. Qu, C. Lin, X. Huang, A. Zhu, Q. Liu, Cellulose nanofiber intermediary to fabricate highly-permeable ultrathin nanofiltration membranes for fast water purification, *Journal of Membrane Science*, 524 (2017) 174–185.

98. L. Chen, J.-H. Moon, X. Ma, L. Zhang, Q. Chen, L. Chen, R. Peng, P. Si, J. Feng, Y. Li, J. Lou, L. Ci, High performance graphene oxide nanofiltration membrane prepared by electrospraying for wastewater purification, *Carbon*, 130 (2018) 487–494.

99. Q. Xie, S. Zhang, H. Ma, W. Shao, X. Gong, Z. Hong, A novel thin-film nanocomposite nanofiltration membrane by incorporating 3D hyperbranched polymer functionalized 2D graphene oxide, *Polymers*, 10 (2018) 1253.

100. W. Ye, H. Liu, F. Lin, J. Lin, S. Zhao, S. Yang, J. Hou, S. Zhou, B. Van der Bruggen, High-flux nanofiltration membranes tailored by bio-inspired co-deposition of hydrophilic g-C3N4 nanosheets for enhanced selectivity towards organics and salts, *Environmental Science: Nano*, 6 (2019) 2958–2967.

101. Q. Bi, C. Zhang, J. Liu, X. Liu, S. Xu, Positively charged zwitterion-carbon nitride functionalized nanofiltration membranes with excellent separation performance of Mg2+/Li+ and good antifouling properties, *Separation and Purification Technology*, 257 (2021) 117959.

102. L.M. Robeson, The upper bound revisited, *Journal of Membrane Science*, 320 (2008) 390–400.

103. Z. Yang, H. Guo, C.Y. Tang, The upper bound of thin-film composite (TFC) polyamide membranes for desalination, *Journal of Membrane Science*, 590 (2019) 117297.

104. R. Zhang, J. Tian, S. Gao, B. Van der Bruggen, How to coordinate the trade-off between water permeability and salt rejection in nanofiltration? *Journal of Materials Chemistry A*, 8 (2020) 8831–8847.

105. H.-F. Xiao, C.-H. Chu, W.-T. Xu, B.-Z. Chen, X.-H. Ju, W. Xing, S.-P. Sun, Amphibian-inspired amino acid ionic liquid functionalized nanofiltration membranes with high water permeability and ion selectivity for pigment wastewater treatment, *Journal of Membrane Science*, 586 (2019) 44–52.

106. R. Reis, L.F. Dumée, B.L. Tardy, R. Dagastine, J.D. Orbell, J.A. Schutz, M.C. Duke, Towards enhanced performance thin-film composite membranes via surface plasma modification, *Scientific Reports*, 6 (2016) 29206.

107. X. Li, S. Chou, R. Wang, L. Shi, W. Fang, G. Chaitra, C.Y. Tang, J. Torres, X. Hu, A.G. Fane, Nature gives the best solution for desalination: Aquaporin-based hollow fiber composite membrane with superior performance, *Journal of Membrane Science*, 494 (2015) 68–77.

108. S. Bano, A. Mahmood, S.-J. Kim, K.-H. Lee, Graphene oxide modified polyamide nanofiltration membrane with improved flux and antifouling properties, *Journal of Materials Chemistry A*, 3 (2015) 2065–2071.

109. W.-F. Chan, H.-y. Chen, A. Surapathi, M.G. Taylor, X. Shao, E. Marand, J.K. Johnson, Zwitterion functionalized carbon nanotube/polyamide nanocomposite membranes for water desalination, *ACS Nano*, 7 (2013) 5308–5319.

110. S. Karan, Z. Jiang, A.G. Livingston, Sub-10 nm polyamide nanofilms with ultrafast solvent transport for molecular separation, *Science*, 348 (2015) 1347–1351.

111. H. Yu, X. Qiu, N. Moreno, Z. Ma, V.M. Calo, S.P. Nunes, K.-V. Peinemann, Self-assembled asymmetric block copolymer membranes: Bridging the gap from ultra- to nanofiltration, *Angewandte Chemie International Edition*, 54 (2015) 13937–13941.

112. Y. Du, Y. Lv, W.-Z. Qiu, J. Wu, Z.-K. Xu, Nanofiltration membranes with narrowed pore size distribution via pore wall modification, *Chemical Communications*, 52 (2016) 8589–8592.

113. M.F. Jimenez Solomon, Y. Bhole, A.G. Livingston, High flux membranes for organic solvent nanofiltration (OSN) – Interfacial polymerization with solvent activation, *Journal of Membrane Science*, 423–424 (2012) 371–382.

114. J.E. Cadotte, R.S. King, R.J. Majerle, R.J. Petersen, Interfacial synthesis in the preparation of reverse osmosis membranes, *Journal of Macromolecular Science: Part A – Chemistry*, 15 (1981) 727–755.

115. J.M. Gohil, P. Ray, A review on semi-aromatic polyamide TFC membranes prepared by interfacial polymerization: Potential for water treatment and desalination, *Separation and Purification Technology*, 181 (2017) 159–182.

116. X. Lu, M. Elimelech, Fabrication of desalination membranes by interfacial polymerization: History, current efforts, and future directions, *Chemical Society Reviews*, 50 (2021) 6290–6307.

117. Z. Tan, S. Chen, X. Peng, L. Zhang, C. Gao, Polyamide membranes with nanoscale Turing structures for water purification, *Science*, 360 (2018) 518–521.

118. Y.-l. Liu, Y.-y. Zhao, X.-m. Wang, X.-h. Wen, X. Huang, Y.F. Xie, Effect of varying piperazine concentration and post-modification on prepared nanofiltration membranes in selectively rejecting organic micropollutants and salts, *Journal of Membrane Science*, 582 (2019) 274–283.

119. Y. Li, E. Wong, Z. Mai, B. Van der Bruggen, Fabrication of composite polyamide/Kevlar aramid nanofiber nanofiltration membranes with high permselectivity in water desalination, *Journal of Membrane Science*, 592 (2019) 117396.

120. S.P. Surwade, S.N. Smirnov, I.V. Vlassiouk, R.R. Unocic, G.M. Veith, S. Dai, S.M. Mahurin, Water desalination using nanoporous single-layer graphene, *Nature Nanotechnology*, 10 (2015) 459–464.

121. M.E.A. Ali, L. Wang, X. Wang, X. Feng, Thin film composite membranes embedded with graphene oxide for water desalination, *Desalination*, 386 (2016) 67–76.

122. Z. Wang, Z. Wang, S. Lin, H. Jin, S. Gao, Y. Zhu, J. Jin, Nanoparticle-templated nanofiltration membranes for ultrahigh performance desalination, *Nature Communications*, 9 (2018) 2004.

123. K. Shen, P. Li, T. Zhang, X. Wang, Salt-tuned fabrication of novel polyamide composite nanofiltration membranes with three-dimensional turing structures for effective desalination, *Journal of Membrane Science*, 607 (2020) 118153.

124. H. Wu, X. Zhang, X.-T. Zhao, K. Li, C.-Y. Yu, L.-F. Liu, Y.-F. Zhou, C.-J. Gao, High flux reverse osmosis membranes fabricated with hyperbranched polymers via novel twice-crosslinked interfacial polymerization method, *Journal of Membrane Science*, 595 (2020) 117480.

125. Y. Liang, Y. Zhu, C. Liu, K.-R. Lee, W.-S. Hung, Z. Wang, Y. Li, M. Elimelech, J. Jin, S. Lin, Polyamide nanofiltration membrane with highly uniform sub-nanometre pores for sub-1 Å precision separation, *Nature Communications*, 11 (2020) 2015.

126. K. Li, B. Lee, Y. Kim, High performance reverse osmosis membrane with carbon nanotube support layer, *Journal of Membrane Science*, 592 (2019) 117358.

5 Nanofiltration and Emerging and Trace Organic Contaminants Wastewater

Míriam Cristina Santos Amaral, Flávia Cristina Rodrigues Costa, and Carolina Rodrigues dos Santos
Universidade Federal de Minas Gerais

CONTENTS

5.1 INTRODUCTION

Trace organic contaminants (TrOCs) include a wide variety of compounds that promote environmental and human health risks due to their physical, chemical, and ecotoxicological characteristics [1]. For example, these compounds can be pharmaceutically active compounds (PhACs), personal care products (PCPs), endocrine-disrupting compounds, pesticides, and organohalogen compounds.

These compounds reach environmental compartments mainly via industrial (such as pharmaceutical and pesticide industries) wastewater, effluents from hospitals, landfill leachate, and runoff from agriculture and animal husbandry, where TrOCs such as steroid hormones and antibiotics are applied to increase production [2,3]. In addition to these sources, domestic wastewater requires great attention.

TrOCs reach municipal wastewater through human urine and feces containing pharmaceuticals, hormones, and their metabolites; improper disposal of pharmaceuticals; and due to routine use of PCPs, for example. The major concern is that most wastewater treatment plants (WWTPs) are not designed to remove these compounds and their eventual by-products [4].

Among conventional biological treatments, active sludge can be highlighted as it is widely used in WWTPs. TrOCs removal mechanisms in active sludge mainly include degradation by microorganisms and sorption on flocs. However, concerning TrOCs, an adequate removal is frequently not

DOI: 10.1201/9781003261827-5

reached, and it is often not possible to meet the discharge limit standards, requiring a coupling to a tertiary treatment [5]. Because many TrOCs are recalcitrant to conventional treatments and remain in the treated effluent [4], they are also found in surface water, groundwater, and even drinking water worldwide [6–8].

One factor that holds back a more vast application of technologies capable of effectively removing TrOCs is that there is still a lack of regulation and monitoring of these compounds in aquatic compartments in most countries. For example, in the European Union (28 countries), Environmental Quality Standards have been defined for 48 priority substances and 8 pollutants in surface waters [9]. However, no TrOC has been directly regulated to date.

In the United States, effluent discharge from WWTPs is regulated by licenses from the National Pollution Discharge Elimination System, but TrOCs are not regulated [9]. In contrast, the National Water Research Institute published a report on public health criteria for direct potable reuse systems. Reference values were given for some substances such as carbamazepine (10 µg/L), cotinine (1 µg/L), primidone (10 µg/L), phenytoin (2 µg/L), meprobamate (200 µg/L), atenolol (4 µg/L), estrone (320 ng/L), triclosan (2,100 µg/L), and N,N-Diethyl-3-methylbenzamide (200 µg/L) [10].

Switzerland is one of the pioneering countries in the protection of the environment and water resources and has an innovative regulatory politics. The country has an environmental program that includes monitoring TrOCs in aquatic environments. Thus, the Quality Monitoring Program (NAWA) promotes regular measurements of pharmaceuticals (atenolol, azithromycin, bezafibrate, carbamazepine, clarithromycin, diclofenac, mefenamic acid, metoprolol, naproxen, sotalol, sulfamethazine, sulfamethoxazole, and trimethoprim) [11]. In addition, there is a commitment to invest in reducing the release of TrOCs into the aquatic environment from WWTPs.

Generally, TrOCs are found in aquatic compartments at low concentrations, in ng/L and µg/L [12]. However, investigations show its risks to aquatic organisms and human health even at low concentrations [13]. The permanence of these compounds in aquatic environments can impact animals and humans that are exposed to them for a long time [14]. For instance, diclofenac (0.5–50 µg/L) can affect gill and kidney tissues of freshwater fish, drug mixtures can inhibit the growth of human embryonic kidney cells [15], and increased incidence of obesity and type 2 diabetes may be related to the presence of TrOCs in the waters [16].

Figure 5.1 illustrates the ecotoxicological risk associated with different TrOCs through the risk quotient (RQ) value. The RQ considers the compound's environmental concentration and the predicted no-effect concentration (PNEC). RQ is considered a reasonable measure of the risk that a compound concentration poses since compounds that cause effects even at very low concentrations are considered [17]. In this way, the importance of removing TrOCs is evidenced, either to ensure safe disposal or to reuse water safely.

5.2 NANOFILTRATION TREATMENT OF TrOCs WASTEWATER

Nanofiltration (NF) is a pressure-driven process in which membranes with pore sizes from 1 to 10 nm are applied. Therefore, NF can retain multivalent ions, suspended solids, viruses, and bacteria. Given the concern about TrOCs present in the environment and their risks, efforts to remove them by the NF process have increased. The rejection of TrOCs by the NF process can occur by some mechanisms represented in Figure 5.2, and these mechanisms are described in the following sections.

5.2.1 PORE SIZE EXCLUSION

The predominant removal mechanism is the rejection of NF membranes due to particle or molecule size compared to membrane pore size (size exclusion). Molecular weight cut-off (MWCO) values of NF membranes are between 200 and 2,000 Da. Generally, NF can reject 98% divalent ions and 30%–85% monovalent ions [27]. Therefore, the size of TrOC is a fundamental parameter to determine its transport across the membrane and retention rate. In this context,

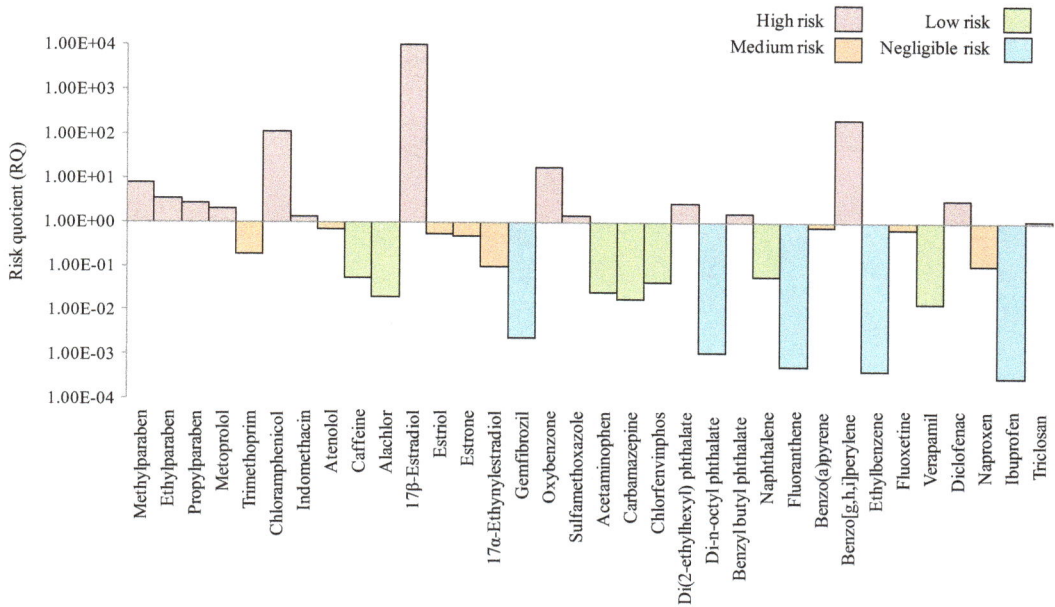

FIGURE 5.1 Ecotoxicological risk (risk quotient (RQ)) for several TrOCs. (References: [18–26, https://pubchem.ncbi.nlm.nih.gov].)

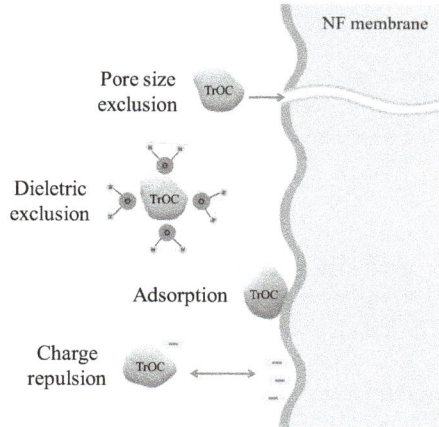

FIGURE 5.2 TrOCs rejection mechanisms by NF membrane.

several parameters have been proposed to represent the TrOC molecular size, including molecular weight (MW), Stokes diameter, molecular width and length, minimum projection area, and van der Waals volume [28].

In general, if TrOC MW is greater than the membrane MWCO, the compound is removed by the size exclusion mechanism. In Figure 5.3, there are several TrOCs that can be removed by NF through the size exclusion mechanism. For example, Ref. [29] found that TrOCs with MW above 200 g/mol had removal efficiency above 90% by a polyamide thin-film composite membrane, with MWCO of 200 Da. Another critical factor in pore size exclusion is the feed solution temperature. Reference [30] showed a considerable reduction in the rejection of TrOCs by the NF membrane with increasing temperature. This result is due to increased membrane pore size, which shows that the operational conditions can harm or favor the rejection by pore size.

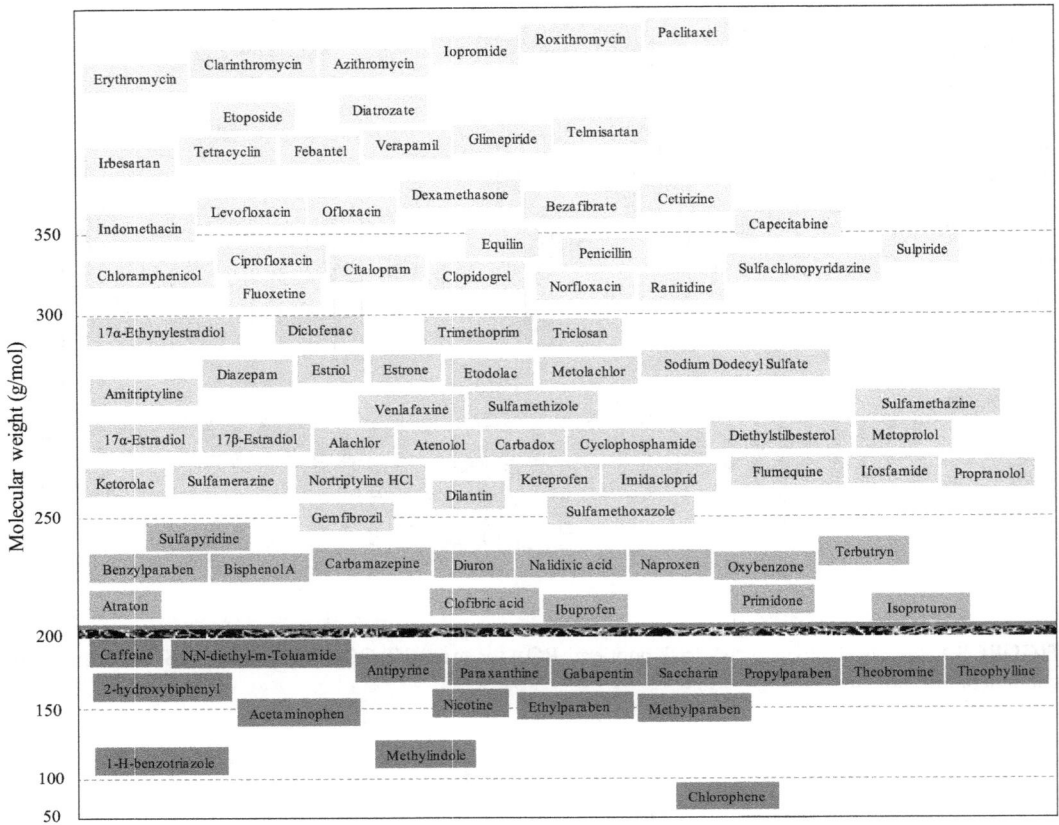

FIGURE 5.3 TrOCs with potential to be removed by NF through size exclusion.

5.2.2 Dielectric Exclusion

The phenomenon of dielectric exclusion occurs when media with different dielectric properties come into contact, as when the NF membrane comes into contact with feed solutions during wastewater treatment. Dielectric exclusion can be based on image force or ion solvation [31].

- **Dielectric exclusion based on image forces**: the intrapore solution of the membrane has a lower dielectric constant than bulk. Due to the generated discontinuity, membrane surface polarization occurs. In this way, the surface polarization charges have the same sign as the ionic charges present in bulk. Consequently, when the ion enters the membrane pore, a fictitious ion is created by the ion itself, inducing its repulsion [31].
- **Dielectric exclusion based on ion solvation**: the nanometer scale of the NF pores introduces significant confinement of the solvent within the pore. Therefore, the solvent interacts with the fixed charge on the membrane and orient itself spatially. This realignment of the solvent causes a change in its molecules' physical and electrical properties and, consequently, affects its viscosity and dielectric constant, leading to greater rejection by the membrane [31].

Dielectric exclusion is a complex mechanism and may vary with the type of solute, the number of ions in contact with the membrane surface, and feed concentration, among other parameters [31]. Thus, dielectric exclusion by NF membranes is less understood and explored than other mechanisms, principally related to TrOCs removal.

5.2.3 CHARGE REPULSION

Charge repulsion occurs due to electrostatic interactions between the polymer matrix of the NF membrane and the charged species in the feed solution. For fixed charge membranes, solutes with the opposite charge to the membrane are attracted, while solutes with the same charge are repelled, increasing their rejection. As a result, the counter-ion concentration (opposite charge from the membrane fixed charge) is higher at the membrane surface than in bulk, while the co-ion concentration (same charge as the fixed charge on the membrane) is lower at the membrane surface. The Donnan effect balances the effect of charged components transport due to membrane charge: when a semi-permeable membrane separates two compartments containing electrolytes, the mobility of these electrolytes is altered by those rejected, aiming at the electrical balance between compartments [32].

Electrostatic interactions between charged TrOCs and the NF membrane are already reported by several studies to improve the rejection of these compounds. For example, Ref. [33] evaluated the removal of 34 TrOCs by NF and observed that all charged TrOCs were highly rejected (>80%) by the membrane. However, atenolol, amitriptyline, and fluoxetine (positively charged) had lower rejections than negatively charged TrOCs with similar MWs. Importantly, although most NF membranes are negatively charged, positively charged NF membranes can be developed for the separation of cations with high efficiency.

In this context, Ref. [29] evaluated the removal of 29 TrOCs by an NF-EMBR (Enzymatic Membrane Bioreactor). They concluded that salicylic acid, atrazine, and N-Diethyl-meta-toluamide – negatively charged at pH between 6.7 and 6.9 – had charge repulsion as the primary removal mechanism. It shows the importance of controlling operating conditions to induce rejection of certain TrOCs via charge repulsion.

5.2.4 ADSORPTION ON THE MEMBRANE SURFACE

Adsorption can be described as the interaction of TrOCs with the membrane surface due to their affinity. Sorption can temporarily promote higher initial TrOC rejections and reduce over time once the sorption capacity of the membrane is exhausted [34]. Mass balance has been an adequate methodology to describe the behavior of sorption rejection in NF membranes. However, membrane and TrOC characteristics, pH, and fouling can influence this process [30].

Reference [35] showed that the estrone removal by adsorption on NF membrane reduced from 90% at the beginning of the experiment to the range of 8.2%–15.3% at the end of 23 hours. This exemplifies the short-term adsorption effect by NF membranes. As the feed solution is continuously filtered through the membrane, the available sites are taken up by the adsorbed estrone. In the same study, the authors observed that at pH 10.4, when 50% of the estrone molecules were dissociated, the final rejection was higher than at pH 4 and 7. This result shows that feed pH can mitigate this effect even with reducing rejection over time.

Reference [36] also observed a decreased concentration of antibiotics in the initial stage of wastewater treatment due to adsorption on the NF membrane. This occurred mainly with fluoroquinolones and macrolides that exhibit high sorption properties due to their high solid-water partition coefficients (Kd). According to the extraction results, pharmaceutical compounds such as azithromycin and roxithromycin concentrations were higher on the membrane surface. The authors attribute this to the interaction between positively charged antibiotics and the polyamine membrane (negatively charged) used in the experiments [36].

5.2.5 FOULING

Fouling is a challenging phenomenon in membrane separation processes that can alter the permeate flux and membrane surface properties. Thus, the effect of fouling on the TrOCs rejection has been increasingly discussed in recent studies. For example, NF membrane fouling can influence the retention of TrOCs via pore restriction, modification of the membrane surface charge and hydrophobicity, and cake-enhanced polarization concentration effect [37].

Reference [37] observed that fouling composed of organic fouling agents, represented by bovine serum albumin, alginate, and humic acid, significantly interfere in NF membrane hydrophobicity. Thus, the presence of these compounds in the fouling promoted the adsorption of the hormone triclosan. It also reduced the diffusive transport of this TrOC across the membrane, consequently increasing its rejection. This study illustrated that the characteristics of the foulant and its interactions with the NF surface considerably influenced the greater or lesser removal of TrOCs.

In addition to the foulants' characteristics, the characteristics of TrOC (mostly load and MW) also significantly influence its removal by fouled membranes. It is known that fouling generally reduces membrane surface charge, and therefore, electrostatic repulsions between membranes and ionic solutes are expected to be smaller. On the other hand, Ref. [38] showed that after organic fouling of NF, the removal of diclofenac increased, probably due to the higher roughness of the fouled membranes. For rough membranes, negatively charged solutes that are forced into the cracks of the fouling layer are surrounded by negative charges, increasing electrostatic repulsion, even though the surface charge is lower [38]. In this sense, Zhao et al. in 2008 showed that sodium alginate scaling on a ceramic NF membrane increased the removal of lower MW TrOCs such as phenacetine, clofibric acid, and primidone due to pore restriction, although it also contributed to the reduction in flux.

Reference [39] obtained similar results. The removal of dimethyl phthalate, diethyl phthalate, dibutyl phthalate, acenaphthylene, phenanthrene, and pyrene by two commercial NF membranes (NF90 and NF270) was evaluated to compare their performance when new and fouled. Removal by clean NF90 ranged from 92.4 to >99.9 and clean NF270 from 50.0 to 92.3, except for dimethyl phthalate (87.9 and 32.1 for NF90 and NF270, respectively). Fouling enhanced all NF270 TrOCs removals, ranging between 59.6 and >99.9. For the N90, while rejections of dibutyl phthalate, pyrene, acenaphthylene, and phenanthrene were practically unchanged, dimethyl phthalate rejection increased to 95.0%, and diethyl phthalate rejection decreased slightly (from 95.3 to 92). The increased rejection of TrOCs by fouled NF270 was attributed to top pore blockage and the higher hydrophobicity of the fouled surface. NF90 was not much affected as it already had smaller pores.

In contrast, fouling can reduce the removal of non-ionic hydrophobic compounds, such as the PhACs risperidone and fluoxetine [40]. It can occur due to a higher density of negative charge in the membrane and the decreased rejection of positively charged solutes.

5.3 EFFECT OF WASTEWATER QUALITY AND OPERATIONAL CONDITION ON TrOCs REJECTION

For choosing the technology to remove TrOCs to enable reuse, one must consider the matrix to be treated. The presence of natural organic matters, for example, impacts membrane rejection of TrOCs and permeate quality [41]. The matrix is also closely related to the fouling potential. Fouling can become another barrier to the penetration of pollutants but can also worsen membrane rejection, reduce flux, and modify the properties of the membrane and its interaction with compounds consequently, as discussed in Section 5.2. Assessing which effect will prevail over the long term is important for choosing the process and optimizing its operation.

Although, intuitively, effluents of worse quality will worsen performance for TrOCs removal, evaluating interactions of the target pollutants with other components present in the matrix is relevant to investigate if unfavorable effects will be predominant or not. For example, Ref. [20] observed that pharmaceutical compounds' removal by NF was greater when treating membrane bioreactor (MBR) effluent than when treating Milli-Q water spiked with the compounds. The greater removal with a complex feed was attributed to the interaction between solutes present in the matrix, which overcame the performance decline after the fouling layer formation.

In addition, several operational conditions can influence the TrOCs removal, which increases removal possibilities but, at the same time, makes the removal process more complex. For example,

Ref. [42] showed that the removal of the TrOCs carbamazepine, estrone, 17ethinylestradiol, bisphenol A, 17estradiol, and diclofenac was significantly higher in higher organic load effluents (0.67 kg chemical oxygen demand (COD)/mg/L), compared to medium (0.37 kg COD/mg/L) and low (0.11 kg COD/mg/L) organic load. When integrated with NF, biological factors such as solids retention time (SRT) can be decisive in removing TrOCs. Membrane rejection efficiency contributes to a higher biomass concentration, greater biodiversity, and greater SRT in biological reactors, which favors the emergence of organisms capable of degrading recalcitrant compounds [43].

Regarding temperature effect on NF TrOCs rejection, Ref. [44] found that, for the range of 5°C–25°C, higher temperatures increased the membrane pore size, permeate flux, and the diffusivity of the compounds. Thus, the rejection of positively charged or neutral TrOCs was reduced at higher operating temperatures. However, the negatively charged compounds were almost unaffected by the temperature change.

Accordingly, effluents with similar characteristics, treated with membranes of the same specifications but in different parts of the world, can generate recovered water with different qualities. Therefore, to reproduce a successful tertiary treatment in a northern European country, a country with a tropical climate must carry out studies on a laboratory and pilot scale with ambient conditions to verify the ability of the NF to generate water with acceptable quality for the proposed reuse.

The relevance of studies that consider the particularities of the places where the reuse will be carried out was evidenced by Ref. [44], who also consider NF a cost-efficient tertiary treatment for micropollutants removal from real wastewater. In a pilot-scale NF study with effluent from a WWTP in China, removals of over 80% for all neutral and positively charged micropollutants were obtained, which was in line with what was requested by Swiss legislation, for example (removals of micropollutants >80% in advanced WWTP effluent treatment). Concerning negatively charged compounds, most of them were removed with efficiency greater than 80%. The suitability of NF to generate a permeate of acceptable quality depends on the compounds present in the feed and the current legislation. However, the authors noted the importance of considering the impact of seasonal variation on removal efficiency: smaller pores at lower temperatures provided greater removals. This is because size exclusion is one of the main mechanisms for removing emerging compounds, as discussed previously in Section 5.2. Furthermore, the authors recalled that it is essential to associate risk assessment studies to the environment of that substance before considering that a process generates safe water for reuse.

The permeate recovery rate can also considerably affect the removal efficiency of TrOCs, and it should be a parameter that is carefully evaluated, especially when the objective is the reuse of effluents treated by NF. For example, Ref. [8] evaluated the fluconazole and betamethasone removal by NF and observed its presence in the permeate from a 40% recovery rate, and from a recovery rate of 10%–70%, the removal was dramatically lower (about 80%–60% for both compounds). Likewise, the transmembrane pressure combined with the type of NF membrane can be a decisive factor in removing TrOCs. Ibuprofen, for example, was maximally retained (98.6% for AFC 40 membrane) by applying transmembrane pressure in a range of 15–20 bar. On the other hand, for AFC 30 membrane, ibuprofen removal increased continuously when transmembrane pressure increased and reached the maximum removal (98.0%) when transmembrane pressure was between 25 and 30 bar.

5.4   NF FOR REMOVING TrOCs FROM REAL EFFLUENTS

Given the different mechanisms of TrOCs rejection by NF and the concern about the presence of these compounds in the environment, the focus of several studies has been the evaluation of NF for the removal of PhACs, pesticides, hormones, and endocrine disruptors, among other TrOCs. Efficiencies of NF to remove TrOCs from real wastewater are presented in Table 5.1 and Figure 5.4, where the comparison between the efficiency of different types of NF membranes in removing TrOCs is shown.

Among the compounds evaluated, PhACs are highlighted since they are the most reported in NF studies. Many of these PhACs are entirely removed by NF alone, especially those with MW greater

TABLE 5.1
NF for Removing TrOCs from Real Wastewater

Compound	Effluent	Removal (%)	Permeate Recovery Rate (%)	Membrane	Scale	Reference
Norfloxacin	WWTP secondary effluent	>98.9	75	NFX membrane	Laboratory-scale	[36]
Ofloxacin		>99.5	75			
Roxithromycin		100	75			
Azithromycin		100	75			
Methylparaben	Effluent from WWTP pre-treated by microfiltration (MF)	53.7	–	MoS$_2$ nanosheets NF membrane	Laboratory-scale	[18]
Ethylparaben		69.1	–			
Propylparaben		79.1	–			
Benzylparaben		91.3	–			
Metoprolol	MBR effluent from WWTP	~85	–	DF30 NF membrane	Laboratory-scale	[19]
Trimethoprim		~85	–			
Chloramphenicol		~85	–			
Indomethacin		~95	–			
Atenolol	MBR effluent from WWTP	~97		NF90 NF membrane	Laboratory-scale	[20]
Diatrizoate		~99				
Capecitabine	WWTP secondary effluent	>99.8	70	5DK NF membrane	Pilot-scale	[46]
Cyclophosphamide		95.2	70			
Ifosfamide		97.3	70			
Paclitaxel	WWTP secondary effluent	99.9	–	5DK NF membrane	Pilot-scale	[56]
Etoposide		98.7	–			
Acetaminophen	WWTP secondary effluent	~90	–	MPS-34 NF membrane		[21]
Caffeine		~85	–			
Triclosan		100	–			
Trimethoprim		~99	–			
Sulfamethoxazole		~97	–			
Naproxen		~99	–			
Ibuprofen		~99	–			
Diclofenac		~99	–			
Diazepam		~99	–			

(Continued)

TABLE 5.1 (Continued)
NF for Removing TrOCs from Real Wastewater

Compound	Effluent	Removal (%)	Permeate Recovery Rate (%)	Membrane	Scale	Reference
Sulfamethoxazole	Pharmaceutical wastewater pre-treated	~98	—	NF90 membrane	Laboratory-scale	[57]
Trimethoprim		~99	—			
Ciprofloxacin		~99	—			
Dexamethasone		~99	—			
Febantel		~99	—			
Caffeine	WWTP secondary effluent	96.6		NF270-2540 membrane	Pilot-scale	[58]
Theobromine		99.4				
Theophylline		88.3				
Amoxicillin		>99.4				
Penicillin		>99.7				
Sulfamethazine	WWTP secondary effluent	100	80	DF30-8040 membrane	Pilot-scale	[45]
Sulfamethoxazole		100	80			
Ofloxacin		100	80			
Ciprofloxacin		83.2	80			
Norfloxacin		100	80			
Nalidixic acid		67.9	80			
Diclofenac acid		70.8	80			
Bezafibrate		100	80			
Clofibric acid		100	80			
Roxithromycin		100	80			
Clarithromycin		100	80			
Propranolol		100	80			
Metoprolol		~98	80			
Carbamazepine		90.7	80			
Sulpiride		100	80			
Estrone		93.5	80			
Bisphenol A		100	80			

(Continued)

TABLE 5.1 (Continued)
NF for Removing TrOCs from Real Wastewater

Compound	Effluent	Removal (%)	Permeate Recovery Rate (%)	Membrane	Scale	Reference
Trimethoprim		100	80			
Caffeine		79.2	80			
N,N-diethyl-m-toluamide		91.1	80			
Alachlor	MBR effluent from WWTP	90.8	—	NF270 membrane	Laboratory-scale	[22]
Atraton		79.8	—			
Caffeine		62.2	—			
Carbadox		57.3	—			
Carbamazepine		71.2	—			
Diethylstilbestrol		59.5	—			
Equilin		66.7	—			
17α-Estradiol		68.1	—			
17β-Estradiol		57.6	—			
Estriol		13.2	—			
Estrone		73.6	—			
17α-Ethynylestradiol		76.7	—			
Gemfibrozil		92.9	—			
Metolachlor		93.4	—			
Oxybenzone		82.4	—			
Sulfachloropyridazine		89.5	—			
Sulfamerazine		88.9	—			
Sulfamethizole		90	—			
Sulfamethoxazole		90.1	—			
Atenolol	Primary effluent from WWTP pre-treated by MBR	62.2	—	NE90 membrane	Laboratory-scale	[23]
Carbamazepine		82.3	—			
Clopidogrel		83	—			
Dilantin		78	—			
Ibuprofen		98.6	—			
Iopromide		74	—			
Glimepiride		88	—			
Sulfamethoxazole		73	—			

(Continued)

TABLE 5.1 (*Continued*)
NF for Removing TrOCs from Real Wastewater

Compound	Effluent	Removal (%)	Permeate Recovery Rate (%)	Membrane	Scale	Reference
Atrazine	MBR (anoxic/aerobic bioreactor and ultrafiltration) effluent	44.8	–	–	Full-scale	[24]
Chlorpyrifos		60.0				
Chlorfenvinphos		–				
Di(2-ethylhexyl) phthalate (DEHP)		<0				
Di-n-octyl phthalate (DNOP)		<0				
Benzyl butyl phthalate (BBP)		9.6				
Nonylphenol		73.8				
Octylphenol		98.4				
Naphthalene		81.8				
Anthracene		<0				
Fluoranthene		27.8				
Benzo[b]fluoranthene		<0				
Benzo[k]fluoranthene		6.8				
Benzo(a)pyrene		3.9				
Indeno[1,2,3-cd]pyrene		26				
Benzo[g,h,i]perylene		27.6				
Benzene		37.9				
cis-1,2-Dichloroethene		–				
1,2-Dichloropropane		<0				
Toluene		<0				
1,2-Dichloroethane		–				
Ethylbenzene		38.7				
1,3-Dichloropropene		0				

(*Continued*)

TABLE 5.1 (*Continued*)
NF for Removing TrOCs from Real Wastewater

Compound	Effluent	Removal (%)	Permeate Recovery Rate (%)	Membrane	Scale	Reference
Xylene		<0				
Chlorobenzene		–				
1,4-Dichlorobenzene		<0				
1,1,2,2-Tetrachloroethane		<0				
Sodium dodecyl sulfate	Real textile wastewater	30	–	NP010 membrane	Laboratory-scale	[59]
Etodolac	Pharmaceutical wastewater after Fenton process	99.5		FM NP010	Laboratory-scale	[60]
Carbamazepine	WWTP effluent	Negligible rejection	–	HydraCoRe70 membrane	Laboratory-scale	[51]
23 pharmaceuticals (not specified)	WWTP effluent NF directly (RI) NF after ozonation (RII):	Average: 32; highest: 89 (RI) :>99 (RII)	98–99	ESNA1-LF-4040 membrane	Pilot-scale	[54]
Anionic surfactant	Carwash station effluent after electrocoagulation	100	75	NF270 and Desal 5DL membranes	Laboratory-scale	[61]
Benzotriazole	Microfiltered water from a water reclamation plant	35	44.2/45.8/48	NF90 membrane	Laboratory-scale	[26]
Carbamazepine		96				
Diclofenac		>93				
Diuron		77				
Gemfibrozil		>95				
Ibuprofen		>90				
Naproxen		>98				
Saccharin		88				
Triclosan		>92				
Trimethoprim		>97				

(*Continued*)

TABLE 5.1 (Continued)
NF for Removing TrOCs from Real Wastewater

Compound	Effluent	Removal (%)	Permeate Recovery Rate (%)	Membrane	Scale	Reference
Acetaminophen Sulfamethoxazole Tetracyclin Terbutryn	WWTP effluent after 10 minutes of Ozonation	>80	80	NF90 membrane	Laboratory-scale	[62]
1-H-benzotriazole	Two effluents from WWTP NF directly (RI): NF after advanced	28.8/29.1 (RI); 28.7–96 (RII)	60	HL membrane provided by GE Osmonics, Inc.	Laboratory-scale	[63]
N,N-diethyl-m-toluamide	oxidative process (RII):	100/96.3 (RI); 100 (RII)				
Chlorophene		100 (RI)				
Methylindole		100 (RI)				
Nortriptyline HCl		100/96.3 (RI); 100 (RII)				
Atenolol	Microfiltered biologically treated sewage effluent collected from a water reclamation plant	<5/58/75.5	-	NP010/NP030/NTR 729HF membranes	Laboratory-scale	[25]
Sulfamethoxazole		17/50/98				
Caffeine		-/-/35				
Trimethoprim		8/8/79				
Carbamazepine		<5/<5/87				
Fluoxetine		13/68/not detected				
Antriptyline		52/74/64				
Primidone		23/39/85				
Verapamil		13/57/66				
Diclofenac		16/29/93				
Naproxen		32/54/95				
Gemfibrozil		16/20/72				

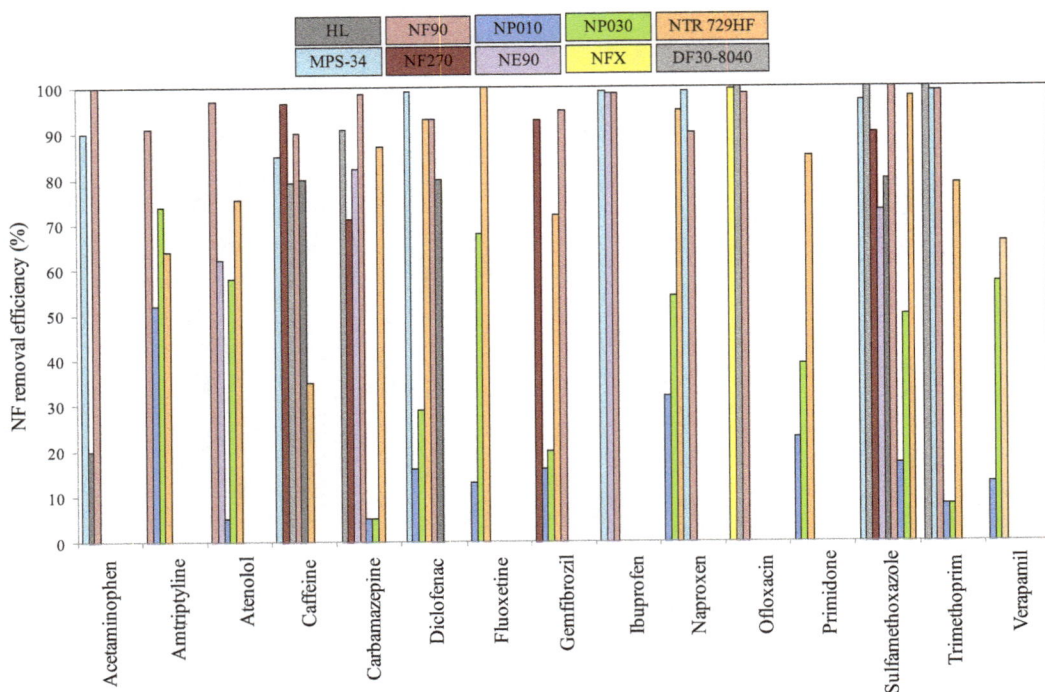

FIGURE 5.4 Comparison of TrOCs removal by different types of NF membranes.

than the MWCO of the evaluated membrane, such as roxithromycin, azithromycin, triclosan, sulfamethazine, sulfamethoxazole, ofloxacin, bezafibrate, norfloxacin, clofibric acid, clarithromycin, propranolol, and trimethoprim [36,45]. On the other hand, some compounds with MW <400 g/mol had low removal efficiency for the NF270 membrane (400 Da) (Table 5.1), such as the veterinary drug carbadox (57.3%) and the hormones diethylstilbestrol (59.5%), 17β-estradiol (57.6), and estriol (13.2%) [22].

Caffeine (MW = 194.19 g/mol) rejection behavior is also an example of the influence of the membrane used and the compound characteristics on the removal efficiency. Using the NF270 membrane, caffeine retention was 62.2% [22], while the NF90 membrane (lower MWCO) had retention of 90% [46], as shown in Table 5.2.

Even nowadays, only a few full-scale TrOCs removal studies have been conducted, and many of them contemplate other matrices than real wastewater, such as surface and groundwater [47]. Reference [48] investigated the performance of a NF plant whose main objectives were to remove hardness and keep the concentration of pesticides found in raw water – atrazine, simazine, and chlorotoluron – below 0.1 µg/L. Groundwater treatment was carried out with DOW NF200 membrane. During the first year of operation, the concentration of pesticides in the permeate was always below the drinking water standards. Full-scale NF has also been successfully applied to remove pesticides in the M&y-Sur-Oise plant (France) [49]. Atrazine, found in raw river water at 850 ng/L, was not detected in NF permeate (concentration below 50 ng/L).

Another study evaluated the removal of several pharmaceuticals (such as ketoprofen, diclofenac, acetaminophen, propyphenazone, carbamazepine, sulfamethoxazole, gemfibrozil, and hydrochlorothiazide) by NF in a full-scale drinking water treatment plant [50]. High rejections (>85%) were observed for several TrOCs because their MW was greater than NF MWCO. Compounds such as ketoprofen, diclofenac, and sulfamethoxazole were removed with efficiencies greater than 95%. The removal of acetaminophen, which is smaller, was around 45%. For gemfibrozil and

TABLE 5.2

Removal of TrOCs from NF Concentrate Treating Real Wastewater

	Nanofiltration			Integrated Process Treating Concentrate				
Compound	Effluent	Removal (%)	Permeate Recovery Rate (%)	Membrane	Process	Removal (%)	Observations	Scale	Reference
4-FAA	Pre-treated (micro-filtration) WWTP effluent	~94[a]	75	NF90-2540 membranes	Ozonation	>99	Only compounds found in concentrations over 100 ng/L in the first set of experiments conducted by authors	Pilot-scale	[47]
Azithromycin		~100[a]							
4-AAA		~74[a]							
Ciprofloxacin		~36[a]							
Erythromycin		~100[a]							
Ofloxacin		~77[a]							
Amitriptyline		~91[a]							
Citalopram		~90[a]							
Paraxanthine		~100[a]							
Nicotine		Not removed							
Atenolol		~61[a]							
Ranitidine		~100[a]							
Sulfamethoxazole		~96[a]							
Trimethoprim		~64[a]							
Venlafaxine		~92[a]							
4-AA		~40[a]							
Carbamazepine		~21[a]							
Ketoprofen		~88[a]							
Antipyrine		~100[a]							
Sulfapyridine		~100[a]							
Clarithromycin		~69[a]							
Caffeine		~90[a]							

(Continued)

TABLE 5.2 (Continued)
Removal of TrOCs from NF Concentrate Treating Real Wastewater

Compound	Nanofiltration				Integrated Process Treating Concentrate				Reference
	Effluent	Removal (%)	Permeate Recovery Rate (%)	Membrane	Process	Removal (%)	Observations	Scale	
Acetaminophen Carbamazepine Atenolol Diatrizoic acid	WWTP effluent	>90	50, 80 and 90	NF90 membrane	Ozonation	90 >90 90 20%–40%	At different times of ozonation and NF recovery rates	NF: Laboratory-scale Ozonation: Pilot-scale	[64]
Norfloxacin Ofloxacin Roxithromycin Azithromycin	WWTP effluent	>98	75	NFX membrane	UV/O$_3$	85%–99% after 5 minutes, >99% after 10 minutes	UV$_{254}$ photolysis was not effective. Removal was 99% with ozonation (10 minutes, norfloxacin required more time)	Laboratory-scale	[36]
Carbamazepine Flumequine Ibuprofen Ofloxacin Sulfamethoxazole	WWTP effluent	>98.5	75	NF90-2540 membranes	Solar photo-Fenton	53–90	Different conditions tested	Pilot-scale	[65]
Acetaminophen Metoprolol Caffeine Antipyrine Sulfamethoxazole Flumequine Ketorolac Atrazine Isoproturon 2-Hydroxybiphenyl Diclofenac	WWTP effluent	>80, except for Acetaminophen (20%)	50	HL membrane provided by GE Osmonics, Inc.	Coagulation (iron and alum), oxidation (chlorine, permanganate, and ozone) and powdered activated carbon (PAC) adsorption	<10 (coagulation); 30.7–100 (PAC: 100 mg/L); 54.2–100 (ozone: 10 mg/L)	–	Laboratory-scale	[66]

(Continued)

TABLE 5.2 (*Continued*)
Removal of TrOCs from NF Concentrate Treating Real Wastewater

Compound	Effluent	Nanofiltration			Integrated Process Treating Concentrate				
		Removal (%)	Permeate Recovery Rate (%)	Membrane	Process	Removal (%)	Observations	Scale	Reference
4-FAA	WWTP effluent	~94[a]	75	NF90-2540 membranes	Solar-assisted anodic oxidation	97	90 minutes of solar-assisted anodic oxidation	Pilot-scale	[67]
4-AAA		~97[a]				98			
Gabapentin		~94[a]				78			
Carbamazepine		~88[a]				44			
Iminostilbene		~95[a]				41			
4-AA		~97[a]				>99			
Imidacloprid		~92[a]				54			
Sulpiride		~96[a]				>99			
Venlafaxine		~97[a]				>99			
Levofloxacin		~98[a]				>99			
Cetirizine		~96[a]				39			
Telmisartan		~93[a]				20			
Irbesartan		~94[a]				42			
Diatrizoic acid		~96[a]				53			
Microcontaminants (<250 ng/L)		~95[a]				77			
Sulfamethazine	WWTP effluent		80	NF90 membrane	Electro-oxidation	>99	–	Laboratory-scale	[68]
Paclitaxel	WWTP effluent	>98	80	Desal 5 Dk membrane	UV, UV/TiO$_2$, and UV/H$_2$O$_2$	>98	–	Laboratory-scale	[69]
Etoposide		>98				>98			
Cyclophosphamide		92±4				Not removed			
Ifosfamide		85±7				Not removed			

[a] Calculated by mass balance from contaminants' concentration in NF feed and concentrate.

mefenamic acid, it was not possible to identify the cause of the relatively low removals (around 50% and 30%). In general, the authors proved the efficiency of the technology for TrOCs removal.

Despite the high efficiencies, it is important to emphasize that more full-scale studies are needed to deal with real effluents since these matrices, even pre-treated, are often more complex and of lower quality than those cited in the studies mentioned here.

5.5 NF AND INTEGRATION WITH OTHER ADVANCED TECHNIQUES

Besides NF is a promising process, it can be improved to overcome challenges concerning TrOCs removal to ensure safe effluents for reuse. Some effluents are very complex and may require several treatment steps. Reference [24] evaluated the removal of micropollutants (pesticides, metals, phthalate esters, phenolic surfactants (alkylphenols), polycyclic aromatic hydrocarbons, and volatile organic compounds) from landfill leachate full-scale plant from Istanbul, Turkey. This type of effluent is complex and requires a combination of technologies for efficient treatment. The full-scale process consists of equalization and settling pond, MBR (anoxic/aerobic bioreactor and ultrafiltration), and NF. Although efficient for removing COD, NH_3, phosphorus, and color (84%, 70%, 82%, and 99%, respectively, only in NF), BTEX compounds (benzene, toluene, ethylbenzene and xylenes) were not mostly removed through membranes. This was expected since the MWs of these compounds were smaller than the pores. The rejection can be increased with the decrease of pores throughout the operation due to the fouling layer formation. Membranes had greater rejection of semi-volatile compounds and metals. Thus, adding one more treatment step to the plant may be necessary for various reuse applications.

From the studies (Tables 5.1), it can be observed the different possibilities and configurations of NF integrated with other processes to remove TrOCs, either after conventional biological processes, membrane bioreactors or integrated with other advanced processes, such as other membrane processes or advanced oxidative processes (AOPs), which has been extensively evaluated. Likewise, integrating other techniques can improve the removal efficiencies of certain compounds, such as carbamazepine, which had negligible rejection using NF alone and 100% removal with NF+thermal Fenton [51].

When the AOP is integrated into the NF as a post-treatment, in this order, it is always very important to verify if the AOP used in the integration can remove the effects associated with the compounds that can compromise the intended reuse, such as toxicity. Reference [47] found increased acute toxicity when using ozonation to treat NF concentrate [52]. Thus, it may be necessary for some reuses to carry out ozonation until complete mineralization or until the formation of easily biodegraded or non-toxic by-products. At the same time, it is necessary to assess whether, for the intended reuse, it is necessary to remove the toxicity to a particular organism.

However, other authors also evaluated the opposite order, applying the AOP before the NF. Reference [53] evaluated the removal of 23 pharmaceuticals with NF and NF after ozonation. With NF alone, the average removal of compounds was 32%. The integrated treatment provided an efficiency greater than 99%. In another study, Ref. [54] evaluated TrOCs removal by NF directly and AOPs (ozone, chlorine, ultraviolet (UV) radiation, UV/H_2O_2, or O_3/H_2O_2) before NF. They observed high removals for NF treatment, except for removing 1-H-benzotriazole (about 29% efficiency). With the application of AOPs before NF, the removal ranged from 28.7% to 96%. As an advantage of the AOP+NF order, it can be mentioned that NF can remove eventual by-products formed during the oxidative process.

5.6 NF CONCENTRATE TREATMENT AND DISPOSAL

Concentrate management and treatment are consistently challenging and deserve attention in all membrane processes, including NF. Concerning TrOCs, they accumulate in the concentrate throughout the operation due to rejection by the NF membrane. Therefore, it is important to remove these compounds at safe levels, including for disposal.

Some authors have evaluated this issue by integrating NF with other processes to concentrate treatment. It is possible to observe that most of the studies evaluated the treatment of the NF concentrate through AOPs (Table 5.2). For example, several TrOCs can be removed (>99%) from NF concentrate using ozonation [47].

An option that may be attractive is treating concentrate, not to the point of complete TrOCs mineralization, which can be very expensive, but until more biodegradable intermediates are generated. It would permit biological treatments, such as concentrate recirculation in biological treatment – often operated before NF. As an example, Ref. [36] treated NF concentrate with O_3/UV. The NF treated the real WWTP effluent. In addition to high antibiotic removal, there was a 58% reduction in acute toxicity and a 40% increase in dissolved organic carbon, and the biochemical oxygen demand/Chemical oxygen demand (BOD/COD) ratio increased by 4.6 times. The authors concluded that the treated NF-concentrated stream could be recirculated in the activated sludge step (the process prior to NF), proposing a route with zero discharge of micropollutants.

In addition to generating a more environmentally safe disposal concentrate, integrating NF with other technologies can be especially attractive when it is possible to reuse both the permeate and the NF concentrate. For the reuse purpose, the integration of NF with other advanced treatment processes can become economically viable, and choosing the most cost-effective combination is logically necessary.

In this sense, Ref. [47] analyzed the economic feasibility of removing micropollutants from real municipal wastewater effluent in Almería (Spain) by a pilot-scale NF combination with solar photo-Fenton and ozonation treating NF concentrate. The experiments were conducted with effluent obtained after secondary treatment. It was compared with the direct treatment with AOPs and effluent treatment with NF, using AOPs as a treatment for the NF concentrate. All configurations were efficient in removing micropollutants. In both cases, integration with NF reduced the total treatment costs, as a smaller volume had to be treated. The authors pointed out that the larger the treatment plant, the more significant are the cost reductions when adding NF as a post-treatment. In another study, Ref. [55] evaluated that using ozonation as a treatment of the NF concentrate and then enabling the reuse of the two NF streams, savings of up to 15.4 k$/year for $125\,m^3/h$ were viable under the conditions evaluated.

5.7 QUALITY OF THE RECOVERED WATER AND REUSE POSSIBILITIES

It has been demonstrated that NF is a technically viable possibility to enable wastewater reclamation [45]. The possibility of applying this promising technology is especially interesting in the context of many locations around the world where tertiary treatment is not yet an ordinary reality. In many countries, WWTPs treat effluents to the secondary level, with biological reactors whose main objectives do not involve the removal of TrOCs commonly found in wastewater. In this way, many treated effluents do not meet the established standards for reuse, even concerning other common pollutants.

For example, Ref. [59] found that the effluent from a WWTP located in Medina Sidonia, Spain, did not meet the regulations for water reuse, even for less-restrictive uses, such as the use of water in industrial applications. This was mainly due to turbidity and *Escherichia coli* values. The WWTP had primary and secondary treatment, which consisted of a biological aeration tank followed by a secondary decanter. The treated effluent was fed into a pilot-scale NF system and was fortified with emerging contaminants. The permeate complied with Spanish legislation for municipal reuse for agriculture, industry, recreational or environmental purposes and was free from emerging contaminants and their associated risks.

Reference [70] considered that water reuse might be an option to alleviate acute water shortages in the City of Cape Town. They evaluated different membrane separation processes on a pilot scale, including NF, for post-treating an effluent from a WWTP that uses MBR, focusing on removing estrogen and testosterone, which is necessary for indirect potable reuse. MBR/NF route was able to meet US EPA and WHO trigger value (<0.7 ng/L) and the PNEC value for fish (1 ng E2/L).

Therefore, when evaluating the options to generate reuse water, it is always imperative to consider a global perspective of the application, considering the required permeate quality/legislation to be complied with, the variation of the concentration of the compounds (including TrOCs) with the season, and the costs involved.

To ensure safe reuse, modeling tools may be of great value to assist in choosing the membrane process to remove TrOCs. Reference [71] sampled the secondary effluent from three Water Resource Recovery Facilities from three different communities that already practice some kind of reuse process, two of which already had plans to enable potable reuse. Of the 63 emerging compounds, most were removed by more than 98% with NF. A multivariate model was developed to predict the rejection of emerging contaminants from a secondary effluent recovered with NF. The model can predict the minimum rejection for recalcitrant residual organic contaminants regardless of the matrix. However, the model did not consider changes in membrane properties throughout the operation due to fouling, for example, which was crucial to obtain information on the viability of NF to maintain the removal of these contaminants and water quality for reuse in the long-term operations.

5.8   REVERSE OSMOSIS (RO) OR NF? PERFORMANCE AND COSTS CONSIDERATIONS

RO can perform superior removals and is chosen in reuse projects regarding TrOCs. For the use of treated sewage in the irrigation of food crops, for example, RO presented itself as a tertiary treatment capable of providing water with the required specificities. At the same time, NF could not remove monovalent ions up to the permitted concentrations [72]. However, it is necessary to consider the lower pressure requirement for the NF, which reduces operating costs compared to RO [56]. In addition, RO permeate had a slightly acidic pH.

Since RO permeate could cause corrosion to downstream fixtures and piping, it would require a post-treatment, including the addition of chemicals to adjust water quality, increasing the costs of RO compared to NF. This issue was studied by Ref. [73], who evaluated the removal of 18 emerging contaminants by NF and obtained results very close to the values reported for RO (82% and 97% of average removal of neutral and ionic compounds, respectively, by NF, against 85% and 99% by RO). Thus, the authors claimed that NF might be a more advantageous option depending on the application.

Reference [56] also defended the greater economic advantage of NF over RO in wastewater reuse, with NF exhibiting a high rejection of organic micropollutants. The authors evaluated that applying NF instead of RO can save 35 k\$/year for 125 m³/h.

Again, it is always crucial to consider the intended reuse water application. Reference [74] evaluated the performance of several NF membranes, selecting one (NF270) to carry out pilot-scale studies at a water reuse facility, evaluating the removal of TrOCs, and comparing it with RO (ESPA2 membrane). The objective was to obtain water for indirect potable reuse. Although more compounds were detected in the NF permeate, the compounds carbamazepine, diethyl-m-toulamide, dilantin, meprobamate, primidone, Tris(1-chloro-2-propyl)phosphate, Tris(1,3-dichloro-2-propyl) phosphate, triclosan, diclofenac, gemfibrozil, ibuprofen, ketoprofen, mecoprop, naproxen, salicylic acid, sulfamethoxazole, and fluoxetine (atenolol and Tris-(2-chloroethyl)phosphate were the exceptions) were found at concentrations lower than 100 ng/L, which the authors observed was lower than any toxicological threshold level. Besides, these concentrations would not be a problem since AOPs already integrated with the membrane system can be another step to remove these contaminants. The authors' economic analysis pointed to annual savings from the application of NF instead of RO. Despite all these favorable results, NF could not remove total organic carbon to satisfactory levels to meet California Department of Public Health regulations and could not be an option in the evaluated conditions.

Reference [75] evaluated the life cycle costs of various advanced large-scale wastewater treatment techniques intended for potable water recovery to identify more economically sustainable options and inform their future development. The study considered capital and operational costs and the possibility of recovering resources and generating waste. Among the technologies evaluated, NF was the second process with the lowest cost, with an approximate value of $0.18 per m^3 of treated effluent. The other processes evaluated were ozonation ($0.15 per m^3), granular activated carbon ($0.27 per m^3), and solar photo-Fenton ($0.31 per m^3). Reference [75] also evaluated different advanced techniques for treating effluents containing pharmaceutical and personal care products. The results showed that, on average, the NF has the lowest impacts for 13 evaluated categories (climate change, fossil depletion, ozone depletion, metal depletion, freshwater eutrophication, marine eutrophication, terrestrial acidification, freshwater ecotoxicity, marine ecotoxicity, human toxicity, urban land occupation, particulate matter formation, and photochemical oxidants formation), considering the environmental impacts of the techniques.

5.9 CHALLENGES TO BE OVERCOME

NF has been widely used to remove TrOCs due to its great application potential and high removals for several TrOCs, although gaps and challenges still exist in the NF process for removing them. The main advantages and challenges of NF for removing TrOCs are represented in Figure 5.5.

Understanding the complex mechanisms of rejection of certain TrOCs by NF is a gap in some cases. In the literature, few reports deepen the study of the mechanism of separation of TrOCs, and thus, the selectivity of NF is still not entirely clear, and consequently, the mathematical models may not be accurate [69]. Studying deeper TrOCs removal mechanisms can help build an important and solid base of knowledge to guide the development of NF technology. Besides, removing many smaller TrOCs by NF still needs more understanding and improvement.

Among the challenges of full-scale application of NF for water recovery to remove TrOCs, the management of contaminated concentrate can be highlighted. Many contaminants are adsorbed on the membrane surface and can be released during variations in feed conditions, such as pH variations. These contaminants can also be released in membrane cleaning solutions. Therefore, the treatment of the concentrate and the cleaning solution must be considered to not dispose of these contaminants in the environment [37].

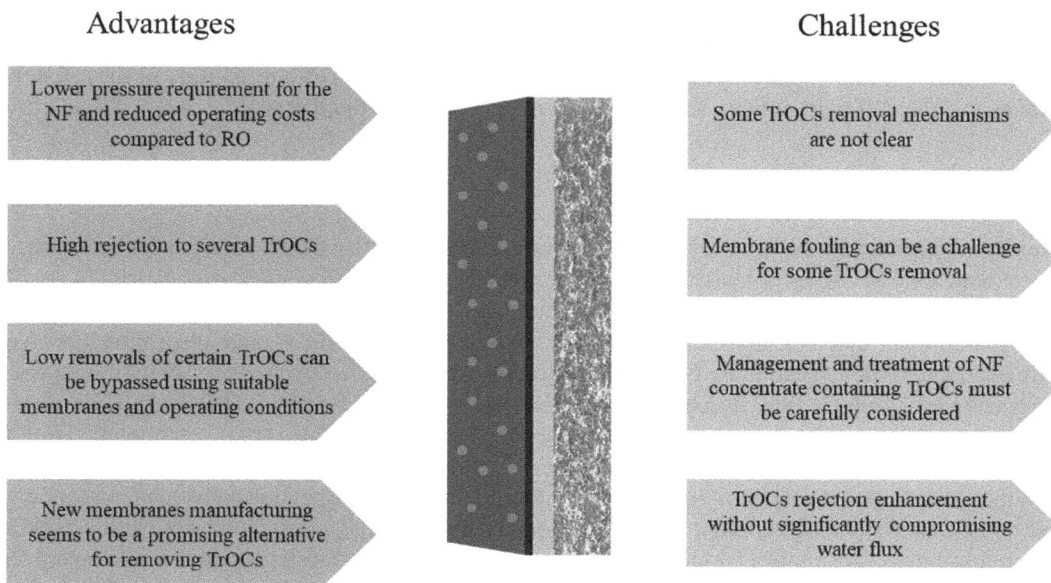

Advantages

- Lower pressure requirement for the NF and reduced operating costs compared to RO
- High rejection to several TrOCs
- Low removals of certain TrOCs can be bypassed using suitable membranes and operating conditions
- New membranes manufacturing seems to be a promising alternative for removing TrOCs

Challenges

- Some TrOCs removal mechanisms are not clear
- Membrane fouling can be a challenge for some TrOCs removal
- Management and treatment of NF concentrate containing TrOCs must be carefully considered
- TrOCs rejection enhancement without significantly compromising water flux

FIGURE 5.5 Advantages and challenges of NF for removing TrOCs.

As previously discussed, despite the fouling layer can increase TrOCs removal, in some cases, it can reduce the removal of trace compounds, as evaluated by Ref. [38]. In these cases, techniques for mitigating fouling are also important for increasing TrOCs removal [76]. In this concern, fouling mitigation techniques have been applied with satisfactory results. The adoption of specific operational parameters (pH, concentration, temperature, and turbulence of the feed flow and applied pressure), for example, can prevent or reduce the presence of fouling [77]. In addition, physical and chemical cleaning of membranes are great allies in reversing this challenge [78], and surface modifications are another way of study to mitigate fouling.

Concerning modifications on surface membranes, they have led to the development of NF membranes with better performance in recent decades. NF membranes have been modified or manufactured with new characteristics to improve TrOCs removal [79–81]. The strategies used can modify the membrane surface, for example, by adding functional materials such as additives or fillers, and constructing new structures to promote an increased rejection by size exclusion or adsorption [28], in addition to the predominant size exclusion mechanism. Reference [82] evaluated the coating of NF (NF90 membrane) with polydopamine to remove 34 organic micropollutants. The coating sealed some defects on the commercial membrane surface, improving the removal of compounds in general. Among the neutral micropollutants, the largest compounds had their transmission to the permeate more reduced. The passage of charged compounds through the original membrane and the coating was small.

It is also possible to add to the membranes the mechanism of catalysis of the compounds. Reference [83] manufactured a biocatalytic NF membrane by modifying the support layer of a commercial NF membrane (NF270), introducing enzymes in its constitution. Evaluating bisphenol A removal by the new membrane, the authors observed that after 36 hours of continuous operation, the micropollutant removal decreased by only 1%, being a stable and promising membrane for removing TrOCs.

Some paradigms may emerge with the modification of NF membranes. For example, Ref. [84] prepared a hydrophilic surface coating with polydopamine on a commercial NF membrane and analyzed its effect on removing TrOCs. The results showed an increase in removing endocrine-disrupting compounds (ethylparaben, propylparaben, and benzylparaben) but a considerable reduction in membrane flux. Thus, new surface modifications still need further development to remove TrOCs from real matrices combined with higher fluxes and salt rejections.

The development of new cost-effective membranes is promising for the competitiveness of NF to be applied in reuses applications [47]. Future studies must also verify the stability of these membranes and the ability to remove TrOCs in the long term in pilot and full scale.

5.10 CONCLUSION

Numerous TrOCs have been detected at concentration from ng/L to μg/L in water bodies and wastewater, and are considered potential environmental threats. NF is a proven technology for the rejection of TrOCs. The predominant removal mechanism is size exclusion. Other than size exclusion effect, additional mechanisms such dielectric exclusion, charge repulsion, and adsorption can also influence the TrOCs rejection by NF. NF rejection is also influenced by the feed solution temperature and membrane fouling. A considerable reduction in the rejection of TrOCs by the NF membrane was related to temperature increase. Membrane fouling can influence the retention of TrOCs via pore restriction, modification of the membrane surface charge and hydrophobicity, and cake-enhanced polarization concentration effect. In addition, the concentrate management and treatment are consistently challenging and deserve attention. Therefore, it is important to remove the TrOCs from concentrate at safe levels, including for disposal. Moreover, future research is warranted to develop a mechanistic rejection of TrOCs by NF and to develop membranes with better performance in TrOCs removal.

REFERENCES

1. Moore, M.R., Vetter, W., Gaus, C., Shaw, G.R., Müller, J.F., 2002. Trace organic compounds in the marine environment. *Mar. Pollut. Bull.* 45, 62–68. https://doi.org/10.1016/S0025-326X(02)00104–2.
2. Luo, Y., Guo, W., Ngo, H.H., Nghiem, L.D., Hai, F.I., Zhang, J., Liang, S., Wang, X.C., 2014. A review on the occurrence of micropollutants in the aquatic environment and their fate and removal during wastewater treatment. *Sci. Total Environ.* 473–474, 619–641. https://doi.org/10.1016/j.scitotenv.2013.12.065.
3. Liu, R., Li, S., Tu, Y., Hao, X., 2021. Capabilities and mechanisms of microalgae on removing micropollutants from wastewater: A review. *J. Environ. Manage.* 285, 112149. https://doi.org/10.1016/j.jenvman.2021.112149.
4. Grandclément, C., Seyssiecq, I., Piram, A., Wong-Wah-Chung, P., Vanot, G., Tiliacos, N., Roche, N., Doumenq, P., 2017. From the conventional biological wastewater treatment to hybrid processes, the evaluation of organic micropollutant removal: A review. *Water Res.* 111, 297–317. https://doi.org/10.1016/j.watres.2017.01.005.
5. Kanaujiya, D.K., Paul, T., Sinharoy, A., Pakshirajan, K., 2019. Biological treatment processes for the removal of organic micropollutants from wastewater: A review. *Curr. Pollut. Reports* 5, 112–128. https://doi.org/10.1007/s40726-019-00110–x.
6. Gu, Y., Huang, J., Zeng, G., Shi, L., Shi, Y., Yi, K., 2018. Fate of pharmaceuticals during membrane bioreactor treatment: Status and perspectives. *Bioresour. Technol.* 268, 733–748. https://doi.org/10.1016/j.biortech.2018.08.029.
7. Eggen, R.I.L., Hollender, J., Joss, A., Schärer, M., Stamm, C., 2014. Reducing the discharge of micropollutants in the aquatic environment: The benefits of upgrading wastewater treatment plants. *Environ. Sci. Technol.* 48, 7683–7689. https://doi.org/10.1021/es500907n.
8. Couto, C.F., Lange, L.C., Amaral, M.C.S., 2019. Occurrence, fate and removal of pharmaceutically active compounds (PhACs) in water and wastewater treatment plants—A review. *J. Water Process Eng.* 32, 100927. https://doi.org/10.1016/j.jwpe.2019.100927.
9. Audenaert, W.T.M., Chys, M., Auvinen, H., Dumoulin, A., 2014. (Future) Regulation of trace organic compounds in WWTP effluents as a driver for advanced wastewater treatment low temperature plasma for applications in medicine view project compartmental modeling of a full scale WRRF view project 269928942.
10. National Water Research Institute (NWRI). 2013. Examining the criteria for direct potable reuse.
11. Miarov, O., Tal, A., Avisar, D., 2020. A critical evaluation of comparative regulatory strategies for monitoring pharmaceuticals in recycled wastewater. *J. Environ. Manage.* 254, 109794. https://doi.org/10.1016/j.jenvman.2019.109794.
12. Calisto, V., Esteves, V.I., 2009. Psychiatric pharmaceuticals in the environment. *Chemosphere.* https://doi.org/10.1016/j.chemosphere.2009.09.021.
13. dos Santos, C.R., Arcanjo, G.S., de Souza Santos, L.V., Koch, K., Amaral, M.C.S., 2021. Aquatic concentration and risk assessment of pharmaceutically active compounds in the environment. *Environ. Pollut.* 290. https://doi.org/10.1016/j.envpol.2021.118049.
14. Deblonde, T., Cossu-Leguille, C., Hartemann, P., 2011. Emerging pollutants in wastewater: A review of the literature. *Int. J. Hyg. Environ. Health* 214, 442–448. https://doi.org/10.1016/j.ijheh.2011.08.002.
15. Alexander, J.T., Hai, F.I., Al-aboud, T.M., 2012. Chemical coagulation-based processes for trace organic contaminant removal: Current state and future potential. *J. Environ. Manage.* 111, 195–207. https://doi.org/10.1016/j.jenvman.2012.07.023.
16. Lyche, J.L., Nourizadeh-Lillabadi, R., Karlsson, C., Stavik, B., Berg, V., Skåre, J.U., Alestrøm, P., Ropstad, E., 2011. Natural mixtures of POPs affected body weight gain and induced transcription of genes involved in weight regulation and insulin signaling. *Aquat. Toxicol.* 102, 197–204. https://doi.org/10.1016/j.aquatox.2011.01.017.
17. Yang, Y., Zhang, X., Jiang, J., Han, J., Li, W., Li, X., Yee Leung, K.M., Snyder, S.A., Alvarez, P.J.J., 2022. Which micropollutants in water environments deserve more attention globally? *Environ. Sci. Technol.* 56, 13–29. https://doi.org/10.1021/acs.est.1c04250.
18. Dai, R., Han, H., Wang, T., Li, X., Wang, Z., 2021. Enhanced removal of hydrophobic endocrine disrupting compounds from wastewater by nanofiltration membranes intercalated with hydrophilic MoS2 nanosheets: Role of surface properties and internal nanochannels. *J. Memb. Sci.* 628, 119267. https://doi.org/10.1016/j.memsci.2021.119267.
19. Xu, M., Huang, H., Li, N., Li, F., Wang, D., Luo, Q., 2019. Occurrence and ecological risk of pharmaceuticals and personal care products (PPCPs) and pesticides in typical surface watersheds, China. *Ecotoxicol. Environ. Saf.* 175, 289–298. https://doi.org/10.1016/j.ecoenv.2019.01.131.

20. Azaïs, A., Mendret, J., Gassara, S., Petit, E., Deratani, A., Brosillon, S., 2014. Nanofiltration for waste-water reuse: Counteractive effects of fouling and matrice on the rejection of pharmaceutical active compounds. *Sep. Purif. Technol.* 133, 313–327. https://doi.org/10.1016/j.seppur.2014.07.007.
21. Garcia-Ivars, J., Martella, L., Massella, M., Carbonell-Alcaina, C., Alcaina-Miranda, M.I., Iborra-Clar, M.I., 2017. Nanofiltration as tertiary treatment method for removing trace pharmaceutically active compounds in wastewater from wastewater treatment plants. *Water Res.* 125, 360–373. https://doi.org/10.1016/j.watres.2017.08.070.
22. Comerton, A.M., Andrews, R.C., Bagley, D.M., Hao, C., 2008. The rejection of endocrine disrupting and pharmaceutically active compounds by NF and RO membranes as a function of compound and water matrix properties. *J. Memb. Sci.* 313, 323–335. https://doi.org/10.1016/j.memsci.2008.01.021.
23. Chon, K., KyongShon, H., Cho, J., 2012. Membrane bioreactor and nanofiltration hybrid system for rec-lamation of municipal wastewater: Removal of nutrients, organic matter and micropollutants. *Bioresour. Technol.* 122, 181–188. https://doi.org/10.1016/j.biortech.2012.04.048.
24. Argun, M.E., Akkuş, M., Ateş, H., 2020. Investigation of micropollutants removal from landfill leach-ate in a full-scale advanced treatment plant in Istanbul city, Turkey. *Sci. Total Environ.* 748. https://doi.org/10.1016/j.scitotenv.2020.141423.
25. Shanmuganathan, S., Vigneswaran, S., Nguyen, T.V., Loganathan, P., Kandasamy, J., 2015. Use of nano-filtration and reverse osmosis in reclaiming micro-filtered biologically treated sewage effluent for irriga-tion. *Desalination* 364, 119–125. https://doi.org/10.1016/j.desal.2014.12.021.
26. Jamil, S., Loganathan, P., Khan, S.J., McDonald, J.A., Kandasamy, J., Vigneswaran, S., 2021. Enhanced nanofiltration rejection of inorganic and organic compounds from a wastewater-reclamation plant's micro-filtered water using adsorption pre-treatment. *Sep. Purif. Technol.* 260, 118207. https://doi.org/10.1016/j.seppur.2020.118207.
27. Koyuncu, I., Sengur, R., Turken, T., Guclu, S., Pasaoglu, M.E., 2015. Advances in water treatment by micro-filtration, ultrafiltration, and nanofiltration. In *Advances in Membrane Technologies for Water Treatment: Materials, Processes and Applications.* https://doi.org/10.1016/B978–1–78242–121–4.00003–4.
28. Shin, M.G., Choi, W., Park, S.J., Jeon, S., Hong, S., Lee, J.H., 2022. Critical review and comprehensive analysis of trace organic compound (TOrC) removal with polyamide RO/NF membranes: Mechanisms and materials. *Chem. Eng. J.* 427, 130957. https://doi.org/10.1016/j.cej.2021.130957.
29. Asif, M.B., Hou, J., Price, W.E., Chen, V., Hai, F.I., 2020. Removal of trace organic contaminants by enzymatic membrane bioreactors: Role of membrane retention and biodegradation. *J. Memb. Sci.* 611, 118345. https://doi.org/10.1016/j.memsci.2020.118345.
30. Dang, H.Q., Nghiem, L.D., Price, W.E., 2014. Factors governing the rejection of trace organic contami-nants by nanofiltration and reverse osmosis membranes. *Desalin. Water Treat.* 52, 589–599. https://doi.org/10.1080/19443994.2013.826851.
31. Oatley, D.L., Llenas, L., Pérez, R., Williams, P.M., Martínez-Lladó, X., Rovira, M., 2012. Review of the dielectric properties of nanofiltration membranes and verification of the single oriented layer approxi-mation. *Adv. Colloid Interface Sci.* 173, 1–11. https://doi.org/10.1016/j.cis.2012.02.001.
32. Schaep, J., Van der Bruggen, B., Vandecasteele, C., Wilms, D., 1998. Retention mechanisms in nanofil-tration. *Chem. Prot. Environ.* 3, 117–125. https://doi.org/10.1007/978-1-4757-9664-3_14.
33. Fujioka, T., Khan, S.J., McDonald, J.A., Nghiem, L.D., 2015. Rejection of trace organic chemicals by a nanofiltration membrane: The role of molecular properties and effects of caustic cleaning. *Environ. Sci. Water Res. Technol.* 1, 846–854. https://doi.org/10.1039/c5ew00170f.
34. Eva, S.D., Eric, L., Martin, R., 2010. Effects of sorption on the rejection of trace organic contaminants during nanofiltration. *Environ. Sci. Technol.* 44, 2592–2598. https://doi.org/10.1021/es902846m.
35. Hu, J.Y., Jin, X., Ong, S.L., 2007. Rejection of estrone by nanofiltration: Influence of solution chemistry. *J. Memb. Sci.* 302, 188–196. https://doi.org/10.1016/j.memsci.2007.06.043.
36. Liu, P., Zhang, H., Feng, Y., Yang, F., Zhang, J., 2014. Removal of trace antibiotics from wastewater: A systematic study of nanofiltration combined with ozone-based advanced oxidation processes. *Chem. Eng. J.* 240, 211–220. https://doi.org/10.1016/j.cej.2013.11.057.
37. Nghiem, L.D., Coleman, P.J., 2008. NF/RO filtration of the hydrophobic ionogenic compound triclosan: Transport mechanisms and the influence of membrane fouling. *Sep. Purif. Technol.* 62, 709–716. https://doi.org/10.1016/j.seppur.2008.03.027.
38. Hajibabania, S., Verliefde, A., Drewes, J.E., Nghiem, L.D., McDonald, J., Khan, S., Le-Clech, P., 2011. Effect of fouling on removal of trace organic compounds by nanofiltration. *Drink. Water Eng. Sci.* 4, 71–82. https://doi.org/10.5194/dwes-4-71-2011.
39. Zhu, L., 2015. Rejection of organic micropollutants by clean and fouled nanofiltration membranes. *J. Chem.* https://doi.org/10.1155/2015/934318.

40. Hajibabania, S., Verliefde, A., McDonald, J.A., Khan, S.J., Le-Clech, P., 2011. Fate of trace organic compounds during treatment by nanofiltration. *J. Memb. Sci.* 373, 130–139. https://doi.org/10.1016/j.memsci.2011.02.040.

41. Liu, S., Ying, G.G., Zhao, J.L., Chen, F., Yang, B., Zhou, L.J., Lai, H., 2011. Trace analysis of 28 steroids in surface water, wastewater and sludge samples by rapid resolution liquid chromatography-electrospray ionization tandem mass spectrometry. *J. Chromatogr. A* 1218, 1367–1378. https://doi.org/10.1016/j.chroma.2011.01.014.

42. Moya-Llamas, M.J., Trapote, A., Prats, D., 2018. Removal of micropollutants from urban wastewater using a UASB reactor coupled to a MBR at different organic loading rates. *Urban Water J.* 15, 437–444. https://doi.org/10.1080/1573062X.2018.1508599.

43. Zhang, Y., Geißen, S.U., Gal, C., 2008. Carbamazepine and diclofenac: Removal in wastewater treatment plants and occurrence in water bodies. *Chemosphere* 73, 1151–1161. https://doi.org/10.1016/j.chemosphere.2008.07.086.

44. Xu, R., Qin, W., Tian, Z., He, Y., Wang, X., Wen, X., 2020. Enhanced micropollutants removal by nanofiltration and their environmental risks in wastewater reclamation: A pilot-scale study. *Sci. Total Environ.* 744, 140954. https://doi.org/10.1016/j.scitotenv.2020.140954.

45. Cristóvão, M.B., Bernardo, J., Bento-Silva, A., Ressureição, M., Bronze, M.R., Crespo, J.G., Pereira, V.J., 2022. Treatment of anticancer drugs in a real wastewater effluent using nanofiltration: A pilot scale study. *Sep. Purif. Technol.* 288. https://doi.org/10.1016/j.seppur.2022.120565.

46. Miralles-Cuevas, S., Oller, I., Agüera, A., Sánchez Pérez, J.A., Sánchez-Moreno, R., Malato, S., 2016. Is the combination of nanofiltration membranes and AOPs for removing microcontaminants cost effective in real municipal wastewater effluents? *Environ. Sci. Water Res. Technol.* 2, 511–520. https://doi.org/10.1039/c6ew00001k.

47. Wang, S., Li, L., Yu, S., Dong, B., Gao, N., Wang, X., 2021. A review of advances in EDCs and PhACs removal by nanofiltration: Mechanisms, impact factors and the influence of organic matter. *Chem. Eng. J.* 406, 126722. https://doi.org/10.1016/j.cej.2020.126722.

48. Wittmann, E., Coté, P., Medici, C., Leech J., Turner, A.G., 1998. Treatment of a hard borehole water containing low levels of pesticide by nanofiltration. *Desalination* 119, 347–352. https://doi.org/10.1016/S0011-9164(98)00180-5.

49. Cyna, B., Chagneau, G., Bablon, G., Tanghe, N., 2002. Two years of nanofiltration at the Méry-sur-Oise plant, France. *Desalination* 147, 69–75. https://doi.org/10.1016/S0011-9164(02)00578-7.

50. Radjenović, J., Petrović, M., Ventura, F., Barceló, D., 2008. Rejection of pharmaceuticals in nanofiltration and reverse osmosis membrane drinking water treatment. *Water Res.* 42, 3601–3610. https://doi.org/10.1016/j.watres.2008.05.020.

51. Minella, M., De Bellis, N., Gallo, A., Giagnorio, M., Minero, C., Bertinetti, S., Sethi, R., Tiraferri, A., Vione, D., 2018. Coupling of nanofiltration and thermal Fenton reaction for the abatement of carbamazepine in wastewater. *ACS Omega* 3, 9407–9418. https://doi.org/10.1021/acsomega.8b01055.

52. Miralles-Cuevas, S., Oller, I., Agüera, A., Llorca, M., Sánchez Pérez, J.A., Malato, S., 2017. Combination of nanofiltration and ozonation for the remediation of real municipal wastewater effluents: Acute and chronic toxicity assessment. *J. Hazard. Mater.* 323, 442–451. https://doi.org/10.1016/j.jhazmat.2016.03.013.

53. Flyborg, L., Björlenius, B., Persson, K.M., 2010. Can treated municipal wastewater be reused after ozonation and nanofiltration? Results from a pilot study of pharmaceutical removal in Henriksdal WWTP, Sweden. *Water Sci. Technol.* 61, 1113–1120. https://doi.org/10.2166/wst.2010.029.

54. Acero, J.L., Benitez, F.J., Teva, F., Leal, A.I., 2010. Retention of emerging micropollutants from UP water and a municipal secondary effluent by ultrafiltration and nanofiltration. *Chem. Eng. J.* 163, 264–272. https://doi.org/10.1016/j.cej.2010.07.060.

55. Mendret, J., Azais, A., Favier, T., Brosillon, S., 2019. Urban wastewater reuse using a coupling between nanofiltration and ozonation: Techno-economic assessment. *Chem. Eng. Res. Des.* 145, 19–28. https://doi.org/10.1016/j.cherd.2019.02.034.

56. Cristóvão, M.B., Torrejais, J., Janssens, R., Luis, P., Van der Bruggen, B., Dubey, K.K., Mandal, M.K., Bronze, M.R., Crespo, J.G., Pereira, V.J., 2019. Treatment of anticancer drugs in hospital and wastewater effluents using nanofiltration. *Sep. Purif. Technol.* 224, 273–280. https://doi.org/10.1016/j.seppur.2019.05.016.

57. Dolar, D., Vuković, A., Ašperger, D., Košutić, K., 2011. Effect of water matrices on removal of veterinary pharmaceuticals by nanofiltration and reverse osmosis membranes. *J. Environ. Sci.* 23, 1299–1307. https://doi.org/10.1016/S1001-0742(10)60545-1.

58. Egea-Corbacho, A., Gutiérrez Ruiz, S., Quiroga Alonso, J.M., 2019. Removal of emerging contaminants from wastewater using nanofiltration for its subsequent reuse: Full–scale pilot plant. *J. Clean. Prod.* 214, 514–523. https://doi.org/10.1016/j.jclepro.2018.12.297.

59. Khosravi, A., Karimi, M., Ebrahimi, H., Fallah, N., 2020. Sequencing batch reactor/nanofiltration hybrid method for water recovery from textile wastewater contained phthalocyanine dye and anionic surfactant. *J. Environ. Chem. Eng.* 8, 103701. https://doi.org/10.1016/j.jece.2020.103701.

60. Vergili, I., Gencdal, S., 2015. Applicability of combined Fenton oxidation and nanofiltration to pharmaceutical wastewater. *Desalin. Water Treat.* 56, 3501–3509. https://doi.org/10.1080/19443994.2014.976772.

61. Gönder, Z.B., Balcıoğlu, G., Vergili, I., Kaya, Y., 2020. An integrated electrocoagulation–nanofiltration process for carwash wastewater reuse. *Chemosphere* 253. https://doi.org/10.1016/j.chemosphere.2020.126713.

62. Yacouba, Z.A., Mendret, J., Lesage, G., Zaviska, F., Brosillon, S., 2021. Removal of organic micropollutants from domestic wastewater: The effect of ozone-based advanced oxidation process on nanofiltration. *J. Water Process Eng.* 39, 101869. https://doi.org/10.1016/j.jwpe.2020.101869.

63. Acero, J.L., Benitez, F.J., Real, F.J., Rodriguez, E., 2015. Elimination of selected emerging contaminants by the combination of membrane filtration and chemical oxidation processes. *Water. Air. Soil Pollut.* 226. https://doi.org/10.1007/s11270-015-2404-8.

64. Azaïs, A., Mendret, J., Petit, E., Brosillon, S., 2016. Influence of volumetric reduction factor during ozonation of nanofiltration concentrates for wastewater reuse. *Chemosphere* 165, 497–506. https://doi.org/10.1016/j.chemosphere.2016.09.071.

65. Miralles-Cuevas, S., Oller, I., Pérez, J.A.S., Malato, S., 2014. Removal of pharmaceuticals from MWTP effluent by nanofiltration and solar photo-Fenton using two different iron complexes at neutral pH. *Water Res.* 64, 23–31. https://doi.org/10.1016/j.watres.2014.06.032.

66. Acero, J.L., Benitez, F.J., Real, F.J., Teva, F., 2016. Micropollutants removal from retentates generated in ultrafiltration and nanofiltration treatments of municipal secondary effluents by means of coagulation, oxidation, and adsorption processes. *Chem. Eng. J.* 289, 48–58. https://doi.org/10.1016/j.cej.2015.12.082.

67. Salmerón, I., Rivas, G., Oller, I., Martínez-Piernas, A., Agüera, A., Malato, S., 2021. Nanofiltration retentate treatment from urban wastewater secondary effluent by solar electrochemical oxidation processes. *Sep. Purif. Technol.* 254, 117614. https://doi.org/10.1016/j.seppur.2020.117614.

68. Du, X., Li, Z., Xiao, M., Mo, Z., Wang, Z., Li, X., Yang, Y., 2021. An electro-oxidation reactor for treatment of nanofiltration concentrate towards zero liquid discharge. *Sci. Total Environ.* 783, 146990. https://doi.org/10.1016/j.scitotenv.2021.146990.

69. Janssens, R., Cristovao, M.B., Bronze, M.R., Crespo, J.G., Pereira, V.J., Luis, P., 2019. Coupling of nanofiltration and UV, UV/TiO$_2$ and UV/H$_2$O$_2$ processes for the removal of anti-cancer drugs from real secondary wastewater effluent. *J. Environ. Chem. Eng.* 7. https://doi.org/10.1016/j.jece.2019.103351.

70. Aziz, M., Ojumu, T., 2020. Exclusion of estrogenic and androgenic steroid hormones from municipal membrane bioreactor wastewater using UF/NF/RO membranes for water reuse application. *Membranes (Basel).* 10. https://doi.org/10.3390/membranes10030037.

71. Jones, S.M., Watts, M.J., Wickramasinghe, S.R., 2017. A nanofiltration decision tool for potable reuse: A new rejection model for recalcitrant CECs. *Water Environ. Res.* 89, 1942–1951. https://doi.org/10.2175/106143017x14902968254629.

72. Hafiz, M.A., Hawari, A.H., Alfahel, R., Hassan, M.K., Altaee, A., 2021. Comparison of nanofiltration with reverse osmosis in reclaiming tertiary treated municipal wastewater for irrigation purposes. *Membranes (Basel).* 11, 1–13. https://doi.org/10.3390/membranes11010032.

73. Yangali-Quintanilla, V., Maeng, S.K., Fujioka, T., Kennedy, M., Li, Z., Amy, G., 2011. Nanofiltration vs. reverse osmosis for the removal of emerging organic contaminants in water reuse. *Desalin. Water Treat.* 34, 50–56. https://doi.org/10.5004/dwt.2011.2860.

74. Bellona, C., Heil, D., Yu, C., Fu, P., Drewes, J.E., 2012. The pros and cons of using nanofiltration in lieu of reverse osmosis for indirect potable reuse applications. *Sep. Purif. Technol.* 85, 69–76. https://doi.org/10.1016/j.seppur.2011.09.046.

75. Tarpani, R.R., Azapagic, A., 2018. Life cycle environmental impacts of advanced wastewater treatment techniques for removal of pharmaceuticals and personal care products (PPCPs). *J. Environ. Manage.* 215, 258–272. https://doi.org/10.1016/j.jenvman.2018.03.047.

76. Abtahi, S.M., Marbelia, L., Gebreyohannes, A.Y., Ahmadiannamini, P., Joannis-Cassan, C., Albasi, C., de Vos, W.M., Vankelecom, I.F.J., 2019. Micropollutant rejection of annealed polyelectrolyte multilayer based nanofiltration membranes for treatment of conventionally-treated municipal wastewater. *Sep. Purif. Technol.* 209, 470–481. https://doi.org/10.1016/j.seppur.2018.07.071.

77. Farrukh, M.A. Nanofiltration, 162p. 2018.10.5772/intechopen.70909.

78. Thombre, N.V., Gadhekar, A.P., Patwardhan, A.V., Gogate, P.R., 2020. Ultrasound induced cleaning of polymeric nanofiltration membranes. *Ultrason. Sonochem.* 62, 104891. https://doi.org/10.1016/j.ultsonch.2019.104891.

79. Guo, H., Deng, Y., Tao, Z., Yao, Z., Wang, J., Lin, C., Zhang, T., Zhu, B., Tang, C.Y., 2016. Does hydrophilic polydopamine coating enhance membrane rejection of hydrophobic endocrine-disrupting compounds? *Environ. Sci. Technol. Lett.* 3, 332–338. https://doi.org/10.1021/acs.estlett.6b00263.

80. Dai, R., Guo, H., Tang, C.Y., Chen, M., Li, J., Wang, Z., 2019. Hydrophilic selective nanochannels created by metal organic frameworks in nanofiltration membranes enhance rejection of hydrophobic endocrine-disrupting compounds. *Environ. Sci. Technol.* 53, 13776–13783. https://doi.org/10.1021/acs.est.9b05343.

81. Medhat Bojnourd, F., Pakizeh, M., 2018. Preparation and characterization of a PVA/PSf thin film composite membrane after incorporation of PSSMA into a selective layer and its application for pharmaceutical removal. *Sep. Purif. Technol.* 192, 5–14. https://doi.org/10.1016/j.seppur.2017.09.054.

82. Huang, S., McDonald, J.A., Kuchel, R.P., Khan, S.J., Leslie, G., Tang, C.Y., Mansouri, J., Fane, A.G., 2021. Surface modification of nanofiltration membranes to improve the removal of organic micropollutants: Linking membrane characteristics to solute transmission. *Water Res.* 203, 117520. https://doi.org/10.1016/j.watres.2021.117520.

83. Zhang, H., Luo, J., Woodley, J.M., Wan, Y., 2021. Confining the motion of enzymes in nanofiltration membrane for efficient and stable removal of micropollutants. *Chem. Eng. J.* 421, 127870. https://doi.org/10.1016/j.cej.2020.127870.

84. Guo, H., Deng, Y., Yao, Z., Yang, Z., Wang, J., Lin, C., Zhang, T., Zhu, B., Tang, C.Y., 2017. A highly selective surface coating for enhanced membrane rejection of endocrine disrupting compounds: Mechanistic insights and implications. *Water Res.* 121, 197–203. https://doi.org/10.1016/j.watres.2017.05.037.

6 Nanofiltration Process in Landfill Leachate Treatment

Ayse Yuksekdag and Sevde Korkut
Istanbul Technical University

Vahid Vatanpour
Istanbul Technical University
Kharazmi University

Ismail Koyuncu
Istanbul Technical University

CONTENTS

6.1   INTRODUCTION

Municipal solid waste (MSW) generation has increased tenfold over the last century as the world's population has expanded, lives in an urban area, and become more consumeristic. It is expected that the world population will have more than doubled by 2025 [1]. The treatment and disposal of solid waste, which is produced rapidly and in large quantities, also cause great costs to the municipal budgets. Incineration, composting, and landfill are commonly used to convert solid waste into stabilized materials. Landfilling has become a widespread technique because of its low cost, with 70% of developing countries and 90% of the world preferring it [2]. Landfill leachate (LL), which generates as MSW degrade in landfills, is a subject that must be addressed since it has the potential to contaminate soil, water, and groundwater.

Leachate is dark-colored wastewater that is extremely difficult to treat. Leachate, which develops as wastes degrade in landfills, is a problem that must be addressed since it has the potential to contaminate soil, water, and groundwater. Numerous factors, including various types of physicochemical reactions, rainfall percolation, and moisture content, all contribute to the formation of LL.

DOI: 10.1201/9781003261827-6

LL often comprises a variety of contaminants that are hazardous to ecosystems, such as organic chemicals, ammoniacal nitrogen (NH_3-N), natural and manufactured ligands, biological micro or macro-organisms, and heavy metals. Uncontrolled LL leak pollutes surface waterways, groundwater, and soil. As a result of leachate migration into groundwater, a large number of organic compounds (about 1,000), >200 of which are regularly found groundwater pollutants, have been discovered [3,4].

Before discharging leachate into receiving waters, treatment systems may be required to avoid contamination and fulfill stringent discharge criteria [5]. Furthermore, effective treatment processes should be devised after completely analyzing the leachate composition [6]. Because the overall composition of LL is so complex and variational according to age, it is challenging to recommend a general treatment process. The technology to be selected should have a low initial investment and operating costs, as well as minimal environmental impact. The removal of NH_3-N, hazardous contaminants, and organic matter in LLs is a critical step in meeting the criteria for safe LL discharge into receiving waters. Different processes have been studied for LL purification and these can be grouped under two headings: biological processes (anaerobic/aerobic treatment, stabilization lagoons, natural water purification) and physicochemical processes (coagulation/flocculation, ion exchange resin, membrane separation processes, air stripping, and advanced oxidation processes [AOPs]).

Since the use of biological processes as a single process in treatment is insufficient, it is supplemented with pretreatment or advanced treatment. Hybrid processes developed by combining conventional and advanced treatment technologies also promise success in the treatment of difficult LL wastewater [4].

In recent studies, the most preferred processes are biological treatment systems such as sequencing batch reactor, an up-flow anaerobic sludge blanket, and lastly membrane bioreactor (MBR) [7]. These treatment technologies are effective in rejecting the high concentration of organic macromolecules and total nitrogen from LL. Coagulation/flocculation might be used as a pretreatment phase before biological processes or as the last step in the removal of organic materials [8]. Adsorption and air stripping methods have some drawbacks such as column fouling and excess sludge production. AOPs are also one of the physicochemical processes in the second stage. AOPs stand out both for their cost-friendliness and for decomposing particularly refractory compounds. Much emphasis has been placed in recent years on the treatment of LLs using hybrid methods of the processes that are mentioned before [4].

Over the last two decades, there has been a lot of interest in the use of membranes in leachate treatment. Low pressure-driven processes are microfiltration (MF), and ultrafiltration (UF), while higher pressures are used in nanofiltration (NF) and reverse osmosis (RO) processes. Hybrid technologies, which are combinations of physicochemical and biological processes, can be developed to overcome the disadvantages and shortcomings of a single process. In particular, the membrane processes undergoing a pretreatment or advanced treatment in LL treatment support the concentrated management as well as the quality effluent.

6.2 SANITARY LANDFILL LEACHATE CHARACTERIZATION

The composition of the leachate can be influenced by a variety of things. Among these include landfill age, rainfall, local meteorological conditions, operational mode, waste type and composition, waste compaction, and other factors [9]. As a result, it is reasonable to assume that the content of leachate changes significantly between landfills and climatic conditions for the same plant [10]. Waste decomposition occurs through a series of biochemical reactions that take place in the following order: initial phase, an acidogenic phase, a methanogenic phase, and a final stabilizing phase [11]. Furthermore, leachate can be classified as young (<5 years), medium (5–10 years), or old (>10 years) based on its age. Table 6.1 shows the range numbers for the key physicochemical properties of leachate based on landfill age. Due to higher biological activity in warmer settings, biodegradability may be even lower in tropical locations than in temperate climates [12].

TABLE 6.1

The Physicochemical Properties of Landfill Leachate as a Function of Landfill Age, and a Comparison with Household Effluent [15]

Parameters	Landfill Age and Leachate Characteristics in Temperate Regions			Landfill Age and Characteristics in Tropical Regions			Domestic Effluent
	0–5 years	5–10 years	10–20 years	0.5–2 years	1.7–2.1 years	7.2–14.4 years	
pH	3–6	6–7	>7.5	7.80–8.50	6.2–8.3	7.3–8.4	7
BOD$_5$ (mg/L)	10,000–25,000	1,000–4,000	50–1,000	275–453	1–7,068	1–12,766	200
COD (mg/L)	15,000–40,000	10,000–20,000	1,000–5,000	1,230–6,027	164–17,440	576–21,137	400
BOD$_5$/COD	0.6–0.7	0.1–0.2	0.1–0.2	–	<0.006–0.3	<0.002–0.3	0.5
Biodegradability	Medium–high	Medium	Low	–	Low	Low	Medium–High
Ammoniacal nitrogen (mg/L)	1,500–4,250	250–700	50–200	526–1,787	21.1–1,120	133–2,808	25
Conductivity (mS/cm)	15–41.5	6.0–14.0	–	8,900–10,872	677–14,590	3,920–25,630	–

The contaminants present in the LL can be divided into four types: dissolved organic components, inorganic macro-components, heavy metals, and xenobiotic organic compounds. In the first group, constituents include carbohydrates, proteins, chemical oxygen demand (COD), total organic carbon (TOC), volatile fatty acids (VFA), biological oxygen demand (BOD), fulvic-like compounds, aliphatic/aromatic/phenolic/esteric molecules, and humic-like compounds [13]. Among the harmful chemicals in the LL, ammonia nitrogen (NH_3-N) compounds constitute one of the greatest threats to aquatic ecological life, because, at high concentrations (100 mg/L), they can impede nitrification, induce toxic effects on living animals, and drive algae development. Furthermore, inorganic ions such as ammonium (NH_4^+), bicarbonate (HCO_3^-), potassium (K^+), calcium (Ca^{2+}), sodium (Na^+), chloride (Cl^-), iron (Fe^{2+}), manganese (Mn^{2+}), magnesium (Mg^{2+}), and sulfate (SO_4^{2-}), make up a significant portion of the LL. HCO_3^-, Cl^-, and SO_4^{2-} ions are detected to be the most common in the LL.

Heavy metals (arsenic, cadmium, chromium, cobalt, copper, lead, mercury, nickel, and zinc) are present in the LL in amounts ranging from 0.50 µg/L to 155 mg/L, depending on the country [14]. In landfills, the rate of solid waste biodegradation reduces when the pH of the LL approaches the acidic region, where heavy metals may simply dissolve and enter the ecosystem. They acquire hazardous properties as a result of their capacity to obtain metal complex compounds, particularly with organic matters. Xenobiotic compounds are a member of organic chemicals derived from industrial and domestic wastes, as well as waste sludges, that contain different monoaromatic and halogenated hydrocarbons [13].

6.3 MEMBRANE TECHNOLOGY IN LANDFILL LEACHATE TREATMENT

Membrane technologies are potential solutions for pollution removal from aqueous matrices. Among them are MF, UF, NF, and RO, which use the membrane as a separating mechanism and hydraulic pressure difference as a driving force. Membranes were first discovered in the 18th century and have survived to the present day. Significant advancements have occurred since then to make membranes more diversified and suited for a global scale of uses. Membrane filtration procedures, including RO, NF, and UF, have been widely used in industrialized nations as leachate treatment technology has advanced and modernized [16–18]. The membrane processes such as RO, NF, and MBR approaches have been used in combination with standard biological treatments and physicochemical methods for treatment of sanitary LL. For example, The Seneca Meadows landfill which has a huge active landfill area in New York State, USA, is an outstanding case of real-scale application of a membrane separation process. The RO process at this site contributed to a contaminant removal effectiveness of more than 95% and the treatment of a considerable volume of LL with a minimal investment cost.

Despite the many membrane separation processes used for leachate treatment, the NF and RO membranes are the most efficient, particularly when combined with additional processes such as MBRs and AOPs. The NF provides a flexible technique that addresses many water quality standards. NF membranes are typically composed of polymeric thin films with molecular weights ranging from 200 to 1,000 Da. When the goal is to remove divalent/multivalent ions and dissolved organic materials, these membranes have successful results such as high flux and rejection rate, but a low rejection rate for sodium (Na^+) and chloride (Cl^-). According to certain research, the treatment effectiveness of COD and NH_3-N in the LL by NF is 65% and 50%, respectively [15].

Membrane-based LL treatment systems have been used at landfills in various European countries (e.g., Germany) since the 1990s [19]. The main disadvantage of membrane-based LL treatment is the formation of more concentrated wastewater that is a dark-colored liquid involving refractory contaminants, high salinity, and high conductivity. Even though concentrate streams from membrane treatment processes of the LL account for 13%–50% of the initial volume of the LL. They contain high levels of refractory organic contaminants such as xenobiotic aromatic compounds, endocrine-disrupting chemicals, long-chain hydrocarbon, halohydrocarbon macromolecules, and inorganic salt molecules [4].

As aforementioned, it is vital to consider that membrane technologies produce a concentrate stream with more pollutants than feed wastewater. This stream can be simply treated by conventional techniques, and because it has a volume that is 70% less than the feed volume, there is an enhancement in efficiency, additionally reducing sludge generation. One option for treating this concentration is to recycle it back to the landfill, a procedure that was first used in Wischhafen (a German landfill) [19]. Furthermore, the high salinity and the presence of the aforementioned refractory contaminants, significantly interrupt the potential of biodegradation of LL membrane concentrates, rendering them non-removable in biological treatment plants.

The raw sanitary LLs are generally treated biologically first. Anaerobic, anoxic, and aerobic processes (e.g., A²O), in which nitrification-denitrification takes place, are often utilized to ensure the removal of ammonia nitrogen as well as organic matter. Conventional biological treatments, on the other hand, are incapable of successfully removing refractory molecules and several micropollutants. It is possible to completely remove these resistant pollutants with the MBRs obtained by adding the membranes to the anaerobic and aerobic treatment plants internally or externally. Since MBRs do not require a final clarifier, they ensure that mixed liquid suspended solids have separated in the biological treatment tank, and less sludge has formed. Also, it has a permanent, stable state, a compact footprint, and high output water quality. As a result of these factors, MBR technology is used in over 50% of new LL treatment operations. Using MBR in LL treatment resulted in BOD_5 removal efficiency of over 99% [20,21]. The external single or two-stage A/O-MBR process is one of the most widely used anoxic/oxic LL purification methods in treatment plants [22]. This MBR method includes anoxic and aeration tanks, followed by a UF process for advanced quality of water. Despite the high removal efficiencies of MBR processes of organic matters (BOD_5, COD), and also NH_3-N, the permeate still fails to fulfill discharge criteria.

To achieve severe effluent discharge criteria, membrane technologies, especially NF and RO, are commonly used in conjunction with biological treatment techniques (Figure 6.1). MBR/NF-RO systems, for example, have evolved as a unique combination procedure using LL treatment [23]. This method successfully separates organic matters and inorganic compounds that could not be destroyed during the biological processes and generates stable residual water. The MBR/NF-RO process has worldwide removal effectiveness of more than 97% for both COD and BOD_5. According to Ahn et al.'s work (2002), LL was treated by the MBR/RO process, increasing the rate of nitrification and biodegradation by MBR. Additionally, the RO process eliminated pollutants that are untreated biologically and inorganic nitrogenous ionic compounds [24]. In Amaral et al.'s work (2015), using NF as a post-treatment to MBR resulted in outstanding treatment performance of LL [25]. Organic matter removal (namely COD) with the MBR/NF process was 80%–96% successful. NH_3-N was rejected with 80%–96% efficiency, while phosphorus removal performance was in the range of 78%–99.8%. The effect of integrating the air stripping to the MBR/NF process as

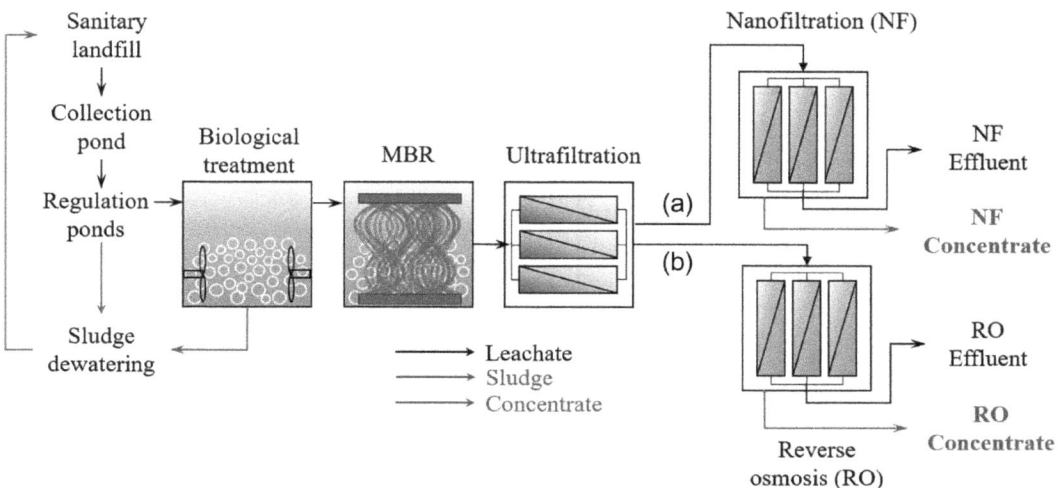

FIGURE 6.1 A diagram depicting the production of membrane concentrates in a typical sanitary landfill leachate treatment process (a: From NF, b: From RO processes) [4].

a pretreatment on performance and treatment efficiency was also investigated. The air stripping/ MBR/NF hybrid system showed performance of 88% in COD removal and 95% in NH_3-N [4].

6.4 STAND-ALONE NANOFILTRATION MEMBRANE

If only the NF process is used in the treatment, the raw leachate is fed directly to the membrane modules. Figure 6.2 shows the stand-alone NF treatment configuration schematically. However, leachate treatment with direct NF is not preferred in real-scale plants due to insufficient treatment, rapid fouling, and lower flux problems. In 1999, the stabilized leachate from France was treated directly with the pilot-scale NF system, and slightly and strongly negatively charged tubular NF membranes were used. Except for Cl^- and SO_4^{2-}, the strongly negative charged NF membrane exhibited better treatment performance. While COD removal efficiencies in slightly and strongly negative charged membranes were 74% and 80%, respectively, nitrogen removal efficiencies were quite insufficient. Total Kjeldahl Nitrogen (TKN) and NH_3-N concentrations in the permeate flow of the slightly negative charged membrane were 415 and 378 mg/L, respectively; these concentrations were observed as 378 and 342 mg/L for the strongly negatively charged membrane. As a result, although the effluent concentrations were low due to the relatively low COD concentrations in stable leachate, adequate treatment for high nitrogen concentrations could not be achieved with NF alone. Therefore, hybrid processes were required for stable leachate with high nitrogen [26]. In another study, lab-scale experiments were conducted to treat leachate directly with NF. As the applied pressure increased from 6 to 12 bar, a decrease was observed in COD, total suspended solids (TSS), nitrate, ammonium nitrogen, and heavy metal removal. As a faster accumulation occurred on the membrane surface with increasing pressure, concentration polarization occurred, and thus, the efficiency was decreased. However, lower fluxes were obtained in the case of operation at low pressure, and the operating time lasted longer. In addition, as in the previous study, nitrate and ammonium nitrogen removal efficiencies of 45% and 20%, respectively, were quite low in this study. In other words, insufficient treatment was obtained when only NF was used [27]. As a result, using NF process only is not well accepted in leachate treatment. Existing studies used lab or pilot scales and were carried out to optimize operating parameters or determine the treatment performance obtained from different membranes. It has been observed that using only the NF process is not sufficient, especially for nitrogenous substance removal.

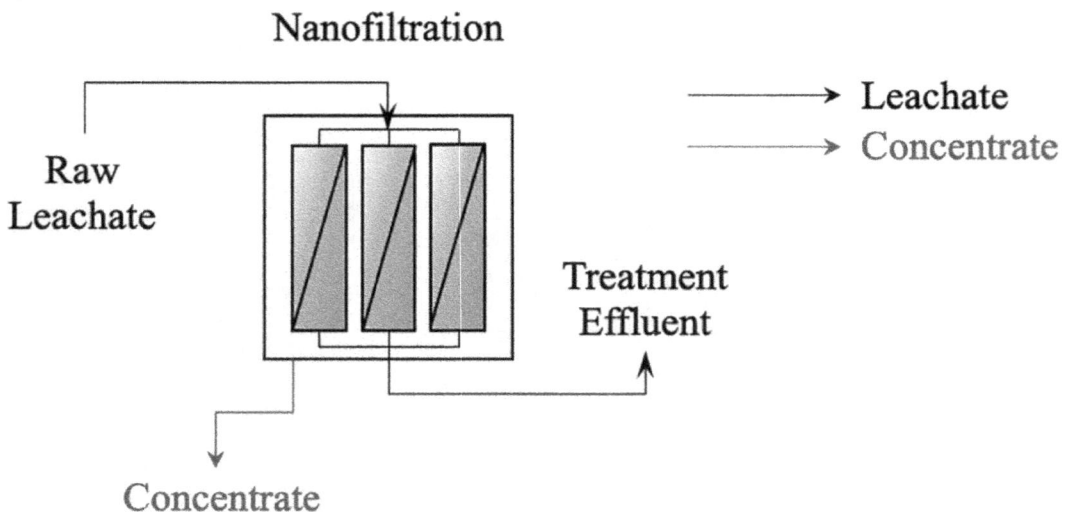

FIGURE 6.2 Schematic representation of direct NF treatment of leachate.

6.4.1   Modification of NF Membranes

LL requires advanced treatment methods because of the high concentrations of ammonium, heavy metals, and refractory materials it contains. One of the most used methods is NF after biological or physicochemical methods. While the NF membrane removes heavy metals and refractory materials such as humic substances (HS) with high efficiency, monovalent salts can pass through the membrane, which leads to the accumulation of high amounts of HSs in the NF concentrate. Although methods such as return to landfill, evaporation, and advanced treatment are applied for concentrate management, special interest is needed for product recovery within the scope of the circular economy. In general, leachate concentration is high in HS and contributes to plant development when employed as fertilizer. However, the heavy metals and salts it contains can be toxic to the plant. Therefore, an NF membrane separating HS and inorganic salts could provide good concentrate management and product recovery [28].

While NF membranes can remove HS, negatively charged commercial NF membranes are tight and thus are able to remove inorganic salts and heavy metals. However, it threatens the valorization of HS in the concentrated phase as fertilizer. Alternative to the tight commercial NF membranes, loose NF membranes can also be used for leachate treatment [29]. Since these membranes have larger pores than tight ones, the electrostatic repulsion forces are lower, and metal ions can pass through the membrane [30]. On the other hand, the removal rate of HS also decreases, and some of these substances can be lost because of the permeation through NF. At this point, it is of great importance to separate HS and inorganic salts between them by modifying the NF surface morphology.

A loose commercial NF membrane surface was modified in a previous study by co-deposition of dopamine and polyethyleneimine (PEI) to fractionate the HS and inorganic salts. The schematic representation of modification is shown in Figure 6.3a. The aim is to decrease the pore diameter while increasing the membrane surface charge. After 90 minutes of coating, the molecular weight cut-off (MWCO) of the membrane decreased to 305 Da. In addition, a decrease in surface negativity was observed by PEI bonding to the surface. Although it is thought that the inorganic salt removal efficiency will increase with the decreasing pore diameter, the electrostatic repulsion has decreased due to the increase in the surface charge. Thus, an enhancement in salt permeation has been obtained. In addition, the removal of HS enhanced with the increasing surface charge. As a result, a modified NF membrane was obtained with increased salt permeability and HS removal. The real leachate filtration experiments demonstrated that the concentration of the HS increased from 1,800 to 17,250 mg/L, while a slight increase was observed in the salt concentration. After NF and diafiltration, a concentrated stream having fertilizer potential containing 99.5% HS was obtained [31].

Another commercial loose NF membrane surface was modified by coating with metal–organic coordination based on tannic acid Fe^{3+} complexes to fractionate the leachate's HS and inorganic salts efficiently. The surface coating of the NF membrane is shown in Figure 6.3b. The membrane surface washed with water is contacted with $FeCl_3$ and tannic acid solutions for a maximum of 1 minute and then washed again with water. The membrane surface was rapidly coated with a homogeneous and stable tannic acid–Fe^{3+} coordination complex. The NF MWCO decreased from 601 to 265 Da with this coating process. Thus, the selectivity between organic compounds and inorganic salts increased. Almost all HS could be retained with the membrane, which was coated for 15 seconds. Moreover, the membrane surface zeta potential decreased after coating. Thus, the electrostatic repulsion effect is alleviated despite the narrowing pore diameter, and inorganic salt removal was reduced. In the case of leachate purification with NF-diafiltration combination, the concentration of HS increased from approximately 1,800 mg/L to almost 14,000 mg/L, and a potential fertilizer containing 98.91% pure HS was obtained. These results are quite similar to the previous study modifying with PEI, considering the importance of loose NF membrane surface modification for sustainable treatment of leachate [32].

Another critical issue to be considered in the fabrication of NF membranes for leachate treatment is the development of innovative anti-fouling membranes. LL, which causes severe fouling on the membrane surface with its complex structure, decreases the membrane flux and causes an increase

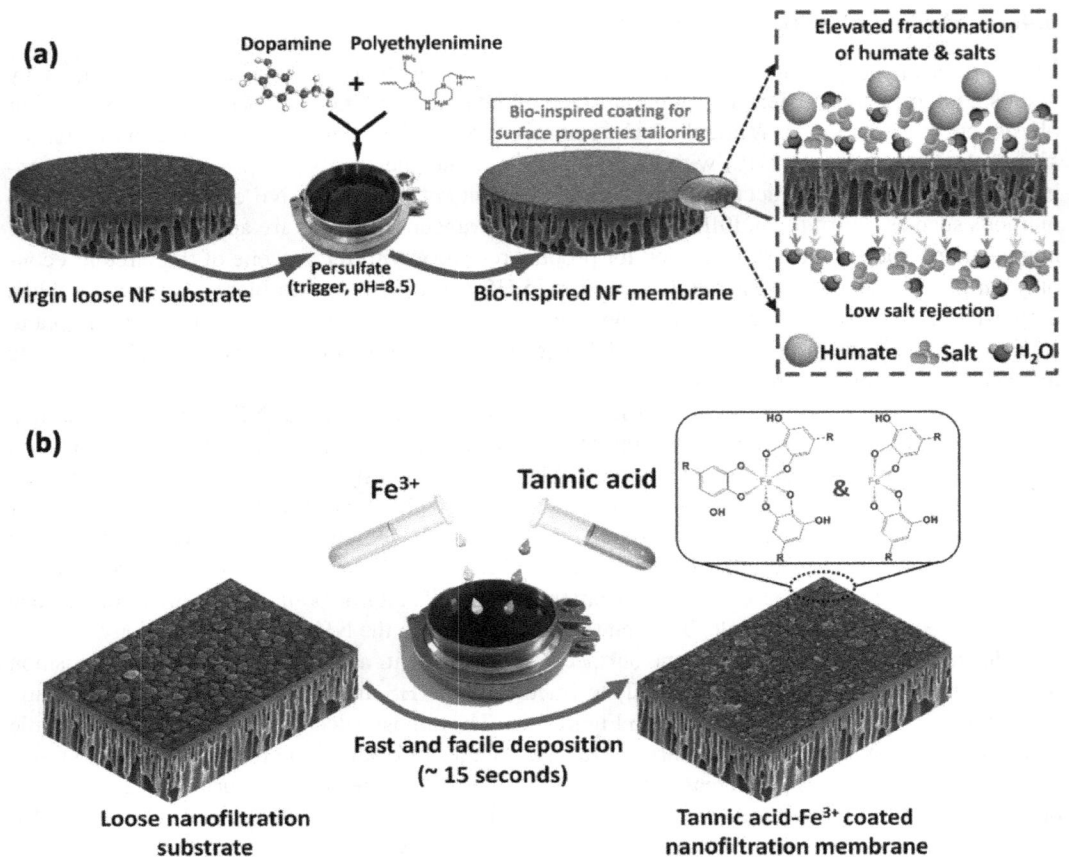

FIGURE 6.3 (a) Dopamine-based NF modification [29] (License number: 5263210050010) and (b) tannic acid-Fe^{3+}-based NF coating [30] (License number: 5263210196613).

in operating costs over time. The fabrication of NF membranes with low fouling tendency, high permeability, and selectivity is an important issue that can reduce operating costs in treating wastewater such as leachate. Graphite oxide (GO) and vanillin are widely investigated for membrane fabrication due to their anti-fouling and flux-increasing properties. The fouling substances generally have nonpolar and hydrophobic structures. On the other hand, GO and vanillin additives are hydrophilic and have anti-fouling properties. If they are used as additives in NF membrane preparation, the hydrophilicity of the membrane could be increased significantly. In a recently published study, GO and vanillin additives were used synergistically in NF membrane fabrication. Vanillin, two-dimensional GO layers, and polysulfone (PSf) were used in the fabrication, as shown in Figure 6.4. In summary, after dispersing the different concentrations of GO in the solvent, vanillin was added and mixed, then PSf was added, and then the polymer solution was mixed continuously for 24 hours. Then, deionized water was used as the coagulation bath, and NF membranes were produced by phase inversion. It was observed that the membrane hydrophilicity increased with increasing GO concentration. In addition, porous networks were formed by GO addition, and polar functional groups were formed on the membrane surface. Thus, an increase in pure water flux was observed. Likewise, with the addition of GO, the interaction of the substances causing the membrane fouling was decreased, and the total, reversible, and irreversible fouling rates were reduced. As a result, the flux recovery rate increased. PSf/GO-vanillin membrane showed better anti-fouling properties compared with PSf-vanillin one. Leachate filtration tests have shown a flux recovery rate of approximately 90%. Finally, GO, and vanillin-doped NF membranes were evaluated as a good alternative for leachate treatment [35].

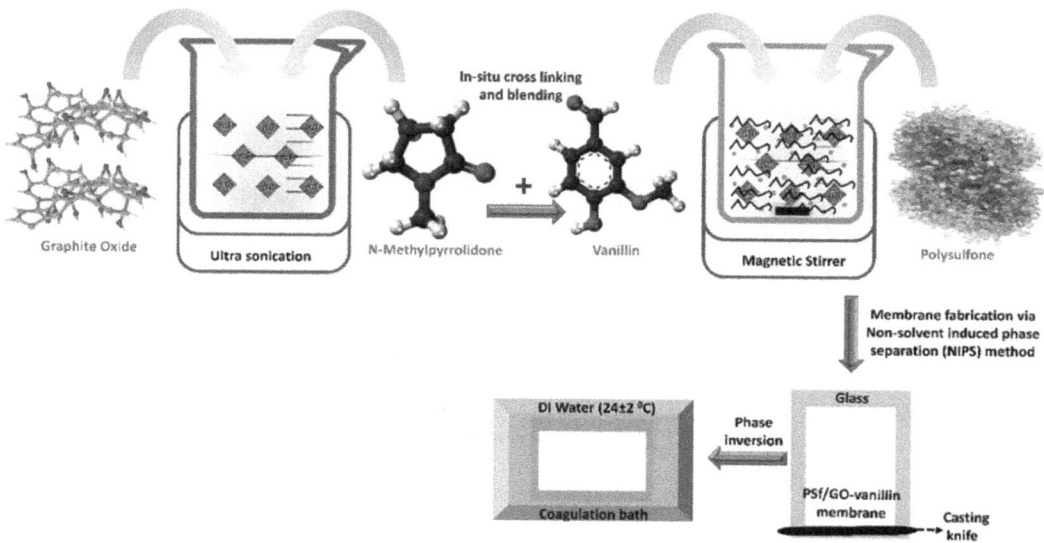

FIGURE 6.4 Fabrication steps of PSf-GO/vanillin composite NF membrane [31] (License number: 5263210326936).

6.5 HYBRID MEMBRANE PROCESSES

Due to having high pollutant concentrations and its toxic nature, leachate treatment is quite complicated and requires multiple stages of processes. For this reason, various physicochemical processes have been researched and developed. Biodegradable organic matter content is high in the leachate of young landfills. Efficient treatment is usually carried out with physical and chemical processes for this type of leachate, followed by biological treatment. However, with the increase in landfill age, the biodegradable organic matter in the leachate is reduced by conversion to methane gas, leaving behind the non-biodegradable organic fraction and high concentrations of nitrogenous compounds. Moreover, increasing heavy metal concentrations may inhibit microbial activity. Thus, hybrid membrane processes are often preferred to treat LL to meet discharge limitations [15]. Combining the NF process with other pressure-driven membrane processes like MBR, MF, UF and RO could be an effective way for further treatment and purification of LL.

6.5.1 MEMBRANE BIOREACTOR/NANOFILTRATION

MBR is compact biological treatment system in which MF or UF membranes separate the biomass from the water, generally after removing nitrogenous and carbonaceous substances in anoxic and aerobic tanks. These membranes can be used in submerged or side-stream configurations. In submerged MBR, modules are placed in the aerobic tank. In the case of the side stream, the membrane modules are located outside the reactors. In this case, a constant rate of the membrane concentrate is recycled to the anoxic tank. Thus, the biomass concentration is kept constant and high in the system. Due to the high biomass concentrations and sludge ages, it is able to provide efficient biodegradation. In addition, it is one of the most preferred biological leachate treatment technologies today [37]. Although it is used in the biological treatment of both young and mature leachates, it cannot provide sufficient treatment alone, especially in the treatment of mature leachates, and generally requires an advanced treatment process to meet the discharge criteria. Especially, MBR applications for LL older than 10 years are found insufficient due to the lower BOD_5/COD and high NH_3-N and COD concentrations and having toxic heavy metals [38]. To achieve a more reliable and stable treatment performance, the NF process is applied to MBR effluent. The NF is the most widely

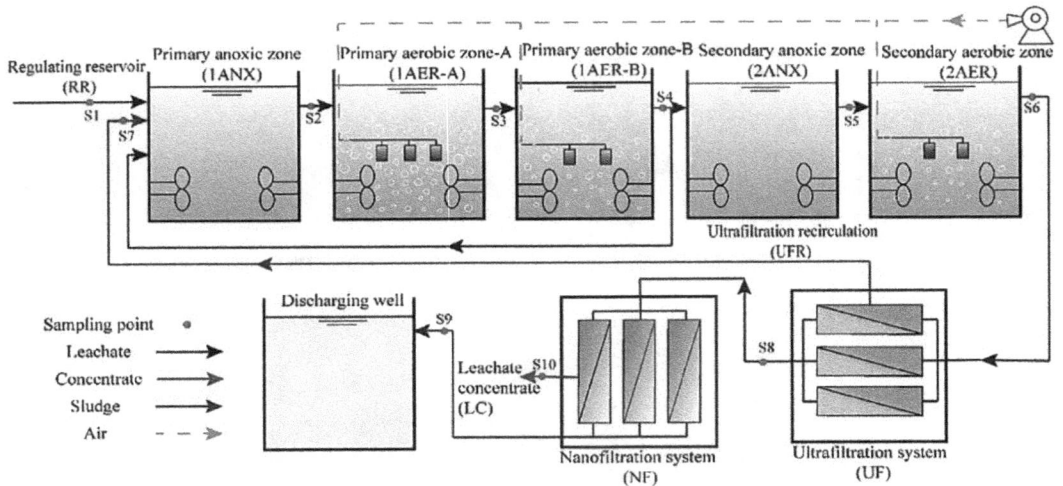

FIGURE 6.5   Schematic diagram of side-stream MBR/NF process [34] (License number: 5263210435297).

used pressure-driven membrane process after MBR due to its ability to remove non-biodegradable organic substances [39]. A typical MBR/NF hybrid process diagram is shown in Figure 6.5 [40].

In parallel with the regulation restriction, MBRs alone have become insufficient for leachate treatment. When MBR and NF processes are applied, sufficient COD and color removals can be obtained due to the high removal efficiency of NF in organic and inorganic substances removal [41]. Although the NF process is integrated into the MBR effluent for further treatment, membrane fouling is still a significant problem in leachate treatment. Therefore, the MBR process must successfully perform an efficient pretreatment for efficient and long-term operation of NF. Wang et al. (2014) investigated the effect of granular activated carbon (GAC) dosage into the aerobic MBR on the removal efficiency of MBR/NF and fouling of membrane processes. For this purpose, an anoxic/aerobic (A/O) GAC-assisted MBR (A/O-GAC-MBR) and A/O-MBR pilot-scale plants were operated comparatively. They observed that GAC addition increased the organic pollutant and heavy metal removal. Moreover, due to the enhancement of bioflocculation, membrane fouling was decreased. By adding GAC in MBR, NF color removal efficiency increased from 41.82% to 93.75% compared to A/O-MBR. Finally, the integrated approach was determined as feasible for large-scale leachate treatment [38].

Toxicity is another crucial threat to biological treatment. To investigate the leachate toxicity removal in MBR/NF, Reis et al. (2017) compared the toxicity removal efficiency of yeast (MBR$_y$) and conventional bacteria MBR (MBR$_b$) integrated with NF. Although COD and color removal efficiencies were observed to be higher in yeast-based MBR than conventional MBR, it was also observed that the effluent toxicity was higher in MBR$_y$. Also, higher nitrite and nitrate concentrations were observed in the MBR$_b$ effluent. The reason for this is that the yeast-based sludge does not provide the necessary conditions for the growth of nitrifying bacteria. Toxicity removal was lower in MBR$_y$ due to the low biodegradation. Finally, they emphasized in their study that the NF process is of great importance because it can purify all or some of the toxic substances in leachate [42]. Dissolved organic matter (DOM) is one of the essential components of LL. Knowing the forms and molecular weight distributions of these DOMs in biological or chemical treatment processes is significant to determine the necessary and sufficient process.

Wang et al. (2020) investigated natural organic matter (NOM) removal and conversion based on a two-stage A/O-MBR-NF process. They reported that biological treatment could readily decompose NOMs with higher H/C and lower O/C ratios. The remaining NOM species were purified efficiently with NF, and thus, they were able to meet their discharge limits with this combination of processes. On the other hand, it has been stated that sulfur-containing compounds are primarily found in the NF concentrate stream. In order to meet the zero waste approach in leachate treatment, they stated that sulfur-oxidizing bacteria or AOPs should be investigated to remove sulfur-containing substances in the NF concentrate stream [43]. In another study, researchers examined a real-scale MBR+NF plant treating mature LL to determine the fate of DOM at the molecular level and determine how much DOM removal was achieved in individual processes. According to the research results, it was stated that if the DOM content of wastewater is high, the anoxic biological process has vital importance for the treatment. On the other hand, if the leachate is rich in lignin/carboxyl-rich alicyclic-like organics, then UF and NF processes will perform adequate treatment. In addition, it was stated that the newly formed nitrogen-containing organics during biological treatment should be taken into account in the treatment process [40].

A general problem of membrane processes is the management and disposal of the concentrated stream. As it is known, NF cannot remove monovalent ions. Besides, divalent and other multivalent ions are rejected, and these ions remain in the concentrated phase. Therefore, it is convenient to separate HS from salts. For example, it is possible to obtain fertilizer-like products from the concentrated streams of the NF process used for MBR effluent. Thus, both concentrate management and product recovery take place simultaneously. In a study, the liquid fertilizer quality of loose NF membrane concentrate after MBR was investigated. It was observed that the germination of green mung bean seed increased when NF concentrate was used as liquid fertilizer. Moreover, no significant phytotoxicity was observed. These results showed that the potential of recovering HS-rich liquid fertilizer from the concentrate stream obtained by applying loose NF membrane processes to the leachate MBR effluent is quite high [29]. In addition to NF concentrate, permeate stream also has great potential for recovery. With the increase in the importance of water recovery due to water scarcity, the adoption of advanced technologies in wastewater treatment is increasing in every field. MBR+NF combination is also required to use the recovered leachate in processes that require lower quality water, such as dust arrestment in landfills [44].

6.5.2 REVERSE OSMOSIS/NANOFILTRATION

RO membranes can be used for advanced treatment and water recovery in leachate. These membranes can treat wastewaters at desired high quality, but the limiting factor is the water recovery rate. The factors limiting the water recovery rate are fouling, scaling, and osmotic pressure. Especially in leachate treatment, pretreatment is essential to reduce the RO fouling and use the membranes longer. NF membranes can be effectively utilized for pretreatment prior to RO. In particular, if the treated leachate is intended to be reused, the NF/RO pressure-driven membrane combination offers the most effective solution. In principle, it is possible to achieve high water recovery rates in leachate treatment with the NF/RO configuration [45]. Since divalent ions and refractory organic materials are rejected by the NF membrane after biological or physicochemical treatment, mostly monovalent ions remain in the permeate stream to RO. Thus, both RO fouling is reduced, and accordingly, it becomes possible to operate the system at higher fluxes. Another approach in the NF/RO configuration is to treat the RO concentrate with NF to increase the water recovery rate. The common goal in both configurations is to reach a higher water recovery rate.

A GAC-assisted A/O-MBR pilot plant was integrated with NF/RO for water recovery purposes in China. After NF, the COD concentration decreased to the range of 220–380 mg/L, while the NH_3-N content was between 24 and 33 mg/L. The total salt removal of NF is between 15% and 20%, while most of the ions in the filtrate are monovalent ions, almost the whole divalent ions were removed. Compared to A/O-MBR, NF provided excellent color removal of approximately 94%. In other words, a very effective pretreatment was carried out for RO. In addition, RO's COD and NH_3-N removal efficiencies were approximately 96% and 90%, respectively, while the salt removal was higher than 95%, and the permeate conductivity value was measured as 134 µS/cm. RO filtration water quality not only met leachate discharge limits but also exceeded Chinese industrial water consumption water quality standards. As a result, with the NF/RO combination, high-quality water was recovered, and the RO membrane fouling was highly alleviated, thus reducing the operating cost [38].

Another study compared two different combinations of the NF/RO configuration [46]. The first alternative is the treatment of the NF filtrate with RO, that is, NF+RO, while the other alternative is the treatment with RO first and then the treatment of the RO concentrate with NF, namely RO+NF. These two alternatives have been tried in mature leachate treatment in the hazardous waste storage area, and the most advantageous process was investigated. Higher fluxes were obtained due to the lower osmotic pressures generated at each stage of the NF+RO operation compared to RO+NF. On the other hand, in the RO+NF configuration, as the osmotic pressure increases continuously throughout the operation, it has increased the operating and maintenance costs. In addition to the fact that RO+NF was a more energy-intensive process, it was also difficult to combine with other physicochemical or biological processes due to the high salinity of RO concentrate. As a result, it was stated that the NF+RO process could be combined with other processes due to the fractionation of monovalent ions, including NH_4^+, and it was a more energy-efficient process [46].

6.5.3 MICROFILTRATION/NANOFILTRATION

MF membranes are generally not used in leachate treatment. However, a pretreatment is essential before NF which is widely used in leachate treatment. For this reason, MF membranes are generally employed as a pretreatment after advanced oxidation (AO), chemical precipitation, adsorption, or biological treatment. Using MF between biological or chemical treatment and NF is not only for higher efficiency pretreatment but also because it requires less footprint and provides a shorter treatment time than settling tanks.

NF membranes can be operated as side-stream mode, while MF membranes can be operated as a side stream or submerged. Since the aeration applied in the submerged mode increases the turbulence, they are more desirable for large-scale applications. Thus, concentration polarization and membrane fouling are decreased [47].

It is also possible to achieve more considerable water recovery with the MF/NF configuration. For example, one study compared two different combinations of MF/NF. The leachate was treated with AO/MF/NF in the first alternative. In the other alternative, the raw leachate was treated with MF/NF, and the concentrate streams of MF and NF were treated with AO/MF/NF. Moreover, permeate of concentrate treatment was fed to NF with MF permeate. When the removal results of both alternatives were compared, similar COD and color removals were obtained. However, in the second alternative, the MF and NF concentrated phases were treated with AO/MF/NF, resulting in higher water recovery [48]. HS have essential importance because they cause fouling of NF membranes. Before further treatment of leachate with NF, the combination of chemical precipitation and MF can also achieve an effective pretreatment, which can remove HS. In a study, a stabilized leachate was precipitated by adding lime and passed through the MF membrane. While 41% HS removal was achieved, the total removal increased to 92% after NF treatment. As a result, chemical precipitation and MF pretreatment were seen to significantly reduce NF fouling, while the importance of NF for satisfactory organic matter removal in stabilized leachate was demonstrated [49].

6.6 INTEGRATION OF NANOFILTRATION PROCESS WITH OTHER CONVENTIONAL TECHNIQUES

6.6.1 COAGULATION/NANOFILTRATION

Besides multi-membrane process combinations, various physicochemical processes are also applied as a pretreatment before NF. One of these pretreatment alternatives is coagulation. The coagulation method can provide an effective and economical pretreatment. Generally, HS are removed by adding iron or aluminum salts. The electrocoagulation (EC) method was derived from coagulation, and it consists of using electrodes of these metals instead of metal salts. Coagulation occurs by releasing the metal ions from the electrode to the wastewater. The efficiency of EC is higher than conventional coagulation, and less sludge is formed. However, it may be more costly due to the need for electricity. Both methods can exhibit substantial organic matter and color removals by

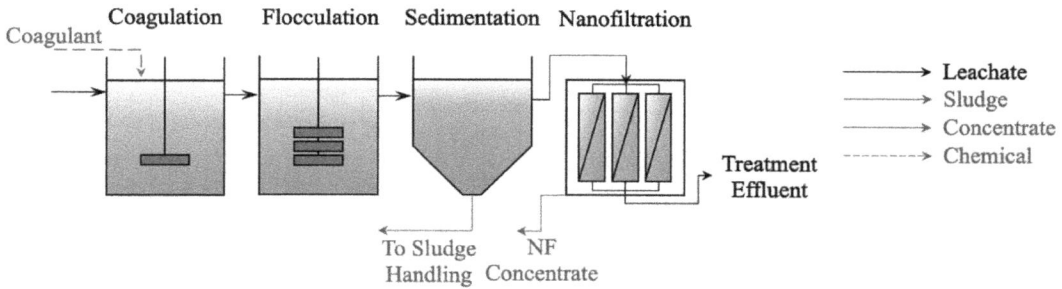

FIGURE 6.6 Schematic diagram of conventional coagulation/NF process.

combining the NF process. A typical conventional coagulation/NF hybrid system consists of four units, as shown in Figure 6.6. The raw leachate is subjected to rapid mixing by adding metal salt and pH adjustment chemicals for coagulation in the first tank. Then, the leachate passing through the flocculation and settling tanks, respectively, is treated by NF. Sometimes, an aeration unit can be placed between the precipitation and NF to strip the ammonia into the air.

In a study, EC/NF and CC/NF processes were compared to investigate the pretreatment efficiency with EC) and classical coagulation (CC). TOC and turbidity removal efficiencies of EC were higher than CC. However, no difference was observed between the total removals of EC/NF and CC/NF processes. This result demonstrated the effectiveness of the NF process in advanced treatment. However, while NF fouling occurred in the CC/NF process, no fouling was observed in the EC/NF hybrid system. Therefore, it was decided that the EC/NF alternative is more suitable than CC/NF one [50]. Although the EC/NF alternative was more effective in this study, it should be noted that the applicability of the EC process for large-scale plants will be more difficult. For example, a coagulation process was applied through the dosing of lime for pretreatment before NF. The precipitation process with slow stirring also aimed to strip the ammonia into the air. According to the results, the optimum removal efficiency was obtained with a lime dose of 10 g/L. Overall, 94% COD, 89% TOC, NH_3-N reductions of approximately 70%, and HS removals of approximately 80% were observed by the CC/NF process. Although the removal of recalcitrant organic matter by coagulation is relatively lower, 54% TOC and 33% COD removals were observed in the coagulation process. Such performance was evaluated as an effective pretreatment to reduce NF fouling [8].

6.6.2 Advanced Oxidation Processes/Nanofiltration

The NF process is widely used after biological or physicochemical treatment to remove refractory organic materials in leachate. However, the NF does not destroy pollutants but concentrates them into smaller volumes like other membrane processes. The result is a concentrated stream with high polluting potential, one of the biggest challenges in membrane processes. In some enterprises, membrane concentrate streams are recycled to landfill storage. This situation may lead to the accumulation of pollutants and salts in the system over time. Evaporation and solidification are alternative concentrate management processes, but these processes are not preferred due to the high operating cost. Another method is to treat NF concentrate streams by AOPs. These AOPs have been researched extensively due to their obvious advantages, such as degradation of refractory and hazardous DOM and increasing the percentage of biodegradable fractions.

As a result of the AOP, the non-biodegradable COD fraction could be converted to a biodegradable form, or mineralization of the refractory organic matter may occur. Mineralization can be achieved by increasing the oxidant concentration or extending the reaction time, but in that case, the operating cost will increase. Instead, when the AOP is stopped before complete mineralization, organics with large aromatic structures become biodegradable, making it more cost-effective to

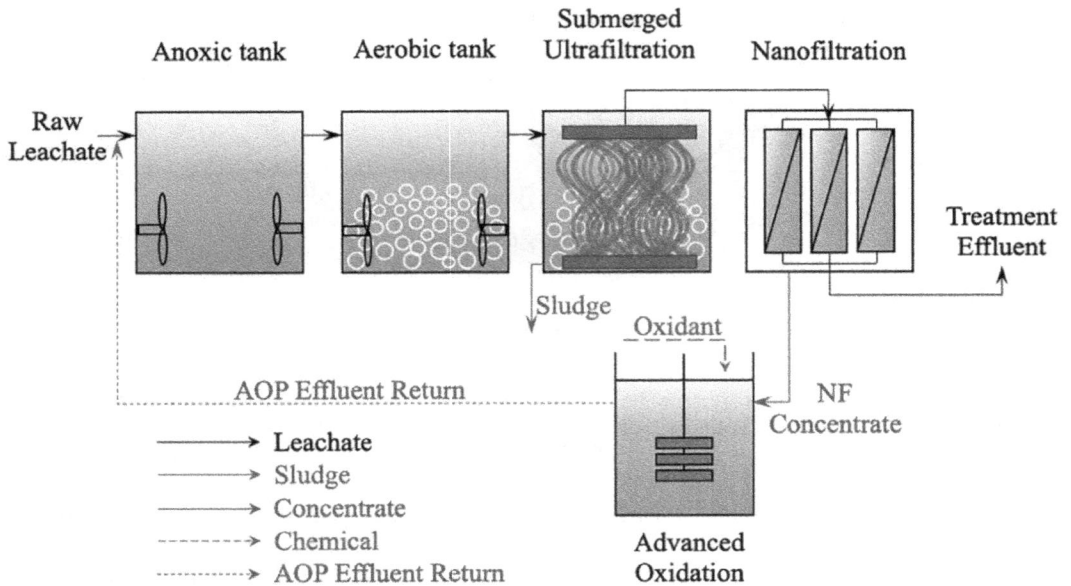

FIGURE 6.7   Schematic diagram of NF/AOP configuration for landfill leachate treatment.

operate. Thus, the effluent of the AOP could be returned to biological treatment. In other words, organic substances that cannot be removed in biological treatment are treated with NF, and concentrated leachate is sent to the AOP. Then, the leachate concentrate, which has become biodegradable, is transferred to biological treatment. Thus, both concentrate management and further organic matter removal take place. The typical NF/AOPs combination based on MBR+NF, one of the most used NF combinations in leachate treatment, is illustrated in Figure 6.7.

There are different AOPs such as Fenton (like) oxidation, electrochemical oxidation, ozone oxidation, and photochemical oxidation. Table 6.2 lists the studies on treating the NF leachate concentrated stream in terms of non-biodegradable DOMs. In a previous study, the electro-Fenton (EF) process, a type of advanced electrochemical oxidation, was implemented to investigate MBR+NF concentrated leachate's mineralization and biodegradability behavior. Raw carbon felt (CF) and $Fe^{II}Fe^{III}$ layered double hydroxides-modified CF ($Fe^{II}Fe^{III}$-LDH modified CF) were utilized as a cathode to understand the effect of homogenous and heterogeneous EF, while a Ti_4O_7 was used as anode in the experiments. Although the highest DOM mineralization efficiency was obtained from the heterogeneous EF process with $8.3\,mA/cm^2$ for 8 hours, the heterogeneous EF process with $4.2\,mA/cm^2$ for 4 hours ($Fe^{II}Fe^{III}$-LDH modified cathode and Ti_4O_7 anode) was defined as the most efficient treatment alternative. As a result of the respirometry analysis, it was also stated that by returning the NF concentrate stream to MBR after partial mineralization, the biodegradable organics would be removed, and thus the process cost would be improved [51].

In another study, leachate concentrate of MBR+NF configuration was targeted. In this study, AO with ozone was applied, and the size of the ozone bubbles on the refractory organic matter removal efficiency was investigated. The $2.4\,g/L$ ozone dosage, pH 9, and the ozonation time of 120 minutes were optimally chosen, and it was observed that the reaction rate constant of the micro-bubble method was three times higher than the medium and large bubbles under these conditions. After micro-bubble ozonation, the concentration of humic acids decreased from 24% to 14% and of fulvic acids from 50% to 25%. In addition, it was observed that the concentration of hydrophilic components increased from 26% to 61%. These results prove that the concentration of biodegradable organic matter is increasing. While a reaction time of at least 45 minutes is required to degrade the organic matters with large molecular weight, it has been evaluated as a promising process for

TABLE 6.2

Applied AOPs to NF Leachate Concentrate Treatment

NF Concentrate Characteristics	Oxidation Process	Removal	Main Findings	Ref.
pH: 8.5 COD: 1,330 mg/L TOC: 715 mg/L BOD: 34 mg/L Cl⁻: 1,690 mg/L SO_4^{-2}:3,409 mg/L	Fenton oxidative coagulation and ultraviolet photo-Fenton	COD: 91% TOC: 93%	– Possible to remove 70% of COD by using low concentration of oxidant for Fenton oxidation – Biodegradability increased – Possible to remove phthalic acid esters and polycyclic aromatic hydrocarbons in the leachate up to 90%	[54]
pH: 7.7 COD: 10,000 mg/L TOC: 3,719 mg/L Cl⁻: 7,200 mg/L	Electro-Fenton	COD: 69% TOC: 60%	– Compared to EC, EF was found more effective in terms of conversion the insoluble COD to soluble COD – EF produced less sludge than EC	[55]
pH: 7.9 COD: 2,900 mg/L Biodegradable COD (bCOD): 260 mg/L Dissolved organic carbon (DOC): 990 mg/L Cl⁻: 230 mg/L SO_4^{-2}: 2,000 mg/L	Electro-Fenton	DOC: 72%–96%	– Ti_4O_7 anode material is significant to improve the biodegradability – Colloidal proteins are defined as the most refractory organic component – Although the acute toxicity of the effluent was not reduced, high biodegradability was achieved after 4 hours of EF process	[56]
pH: 7.3 COD: 5,846 mg/L TOC: 2,082 mg/L	Ozonation	COD: 43% TOC: 5.6%	– 100% removal of non-biodegradable humic acids – Organic matter with high molecular weight (MW) was converted to low MW (<10kDa) – Possible to treat the oxidized concentrated leachate biologically	[57]
pH: 8.2 COD: 2,930 mg/L BOD: 131 mg/L TOC: 1,129 mg/L	Micro-bubble ozonation	COD: 76% TOC: 70%	– Reaction rate was three times higher when compared to normal ozonation – Concentration of organic matter with molecular weight lower than 2,000Da increased from 0% to 63%	[58]
pH: 8 COD: 4,700 mg/L Cl⁻: 10,458 mg/L	Coagulation combined with (I) Fenton oxidation (II) photoelectrooxidation (PEO)	(I) COD: 77% (II) COD: 86%	– Compared to direct Fenton oxidation, the combination of coagulation and Fenton oxidation exhibited a higher COD removal efficiency – Coagulation and PEO treatment was found more effective – All the halogenated compounds could be eliminated by PEO	[59]
pH: 6–8 COD: 2,088 mg/L BOD: 25 mg/L Cl⁻: 7,100 mg/L	Electrocatalytic oxidation	COD: 99% TOC: 57%	– Dissolved organic matter could effectively be degraded – Degradation order was found as: unsaturated double bond, fulvic-like component, and terrestrial humic-like component	[60]

NF concentrate management since it does not create by-products compared to other physicochemical oxidation processes [52]. Another study investigated the electrooxidation (EO) and photoelectrooxidation (PEO) processes after NaClO enhanced Fe^{2+} coagulation to treat organic matter and NH_4^+-N in concentrated leachate obtained from the MBR+NF process. Higher COD was removed with Fenton oxidation combined with coagulation compared to Fenton oxidation. On the other hand, COD removal increased by 10% when PEO was used instead of Fenton oxidation. However, no difference was observed in UV_{254}, NH_4^+-N, and color removals. Unlike other studies, the formation and degradation of chlorinated intermediate compounds formed during AO and NaClO enhanced Fe^{2+} coagulation were also investigated. Trihalomethane concentration increased to 10.5 mg/L at the end of 60 minutes and decreased to zero at 120 minutes by electrochemical oxidation process (EOP). Likewise, the concentrations of haloacetic acids and haloacetonitrile increased over time and decreased to zero at the end of 180 minutes. However, the formation and degradation of chlorinated intermediates in EO did not follow the same trend. After 180 minutes, many chlorinated intermediates remained without degradation. Therefore, the PEO process was considered more suitable than EO [53].

6.7 CONCLUSION

LL is a complex type of wastewater that is difficult to treat due to its refractory organic materials, high salinity, toxicity, and heavy metal content. In addition, leachate characteristics do not remain constant over time but change significantly. Treatment becomes even more difficult, especially as mature leachate loses biodegradability. Thus, it has become inevitable to use advanced treatment technologies to meet the tight discharge limitations. Due to the increasing water scarcity, water recovery has become a new requirement for every treatment process. Membrane processes are state-of-the-art treatment technology that can meet both advanced treatment and water recovery targets. However, a single process is not sufficient for the treatment of leachate. Generally, the desired effluent quality is achieved by combining various physicochemical and biological processes with pressure-driven membrane processes.

NF is more advantageous among pressure-driven membrane processes than RO because it requires lower energy and higher flux. Moreover, NF is frequently preferred due to its satisfactory organic matter and color removal. However, one of the most significant challenges of membrane processes is the fouling problem. Therefore, if NF is used directly, rapid fouling will occur, and thus operating and maintenance costs will increase. Various physicochemical or biological pretreatments are used before NF to reduce the energy requirement and use the membranes longer. Examples of hybrid NF processes are MBR/NF, MF/NF, and coagulation/NF. The MBR+NF hybrid treatment system is the most preferred process, including real applications. While MBR can treat biologically degradable organic matters, refractory organic materials, divalent inorganic salts, and color parameters can be effectively removed by NF to obtain the desired effluent quality. Another hybrid NF process is the MF/NF configuration. Before NF, direct MF is not used very often. The purpose of MF usage before NF is to provide a pretreatment by combining it with physicochemical or biological treatment. MF membranes are generally used as pretreatment after coagulation and chemical precipitation processes, with the advantages of superior liquid–solid separation and less footprint requirement, and are preferred to reduce fouling in the NF process. In the coagulation/NF hybrid system, both conventional coagulation/NF and EC/NF processes can be used for leachate treatment. Since no external chemical is added in EC, less sludge is generated, while the energy requirement is higher than the conventional one. The overall aim of these hybrid processes is to provide more effective removal of DOM in mature leachate having reduced biodegradable fraction. Besides, when the leachate matures, refractory matter concentrations in the NF concentrate phase increase, which is another major challenge of the pressure-driven membrane processes in concentrate management. Various techniques such as returning the concentrated phase to the landfill area, distillation, evaporation, or AO methods are applied in the landfill operation. Distillation and evaporation are relatively energy-intensive processes. Although the return to the landfill is mainly applied, new methods are needed due to this method causes an increase in the salinity and toxicity of the leachate over time. On the other hand, AO methods can be preferred for effective concentrate management. With the AOP, more organic matter removal can be achieved when the refractory materials accumulated in the NF concentrate phase are degraded and returned to biological treatment.

In addition to water recovery from leachate treatment, which requires the hybrid treatment processes, product recovery is also possible at the same time. The highly concentrated HS in the NF concentrate phase have fertilizer value for plants. However, since heavy metals are concentrated simultaneously, they may have a toxic effect on plants. By operating loose NF membranes, heavy metal removal could be reduced, and HS can be concentrated with higher purity. Another current approach is manufacturing innovative NF membranes to fractionate HS and inorganic salts. With the studies carried out in recent years, the surfaces of loose NF membranes have been modified, and the NF concentrate phase having fertilizer value has been successfully obtained by separating HS and inorganic salts from each other. As a result, it is seen that the NF process is the essential membrane process for leachate treatment due to both obtaining high-quality effluent and being energy-efficient, and studies showed that product recovery is also possible in recent years.

REFERENCES

1. D. Hoornweg, P. Bhada-Tata, C. Kennedy, *Nature* 502 (2013) 615–617.
2. M.R. Sabour, A. Amiri, *Waste Management* 65 (2017) 54–62.
3. H. Luo, Y. Zeng, Y. Cheng, D. He, X. Pan, *Science of the Total Environment* 703 (2020) 135468.
4. R. Keyikoglu, O. Karatas, H. Rezania, M. Kobya, V. Vatanpour, A. Khataee, *Separation and Purification Technology* 259 (2021) 118182.
5. S. Renou, J.G. Givaudan, S. Poulain, F. Dirassouyan, P. Moulin, *Journal of Hazardous Materials* 150 (2008) 468–493.
6. A.M.H. Shadi, M.A. Kamaruddin, N.M. Niza, M.I. Emmanuel, M.S. Hossain, N. Ismail, *International Journal of Environmental Science and Technology* 18 (2021) 2425–2440.
7. W. Chen, Z. Gu, G. Ran, Q. Li, *Waste Management* 121 (2021) 127–140.
8. R. de Almeida, A. Moraes Costa, F. de Almeida Oroski, J. Carbonelli Campos, *Journal of Environmental Science and Health, Part A* 54 (2019) 1091–1098.
9. M.A. Kamaruddin, M.S. Yusoff, L.M. Rui, A.M. Isa, M.H. Zawawi, R. Alrozi, *Environmental Science and Pollution Research* 24 (2017) 26988–27020.
10. D. Kulikowska, K. Zielińska, K. Konopka, *International Journal of Environmental Science and Technology* 16 (2019) 423–430.
11. P. Mandal, B.K. Dubey, A.K. Gupta, *Waste Management* 69 (2017) 250–273.
12. P.H. Chen, *Environment International* 22 (1996) 225–237.
13. C. Xiaoli, L. Guixiang, Z. Xin, H. Yongxia, Z. Youcai, *Waste Management* 32 (2012) 438–447.
14. D.L. Baun, T.H. Christensen, *Waste Management and Research* 22 (2004) 3–23.
15. Y.A.R. Lebron, V.R. Moreira, Y.L. Brasil, A.F.R. Silva, L.V. de S. Santos, L.C. Lange, M.C.S. Amaral, L.V. de Souza Santos, L.C. Lange, M.C.S. Amaral, *Journal of Environmental Management* 288 (2021) 112475.
16. A. Karimi, A. Khataee, M. Safarpour, V. Vatanpour, *Separation and Purification Technology* 237 (2020) 116358.
17. V. Vatanpour, S.S. Madaeni, A.R. Khataee, E. Salehi, S. Zinadini, H.A. Monfared, *Desalination* 292 (2012) 19–29.
18. Y. Deng, J.D. Englehardt, *Water Research* 40 (2006) 3683–3694.
19. A.H. Robinson, *Membrane Technology* 2005 (2005) 6–12.
20. C. Chiemchaisri, W. Chiemchaisri, P. Nindee, C.Y. Chang, K. Yamamoto, *Water Science and Technology* 64 (2011) 1064–1072.
21. Y. Xu, Y. Zhou, D. Wang, S. Chen, J. Liu, Z. Wang, *Journal of Environmental Sciences* 20 (2008) 1281–1287.
22. J. Liu, H. Zhang, P. Zhang, Y. Wu, X. Gou, Y. Song, Z. Tian, G. Zeng, *Bioresource Technology* 243 (2017) 738–746.
23. F.N. Ahmed, C.Q. Lan, *Desalination* 287 (2012) 41–54.
24. W.-Y. Ahn, M.-S. Kang, S.-K. Yim, K.-H. Choi, *Desalination* 149 (2002) 109–114.

25. M.C.S. Amaral, W.G. Moravia, L.C. Lange, M.M.Z. Roberto, N.C. Magalhaes, T.L. dos Santos, *Desalination and Water Treatment* 53 (2015) 1482–1491.
26. D. Trebouet, J.P. Schlumpf, P. Jaouen, J.P. Maleriat, F. Quemeneur, *Environmental Technology* 20 (1999) 587–596.
27. A.W. Mohammad, N. Hilal, L.Y. Pei, *International Journal of Green Energy* 1 (2004) 251–263.
28. W. Ye, H. Liu, M. Jiang, J. Lin, K. Ye, S. Fang, Y. Xu, S. Zhao, B. Van der Bruggen, Z. He, *Water Research* 157 (2019) 555–563.
29. W. Ye, H. Liu, M. Jiang, J. Lin, K. Ye, S. Fang, Y. Xu, S. Zhao, B. Van der Bruggen, Z. He, *Water Research* 157 (2019) 555–563.
30. T. Tavangar, M. Karimi, M. Rezakazemi, K.R. Reddy, T.M. Aminabhavi, *Chemical Engineering Journal* 385 (2020) 123787.
31. W. Ye, R. Liu, F. Lin, K. Ye, J. Lin, S. Zhao, B. Van der Bruggen, *Chemical Engineering Journal* 388 (2020) 124200.
32. J. Lin, Q. Chen, R. Liu, W. Ye, P. Luis, B. Van der Bruggen, S. Zhao, *Water Research* 204 (2021) 117633.
33. W. Ye, R. Liu, F. Lin, K. Ye, J. Lin, S. Zhao, B. Van der Bruggen, *Chemical Engineering Journal* 388 (2020) 124200.
34. J. Lin, Q. Chen, R. Liu, W. Ye, P. Luis, B. Van der Bruggen, S. Zhao, *Water Research* 204 (2021) 117633.
35. S. Yadav, I. Ibrar, A.K. Samal, A. Altaee, S. Déon, J. Zhou, N. Ghaffour, *Journal of Hazardous Materials* 421 (2022) 126744.
36. S. Yadav, I. Ibrar, A.K. Samal, A. Altaee, S. Déon, J. Zhou, N. Ghaffour, *Journal of Hazardous Materials* 421 (2022) 126744.
37. G. Wang, Z. Fan, D. Wu, L. Qin, G. Zhang, C. Gao, Q. Meng, *Desalination* 349 (2014) 136–144.
38. G. Wang, Z. Fan, D. Wu, L. Qin, G. Zhang, C. Gao, Q. Meng, *Desalination* 349 (2014) 136–144.
39. B.G. Reis, A.L. Silveira, L.P. Tostes Teixeira, A.A. Okuma, L.C. Lange, M.C.S. Amaral, *Waste Management* 70 (2017) 170–180.
40. L. Shao, Y. Deng, J. Qiu, H. Zhang, W. Liu, K. Bazienė, F. Lü, P. He, *Water Research* 195 (2021) 117000.
41. G. Li, W. Wang, Q. Du, *Journal of Applied Polymer Science* 116 (2010) 2343–2347.
42. B.G. Reis, A.L. Silveira, L.P. Tostes Teixeira, A.A. Okuma, L.C. Lange, M.C.S. Amaral, *Waste Management* 70 (2017) 170–180.
43. H. Wang, Z. Cheng, Z. Sun, N. Zhu, H. Yuan, Z. Lou, X. Chen, *Chemosphere* 243 (2020) 125354.
44. M.C.S. Amaral, W.G. Moravia, L.C. Lange, M.R. Zico, N.C. Magalhães, B.C. Ricci, B.G. Reis, *Journal of Environmental Science and Health, Part A* 51 (2016) 1–10.
45. R. Rautenbach, *Journal of Membrane Science* 174 (2000) 231–241.
46. S. Ramaswami, J. Behrendt, R. Otterpohl, *Membranes* 8 (2018) 17.
47. A.V. Santos, L.H. de Andrade, M.C.S. Amaral, L.C. Lange, *Environmental Technology* 40 (2019) 2897–2905.
48. A.V. Santos, L.H. de Andrade, M.C.S. Amaral, L.C. Lange, *Environmental Technology* 40 (2019) 2897–2905.
49. M.C.S. Amaral, H. V. Pereira, E. Nani, L.C. Lange, *Water Science and Technology* 72 (2015) 269–276.
50. T. Mariam, L.D. Nghiem, *Desalination* 250 (2010) 677–681.
51. M. El Kateb, C. Trellu, A. Darwich, M. Rivallin, M. Bechelany, S. Nagarajan, S. Lacour, N. Bellakhal, G. Lesage, M. Héran, M. Cretin, *Water Research* 162 (2019) 446–455.
52. H. Wang, Y. Wang, Z. Lou, N. Zhu, H. Yuan, *Waste Management* 69 (2017) 274–280.
53. M. Qiao, X. Zhao, X. Wei, *Scientific Reports* 8 (2018) 12525.
54. J. Li, L. Zhao, L. Qin, X. Tian, A. Wang, Y. Zhou, L. Meng, Y. Chen, *Chemosphere* 146 (2016) 442–449.
55. S. Yazici Guvenc, K. Dincer, G. Varank, *Journal of Water Process Engineering* 31 (2019) 100863.
56. M. El Kateb, C. Trellu, A. Darwich, M. Rivallin, M. Bechelany, S. Nagarajan, S. Lacour, N. Bellakhal, G. Lesage, M. Héran, M. Cretin, *Water Research* 162 (2019) 446–455.
57. H. Wang, Y. nan Wang, X. Li, Y. Sun, H. Wu, D. Chen, *Waste Management* 56 (2016) 271–279.
58. H. Wang, Y. Wang, Z. Lou, N. Zhu, H. Yuan, *Waste Management* 69 (2017) 274–280.
59. M. Qiao, X. Zhao, X. Wei, *Scientific Reports* 8 (2018) 1–9.
60. Z. Zhang, C. Teng, K. Zhou, C. Peng, W. Chen, *Chemosphere* 255 (2020) 127055.

7 Nanofiltration Membranes for Textile Industry
Water Reclamation and Resource Recovery

Nadiene Salleha Mohd Nawi, Mei Qun Seah,
Woei Jye Lau, Pei Sean Goh, and Ahmad Fauzi Ismail
Universiti Teknologi Malaysia

CONTENTS

7.1 INTRODUCTION

Fast fashion is one of the major reasons contributing to the rise of fashion and textile industry. The textile industry was estimated to support about 120 million employments and valued around USD 2 trillion worldwide [1], making it one of the largest industries globally. However, this industry is notorious for its large water consumption (1.5 trillion litres per year) and is responsible for ~20% of industrial water pollution from its wastewater produced [2]. Currently, China and India are the two major players of textile industry and annually, they would consume ~8.65 and ~0.83 billion m³ of groundwater for textile production, respectively [3,4]. In terms of the quantity of wastewater discharged, both countries generated up to ~2.5 billion m³/year [3,4].

Typically, large quantity of water is used at the wet stage of textile manufacturing such as sizing and desizing, scouring, bleaching, mercerization, dyeing/printing and finishing [5]. Sizing refers to the application of binders or adhesives such as polyvinyl alcohol and polycyclic acids to strengthen and improve fibre weavability. Desizing meanwhile removes sizing materials using enzymes and mineral acids. This is followed by scouring process where alkali agents (e.g., glycerol, sodium hydroxide and detergents) are used to remove impurities (e.g., waxes and surfactants) from the fibres. Bleaching is later performed using hydrogen peroxide (H_2O_2) to enhance fibre whiteness, while mercerization aims to improve fibre strength, lustre and dye affinity through caustic sodium hydroxide (NaOH) treatment. Dyeing/printing is then carried out using a wide variety of dyes in the presence of salts to colour the fibres. Finally, finishing uses protection and maintenance chemicals such as biocides and waxes to improve fibre properties such as stain and water proofing, microbial protection and ultraviolet (UV) damage.

Since many different chemicals/materials are used during textile manufacturing, a proper treatment of textile effluent discharge is compulsory to remove hazardous/toxic pollutants before being released into environment. Improper wastewater treatment could potentially contribute to ecosystem degradation. For instance, a study showed that a chemically stable fluoropolymers used in textile waterproofing was found in remote Arctic locations [6]. Persistent perfluoroalkyl substances (also known as forever chemicals), which are used in anti-staining fabrics, were previously found in urban rainwater [7]. A 2014 Swedish report revealed that out of the 2,450 textile manufacturing-related chemicals investigated, 10% were identified to have potential risk to human health [8]. These chemicals include fragrances, azo dyes, reproductive toxins such as brominated flame retardant, highly fluorinated water and phthalates as well as antibacterial agents. Bioaccumulation of textile wastewater contaminants in the food chain could result in severe toxic hazard to human health, as these pollutants may have carcinogenic, mutagenic, allergenic and/or endocrine disruptive properties [1].

In view of this, cleaner production should be adopted to counter large amount of hazardous emissions from the textile industry in order to achieve the United Nation's Sustainable Development Goals 9, 12 and 15 [9]. The implementation of cleaner production practices includes the use and development of alternative ecological materials (e.g., bamboo fibres and natural dyes) and less toxic chemical products [10], in addition to technical and manufacturing process improvements. With this in mind, research on water reclamation and resource recovery (e.g., concentrated salts and spent dyes) from textile effluents is gaining immense traction [11–13]. Recently, the application of membrane technology in saline wastewater treatment and resource recovery was highlighted, and the review detailed the advantages of membranes over conventional physicochemical treatment methods [14].

Textile effluents are commonly treated via physical (e.g., adsorption, coagulation and flocculation), chemical (oxidation and photocatalytic degradation using H_2O_2, UV, ozone and NaOCl), electrochemical and biological (bioremediation) means. These primary and secondary treatment methods mainly

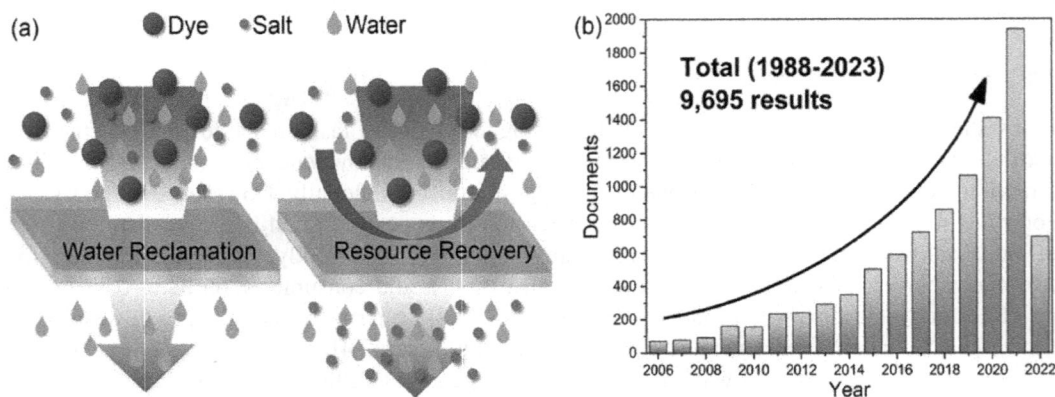

FIGURE 7.1 (a) Schematic diagram of water reclamation and resource recovery using different types of NF membranes and (b) number of documented papers related to NF membranes for textile wastewater treatment. (Data were retrieved from the Scopus based on the keyword 'nanofiltration AND membrane AND textile AND wastewater' (Accessed on 14 March 2022).)

TABLE 7.1

Contaminants Discharged from Each Wet Processing Step and the Properties of Their Effluents [16,18]

Wet Process	Contaminants/Pollutants	Effluent Characteristics
Sizing and desizing	Sizes, enzymes, starch, waxes and lubricants	High biochemical oxygen demand (BOD)
Scouring	Sodium hydroxide, disinfectant/insecticide residue, surfactants, detergent, fats, lubricants, oils and waxes	High pH
Bleaching	Hydrogen peroxide, sodium silicate and organic stabilizer	High pH (alkaline conditions)
Mercerizing	Sodium hydroxide	High pH
Dyeing	Heavy metals, metals, salts, surfactants and dyes	High BOD, high/low pH (alkaline/acidic conditions)
Printing	Suspended solids, metals, dyes, urea, formaldehyde and solvents	High BOD
Finishing	Suspended solids, softeners, solvents/spent solvents, resins and waxes	High BOD and COD

aim to remove colour/dissolved solids and degrade toxic dyes/chemicals [15], but could not recycle valuable resources such as dyes, acids/bases and water from the effluent. To address the limitations of these methods, nanofiltration (NF) membrane with unique properties, targeting either for pure water production or salt recovery, is the most befitting class for water reclamation and resource recovery applications (see Figure 7.1a). Figure 7.1b shows that NF membranes have been increasingly reported as an efficient water reclamation and resource recovery tool for textile effluents.

7.2 PROPERTIES OF TEXTILE EFFLUENT

Among the stages of textile fabrication, wet processing is the stage that produces the highest volume of toxic and hazardous effluents due to the substantial amount of water consumed [16,17]. In terms of the textile effluent characteristics, Yaseen and Scholz [18] highlighted its unpredictability when they compared the properties and chemical constituents of nine different effluents obtained from various sources, and wildly fluctuating properties were recorded. Nevertheless, the typical pollutants discharged from each wet processing stage could still be identified and are outlined in Table 7.1.

Textile effluent discharged from factories is largely made up of dyes, salts, metals/heavy metals, volatile organic compounds and other pollutants [1,18]. Among them, synthetic dyes such as reactive, direct, basic and acid dyes are the most prominent and persistent pollutants in the textile effluents due to their high water solubility and difficulty of being removed by primary or secondary wastewater treatments [19]. The presence of pigments in textile effluent is also found to be an environmental issue as the deep-coloured effluent could impair the photosynthesis of underwater flora [20,21]. In addition to disrupting aquatic photosynthesis, organic dyes are also toxic, non-biodegradable and high in chemical oxygen demand (COD). Meanwhile, the existence of high salt content decreases the efficiency of biological treatments due to lowered metabolic functions of activated sludge microorganisms [22].

7.3 TYPES OF NANOFILTRATION MEMBRANES

In the current market, polymeric NF membranes are sold in either hollow fibre or spiral wound configurations. However, spiral wound module with polyamide (PA) thin-film composite (TFC) membrane dominates the NF market owing to its high water flux and salt rejection compared to the asymmetric hollow fibre membrane which is manufactured using phase inversion technique. In this chapter, we will cover the studies not only related to the PA TFC membranes [or TFC membrane modified by nanoparticles, also known as thin-film nanocomposite (TFN)] but also asymmetric membranes in both flat sheet and hollow fibre configuration. Generally, the TFC membrane is composed of three main layers, i.e., top thin PA selective layer, intermediate microporous substrate and bottom non-woven fabric. The PA selective layer is the most important layer of the TFC membrane as it determines the filtration performance and the antifouling properties. Asymmetric NF membranes meanwhile are the membranes made of either polymers or polymer/inorganic nanomaterials using phase inversion technique.

In this chapter, two main applications of NF membranes will be reviewed. The first NF application is aimed at achieving promising separation during dyeing solution (or textile effluent) treatment so as the treated permeate can either meet the discharge standard or fulfil the requirements of reclamation. The second application is to selectively separate the components from the feed using NF membrane with the purpose of recovering resources. The properties of NF membranes employed for these applications can be flexibly designed by optimizing the conditions during synthesis process.

7.4 NANOFILTRATION MEMBRANE FOR WATER RECLAMATION

Textile industry is one of the water-intensive industries that contribute majorly towards saline wastewater production. Faced with more stringent discharge standards, effective yet affordable treatment strategies are urgently required. As such, water reclamation has become a major topic in the textile industry. Besides aiming to meet the local standard of discharge, the treated wastewater (with high quality) can be possibly reused to reduce not only the freshwater demand but also the operating cost.

The development of highly efficient NF membranes with effective removal of organic compounds and salts as well as antifouling property is paramount for water reclamation. With its high removal efficiencies for a wide range of contaminants, NF membranes has been recognized as an effective tool to treat textile effluent and could produce high-quality effluents that could be suitable for reuse. Parameters such as permeate flux and pollutants removal rate (e.g., dye and dissolved ions) are the key factors affecting the reclamation. In view of this, the following section would discuss the properties of NF membranes for the water reclamation of textile (or dyeing) wastewater. Table 7.2 summarizes the key findings of studies that employed NF membranes in water reclamation of textile wastewater.

7.4.1 THIN-FILM COMPOSITE MEMBRANES

Liu [23] studied the effect of hydrophilic diethanolamine (DEA) as amine monomer on the PA properties of TFC membrane and used the membrane for textile wastewater treatment. Compared to the neat TFC membrane fabricated using piperazine (PIP) and trimesoyl chloride (TMC), the DEA-modified TFC membrane exhibited 26.8% higher pure water permeability (PWP) (increased from 13.4 to 17.0 L/m^2·h·bar) while maintaining similar rejection rates against different salts. The significant flux enhancement is due to the grafting of DEA onto PA layer that improved surface hydrophilicity without affecting its thickness and compactness. As shown in Figure 7.2a, the DEA-modified TFC membrane exhibited the rejection in the order of methyl blue (MYB, 99.8%) > Congo red (99.6%) > sunset yellow (97.6%) > neutral red (80.6%) during filtration of 50 mg/L dye solutions. The modified membrane also demonstrated smaller degree of flux decline after a simple hydraulic washing, which indicated its improved antifouling property for wastewater treatment.

To improve the permselectivity of TFC membrane, Lü [39] utilized tannic acid (TA) to produce an additional skin layer atop existing PA layer. Through optimization, they reported that the best performing NF membrane could be synthesized by cross-linking 0.25% PIP with 0.10% TMC, followed by coating using 0.08% TA. Although the PWP of the TFC membrane was affected (decreased from 25 to 10.32 L/m^2·h·bar) upon TA coating, the NaCl and Na$_2$SO$_4$ rejections of coated membrane was found to improve by 64.71% and 9.28%, respectively, reaching 71.2% and 99.4%.

TABLE 7.2

NF Membranes Employed in Water Reclamation of Saline Textile Wastewater

Type	Selective Layer (Nanomaterial)	Substrate	Dye Rejection (%)	Salt Rejection (%)	Permeability (L/m²·h·bar)	Reference
TFC	PIP+TMC	PSf	99.8% Methylene Blue	98.5% Na$_2$SO$_4$	17	[23]
			99.6% Congo Red	88.3% MgSO$_4$		
			97.6% Sunset Yellow	50.6% NaCl		
			80.6% Neutral Red	31.2% MgCl$_2$		
	PIP+TMC	PMIA	97% Chromotrope FB	98.45% Na$_2$SO$_4$	17.36	[24]
			98% Reactive Yellow 3	97.98% MgSO$_4$		
			99% Direct Fast Blue	95.74% MgCl$_2$		
	PIP+TMC	PSf	99.9% Sunset Yellow	71.2% NaCl	10.32	[25]
			99.9% Congo Red	99.4% Na$_2$SO$_4$		
	PIP+TMC	PPSU	86% Rhodamine B	Not available	12.7	[26]
			80% Methyl Orange			
TFN	PIP+TMC	PSf	99.7% Setazol Red	49% NaCl	4.22	[27]
			99.7% Reactive Orange	94.4% MgSO$_4$		
	PIP+TMC	PSf	99.95% Rose Bengal	95.8% Na$_2$SO$_4$	4.45	[28]
			98.72% Reactive Black 5	97.7% MgSO$_4$		
			96.57% Methyl Blue			
	PIP+TMC	PES	100% Reactive Blue 4	Not available	8.8	[29]
			100% Acid Blue 193			
			100% Methyl Violet			
	PIP+TMC	PSf	99.73% Reactive Black 5	92.4% Na$_2$SO$_4$	20.8	[30]
LbL	PA+ZIF-8	PSf	99.8% Congo Red	40% MgSO$_4$	2.01	[31]
	PDDA+PAMPS	PSf	95.5% Acid Blue	73.98% Na$_2$SO$_4$	20.5	[32]
	PAA+PEI	Isopore polycarbonate	96.3% Congo Red	88.2% MgSO$_4$	166	[33]
				86.7% MgCl$_2$		
	TA+Jeffamine	PAN	99.5% Rose Bengal	Not available	37	[34]
			99% Rhodamine B			
			98% Methylene Blue			
MMMs	MMGO	PES	99.9% Direct Red 16	81% Na$_2$SO$_4$	36.1	[35]
				15% NaCl		
	C-MWCNT	PMIA	98% Congo Red	80.7% Na$_2$SO$_4$	5.1	[36]
			98% Direct Red 23			
	ACF-MWCNT	PVDF	97.13% Reactive Green 19	Not available	46	[37]
	Brij-58+PEG 400	PVDF	90% Reactive red 141	Not available	31.2	[38]

[a] PIP, piperazine; TMC, trimesoyl chloride; PA, polyamide; ZIF-8, zeolitic imidazolate framework; PDDA, poly(diallyldimethylammonium chloride); PAMPS, poly(2-acrylamido-2-methylpropanesulfonate); PAA, polyacrylic acid; PEI, polyetherimide; TA, tannic acid; MMGO, magnetic graphene-based composite; C-MWCNT, carboxylic multiwalled carbon nanotube; ACF-MWCNT, carboxyl-functionalized multiwalled carbon nanotubes; PEG, polyethylene glycol; PSf, polysulfone; PMIA, poly(m-phenylene isophthal amide); PPSU, polyphenylsulfone; PES, polyethersulfone; PAN, polyacrylonitrile; PVDF, polyvinylidene fluoride.

The reduced water flux can be attributed to the thicker and more compact selective layer formed due to deposition of TA layer atop PA, which resulted in increased water resistance. Further comparison showed that while both the coated and the pristine TFC membranes were able to reject >99.9% of sunset yellow and Congo red, the coated membrane demonstrated lower flux decline owing to its improved antifouling property and higher negative membrane surface charge.

In another study, Liu [26] synthesized two sulfonated diamine monomers, i.e., 2,5-bis(4-amino-2-trifluoromethyl-phenoxy)benzenesulfonic acid (6FAPBS) and 4,4′-bis(4-amino-2-trifluoromethyl-phenoxy)biphenyl-4,4′-disulfonic acid (6FBABDS) and used them to prepare respective TFC membranes in the presence of PIP and TMC. The introduction of 0.30% 6FAPBS and 6FBABDS into PA layer resulted in increased membrane hydrophilicity (with contact angle dropped from 72° to 55° and 47°, respectively), which in turn facilitated water transport through the membranes (increased from 9.24 to 12.7 and 14.26 L/m^2·h·bar, respectively) due to the presence of sulfonic groups that enhanced surface wettability. The improved surface chemistry also helped in rejecting ~86% of Rhodamine B and ~77% of methyl orange. As shown in Figure 7.2b, it was observed that the water fluxes of modified membranes were maintained at ~52.5 L/m^2·h (TFC/6FAPBS) and >60 L/m^2·h (TFC/6FBABDS) with only slight reduction over the testing duration. The water fluxes of both modified membranes were obviously higher compared to the control PIP-TMC membrane (~45 L/m^2·h).

Using hollow fibre substrate made of poly(m-phenylene isophthalamide) (PMIA), Wang [24] fabricated PIP-TMC-based TFC membrane and reported that the resultant membrane exhibited promising permeability and Na$_2$SO$_4$ rejection performance of 17.36 L/m^2·h·bar and 98.45%, respectively. As illustrated in Figure 7.2c, the formation of 'ridge and valley' structure on the membrane surface indicated the successful cross-linking of two active monomers in producing PA layer. When tested against three different types of dyes (Chromotrope FB, Reactive Yellow 3 and Direct Fast Blue B2RL) at solution pH ranging between 3 and 11, the authors observed an increased membrane flux at higher pH value owing to negatively charged membrane surface that effectively hindered dye adhesion onto membrane. The dye removal rates by this membrane were recorded high at 97%–99% regardless of pH and dyes.

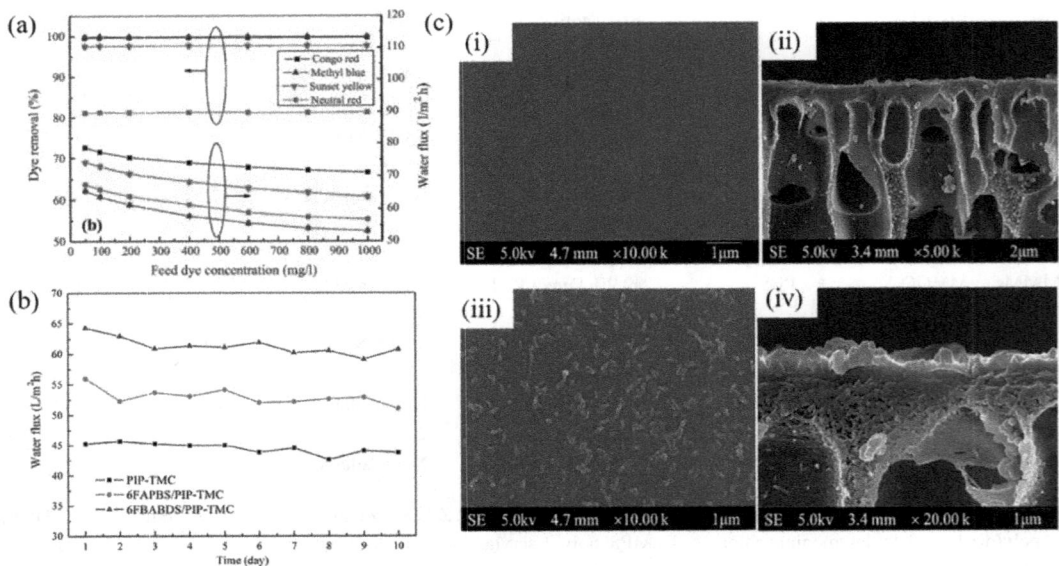

FIGURE 7.2 (a) Impact of dye concentration on dye removal and water flux of DEA-modified TFC membrane tested at 5 bar [23], (b) long-term stability of TFC membranes modified with 6FAPBS and 6FBABDS during filtration of 50-ppm methyl orange solution at 5 bar [26] and (c) scanning electron microcopy (SEM) surface and cross-sectional images of (i, ii) PMIA substrate and (iii, iv) PMIA-based TFC membrane [24].

7.4.2 THIN-FILM NANOCOMPOSITE MEMBRANES

Concerted efforts have been made over the past decade in addressing the limitations of TFC membranes such as water flux/rejection trade-off relation and surface fouling. This led to significant growth in the number of articles related to the use of nanomaterials in modifying the characteristics of pristine PA layer during interfacial polymerization (IP) process to produce TFN membranes with improved filtration performance.

The earliest work of TFN membranes was reported by Jeong et al. [40] in which they introduced zeolites into PA layer of TFC membrane and achieved higher water permeability and solute selectivity for reverse osmosis process. Since then, this type of membrane has received great attention from the researchers worldwide for various applications including dye removal. Ghaemi and Safari [29] prepared new type of TFN membranes incorporating silicoaluminophosphate nanoporous zeolite (SAPO-34) and reported that it could completely remove all dye compounds from the feed (initial dye concentration: 50 mg/L) while maintaining consistent flux for a period of 240 minutes. The improved surface hydrophilicity along with creation of nanochannels due to SAPO-34 addition has made the TFN membrane to exhibit much higher water permeance (8.8 L/m²·h·bar) compared to the control TFC membrane (2.73 L/m²·h·bar).

Wu et al. [30], on the other hand, incorporated ZIF-8 nanoparticles into PA layer via two different strategies, i.e., in-situ growth and one-step blending IP methods, as depicted in Figure 7.3a. The authors found that the in-situ growth of ZIF-8 on polysulfone (PSf) substrate was more effective than blending method that introduced ZIP-8 into PIP solution. When ZIF-8 was present in the PIP solution, it tended to decrease the diffusion of PIP monomer from the aqueous to the organic phase, reducing cross-linking degree of the PA layer. Using the in-situ growth method for TFN membrane fabrication, the authors were able to achieve considerably high PWP (20.8 L/m²·h·bar) with Na_2SO_4 and Reactive Black 5 rejection recorded at 92.4% and 99.73%, respectively.

Carbon-based nanomaterials such as graphene oxide (GO) nanosheets are also popularly used in modifying TFC membrane over the years. To develop TFN membrane, Lai et al. [28] first deposited GO nanosheets on the surface of PSf substrate via vacuum filtration method. The GO-deposited substrate was then subjected to IP process by filtering 2 wt% PIP solution followed by 0.2 wt% TMC solution. The novel method produced TFN membrane with even nanomaterials distribution

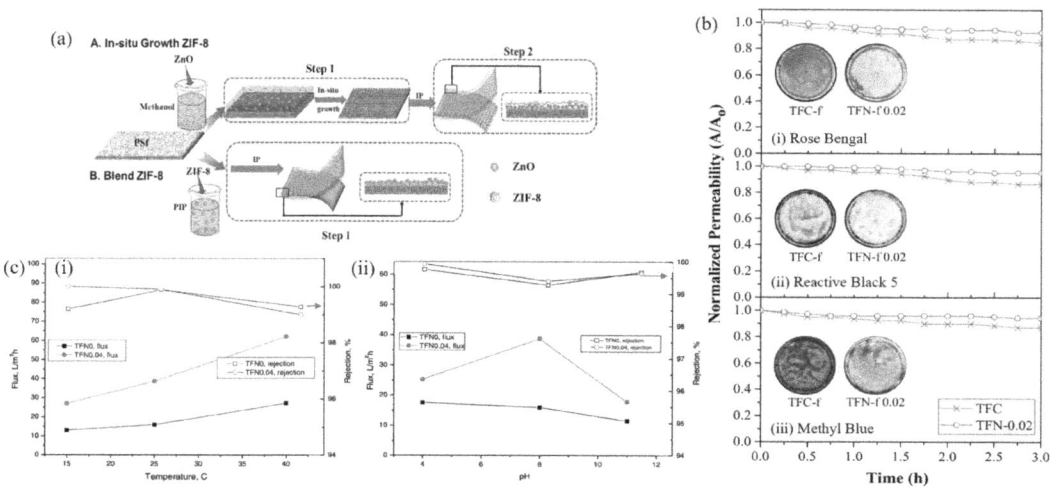

FIGURE 7.3 (a) Illustration of ZIF-8 deposition on TFN membranes with two different strategies [30], (b) normalized permeability of TFC and GO-incorporated TFN membranes during filtration of three different dyeing solutions [28] and (c) flux and setazol red rejection of TFC and HNT-incorporated TFN membranes as a function of (i) temperature and (ii) solution pH [27].

without having any loss of precious nanomaterials. With respect to performance, the TFN membrane recorded 43% flux enhancement compared to the control TFC membrane without compromising MgSO$_4$ rejection (97.7%). Further evaluation indicated that besides demonstrating excellent dye removal (>96%), the TFN membranes experienced lower flux decline (5%–8%) compared to the TFC membrane (12%–15%) for a testing period of 3 hours (Figure 7.3b). This clearly revealed the important role of GO nanosheets in improving fouling resistance of membrane.

Ormanci-Acar [27] modified the selective layer of membrane with halloysite nanotubes (HNTs) at various loadings (0%–0.06%) and found that 0.04% was the most ideal nanoparticle loading to achieve the best performance. At this loading, the resultant membrane demonstrated water permeability of 4.22 L/m^2·h·bar, MgSO$_4$ rejection of 94.4% and almost complete rejection of dye (~99.7%). This membrane was further evaluated under different operating temperatures (15°C, 25°C and 40°C) and pH (4, 8 and 11) and the results are presented in Figure 7.3c. As shown in the figure, the membrane flux increased but dye rejection decreased slightly with rising temperature. Higher temperature could decrease the solution viscosity which led to flux improvement but slight reduction in membrane selectivity. Separately, pH alteration was found to be insignificant on dye rejection but alkaline conditions resulted in flux decline due to dye aggregation.

7.4.3 Layer-by-Layer Membranes

Layer-by-layer (LbL) assembly is another approach commonly used by researchers to synthesize NF membranes by depositing oppositely charged polyelectrolytes on a substrate membrane. This technique allows control over film thickness and its composition at molecular level while maintaining membrane high selectivity [41].

Taking the advantages of the LbL approach, Wang et al. [42] deposited an interlayer of ZIF-8 (up to four bilayers) via in-situ growth atop PSf substrate prior to the IP process. The authors found that the characteristic peaks at 1,660 and 1,550 cm^{-1}, which are due to the typical amide structures, became weaker with increasing bilayers of ZIF-8 (Figure 7.4a). They attributed this to the negative impact of ZIF-8 on the degree of cross-linking between MPD and TMC which decreased PA thickness. Although the PA cross-linking degree was reduced, the water permeance of the resultant membrane experienced two times higher flux compared to neat membrane, reaching 2.01 L/m^2·h·bar with a high rejection against Congo red (99.8%).

Ahmadian-Alam et al. [32] developed a hybrid multilayer NF membrane by incorporating hexagonal boron nitride (HBN) nanosheets within the polyelectrolyte solutions of poly(diallyldimethylammonium chloride) (PDDA) and poly(2-acrylamido-2-methylpropanesulfonate), aiming to improve the membrane permeation rate. The formation of dense bilayers on the substrate could be successfully seen in Figure 7.4b(i). In terms of surface roughness, the HBN-incorporated multilayer membrane exhibited rougher surface compared to the pristine multilayer membrane due to the presence of nanosheets on the membrane surface (Figure 7.4b(ii)). It must be noted that owing to the hydrophilicity of the HBN, the resultant membrane was reported to achieve 19% higher flux than that of pristine membrane. Although the HBN-incorporated multilayer membrane was able to retain up to 95.5% acid blue, its separation rate against Na$_2$SO$_4$ was considerably low at 73.98%.

Lin et al. [33], on the other hand, reported a novel strategy to develop multilayer membrane by depositing polyacrylic acid (PAA) and polyetherimide (PEI) via LbL self-assembly with nickel hydroxide (Ni(OH)$_2$) nanosheet sacrificial layer, as illustrated in Figure 7.4c. The sacrificial layer aided in depositing the polyelectrolyte layers onto substrate before being dissolved with hydrochloric acid. The results showed that the fabricated multilayer membrane displayed dense, smooth and compact structure with controllable thickness that increased linearly by ~70 nm per layer. The PWP and MgCl$_2$ rejection of the optimized membrane (seven layers) were able to reach 166 L/m^2·h·bar and ~86%, respectively. Moreover, the negatively charged membrane played a significant role in improving membrane dye filtration by rejecting 96.3% of Congo red.

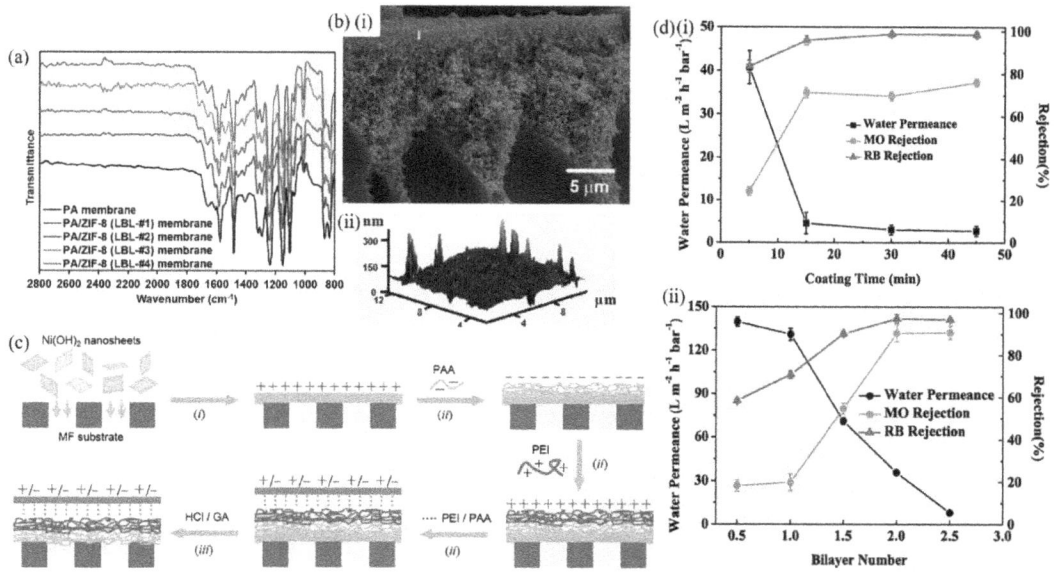

FIGURE 7.4 (a) Fourier transform Infrared (FTIR) spectra of PA membrane (pristine) and PA/ZIF-8 membranes with different bilayers [31], (b) morphology of HBN-incorporated multilayer membrane, (i) field emission scanning electron microscopy (FESEM) and (ii) 3D AFM [32], (c) schematic illustration of LbL assembly using PAA and PEI for multilayer NF membrane fabrication [33] and (d) the effect of coating time and bilayer number on the water flux and dye rejection of TA/JA multilayer NF membrane [34].

Based on the work done by Guo et al. [34], NF membrane with a thin polyelectrolyte layer was developed by depositing TA and Jeffamine (JA) via LbL assembly on polyacrylonitrile (PAN) support. It was reported that the increase in bilayer number and deposition time resulted in higher dye removal rate but reduced water permeability owing to increased top layer thickness (i.e., corresponded to greater water resistance), as shown in Figure 7.4d. At 15-minute deposition and two bilayers, the authors were able to produce the best performing NF membrane, achieving 37 L/m^2·h·bar PWP and >91% rejection against various dyes. Furthermore, this membrane was reported to exhibit promising antifouling properties based on its high flux recovery rate (FRR, ~98%) after subjecting to bovine serum albumin (BSA) fouling test followed by simple cleaning using water.

7.4.4 ASYMMETRIC MEMBRANES

Compared to the widely used TFC membranes, the asymmetric membrane made using non-solvent induced phase separation method offers several advantages such as simple fabrication process and less susceptible to chlorine attack. In an attempt to enhance NF performance of asymmetric membrane, Abdi et al. [35] used magnetic graphene-based composite (MMGO) to modify properties of polyethersulfone (PES)-based membrane. Figure 7.5a compares the cross-section structure of the membranes with and without MMGO addition. Both membranes displayed typical asymmetrical structure with a dense skin layer and porous sublayer. Longer finger-like structure, however, could be observed in MMGO-modified membrane due to faster exchange between solvent and non-solvent caused by the addition of particles. This membrane showed 2.5 times higher pure water flux than the bare PES membrane, recording 8.25 L/m^2·h·bar. Besides, it achieved >99% and 81% rejection against direct red 16 and Na_2SO_4, respectively.

Zhao et al. [36] introduced carboxylic multiwalled carbon nanotube (C-MWCNT) into PMIA-based hollow fibre membranes in order to improve the membrane performance for dye removal.

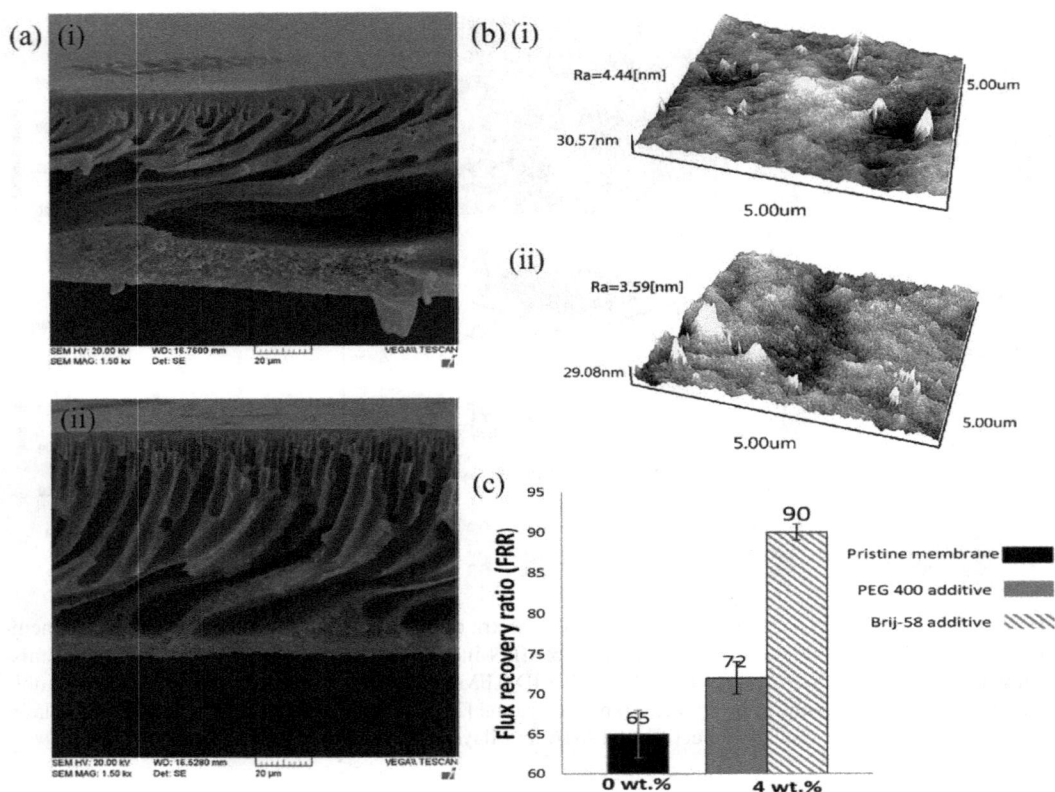

FIGURE 7.5 (a) Cross-sectional SEM images of (i) PES and (ii) 0.5 wt% MMGO-modified PES membrane [35], (b) AFM images of PMIA membranes (i) without and (ii) with C-MWCNT [36] and (c) FRR of different PVDF membranes [38].

With the incorporation of only 0.10 wt% C-MWCNT, the modified membrane showed 70% greater water flux (5.1 L/m^2·h·bar) than pure PMIA membrane (2 L/m^2·h·bar). In addition, the presence of nanoparticles in the PMIA membrane was found to have slightly smoother surface when examined from atomic force microscopy (AFM) image (Figure 7.5b). When tested against 1,000 mg/L Na$_2$SO$_4$ at 10 bar, the salt rejection of PMIA membrane increased from 76.6% to 80.7% upon the addition of 0.10 wt% C-MWCNT. In terms of dye removal, the modified PMIA membrane showed at least 98% removal against Congo red and direct red, owing to its surface negative charge that repelled the pigments more effectively.

Recently, Gholami et al. [37] embedded carboxyl-functionalized multiwalled carbon nanotubes (ACF-MWCNT) into polyvinylidene fluoride (PVDF) membrane and used it to separate dye from feed solution at low operating pressure (3 bar). It must be noted that the presence of 0.1 wt% ACF-MWCNT could improve the pristine membrane's water flux by 17% without compromising reactive green rejection (97.13%), revealing the potential of this material in overcoming the trade-off effect between permeability and rejection. The researchers also found the modified membrane exhibited higher FRR (96.6%) than the pristine PVDF (~60%) after being tested with 500 ppm BSA solution owing to this improved surface hydrophilicity upon modification.

To study the effectiveness of Brij-58 surfactant as a hydrophilic additive, Nikooe and Saljoughi [38] introduced it into PVDF membrane and compared its performance with the PVDF membrane containing common hydrophilic additive, i.e., polyethylene glycol (PEG) 400. Although both modified membranes exhibited comparable PWP (~30 L/m^2·h·bar) that was 33% higher than the pure

PVDF membrane, they suffered from lower dye rejection as a result of formation of more porous structure. Nevertheless, the incorporation of hydrophilic additive played a main role in enhancing membrane antifouling performance in which the Brij-58-modified membrane recorded the highest FRR (90%), followed by PEG 400-modified membrane (72%) and pristine membrane (65%), as shown in Figure 7.5c. The superior performance of Brij-58 in comparison to PEG 400 was due to its higher molecular weight and having a hydrophobic tail which improved its interaction with PVDF membrane.

7.5 NANOFILTRATION MEMBRANES FOR RESOURCE RECOVERY

The dyeing process in the textile industry often uses a large amount of salts such as NaCl and Glauber's salt ($Na_2SO_4 \cdot 10H_2O$) to maximize exhaustion of dye molecules. Salt is necessary to drive dyes into textile during the dyeing process, acting as an electrolyte for dye migration, adsorption and fixation onto the cellulosic material. Recent paradigm shifts towards resource recovery from wastewaters [43] prompt the application of 'loose' NF membranes, aiming to recover either dyes or salts from effluent [11,44,45]. Typical NF membranes are less effective for this application due to their dense selective layer which results in high salt rejection. The fractionation process, which is known as dye desalination (removing salts from salt/dye mixtures) or dye/salt separation, could result in a high permeation of salty water while retaining high rejection of dye molecules even at low operation pressure, effectively separating dyes from salts. Table 7.3 summarizes the performances of different types of NF membranes in recovering salts from dye/salt mixture.

7.5.1 Thin-Film Composite Membrane

TFC membranes have been widely applied in NF application [61] and most recently applied in resource recovery from wastewaters. Wang et al. [13], for instance, synthesized TFC hollow fibre membrane by cross-linking PIP with TMC on the surface of PVDF hollow fibre. By optimizing the fabrication parameters, the authors were able to produce a TFC membrane with stable performance (up to 72 hours) that could effectively separate low-molecular-weight Methylene Blue (MB) from NaCl with a 9,780 NaCl/MB selectivity (2.2% NaCl and 99.99% MB rejection). Its PWP was reported to be relatively high at 10.2 L/m²·h·bar. However, it should be noted that such conventional PIP/TMC-based PA layer in TFC flat sheet configuration typically results in a selective layer that is far too dense for resource recovery (see Section 7.4). Hence, additives such as LiCl (a common pore forming agent) have been employed to loosen the NF membranes [62]. Li et al. [63] showed that by adding TA in the aqueous solution, the IP between TMC and PIP/TA mixed monomer could enlarge pores within the PA layer, leading to promising water flux of 32.57 L/m²·h·bar. It also demonstrated potential for salt/dye separation due to its extremely low NaCl rejection (2.25%) but excellent dyes rejection (>99%).

Novel active monomers were also used to develop 'loose' NF membranes [64]. By substituting typical PIP with a zwitterionic 3,3′-diamino-N-methyldipropylamine (ZDNMA) monomer, Mi et al. [65] formed a ZDNMA/TMC-based membrane with an extremely high NaCl/MYB selectivity factor, i.e., 857. The authors attributed the high NaCl passage and elevated PWP (10.67 L/m²·h·bar) of the TFC membrane to the zwitterionic features of ZDNMA. Textile wastewaters are known to contain various foulants; hence, besides high selectivity, antifouling properties are also crucial for NF membranes treating this effluent. The ZDNMA/TMC membrane recorded a higher 94.9% FRR when compared to the conventional PIP/TMC TFC membrane (87.7%) due to its enhanced hydrophilicity. Meanwhile, sulfonated polyethylenimine (SPEI) was also shown to be effective in replacing traditional PIP monomer for preparing loose NF membranes [46]. The restrained IP effect due to the sulfonic acid groups of SPEI (see Figure 7.6a) helped to create a loose selective layer with promising PWP (41.1 L/m²·h·bar) and excellent dye/salt fractionation. The developed membrane

TABLE 7.3

NF Membranes Employed in Dye/Salt Separation (Dye Desalination) of Saline Textile Wastewater

Type	Selective Layer (Nanomaterial)	Substrate	Dye Rejection (%)	Salt Rejection (%)	Salt/Dye Selectivity $\left(\alpha = \dfrac{1-R_S}{1-R_D}\right)$	Permeability (L/m²·h·bar)	Reference
TFC	PIP+TMC	PVDF	99.99% Methylene Blue	2.2% NaCl	9,780	10.2	[13]
			100% Congo Red	6.2% NaCl	9,380		
	SPEI+TMC	PSf	97% Congo Red	3.2% NaCl	31.3	41.1	[46]
	Su+TMC	PES	99.4% Congo Red	3.3% NaCl	161.2	52.4	[47]
	βCD+TMC	PES	99.0% Congo Red	10.6% NaCl	89.4	103.9	[48]
	PEI+TA	PES	99.8% Congo Red	8.4% NaCl	458	41.5	[49]
TFN	PIP/BP+TMC (GQDs)	PES	99.8% Direct Red 80	8.8% NaCl	456	59.6	[50]
	PEI (PDA-CNCs)	ENMs	99.91% Congo Red	1.14% NaCl	1,098	128.4	[51]
				~2% Na₂SO₄	~1,088		
	PDA (HGO)	PES	99.0% Direct Red 80	10.8% Na₂SO₄	89.2	120	[52]
	PIP+TMC (lipid nanovesicles)	PES	~99.8% Direct Red 23	~7% NaCl	464	20.6	[53]
	PVA+GA [MOF CuTz-1/GO]	HPAN	99.4% Congo Red	0.3% NaCl	161	40.2	[54]
LbL	PVIm-Me+PStSO₃Na	PVDF-NMe₃	99.9% Rose Bengal	~30% MgCl₂	~720	~64	[55]
				~18% NaCl	~820		
MMMs	TA+PEI	PES/Fe	98.5% Congo Red	2.4% Na₂SO₄	65	62.3	[56]
	CS+NaAlg	HPAN	98.8% Congo Red	12.1% NaCl	73	54.8	[57]
	PEI [MOF BUT-8(A)]	HPAN	99.8% Congo Red	9.4% NaCl	453	68.3	[58]
	PES (PDA-MoS₂)	Non-woven fabric	99.88% Janus Green B	1.58% NaCl	820	42	[59]
	PSf-COOH (PSf-COOK)	–	95.5% Congo Red	5.6% Na₂SO₄	21	153.8	[60]

PIP, piperazine; TMC, trimesoyl chloride; SPEI, sulfonated polyethylenimine; Su, sucrose; βCD, β-cyclodextrin; PEI, polyethylenimine; TA, tannic acid; BP, 4,4′-bipiperidine; GQDs, graphene quantum dots; PDA, polydopamine; CNC, cellulose nanocrystals; HGO, holey graphene oxide; PVA, polyvinyl alcohol; GA, glutaraldehyde; CuTz, copper-triazolate; GO, graphene oxide; PVIm-Me, poly(vinyl imidazole); PStSO3Na, poly(styrene sodium)sulfonate; CS, chitosan; NaAlg, sodium alginate; BUT-8(A), (4,8-disulfonaphthalene-2,6-dicarboxylate); MoS2, molybdenum disulfide; PSf-COOH, carboxylated polysulfone; PSf-COOK, carboxyl potassium salt polysulfone; PVDF-NMe3, cationic polyvinylidene fluoride; PSf, polysulfone; PES, polyethersulfone; ENMs, electrospun nanofibre mats; HPAN, hydrolyzed polyacrylonitrile; Fe, iron metal ion.

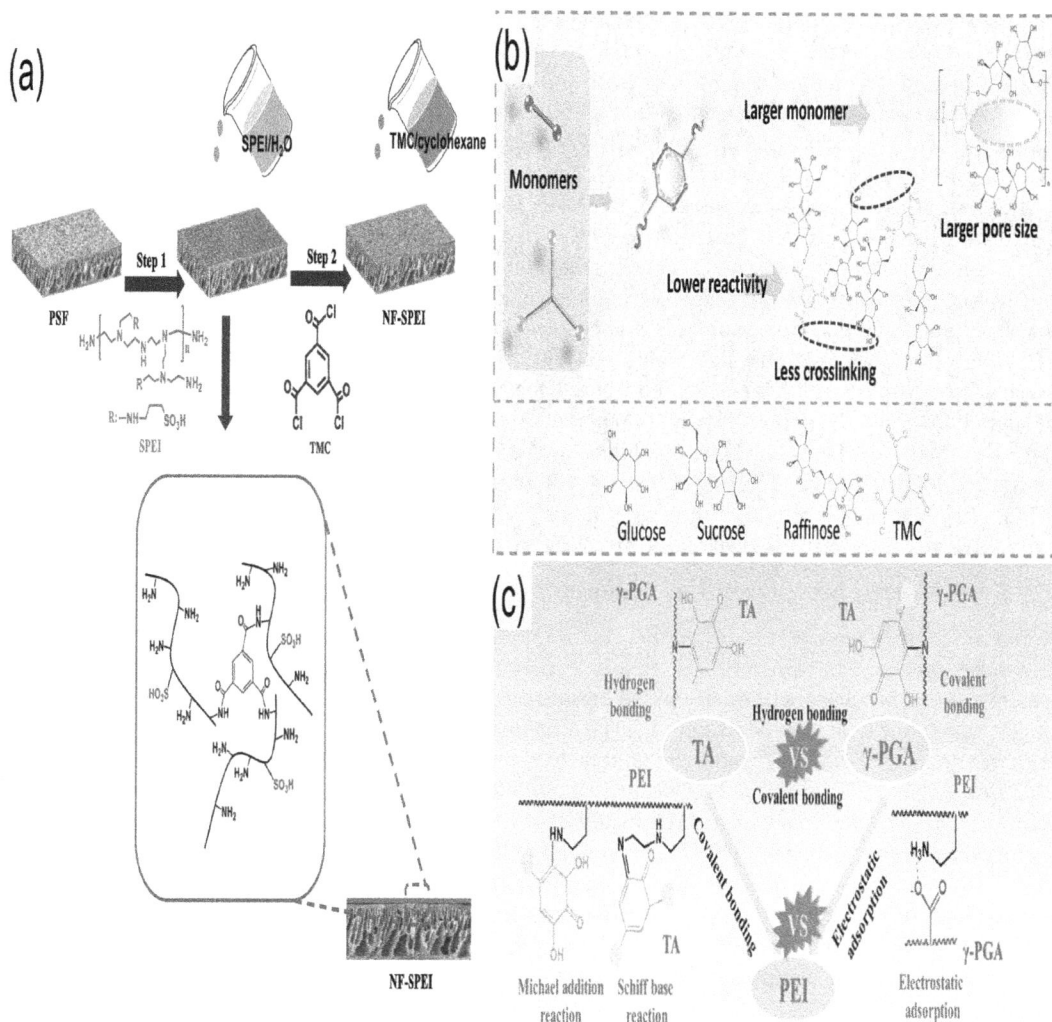

FIGURE 7.6 (a) Schematic diagram of SPEI NF membrane fabrication [46], (b) mechanism of developing loose NF membrane via IP between natural carbohydrates and TMC [47] and (c) proposed competitive reaction mechanisms of PEI, TA and γ-PGA [69].

achieved >97% dye rejections for Victoria blue B, MYB and Reactive red 120 but low salt rejections (8.4% and 3.2% for Na_2SO_4 and NaCl, respectively).

Moreover, research towards green chemistry and sustainable development has always been trending. For instance, instead of using synthetic chemicals, bio-inspired active monomers such as simple sugars [47], sugar alcohols (erythritol) [66] and oligosaccharides (β-cyclodextrin) [48,67], as well as chitosan biopolymer [68], were employed in NF membrane fabrication for dye/salt separation. Jin et al. [47] studied the feasibility of using inexpensive natural carbohydrate-derived sugars [sucrose (Su), glucose (Gl) and raffinose (Ra)] in replacing PIP monomers (see Figure 7.6b). The large size and low reactivity of these sugars produced a looser polyester structure while maintaining a competitive dye/salt separation performance, achieving an excellent PWP of 52.4 L/m²·h·bar and NaCl/CR selectivity of ~160 (for Su/TMC membrane). The authors also demonstrated that the Su/TMC membrane had extraordinary tolerance to weakly acidic conditions (4.5 pH) for up to 75 hours, even after a 10-day immersion in weakly acidic water.

Besides the employment of novel monomers/polymers, solvent-free membrane fabrication procedures are also gaining research interest. Currently, the usage of solvents in membrane fabrication (mainly as an intermediate for TMC deposition) is still prevalent, although these organic solvents have been associated with harmful side effects. Li et al. [49] developed a solvent-free green rapid coating process by cross-linking PEI with TA to form the selective layer of the TFC membrane. When tested with solution containing high salt content (60 g/L), the optimized membrane could still record a 97.6% CR rejection and 2.1% NaCl rejection, exhibiting promising potential in dye/salt separation. Further tests showed that after pre-treatment with citric acid (pH 2) or NaOH (pH 12), the membrane demonstrated stable performance by maintaining the efficiency in fractioning salt from dye/salt mixture. Building on this, Cao et al. [69] created a dual-charged membrane via a negatively charged co-deposition of TA and poly-γ-glutamic acid (γ-PGA) atop positively charged PEI layer. The loose dual-charged structure allowed for a high water permeability (36.9 L/m²·h·bar), low salt rejections (11.1% for Na_2SO_4 and 14.6% for NaCl) and high rejections to both positively and negatively charged dyes (96% methyl orange and 86% MB). The authors explained that the loosening of the selective layer could be attributed to the interactions between TA, γ-PGA and PEI that restricted the self-polymerization of TA as well as the competitive pre-reactions between TA and γ-PGA that effectively hindered their interactions with PEI (see Figure 7.6c).

7.5.2 THIN-FILM NANOCOMPOSITE MEMBRANE

Rapid research on nanomaterials and their incorporation into TFC membranes have resulted in extensive development of TFN membranes. In this section, the TFN membranes employed for textile wastewater resource recovery will be discussed based on the dimension of the nanomaterials (0D, 1D, 2D or 3D) used. For example, Metecan et al. [70] compared the chelating ability of PEI-alginate (Alg) with 0D Zn or Fe metal ions, and found that the positively charged PEI-Alg-Fe^{3+} TFN membrane with 26.3 L/m²·h·bar PWP performed the best when tested using real textile wastewater, with 92.4% colour removal and 8% conductivity (salt) removal. Other 0D nanoparticles that can be used to modify NF membrane are metal oxides (TiO_2 [71] and CuO [72]) and graphene quantum dots. Song et al. [50] have also found their way into similar applications, although these nanoparticles were generally incorporated into the selective layer via relatively facile nanoparticle deposition method.

Examples of 1D nanomaterials in TFN membranes for resource recovery application are rod-like structures such as cellulose nanocrystals (CNCs) [51,73] and Ag nanorods [74]. Yang et al. [51] fabricated TFN membranes through the surface coating of polydopamine (PDA)-modified CNCs onto the substrate, followed by PEI cross-linking. The optimized membrane showed superior dye rejection (99.91%) and PWP (128.4 L/m²·h·bar) with impressive selectivity factor for CR/NaCl mixture (1,098). Meanwhile, Istirokhatun et al. [74] manipulated the decomposition of Ag nanorods (see Figure 7.7a) to release 0D Ag nanoparticles within the PA matrix, endowing the TFN membrane with enhanced antifouling/antibiofouling properties towards model foulants while retaining high PWP of 10.4 L/m²·h·bar and <20% NaCl rejection.

In terms of 2D nanomaterials, the large surface area of these sheet-like structures enables easier functionalization and nanochannel formation. GO nanosheet is the 2D nanomaterial that is widely used in modifying membranes for water treatment processes [75,76]. However, Chen et al. [52] reported that they are often cross-linked or reduced to improve membrane stability, resulting in reduced water permeance. Hence, the authors performed chemical etching using H_2O_2 on GO to form holey GO, shortening the water diffusion path due to the formation of nanopores. The optimized membranes exhibited water permeance between 70 and 120 L/m²·h·bar with <20% Na_2SO_4 rejection and >98.5% CR rejection. When the membrane with 70 L/m²·h·bar PWP was subject to a 3-day continuous filtration test, the membrane demonstrated high stability and retained a high CR rejection (99.3%). It is interesting to note that when the etching time increased from 2 to 4 hours, the membrane PWP increased to an astounding 460 L/m²·h·bar with ~0% Na_2SO_4 rejection and 94.4% Direct Red 80 rejection. Molybdenum disulfide (MoS_2) nanosheets were also shown to be effective

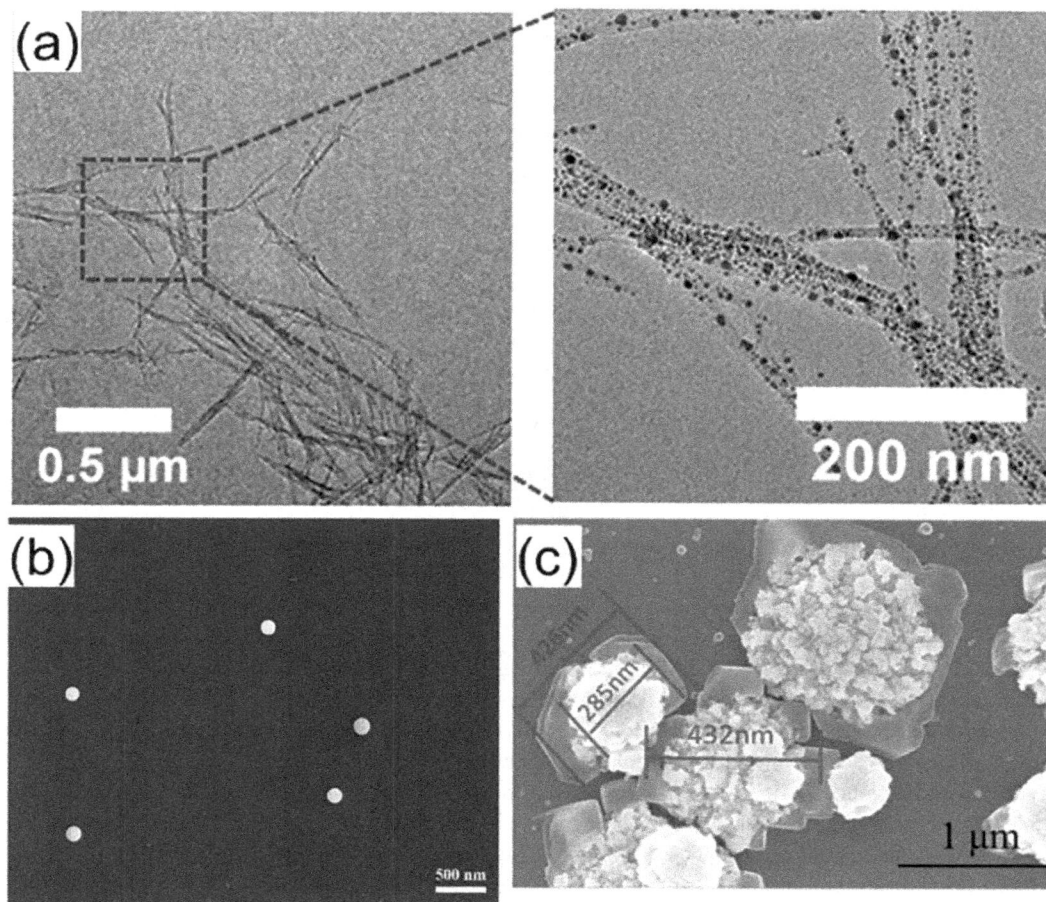

FIGURE 7.7 Morphology of different nanomaterials used for TFN membrane fabrication: (a) transmission electron microscopy (TEM) images of Ag nanorods prior to decomposition [74], (b) SEM image of lipid nanovesicles [53] and (c) SEM image of VES-AgCl [78].

in forming high-throughput TFN membrane for dye desalination [77]. This developed membrane could reach up to 262 L/m^2·h·bar PWP, 99.8% MB rejection and 12.9% Na$_2$SO$_4$ rejection.

On the other hand, 3D nanomaterials with all three dimensions (length, width and height) each exceeding 100 nm have also been incorporated into TFC membranes for dye/salt separation. Zhou et al. [53] incorporated positively charged lipid nanovesicles (~130 nm) into the PA layer (see Figure 7.7b) to tune its surface charge density and pore size for efficient separation of dye and salt at high salinity (60 g/L). In another study, Liu et al. [78] initially formed a Ag$^+$-PEI chelate complex layer on the hydrolyzed PAN substrate. Then, the Ag$^+$ from the complex layer reacted with vitamin E succinate (VES)/NaCl solution to in-situ synthesize the VES-functionalized AgCl nanoparticles. Subsequently, flower-like nanoparticles of about 430 nm were formed on the membrane surface, consisting of sphere-like amphiphilic VES aggregates on top of plate-like AgCl nanoparticles (see Figure 7.7c). The VES/AgCl nanoparticles provided additional water transport pathways, achieving 20%–50% higher permeability compared to the Ag$^+$-PEI membrane, with low rejections for both NaCl (9.6%) and Na$_2$SO$_4$ (12.8%). Since the nanoparticles were synthesized in situ, they possessed uniform distribution and good compatibility within the polymer layer, resulting in a stable performance during a 60-hour dye/salt mixture filtration.

Meanwhile, other studies integrated more than one type of nanomaterials concurrently [e.g., 3D metal-organic frameworks (MOFs) and 2D GO] into the TFN membranes to exploit the interesting properties of each nanomaterial [54]. Kang et al. [79] developed a TFN membrane with reduced nanomaterial agglomeration by incorporating magnetite decorated sulfonated-GO (SGO) nanosheets in the substrate layer. The substrate layer with SGO achieved increased porosity and hydrophilicity which aided in the formation of a loose NF membrane, while the high TFN membrane dye rejection was attributed to the high adsorption capacity of the magnetite nanoparticles. Hence, the resultant TFN membrane recorded a 44.4 $L/m^2 \cdot h \cdot bar$ PWP, 3.9% NaCl rejection and 99% CR rejection. Separately, Xu et al. [80] constructed a TFN membrane with positively charged surface (PDA/PEI) and negatively charged subsurface (GO/dopamine-modified TiO_2), with water permeance as high as 41.6 $L/m^2 \cdot h \cdot bar$ and Eriochrome black T (EBT)/Na_2SO_4 selectivity of 47.6. Diafiltration process using the best membrane showed a ~15-fold EBT concentration with Na_2SO_4 reduction to 0.36%. The results demonstrated superior dye concentration and salt-removing ability of the TFN membrane.

7.5.3 Layer-by-Layer Membrane

In general, polyelectrolyte is one of the most prevalent materials used for the fabrication of LbL membranes [55]. Joshi et al. [55] fabricated LbL NF membranes using cationic PVIm-Me and anionic $PStSO_3Na$ polyelectrolytes on the cationic PVDF-NMe_3 substrate. The membrane with five layers (2.5 bilayer) performed remarkably well with 71 $L/m^2 \cdot h \cdot bar$ water permeance and ~164 $MgCl_2$/RB selectivity factor. $MgCl_2$/RB selectivity as high as 720 could be achieved using the membrane with seven layers, but the membrane water flux was compromised (~11 $L/m^2 \cdot h \cdot bar$). Another example is the alternate self-assembly of GO and oxidized carbon nanotubes with PDDA as the connector between the two oppositely charged layers [81].

Other studies merged the LbL self-assembly approach with cross-linking mechanisms to develop a membrane with greater stability [82]. For example, Liu et al. [56] grafted TA and PEI onto the PES/Fe substrate via coordination assembly and Michael addition strategy. This approach produced a loose NF membrane with 62.3 $L/m^2 \cdot h \cdot bar$ PWP, >98% dye rejections (CR, EBT and Alcian blue 8GX) and <5% NaCl rejection. Similarly, Vatanpour and Paziresh [83] fabricated loose NF membranes through the assembly of PEI and PEI-PO_3Na polyelectrolyte coupled with glutaraldehyde cross-linking. It is interesting to note that natural biopolymers such as chitosan (CS) and sodium alginate (NaAlg) have been used in such cross-linking-assisted LbL self-assembly method [57]. The authors used CS and NaAlg as the cationic and anionic polyelectrolytes, respectively, while $CaCl_2$ was used as the ionic cross-linker based on the chelating ability of Ca^{2+} ions with the biopolymers (see Figure 7.8a). These findings revealed the potential of LbL membranes for effectively separating dyes and salts from saline textile wastewater for resource recovery.

7.5.4 Asymmetric Mixed Matrix Membranes

Asymmetric mixed matrix membranes (MMMs) are known for their simple fabrication procedure based on phase separation process. However, the inherently dense MMMs prompt researchers to incorporate additives such as porous MOFs and covalent-organic frameworks into the membrane, especially for saline wastewater recovery application which requires high water and salt permeation [84,85]. The incorporation of hydrolytic-stable BUT-8(A) (4,8-disulfonaphthalene-2,6-dicarboxylate) MOF in the PEI matrix has been shown to be capable of producing a MMM with 68.3 $L/m^2 \cdot h \cdot bar$ water permeability, 99.8% CR rejection and >90% NaCl permeation [58]. Such high water permeance was attributed to the BUT-8(A) channels, which facilitated molecule transport. The authors also reported a thin and uniform active layer (~140 nm) as a result of excellent MOF dispersibility in water (see Figure 7.8b) and good MOF-polymer compatibility.

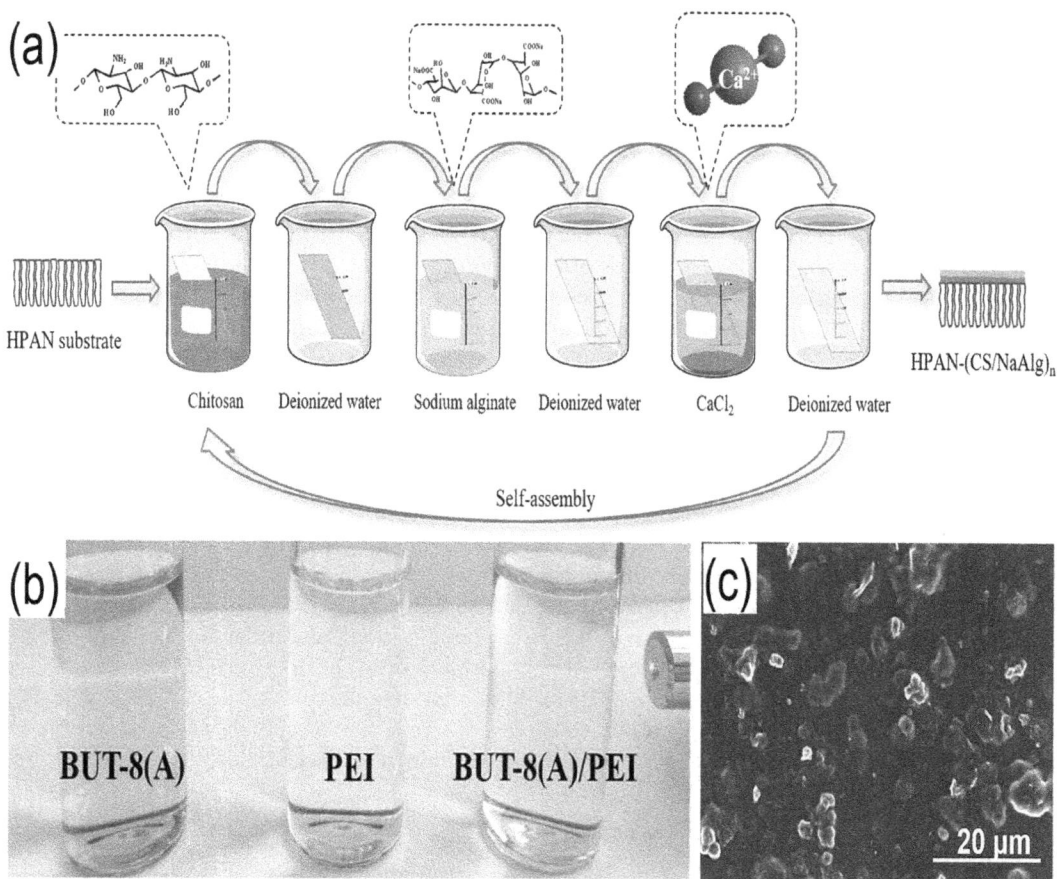

FIGURE 7.8 (a) Schematic diagram of CS/NaAlg LbL membrane fabrication [57], (b) Tyndall effect showing the good dispersibility of BUT-8(A) MOFs in aqueous solution [58] and (c) SEM image of MMMs incorporated with mesoporous silica thin films [86].

Besides organic frameworks, sheet-like additives could also produce nanochannels within the MMM which allow water and salt permeation while maintaining high dye rejections. For example, Lai et al. [86] and Lessan et al. [87], respectively, incorporated mesoporous silica thin films (see Figure 7.8c) and cellulose nanosheets into their MMMs to enhance membrane dye desalination performance. Besides enhancing dye rejection, the addition of PDA-modified MoS_2 nanosheets in membrane matrix improved MMM's stability in alkaline conditions (pH 11). This modified membrane achieved higher Janus Green B rejection (88.6%) compared to the pristine membrane (77.6%) [59]. Other additives such as carboxyl potassium salt-polysulfone (PSf-COOK) [60], PDA particles [88] and wood-based biochar [89] were also shown to improve MMMs dye/salt separation ability.

Liu et al. [60] tested the dye/salt separation performance of MMMs fabricated with different ratios of PSf-COOK60% additive and carboxylated PSf (PSf-COOH60%). Since –COOK was easily ionized compared to –COOH, the MMM with a higher additive ratio exhibited the highest hydrophilicity and electronegativity (~−30 mV). As a result, the optimum membrane offered ~154 $L/m^2 \cdot h \cdot bar$ PWP, 95.5% CR rejection and organic salt (NaCl and Na_2SO_4) rejection as low as 5.6%. Ang et al. [88] added 0.5% PDA particles (~156 nm diameter) into 19% PES solution, boosting the pure water flux of the MMM without altering Na_2SO_4 (<10%) and MB rejection (99.9%). Compared to the 17% PES solution, the presence of PDA particles in high PES concentration (19%) resulted in a lower total solid content. This increased the membrane porosity, leading to an enhanced water flux (~3.5 folds). However, when PDA particles increased above 0.5%, the obstruction of the PES pores led to a decreased water permeability.

7.6 CONCLUSION

Since the 1980s, NF membranes have experienced massive development and have emerged as one of the most promising alternative membrane-based technologies for water and wastewater treatment. With the current progress, NF membranes offer numerous advantages over conventional methods for the textile wastewater treatment. The growing demand of freshwater coupled with the need of resource recovery calls for water reclamation and salt/dye recovery from the water-intensive textile industries. The unique characteristic of NF membranes that allows selective separation of targeted species has proven to be useful for water reuse and resource recovery.

This chapter reviewed various types of NF membranes, classified according to TFC/TFN membranes, LbL membrane and asymmetric membrane made of phase inversion technique. Our analysis revealed that most of NF membranes were reported to have great permeability and selectivity for textile wastewater treatment. With a simple fabrication process, asymmetric NF membranes have demonstrated vast potential in water reclamation by exhibiting high permeate flux. NF membrane surface charge tailoring by LbL deposition also has a profound significance in textile wastewater treatment as fabricated membrane displayed enhanced fouling resistance. Meanwhile, for the past three decades, the use of TFC membranes has been found to dominate the industrial applications of NF in textile industry. The incorporation of nanoparticles into TFC to produce TFN membranes opens a new chapter for NF technology, as it is able to overcome the drawbacks of typical selective layer and has huge potential in applications involving water reuse and separation of resources from textile wastewater. Currently, there is one membrane manufacturer producing TFN membranes for desalination process.

REFERENCES

1. Kishor R, Purchase D, Saratale GD, Saratale RG, Ferreira LFR, Bilal M, et al. Ecotoxicological and health concerns of persistent coloring pollutants of textile industry wastewater and treatment approaches for environmental safety. *J Environ Chem Eng.* 2021;9(2):105012.
2. Niinimäki K, Peters G, Dahlbo H, Perry P, Rissanen T, Gwilt A. The environmental price of fast fashion. *Nat Rev Earth Environ [Internet].* 2020;1(4):189–200. Available from: https://doi.org/10.1038/s43017-020-0039-9.
3. Jegatheesan V, Pramanik BK, Chen J, Navaratna D, Chang CY, Shu L. Treatment of textile wastewater with membrane bioreactor: A critical review. *Bioresour Technol.* 2016;204:202–12.
4. Li K, Liu Q, Fang F, Wu X, Xin J, Sun S, et al. Influence of nanofiltration concentrate recirculation on performance and economic feasibility of a pilot-scale membrane bioreactor-nanofiltration hybrid process for textile wastewater treatment with high water recovery. *J Clean Prod.* 2020;261:121067.
5. Senthil Kumar P, Grace Pavithra K. Water and textiles. *Water Text Fash.* 2019;21–40.
6. Peters G, Granberg H, Sweet S. The role of science and technology in sustainable fashion. 2014.
7. Evich MG, Davis MJB, McCord JP, Acrey B, Awkerman JA, Knappe DRU, et al. Per- and polyfluoroalkyl substances in the environment. *Science (80-) [Internet].* 2022;375(6580):eabg9065. Available from: https://doi.org/10.1126/science.abg9065.
8. Posner S, Jönsson C. *Chemicals in Textiles – Risks to Human Health and the Environment : Report from a Government Assignment.* Swedish Chemicals Agency; 2014.
9. de Oliveira Neto GC, Ferreira Correia JM, Silva PC, de Oliveira Sanches AG, Lucato WC. Cleaner production in the textile industry and its relationship to sustainable development goals. *J Clean Prod.* 2019;228:1514–25.
10. Cesar da Silva P, Cardoso de Oliveira Neto G, Ferreira Correia JM, Pujol Tucci HN. Evaluation of economic, environmental and operational performance of the adoption of cleaner production: Survey in large textile industries. *J Clean Prod.* 2021;278:123855.

11. Guo S, Wan Y, Chen X, Luo J. Loose nanofiltration membrane custom-tailored for resource recovery. *Chem Eng J*. 2021;409:127376.

12. Soyekwo F, Liu C, Wen H, Hu Y. Construction of an electroneutral zinc incorporated polymer network nanocomposite membrane with enhanced selectivity for salt/dye separation. *Chem Eng J*. 2020;380:122560.

13. Wang C, Chen Y, Hu X, Feng X. In-situ synthesis of PA/PVDF composite hollow fiber membranes with an outer selective structure for efficient fractionation of low-molecular-weight dyes-salts. *Desalination*. 2021;503:114957.

14. Goh PS, Wong KC, Ismail AF. Membrane technology: A versatile tool for saline wastewater treatment and resource recovery. *Desalination*. 2022;521:115377.

15. Kiran S, Adeel S, Nosheen S, Hassan A, Usman M, Rafique MA. Recent Trends in Textile Effluent Treatments: A Review [Internet]. *Adv Mater Wastewater Treat*. 2017:29–49. (Wiley Online Books). Available from: https://doi.org/10.1002/9781119407805.ch2.

16. Madhav S, Ahamad A, Singh P, Mishra PK. A review of textile industry: Wet processing, environmental impacts, and effluent treatment methods. *Environ Qual Manag [Internet]*. 2018;27(3):31–41. Available from: https://doi.org/10.1002/tqem.21538.

17. Stone C, Windsor FM, Munday M, Durance I. Natural or synthetic – how global trends in textile usage threaten freshwater environments. *Sci Total Environ*. 2020;718:134689.

18. Yaseen DA, Scholz M. Textile dye wastewater characteristics and constituents of synthetic effluents: A critical review. *Int J Environ Sci Technol [Internet]*. 2019;16(2):1193–226. Available from: https://doi.org/10.1007/s13762-018-2130-z.

19. Bhatia D, Sharma NR, Singh J, Kanwar RS. Biological methods for textile dye removal from wastewater: A review. *Crit Rev Environ Sci Technol [Internet]*. 2017;47(19):1836–76. Available from: https://doi.org/10.1080/10643389.2017.1393263.

20. Hassan MM, Carr CM. A critical review on recent advancements of the removal of reactive dyes from dyehouse effluent by ion-exchange adsorbents. *Chemosphere*. 2018;209:201–19.

21. Qamar SA, Ashiq M, Jahangeer M, Riasat A, Bilal M. Chitosan-based hybrid materials as adsorbents for textile dyes–A review. *Case Stud Chem Environ Eng*. 2020;2:100021.

22. Mirbolooki H, Amirnezhad R, Pendashteh AR. Treatment of high saline textile wastewater by activated sludge microorganisms. *J Appl Res Technol*. 2017;15(2):167–72.

23. Liu M, Chen Q, Lu K, Huang W, Lü Z, Zhou C, et al. High efficient removal of dyes from aqueous solution through nanofiltration using diethanolamine-modified polyamide thin-film composite membrane. *Sep Purif Technol*. 2017;173:135–43.

24. Wang T, He X, Li Y, Li J. Novel poly(piperazine-amide) (PA) nanofiltration membrane based poly(m-phenylene isophthalamide) (PMIA) hollow fiber substrate for treatment of dye solutions. *Chem Eng J*. 2018;351:1013–26.

25. Lü Z, Hu F, Li H, Zhang X, Yu S, Liu M, et al. Composite nanofiltration membrane with asymmetric selective separation layer for enhanced separation efficiency to anionic dye aqueous solution. *J Hazard Mater*. 2019;368:436–43.

26. Liu Y, Zhang S, Zhou Z, Ren J, Geng Z, Luan J, et al. Novel sulfonated thin-film composite nanofiltration membranes with improved water flux for treatment of dye solutions. *J Memb Sci [Internet]*. 2012 [cited 2022 Mar 28];394–395:218–29. Available from: https://linkinghub.elsevier.com/retrieve/pii/S037673881100963X.

27. Ormanci-Acar T, Celebi F, Keskin B, Mutlu-Salmanlı O, Agtas M, Turken T, et al. Fabrication and characterization of temperature and pH resistant thin film nanocomposite membranes embedded with halloysite nanotubes for dye rejection. *Desalination*. 2018;429:20–32.

28. Lai GS, Lau WJ, Goh PS, Ismail AF, Tan YH, Chong CY, et al. Tailor-made thin film nanocomposite membrane incorporated with graphene oxide using novel interfacial polymerization technique for enhanced water separation. *Chem Eng J [Internet]*. 2018 [cited 2021 Oct 18];344:524–34. Available from: https://linkinghub.elsevier.com/retrieve/pii/S1385894718304741.

29. Ghaemi N, Safari P. Nano-porous SAPO-34 enhanced thin-film nanocomposite polymeric membrane: Simultaneously high water permeation and complete removal of cationic/anionic dyes from water. *J Hazard Mater*. 2018;358:376–88.

30. Wu X, Yang L, Meng F, Shao W, Liu X, Li M. ZIF-8-incorporated thin-film nanocomposite (TFN) nanofiltration membranes: Importance of particle deposition methods on structure and performance. *J Memb Sci*. 2021;632:119356.

31. Wang L, Fang M, Liu J, He J, Li J, Lei J. Layer-by-layer fabrication of high-performance polyamide/ZIF-8 nanocomposite membrane for nanofiltration applications. *ACS Appl Mater Interfaces [Internet]*. 2015;7(43):24082–93. Available from: https://doi.org/10.1021/acsami.5b07128.

32. Ahmadian-Alam L, Mahdavi H, Mousavi Davijani SM. Influence of structurally and morphologically different nanofillers on the performance of polysulfone membranes modified by the assembled PDDA/PAMPS-based hybrid multilayer thin film. *J Environ Manage*. 2021;300:113809.

33. Lin Z, Zhang Q, Qu Y, Chen M, Soyekwo F, Lin C, et al. LBL assembled polyelectrolyte nanofiltration membranes with tunable surface charges and high permeation by employing a nanosheet sacrificial layer. *J Mater Chem A [Internet]*. 2017;5(28):14819–27. Available from: http://dx.doi.org/10.1039/C7TA03183A.

34. Guo D, Xiao Y, Li T, Zhou Q, Shen L, Li R, et al. Fabrication of high-performance composite nanofiltration membranes for dye wastewater treatment: Mussel-inspired layer-by-layer self-assembly. *J Colloid Interface Sci [Internet]*. 2020 [cited 2022 Mar 27];560:273–83. Available from: https://linkinghub.elsevier.com/retrieve/pii/S0021979719312664.

35. Abdi G, Alizadeh A, Zinadini S, Moradi G. Removal of dye and heavy metal ion using a novel synthetic polyethersulfone nanofiltration membrane modified by magnetic graphene oxide/metformin hybrid. *J Memb Sci*. 2018;552:326–35.

36. Zhao C, Yang B, Han J, Meng Y, Yu L, Hou D, et al. Preparation of carboxylic multiwalled-carbon-nanotube–modified poly(m-phenylene isophthalamide) hollow fiber nanofiltration membranes with improved performance and application for dye removal. *Appl Surf Sci*. 2018;453:502–12.

37. Gholami S, Llacuna JL, Vatanpour V, Dehqan A, Paziresh S, Cortina JL. Impact of a new functionalization of multiwalled carbon nanotubes on antifouling and permeability of PVDF nanocomposite membranes for dye wastewater treatment. *Chemosphere*. 2022;294:133699.

38. Nikooe N, Saljoughi E. Preparation and characterization of novel PVDF nanofiltration membranes with hydrophilic property for filtration of dye aqueous solution. *Appl Surf Sci*. 2017;413:41–9.

39. Lü Z, Hu F, Li H, Zhang X, Yu S, Liu M, et al. Composite nanofiltration membrane with asymmetric selective separation layer for enhanced separation efficiency to anionic dye aqueous solution. *J Hazard Mater*. 2019;368:436–43.

40. Jeong BH, Hoek E, Yan Y, Subramani A, Huang X, Hurwitz G, et al. Interfacial polymerization of thin film nanocomposites: A new concept for reverse osmosis membranes. *J Memb Sci*. 2007;294:1–7.

41. Khoo YS, Lau WJ, Liang YY, Yusof N, Fauzi Ismail A. Surface modification of PA layer of TFC membranes: Does it effective for performance Improvement? *J Ind Eng Chem*. 2021;102:271–92.

42. Wang L, Fang M, Liu J, He J, Li J, Lei J. Layer-by-layer fabrication of high-performance polyamide/ZIF-8 nanocomposite membrane for nanofiltration applications. *ACS Appl Mater Interfaces [Internet]*. 2015;7(43):24082–93. Available from: https://doi.org/10.1021/acsami.5b07128.

43. Goh PS, Wong KC, Ismail AF. Membrane technology: A versatile tool for saline wastewater treatment and resource recovery. *Desalination*. 2022;521:115377.

44. Ahmad NNR, Ang WL, Teow YH, Mohammad AW, Hilal N. Nanofiltration membrane processes for water recycling, reuse and product recovery within various industries: A review. *J Water Process Eng*. 2022;45:102478.

45. Feng X, Peng D, Zhu J, Wang Y, Zhang Y. Recent advances of loose nanofiltration membranes for dye/salt separation. *Sep Purif Technol*. 2022;285:120228.

46. Ding J, Wu H, Wu P. Preparation of highly permeable loose nanofiltration membranes using sulfonated polyethylenimine for effective dye/salt fractionation. *Chem Eng J*. 2020;396:125199.

47. Jin P, Chergaoui S, Zheng J, Volodine A, Zhang X, Liu Z, et al. Low-pressure highly permeable polyester loose nanofiltration membranes tailored by natural carbohydrates for effective dye/salt fractionation. *J Hazard Mater*. 2022;421:126716.

48. Liu L, Yu L, Borjigin B, Liu Q, Zhao C, Hou D. Fabrication of thin-film composite nanofiltration membranes with improved performance using β-cyclodextrin as monomer for efficient separation of dye/salt mixtures. *Appl Surf Sci*. 2021;539:148284.

49. Li Q, Liao Z, Fang X, Wang D, Xie J, Sun X, et al. Tannic acid-polyethyleneimine crosslinked loose nanofiltration membrane for dye/salt mixture separation. *J Memb Sci*. 2019;584:324–32.

50. Song Y, Sun Y, Zhang N, Li C, Hou M, Chen K, et al. Custom-tailoring loose nanocomposite membrane incorporated bipiperidine/graphene quantum dots for high-efficient dye/salt fractionation in hairwork dyeing effluent. *Sep Purif Technol*. 2021;271:118870.

51. Yang L, Liu X, Zhang X, Chen T, Ye Z, Rahaman MS. High performance nanocomposite nanofiltration membranes with polydopamine-modified cellulose nanocrystals for efficient dye/salt separation. *Desalination*. 2022;521:115385.

52. Chen X, Deng E, Lin X, Manoj Tandel A, Rub D, Zhu L, et al. Engineering hierarchical nanochannels in graphene oxide membranes by etching and polydopamine intercalation for highly efficient dye recovery. *Chem Eng J*. 2022;433:133593.

53. Zhou H, Li X, Li Y, Dai R, Wang Z. Tuning of nanofiltration membrane by multifunctionalized nanovesicles to enable an ultrahigh dye/salt separation at high salinity. *J Memb Sci*. 2022;644:120094.

54. Zhou S, Feng X, Zhu J, Song Q, Yang G, Zhang Y, et al. Self-cleaning loose nanofiltration membranes enabled by photocatalytic Cu-triazolate MOFs for dye/salt separation. *J Memb Sci*. 2021;623:119058.

55. Joshi US, Bhalani D V., Chaudhary A, Jewrajka SK. Multipurpose tight ultrafiltration membrane through controlled layer-by-layer assembly for low pressure molecular separation. *J Memb Sci*. 2022;641:119908.

56. Liu S, Fang X, Lou M, Qi Y, Li R, Chen G, et al. Construction of loose positively charged NF membrane by layer-by-layer grafting of polyphenol and polyethyleneimine on the PES/Fe Substrate for dye/salt separation. *Membranes (Basel) [Internet]*. 2021;11(9). Available from: https://www.mdpi.com/2077-0375/11/9/699.

57. Wang X, Dong S, Qin W, Xue Y, Wang Q, Zhang J, et al. Fabrication of highly permeable CS/NaAlg loose nanofiltration membrane by ionic crosslinking assisted layer-by-layer self-assembly for dye desalination. *Sep Purif Technol*. 2022;284:120202.

58. Meng Y, Shu L, Liu L, Wu Y, Xie LH, Zhao MJ, et al. A high-flux mixed matrix nanofiltration membrane with highly water-dispersible MOF crystallites as filler. *J Memb Sci*. 2019;591:117360.

59. Tian H, Wu X, Zhang K. Polydopamine-assisted two-dimensional molybdenum disulfide (MoS2)-modified PES tight ultrafiltration mixed-matrix membranes: Enhanced dye separation performance. *Membranes (Basel) [Internet]*. 2021;11(2). Available from: https://www.mdpi.com/2077-0375/11/2/96.

60. Liu Z, Wang L, Mi Z, Jin S, Wang D, Zhao X, et al. A carboxyl potassium salt polysulfone (PSF-COOK)-embedded mixed matrix membrane with high permeability and anti-fouling properties for the effective separation of dyes and salts. *Appl Surf Sci*. 2019;490:7–17.

61. Lau WJ, Ismail AF, Misdan N, Kassim MA. A recent progress in thin film composite membrane: A review. *Desalination*. 2012;287:190–9.

62. Wang C, Chen Y, Hu X, Guo P. Scalable dual-layer PVDF loose nanofiltration hollow fiber membranes for treating textile wastewater. *J Water Process Eng*. 2022;46:102579.

63. Li Q, Liao Z, Fang X, Xie J, Ni L, Wang D, et al. Tannic acid assisted interfacial polymerization based loose thin-film composite NF membrane for dye/salt separation. *Desalination*. 2020;479:114343.

64. Zheng J, Zhao R, Uliana AA, Liu Y, de Donnea D, Zhang X, et al. Separation of textile wastewater using a highly permeable resveratrol-based loose nanofiltration membrane with excellent anti-fouling performance. *Chem Eng J*. 2022;434:134705.

65. Mi YF, Xu G, Guo YS, Wu B, An QF. Development of antifouling nanofiltration membrane with zwitterionic functionalized monomer for efficient dye/salt selective separation. *J Memb Sci*. 2020;601:117795.

66. Jin P, Zhu J, Yuan S, Zhang G, Volodine A, Tian M, et al. Erythritol-based polyester loose nanofiltration membrane with fast water transport for efficient dye/salt separation. *Chem Eng J*. 2021;406:126796.

67. Li J, Gong JL, Zeng GM, Song B, Cao WC, Fang SY, et al. Thin-film composite polyester nanofiltration membrane with high flux and efficient dye/salts separation fabricated from precise molecular sieving structure of β-cyclodextrin. *Sep Purif Technol*. 2021;276:119352.

68. Halakarni M, Mahto A, Aruchamy K, Mondal D, Nataraj SK. Developing helical carbon functionalized chitosan-based loose nanofiltration membranes for selective separation and wastewater treatment. *Chem Eng J*. 2021;417:127911.

69. Cao Y, Zhang H, Guo S, Luo J, Wan Y. A robust dually charged membrane prepared via catechol-amine chemistry for highly efficient dye/salt separation. *J Memb Sci*. 2021;629:119287.

70. Metecan A, Cihanoğlu A, Alsoy Altinkaya S. A positively charged loose nanofiltration membrane fabricated through complexing of alginate and polyethyleneimine with metal ions on the polyamideimide support for dye desalination. *Chem Eng J*. 2021;416:128946.

71. Yan XY, Wang Q, Wang Y, Fu ZJ, Wang ZY, Mamba B, et al. Designing durable self-cleaning nanofiltration membranes via sol-gel assisted interfacial polymerization for textile wastewater treatment. *Sep Purif Technol*. 2022;289:120752.

72. Waheed A, Baig U, Ansari MA. Fabrication of CuO nanoparticles immobilized nanofiltration composite membrane for dye/salt fractionation: Performance and antibiofouling. *J Environ Chem Eng*. 2022;10(1):106960.

73. Bai L, Liu Y, Ding A, Ren N, Li G, Liang H. Fabrication and characterization of thin-film composite (TFC) nanofiltration membranes incorporated with cellulose nanocrystals (CNCs) for enhanced desalination performance and dye removal. *Chem Eng J.* 2019;358:1519–28.

74. Istirokhatun T, Lin Y, Shen Q, Guan K, Wang S, Matsuyama H. Ag-based nanocapsule-regulated interfacial polymerization Enables synchronous nanostructure towards high-performance nanofiltration membrane for sustainable water remediation. *J Memb Sci.* 2022;645:120196.

75. Zhang H, Bin Li, Pan J, Qi Y, Shen J, Gao C, et al. Carboxyl-functionalized graphene oxide polyamide nanofiltration membrane for desalination of dye solutions containing monovalent salt. *J Memb Sci.* 2017;539:128–37.

76. Seah MQ, Lau WJ, Goh PS, Ismail AF. Greener synthesis of functionalized-GO incorporated TFN NF membrane for potential recovery of saline water from salt/dye mixed solution. *Desalination [Internet].* 2022 Feb 1 [cited 2021 Nov 14];523:115403. Available from: https://linkinghub.elsevier.com/retrieve/pii/S0011916421004744.

77. Zhang M, Gao J, Liu G, Zhang M, Liu H, Zhou L, et al. High-throughput Zwitterion-modified MoS$_2$ membranes: Preparation and application in dye desalination. *Langmuir [Internet].* 2021;37(1):417–27. Available from: https://doi.org/10.1021/acs.langmuir.0c03068.

78. Liu S, Wang Z, Ban M, Song P, Song X, Khan B. Chelation–assisted in situ self-assembly route to prepare the loose PAN–based nanocomposite membrane for dye desalination. *J Memb Sci.* 2018;566:168–80.

79. Kang Y, Jang J, Lee Y, Kim IS. Dye adsorptive thin-film composite membrane with magnetite decorated sulfonated graphene oxide for efficient dye/salt mixture separation. *Desalination [Internet].* 2022 [cited 2021 Nov 29];524:115462. Available from: https://linkinghub.elsevier.com/retrieve/pii/S0011916421005336.

80. Xu Y, Peng G, Liao J, Shen J, Gao C. Preparation of molecular selective GO/DTiO2-PDA-PEI composite nanofiltration membrane for highly pure dye separation. *J Memb Sci.* 2020;601:117727.

81. Liu L, Kang H, Wang W, Xu Z, Mai W, Li J, et al. Layer-by-layer self-assembly of polycation/GO/OCNTs nanofiltration membrane with enhanced stability and flux. *J Mater Sci.* 2018;53.

82. Yang Y, Lan Q, Wang Y. Gradient nanoporous phenolics as substrates for high-flux nanofiltration membranes by layer-by-layer assembly of polyelectrolytes. *Chinese J Chem Eng.* 2020;28(1):114–21.

83. Xiong S, Han C, Phommachanh A, Li W, Xu S, Wang Y. High-performance loose nanofiltration membrane prepared with assembly of covalently cross-linked polyethyleneimine-based polyelectrolytes for textile wastewater treatment. *Sep Purif Technol.* 2021;274:119105.

84. Vatanpour V, Paziresh S. A melamine-based covalent organic framework nanomaterial as a nanofiller in polyethersulfone mixed matrix membranes to improve separation and antifouling performance. *J Appl Polym Sci [Internet].* 2022;139(1):51428. Available from: https://doi.org/10.1002/app.51428.

85. Ruan H, Guo C, Yu H, Shen J, Gao C, Sotto A, et al. Fabrication of a MIL-53(Al) nanocomposite membrane and potential application in desalination of dye solutions. *Ind Eng Chem Res [Internet].* 2016;55(46):12099–110. Available from: https://doi.org/10.1021/acs.iecr.6b03201.

86. Lai YS, Yang J, Mou CY. Mesoporous silica thin films incorporated chitosan mixed matrix nanofiltration membranes for textile wastewater treatment. *J Chinese Chem Soc [Internet].* 2021;68(3):451–61. Available from: https://doi.org/10.1002/jccs.202000495.

87. Lessan F, Karimi M, Arami M. Tailoring the hierarchical porous structure within polyethersulfone/cellulose nanosheets mixed matrix membrane to achieve efficient dye/salt mixture fractionation. *J Polym Res [Internet].* 2016;23(9):171. Available from: https://doi.org/10.1007/s10965-016-1034–1.

88. Ang MBMY, Maganto HLC, Macni CRM, Caparanga AR, Huang SH, Lee KR, et al. Effect of introducing varying amounts of polydopamine particles into different concentrations of polyethersulfone solution on the performance of resultant mixed-matrix membranes intended for dye separation. *J Polym Res [Internet].* 2020;27(8):196. Available from: https://doi.org/10.1007/s10965-020-02174–6.

89. Gu S, Li L, Liu F, Li J. Biochar/Kevlar nanofiber mixed matrix nanofiltration membranes with enhanced dye/salt separation performance. *Membranes (Basel) [Internet].* 2021;11(6). Available from: https://www.mdpi.com/2077-0375/11/6/443.

8 The Application of Nanofiltration for Water Reuse in the Hybrid Nanofiltration-Forward Osmosis Process

Zhiyuan Zong, Nick Hankins, and Fozia Parveen
University of Oxford

CONTENTS

8.1 INTRODUCTION

With global economic development and an increase in population, the demand on water resources is increasing dramatically. The shortage of freshwater resources, which contribute only 2.5% of the world's total water, and the resource inefficiency of wastewater treatment have led to an increasing scarcity of utilizable water resources [1]. To this end, efforts are ongoing to try and utilize ocean water resources and to optimize the reuse of wastewater. Regardless of the direction in water treatment research, researchers are focused on trying to improve its efficiency and sustainability in order to obtain potable clean water with reduced energy and chemical consumption.

In the past, the primary way to treat raw freshwater was to filter it coarsely and to add chemicals to remove pathogens and harmful substances. However, society has become more skeptical of chemical water treatment, especially in view of the environmental consequences and the fact that outbreaks of disease can still occur. In the meantime, the use of membrane technology has gradually assumed a greater importance in mainstream water treatment [2].

The nanofiltration (NF) membrane is one of the most reliable membrane technologies to emerge for water processing since the 1980s. The membrane typically exhibits a pore size around 1 nm and is able to separate out multivalent inorganic salts and low- to medium- molecular-weight organic molecules [3]. However, there are a few drawbacks and limitations worthy of note in terms of

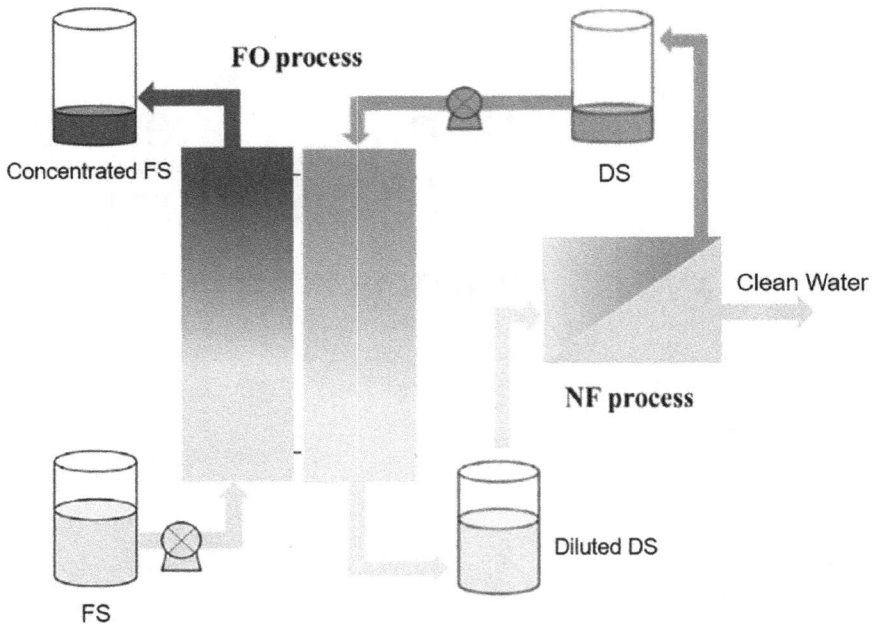

FIGURE 8.1   Schematic of a hybridized and continuous NF-FO process. (Adapted from Ref. [5].)

insufficient separation for species with a wide size distribution, lower efficiency for the rejection of low-concentration individual compounds, and so forth [4]. As an existing mature technology, NF offers a great potential for hybridization with other water processing technologies to act as a downstream treatment technology and produce a better treatment performance. Among all the possibilities for hybridization, forward osmosis (FO) has shown a particularly good potential, especially for seawater desalination and wastewater treatment.

FO holds many unique advantages, such as low energy consumption, efficient contaminant separation, and low materials cost. Unlike reverse osmosis, FO benefits directly from the natural osmotic driving force. By utilizing an osmotic pressure difference, pure water can reach the draw solution (DS) side from the feed solution (FS) side through a selectively permeable membrane. In order to acquire clean water from the diluted DS, a membrane-based technology among others can be applied. In particular, NF is qualified as an alternative process to separate and eventually recycle the DS. Figure 8.1 shows the schematic of a hybrid NF-FO system [5].

This combined system includes a conventional FO unit as the upstream treatment process, and an NF membrane system as a downstream post-treatment process, to enable draw water circulation, extract clean water, and enhance sustainability. The filtered, concentrated DS from the NF cell goes back to the FO system for reuse; the molecular shape and weight of the draw solute has been selected such that it can be separated by NF, such as a multivalent salt or a polyelectrolyte. The clean water permeate from the NF cells could now be used for specific purposes. In this fashion, NF is considered as a suitable candidate for concentrating the diluted DS. With an appropriate choice of draw solute, the total dissolved solids (TDS) in the produced clean water can reach a level lower than 1,000 mg/L, which satisfies the drinking water standard for most countries in the world [5].

A real hybrid cell is usually more complex than the block flow diagram shown in Figure 8.1. Figure 8.2 shows the details of a combination of NF and FO with a more complex unit design, for the treatment of microfiltered coke wastewater. The applied pressure in the NF section within this hybrid system is usually over 20 bars. In order to satisfy this demand, some pressure-boosting devices are usually required, such as high-pressure pumps. A pH probe will be applied in the DS tank to ensure that the pH of the dilute/concentrated DS lies within a suitable range and is not

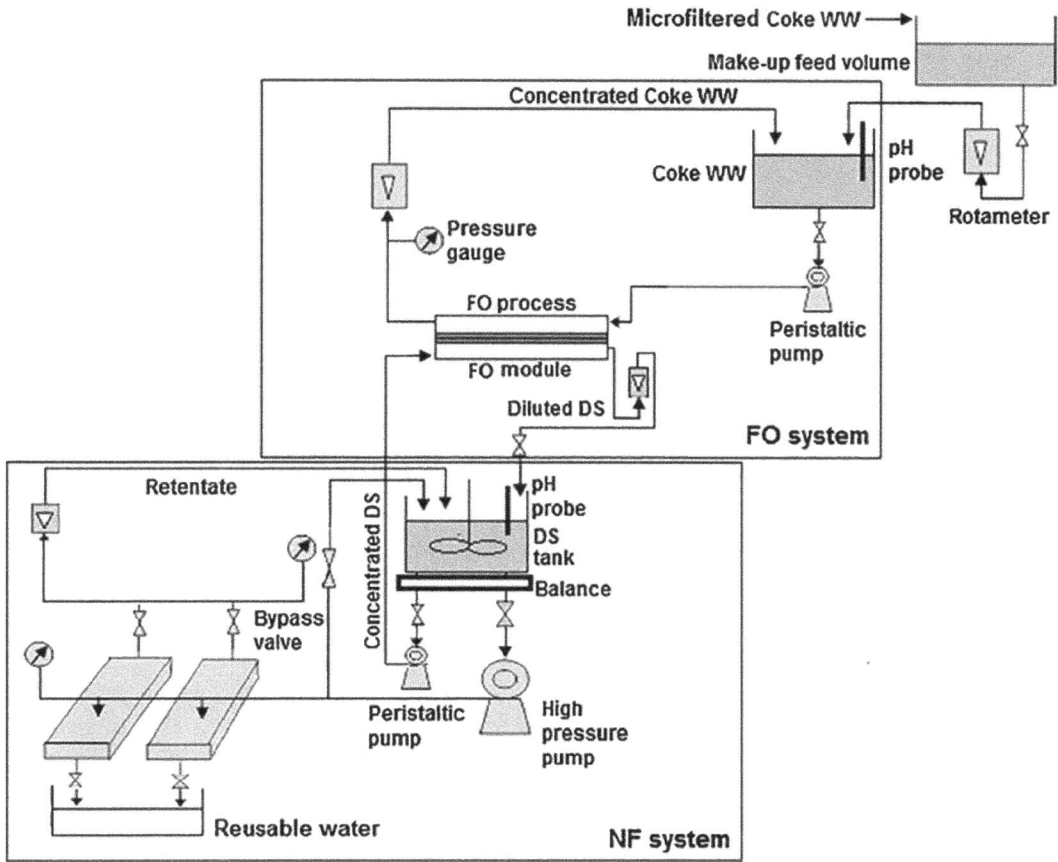

FIGURE 8.2 A detailed schematic diagram for a hybrid NF-FO system in coke-oven wastewater treatment. (Adapted from Ref. [7].)

potentially damaging to the NF membrane. An optimal hybrid NF-FO system could provide a water recovery rate of over 85% from the feed wastewater [6].

Another possible hybrid NF-FO configuration is when the NF membrane is itself incorporated into the FO system. The semi-batch, co-current process shown in Figure 8.3 involves only one circulation loop. The membrane used in the FO cell is a hollow fiber NF membrane instead of a conventional FO membrane, of which the latter tends to be close in nature to an RO membrane. The entire process is acting as an FO process with a circulating DS. An effective NF membrane could provide a reasonable water flux and exceptional draw solute selectivity from its tightly distributed narrow pore sizes. In this manner, the NF membrane could also be an excellent candidate for an FO membrane applied in the FO module.

In principle, there are two major potential opportunities to combine NF technology with the FO system. From a process design standpoint, NF could be a reliable candidate to act as either a downstream treatment to concentrate the diluted DS from the upstream FO [5–7,9–19], or it could be a pre-treatment technology prior to the FO system in order to improve the overall treatment efficiency [20,21]. From a standpoint of detailed unit design, hollow fiber NF membranes are available to be employed in the FO unit. The NF-FO technology integration within the FO cell dramatically enhances the water recovery rate and energy efficiency compared to conventional FO technology [8,22–31].

In order to achieve the potential of the hybrid system in a water reuse process, optimization of the hybrid system design is necessary. There are two key aspects of this combined NF-FO technology

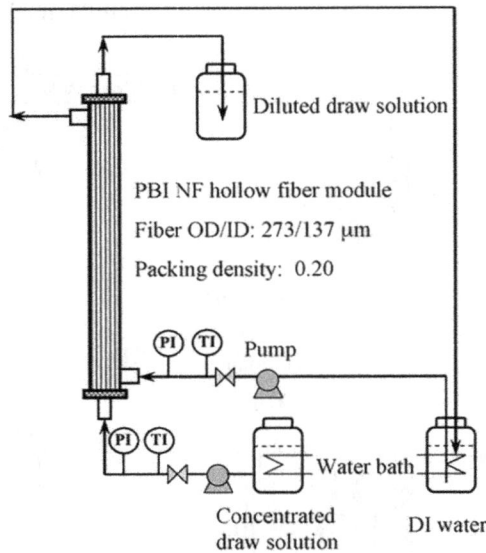

FIGURE 8.3 Schematic diagram of a lab-scale FO cell with NF hollow membrane utilized. (Adapted from Ref. [8].)

which are worthy of more detailed discussion: the selectively permeable NF membrane, made of nanomaterials, and the DS. This chapter will examine the most crucial features of those two key aspects of the NF-FO system, along with some industrial applications that have been proposed.

8.2   DEVELOPMENT OF DSs FOR THE HYBRID NF-FO SYSTEM

As we have seen, the NF membrane can be used either as an upstream/downstream technology attached to the FO unit or applied directly as the working membrane in the FO cells. In either case, the appropriate use of a DS will be a crucial point for discussion. Researchers are trying to develop a DS in the NF-FO system that offers effective contaminant removal performance at the same time as a competitive energy cost for replenishment. A suitable DS will also need to be non-toxic, provide high osmotic pressure, exhibit low viscosity for ease of circulation, and be environmentally friendly. Especially for the hybrid system applied to seawater desalination, the minimum osmotic pressure will be 25–33 atm [32].

Besides the high osmotic pressure of the draw solute, another key feature is to minimize the reverse solute transport. Various types of DSs have been developed and optimized. The DSs suitable for the FO system include (but are not limited to) inorganic salts such as $MgSO_4$, KNO_3, NH_4HCO_3 [33,34], sugars, and proteins such as sucrose and albumin [35,36], hydrophilic nanoparticles such as polyacrylic acid and triethylene glycol-coated nanoparticles [37], magnetic nanoparticles coated with osmotic agents such as poly-sodium-acrylate which can be separated by a magnet [38], stimuli-responsive polymer hydrogels [39], and so on. The use of DSs in the hybrid NF-FO system has been well developed over the past 10 years.

8.2.1   INORGANIC SALT DSs

The most popular applied DS is the inorganic metal salt solution, such as NaCl. The effective flux in a hybrid NF-FO system is affected by both water flux (J_w) and reverse solution transport (J_s). However, a monovalent DS would not be suitable for use in direct contact with an NF membrane as it would not be retained. Compared to a monovalent DS, divalent solutions with a larger hydrated

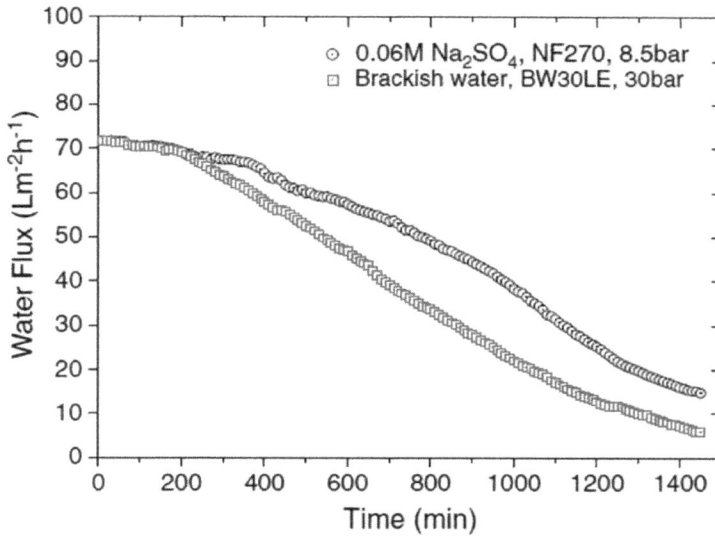

FIGURE 8.4   Water flux decline ratio for stand-alone RO and hybrid NF-FO system with Na_2SO_4 DS over 24 hours. (Adapted from Ref. [18].)

radius are suitable for an NF membrane and could also mitigate reverse solute transport to some extent [40]. In 2014, divalent DSs were proposed, such as Na_2SO_4 and $MgSO_4$ [18].

Researchers compared the water treatment performance between stand-alone reverse osmosis and the hybrid NF-FO system (where the NF is used as a downstream post-treatment to regenerate the DS). It was shown that the operating hydraulic pressure of a hybrid NF-FO system could potentially be 70% lower than the equivalent stand-alone reverse osmosis system. Even though both systems are operated to start with the same water flux, the hybrid NF-FO system shows a much slower decline ratio. Figure 8.4 indicates the water flux decline for both systems over a 24-hour time range. Using $0.06\,M$ Na_2SO_4 as DS, the NF-FO system presents a significantly lower flux decline because of less membrane fouling.

Another divalent inorganic DS, namely $MgCl_2$, was applied in the NF-FO system by Dutta et al. Some interesting findings were reported [23]. From an economic standpoint, compared with the Na-salt–based DS, the Mg-based DS has a higher capital cost [5]. However, increasing the applied hydraulic pressure in the NF cells could help to increase the permeate flux and ultimately break even on capital costs by facilitating the reuse of the expensive draw solute. For the hybrid system with the $MgCl_2$ DS mentioned above, the permeate flux increases about four times as pressure increases from 3 to 10 bar [23]. In some situations, double NF cells could also be applied to recover the precious DSs still further.

Within the FO cell, the DS concentration will affect both the water flux and the reverse solution transport. Figure 8.5 suggests that a higher DS concentration could enhance the water flux. This result also agrees with other studies of a similar nature [8,29]. However, the reverse solute transport increases dramatically with concentration and causes an upward trend in the J_s/J_w ratio. This result has also been exhibited for some other hybrid NF-FO systems with different DSs, such as sucrose and glucose [31]. It implies that optimizing a suitable concentration of DS in different situations will be necessary to achieve the most efficient process. In most cases, the higher the concentration of DS, the higher the reverse solute transport. A similar result has also been reported by a hybrid NF-FO bioreactor with a high-molecular-weight organic DS. It has also been observed that the reverse solute transport can decline over time. One possible explanation for this is that the formation of a viscous layer on the NF membrane by the DS impedes further reverse solute transport [22].

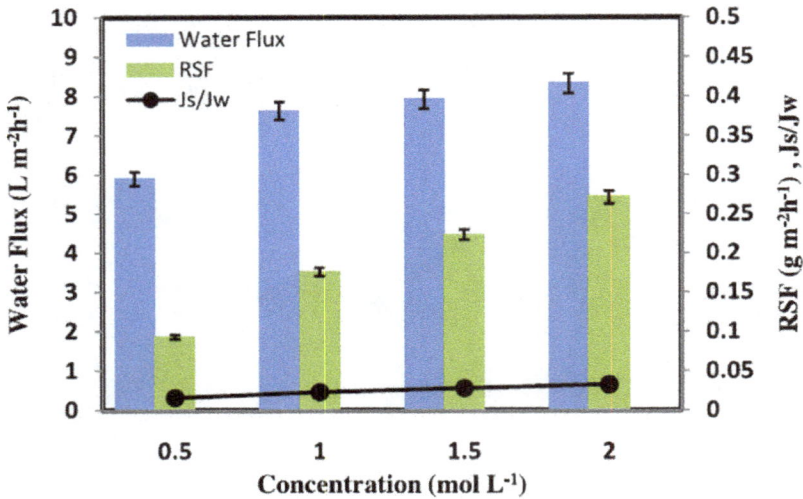

FIGURE 8.5 Water flux and reverse solute transport in the FO system directly using NF membrane with MgCl$_2$ as DS at various concentrations. (Adapted from Ref. [23].)

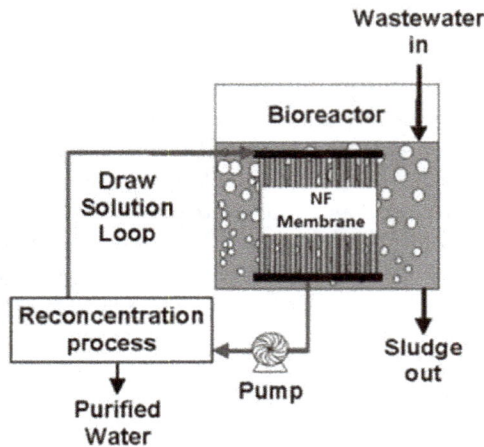

FIGURE 8.6 Schematic diagram of an FO-MBR with NF membrane directly installed. (Adapted from Ref. [41].)

8.2.2 Polymeric and Surfactant-Based DSs

The NF membrane can be used directly in the FO bioreactor. With some recently developed and novel DSs, the overall performance of the FO bioreactor can be boosted significantly. Compared with low-molecular-weight inorganic salt DSs, a DS with a higher molecular weight solute, such as a polymer or surfactant, is usually favored in this hybrid NF-FO bioreactor system. Figure 8.6 presents a simple FO-MBR process with the NF membrane directly installed.

A recent study by Parveen and Hankins [22] applied a single active-layered NF membrane directly inside a membrane bioreactor (MBR) with a diverse range of DS. Both deionized water and biologically active wastewater were tested as FSs in order to compare against each other. Many aspects were tested and reported, such as overall flux rate and percentage decline in water flux, with four distinct DSs applied including surfactant (0.5 M sodium dodecyl sulfate – SDS, 0.5 M

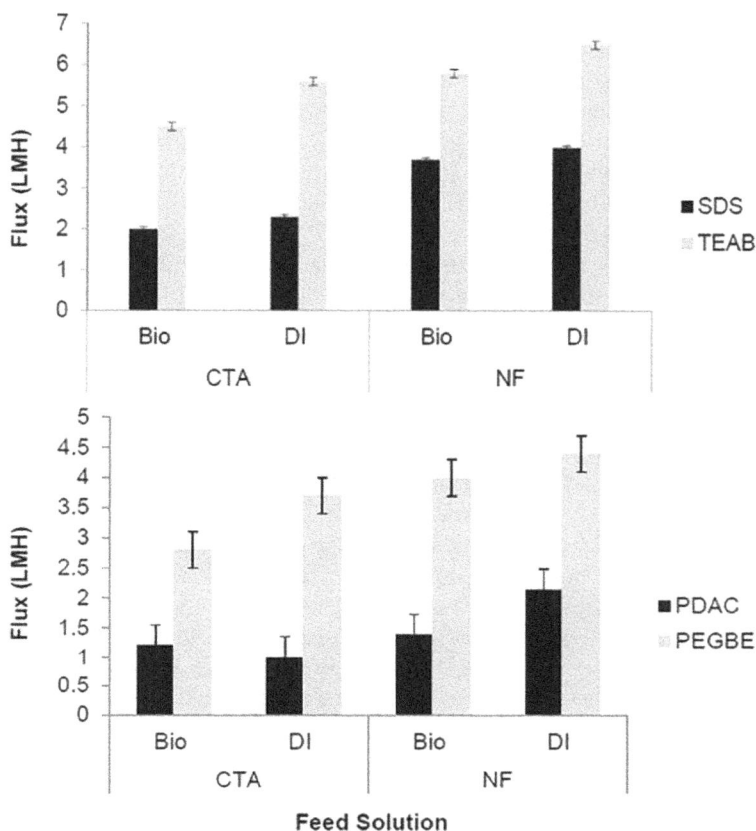

FIGURE 8.7 Initial fluxes for different DSs with the NF-FO hybrid bioreactor and benchmark FO bioreactor (CTA membrane) using both DI water and biologically active wastewater as feed. (Adapted from Ref. [22].)

tetraethylammonium bromide – TEAB) and polyelectrolyte (0.44 M poly-diallyldimethylammo-nium chloride – PDADMAC, 0.67 M poly(ethylene glycol) butyl ether – PEGBE). The use of an NF membrane in the FO system provided one order of magnitude higher water permeability than a conventional Cellulose triacetate (CTA)-based FO membrane. In Figure 8.7, the study indicated that the hybrid NF-FO system produces a significantly higher initial water flux for all studied DSs than a conventional (CTA) FO membrane with the same DS. The enhancement factor varies, but in some cases, the flux is doubled. Within these four DSs, TEAB has the highest osmotic pressure (1.61 MPa), the lowest viscosity (1.2 cP), and, unsurprisingly, the highest water flux (5.6 LMH).

Compared to a polyelectrolyte DS, a surfactant DS has a relatively higher diffusivity and a lower viscosity, both of which favor a higher water flux A surfactant draw solute also shows excellent resistance to reverse solute transport. A recent study by Parveen and Hankins [42] demonstrated that a surfactant DS could reduce the percentage of DS loss in 1 day of operation by more than 50%, compared to a conventional inorganic salt DS. The surfactant DS could save around 26% in draw solute costs due to lower RST loss. As such, surfactant DSs demonstrate great potential for coupling with other inorganic DSs to reduce the RST. Nguyen et al. conducted a series of experiments that coupled a conventional inorganic salt DS, namely Na_3PO_4, with three different kinds of surfactants in an NF-FO system [19]. Compared to the stand-alone Na_3PO_4 DS, the RST was reduced by around 70% and 85% by coupling Na_3PO_4 with SDS and polyethyleneglycol tert-octylphenyl ether, respectively.

The molecular weight of polyelectrolyte can also affect the water flux. Usually, a given polyelectrolyte with a higher molecular weight seems to provide a higher water flux. A possible explanation of this is that, although the number of ionizable groups (and hence the osmotic pressure) when normalized to the weight concentration may be fairly constant, a larger molecule will reduce the reverse solute transport. On the other hand, a larger molecule will diffuse more slowly and contribute more easily to internal and external concentration polarization. A high-molecular-weight PAM (3,000,000 Da) can yield 14–17 LMH [43]. In the study by Parveen and Hankins, the water flux of PEGBE (up to 400,000 Da) was around 4.5 [22]. In a previous study by Ju and Kang, zwitterionic homopolymer polysulfobetaine (PBET) DS was applied in the FO system. The molecular weight was around 18,000 Da, and the water flux reached 3.22 LMH [44]. For some polymers with lower molecular weight, the flux performance can be disappointingly poor. For instance, the 5% PEI solution with ~1,000 Da studied by Jun et al. could only yield around 1 LMH water flux [17].

As already noted, in some situations, the water flux in hybrid systems can be doubled when compared to conventional FO. For instance, with deionized water feed and PDADMAC as the DS, the water flux was elevated from 1 LMH (L/m h) to over 2 LMH by the hybrid system. A similar observation was made by using SDS as the FS [22]. However, the NF-FO hybrid system exhibits an issue of minor concern, regarding a slightly higher reverse solute transport when compared to conventional FO. The work by Parveen and Hankins [22] showed that this had no toxicity effect within the bioreactor, regardless of the draw solute employed.

Parveen and Hankins [22] also pointed out that the decline of water flux in the hybrid NF-FO system is heavily affected by both the type of DS and the nature of the FS. Sometimes, the DS with a higher viscosity demonstrated a lower water flux. The inverse relationship between fluid velocity and viscosity was explained by Poiseuille's law in fundamental fluid mechanics, where the lower velocity leads to a lower water flux. It partly explained why a high viscosity (≥2 cP) surfactant DS exhibits a lower water flux than a moderate viscosity electrolyte DS. On the other hand, increasing the molar concentration of the DS could help to improve water flux to some extent. Clearly, there exists a trade-off. Hence, there is a need to find an optimum concentration for the DSs [22].

During an 8-hour operation, the following three DSs, namely SDS, PDADMAC, and PEGBE, exhibited a lower flux decline with DI water feed than with biologically active feed. Their flux decline was 12.5%, 57%, and 20.45%, respectively, for DI water, and 34%, 63%, and 25%, respectively, for biologically active feed. In contrast, TEAB was more suitable with biological feed than with DI water feed, with a 22.41% decline in water flux with biologically active feed compared to 41% with DI water feed over the same operating time. This result also implies that the DS used in the hybrid NF-FO system should be matched correspondingly for different feed waters. In other words, there is no perfect DS in the NF-FO system that can adapt to all possible wastewater treatment scenarios, and instead an appropriate choice should be made for each particular case.

Toxicity to the biomass in FO-MBR and all FO-associated systems should be considered. Usually, if the reverse solution transport across the membrane is above a certain level, the DS might cause toxicity and harm the bacteria in the bioreactor, lowering the treatment efficiency and polluting their aquatic environment [45,46]. Fortunately, the hybrid NF-FO bioreactor presented no toxicity to bacteria, such as *Bacillus subtilis*. The bacteria grew under all the types of DS tested, in a concentration range from 0.005 to 0.05 M, including with inorganic salts, polyelectrolytes, and surfactants. Measurement of the reverse solute transport indicated that the rates were small or negligible, particularly for the larger molecular weight draw solutes. Although the surfactant SDS showed the highest toxicity to bacteria among tested draw solutes, it still allowed bacteria to grow without hindrance.

8.3 THE NF MEMBRANE IN AN FO MODULE

8.3.1 PROPERTY AND MORPHOLOGY

There are urgent needs for developing the membrane in an FO module which is made of nanomaterials and can be rapidly applied with an environmentally benign and long-life cycle. Ideally, the membrane applied in the FO module is selectively permeable and hydrophilic. The nanomaterial membrane's potential and ability are apparent for many application examples during a water treatment process. Nanomaterials have a myriad of chemical compositions and synthesis routes, and these bestow widely different characteristics in terms of antibacterial activity, adsorption of heavy metal, and so on.

The FO cell membrane acts as an interface barrier between the DS and FS. The vast majority of the membranes currently applied in commercial FO cells are composite RO membranes [47,48]. With the development of NF membranes, some new advantages have been gradually realized by researchers. In a pressure-driven process, the nano-scale active pores in the NF membrane usually have a higher water permeation and a lower salt rejection than the RO membrane. Compared to a UF membrane being used in FO, the NF membrane exhibits a much higher salt rejection ratio. In particular, a typical NF membrane has a better ability to reject divalent solutes than monovalent solutes [49]. With an NF membrane incorporated, FO demonstrates an enhanced ability to remove heavy metal ions from wastewater [50].

As a result of the particular membrane composition and synthesis route, parameters such as membrane permeability, selectivity, fouling propensity, and stability will be of key importance in the hybrid NF-FO system. Applying the appropriate nanomaterial during the membrane fabrication process and tuning the synergy between the support and active layers to an optimum level is important. With this in mind, the morphologies of the nanostructured NF membranes are worthy of further discussion to understand their role.

The NF membrane usually consists of a selective layer and a porous support layer with a certain mechanical strength. Basically, there are two major methods to synthesize NF membranes; simultaneously fabricating an active layer and a support layer (e.g., cellulose acetate [CA] based), and making the selective layer and the support layer separately based on the composite membrane concept [28].

In terms of structure, most of the NF membranes applied in the FO system have a hollow fiber structure (Figure 8.8). This structure is relatively easy to fabricate at an affordable cost on an industrial scale. The hollow fiber structure creates a large membrane area to volume ratio and is usually self-supporting. Hence, commercial membranes are increasingly fabricated in this structure [52]. Figure 8.8 shows a few different NF membranes applied in the FO module, with such a hollow fiber configuration and made of various materials.

In many past works, CA has proven to be a reliable and affordable material to construct NF membranes [53–55]. The first use of a CA-based membrane in the FO system can be traced back to 1975 [56]. In 2010, Su et al. first proposed using a CA-based hollow fiber NF membrane for an FO application [29]. Their study synthesized the CA-based NF membrane by the dry-jet wet-spinning method, with additional heat treatment by immersion in a hot water bath [51]. The heat treatment basically modifies the sublayer structure and affects the surface morphology simultaneously. The thinnest outer layer becomes thicker and denser after heat treatment, which then acts more

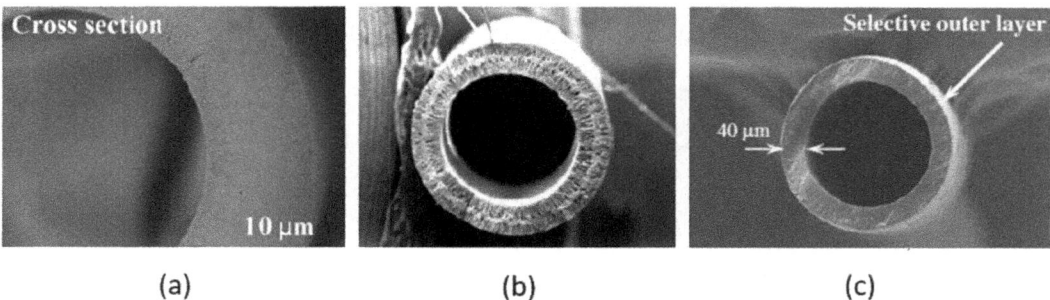

| (a) | (b) | (c) |

FIGURE 8.8 Morphology of NF membrane used in FO system with hollow fiber configuration. (Adapted from: (a) cellulose acetate, Ref. [29], (b) Poly(amide-imide), Ref. [28], (c) Modified Polybenzimidazole, Ref. [24].)

effectively as a semi-permeable barrier. With the heat treatment, the $MgCl_2$ and NaCl rejection rates were elevated from 9.44% and 24.61% to 96.67% and 90.17%, respectively.

The CA-based membrane is more effective for low-salinity feed water than for high-salinity feed water. When the feed concentration increased from 0.05 to 0.6 M, the initial osmotic pressure gradient dropped around 25%. The water flux then dropped from 3.9 to 1.0 LMH, due to concentration polarization taking place at the active layer (selective) side. With the CA-based NF membrane and 0.5 M $MgCl_2$ as the DS, the water flux could reach 2.7 LMH with low salt leakage [51]. A CA-based NF membrane shows strong potential when applied in an FO module because of high rejection to NaCl and $MgCl_2$.

8.3.2 THE NF MEMBRANE WITH SURFACE MODIFICATION

The NF membrane modification process has become a popular procedure to enhance the performance of the hybrid NF-FO system. The surface modification can be either doped or grafted by use of another solution, such as PEI and surfactant. The modification process can tailor the surface charge, and the choice of modification solution depends on the treatment target of the prepared NF membrane.

Poly(amide-imide) (PAI) is a notable material for NF membrane fabrication. PAI is also known as Torlon, a commercially available material that can be easily acquired. It can withstand a wide range of pH and is able to resist many organic solvents [57].

In this case (see Figure 8.9), Setiawan et al. fabricated NF membranes using PAI and doped with PEI at 60°C–80°C, the latter modification further enhancing its performance for FO applications [28]. The wall thickness of the hollow fiber varies from 140 to 200 µm. The amine functional group in PEI bonds with the opened imide ring from the PAI. After doping, the surface of the membrane exhibits more amine groups, and this brings about a positive change over a wider pH range. Those positive charges could help with salt rejection. Compared to the stand-alone PAI, the salt rejection rate for both NaCl and $MgCl_2$ is boosted by a factor of eight times with the PEI-doped membrane. Under a 0.5 M $MgCl_2$ DS, the maximum water flux could reach 9.47 LMH. For the same type and concentration of DS, the doped PAI membrane significantly improves performance over both undoped and CA-based membrane.

The potential to use polybenzimidazole (PBI)-based NF membranes in the FO system is also attractive. Due to its great chemical and thermal stability, PBI has shown a great performance in many applications, such as in ion-exchange membranes [58]. Its highly hydrophilic property gives the membrane a marked ability to avoid fouling, thus being very suitable for water treatment applications.

In 2007, Wang et al. first applied the PBI-based hollow fiber NF membrane in the FO system [8]. With additional chemical cross-linking modification by flowing p-xylylene dichloride, the PBI membrane exhibited a decreased pore size and wall thickness (from 136 to 40 µm) and a narrower pore size

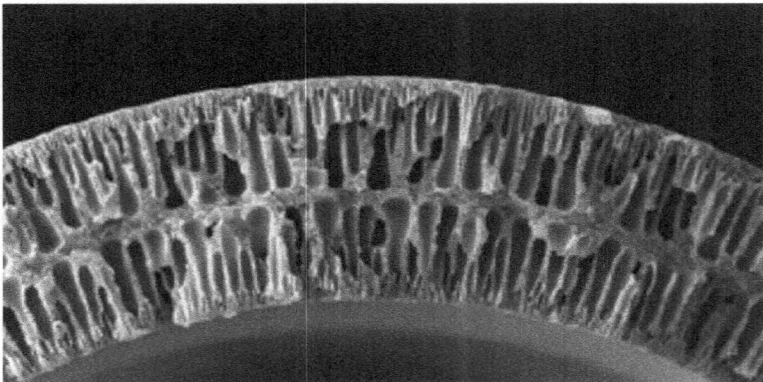

FIGURE 8.9 Morphology of PEI-doped PAI NF membrane. (Adapted from Ref. [28].)

distribution [24]. It helped increase the salt rejection by around 30% and tripled the water flux. The cross-linking process could also mitigate the decline of salt selectivity caused by increased temperature. When the temperature increased from 22°C to 50°C, the non-modified membrane showed a decline in salt selectivity to a value lower than 96%, whereas the cross-linked membrane maintained a salt selectivity above 99.5%. In addition, a longer modification period brought about a greater improvement.

The cross-linking modification method is also available for polyamide (PA) membranes. As before, the higher concentration of amine groups present on the membrane surface usually brings a positive charge over a broader pH range. When the pH is below the membrane's isoelectric point, amine protonation causes the membrane to be positively charged. After a suitable cross-linking modification, more amine groups attached on the membrane surface help to shift the isoelectric point from 4.4 to 10.4 [28]. The positively charged surface helps in turn to repel positively charged species, such as ammonia and heavy metal ions, and ultimately benefits the selectivity and water flux.

Usually, the PA membrane is synthesized based on the TFC (thin-film composite membrane) concept [59,60]. The membrane surface can be modified by many different solutions, such as PEI. In a recent study from 2019 (Figure 8.10), a PA TFC membrane was grafted by PEI, using dicyclo-hexylcarbodiimide (DCC) as an agent [19].

This places more amine groups on the top of the PA surface and provides a positive charge. The grafted PEI thus helps to elevate the critical pH point for neutral charge (the point of zero charge) from pH 3.8 to pH 5.5. It can be further improved by increasing the PEI concentration (from 0.2% to over 1%) and eventually reaching pH 7. Grafting with over 1% PEI means that the PA membrane surface will remain positively charged when the pH is less than 7. The optimal PEI concentration is 3%, providing both the highest water flux and the highest salt selectivity.

Contemporaneously, some other studies have utilized surfactants such as SDS, dodecyltrimethylammonium bromide, and sorbitan monolaurate [27], as well as UV-photografting techniques, to modify NF membranes for FO application [61].

8.3.3 Membrane Fouling and Further Improvements

The perfect FO process with a hydrophilic membrane should have low fouling. However, fouling and internal concentration polarization are generally still observed, and caused mainly by reverse solute transport [62,63].

This issue could potentially be overcome by an appropriate design of the selective membrane. The NF membrane discussed above with a hollow fiber structure exhibits a better fouling resistance

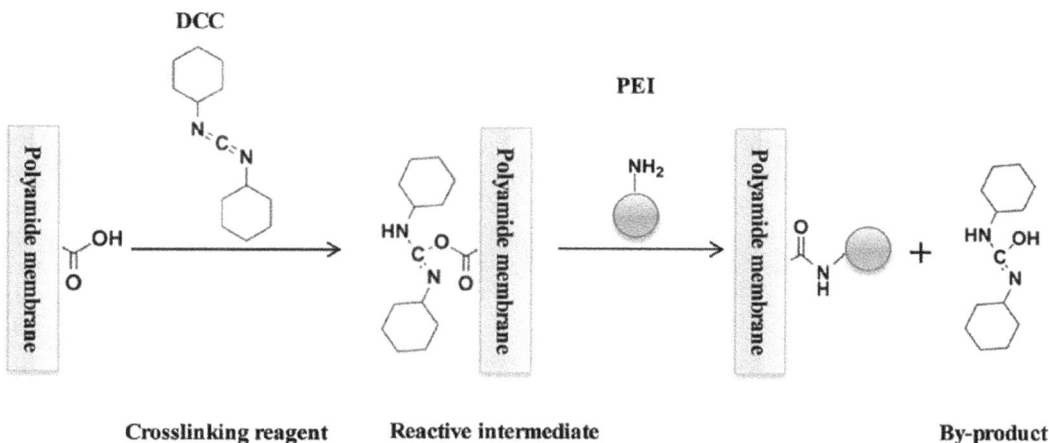

Crosslinking reagent   **Reactive intermediate**   **By-product**

FIGURE 8.10 Surface modification mechanism of PA membrane by PEI with DCC as an agent. (Adapted from Ref. [26].)

than other membrane structures. With a proper membrane fabrication process, the material with a higher hydrophilicity, such as PBI, could help with antifouling to some extent. Liu et al. proposed a novel membrane synthesis process that benefits by overcoming fouling in both NF and FO systems [30]. AgNPs were incorporated into the membrane with a novel layer-by-layer assembly to decrease the contact angle on the surface. The lower contact angle made the surface more hydrophilic and enhanced the antifouling ability.

The NF membrane modification step also tailors the surface charge of the membrane, which plays a significant role in fouling (as mentioned earlier). The positively charged membrane surface can repel positively charged foulants but shows less antifouling propensity to negatively charged foulants, and vice versa. For instance, a membrane surface with amine groups attached as a result of PEI grafting shows a higher fouling propensity when dealing with a negatively charged organic feed but performs better against positively charged ammonium ions [6,26].

Fouling in the NF-FO system could potentially be mitigated by optimizing the FO process design and modifying the NF membrane. Optimization of operating parameters, such as initial flux and crossflow rate, also affects the fouling control. With a high initial flux and a low crossflow rate, the rejected metal ions are more likely to accumulate on the membrane surface by concentration polarization and cause further fouling [64]. In addition, matching of the draw solute and the membrane surface is essential. If a polymer solute has high hydrophilicity for enhanced solubility, then it is best to employ a hydrophobic membrane and vice versa. Both tuning of the operational parameters and applying an appropriate draw solute that provides higher flux are crucial. The application of chemical cleaning agents such as HCl, NaOH, citric acid and EDTA are usually practical to deal with a small amount of fouling [7].

8.4 INDUSTRIAL APPLICATIONS OF THE HYBRID NF-FO SYSTEM

8.4.1 Brackish Water Desalination

Water shortage is a severe problem in arid regions, such as in the Gulf States. Because of its high salinity, seawater desalination usually relies on energy-intensive technologies such as RO [16]. On the other hand, brackish water, such as within an estuary, has a salinity between that of fresh water and that of seawater [65]. The development of a suitable technology to utilize brackish water to produce water for either drinking or fertigation has received much attention. This makes it attractive for desalination by a hybrid NF-FO process, given that the osmotic pressure required for the DS and the reverse solute transport are both relatively modest, while the flux rate is relatively high. Figure 8.11 shows an example of desalinating brackish water for direct fertigation.

In 2011, a hybrid NF-FO system was applied to treat real brackish water as an FS [18]. A loose, flat-sheet cellulose triacetate NF membrane (NF270 from Dow FilmTech) was selected and installed in the FO system. This particular type of NF membrane is normally used to remove organic compounds and also shows great potential for brackish water desalination. With Na_2SO_4 as a DS, the hybrid NF-FO system provided a four times higher initial water flux than a stand-alone RO system. Comparing two frequently discussed DSs, $MgCl_2$ and Na_2SO_4, in brackish water desalination applications, Na_2SO_4 usually provides a higher water flux, while $MgCl_2$ exhibits a better TOC (total organic compound) removal rate [16]. After water cleaning of the membrane, the NF-FO system also has great durability with a flux recovery over 90%. However, the total dissolved solids can remain over the required standard with this NF-FO system, given that an NF membrane can provide a less effective barrier than an RO membrane to monovalent solutes.

To design a more effective NF-FO hybrid system to improve agricultural water standards, some modifications have been made in the process design [15]. NF was used as either pre-treatment or post-treatment with various 1 M fertilizer DSs. NH_4Cl stands out, with a few advantages in terms of providing the highest water flux (over 10 LMH) and complete exclusion of potassium and phosphate ions in brackish water. In order to use the treated water for direct agriculture fertigation (e.g., tomato

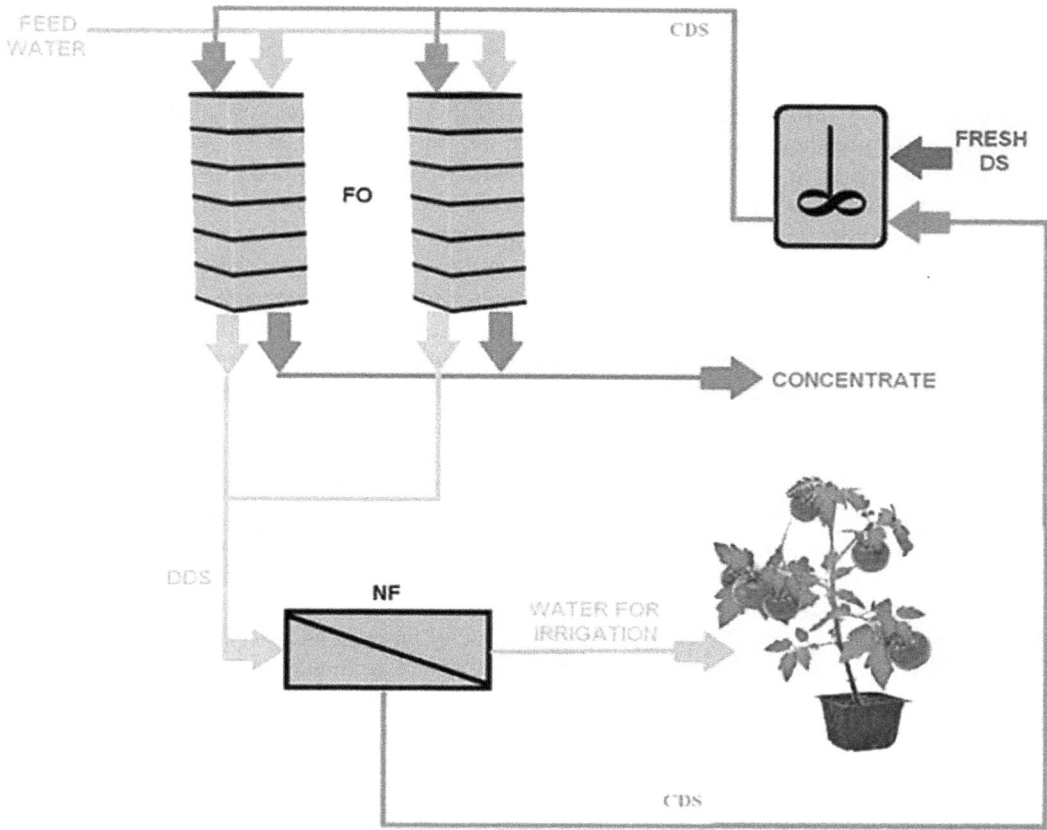

FIGURE 8.11 Brackish water desalination by the NF-FO system for direct fertigation. (Adapted from Ref. [10].)

cultivation), a certain water standard is required. Adequate processes reported include using SOA (sulfate of ammonia) as DS and NF as post-treatment, and the use of $NaNO_3$ or NH_4Cl as DS and NF as pre-treatment.

A long-term evaluation of NF-FO when used in agriculture was conducted by Corzo et al. [10]. The hybrid NF-FO system ran for 480 days, and the study established that it is feasible to use the hybrid system for such a scenario. In addition, the RO process could be attached to the NF module to yield an FO-NF-RO three-in-one system to improve treatment efficiency [5].

8.4.2 Wastewater Treatment

The NF-FO system is also an excellent candidate for wastewater treatment from many sources, and indeed for water recycle. These sources include but are not limited to industrial wastewater such as from the leather industry, agricultural wastewater during fertigation, and effluents requiring the removal of overall or specific dissolved species such as TDS and metallic ions.

During the past few years, several processes that use NF-FO were proposed to treat industrial wastewater. The wastewater from the leather industry usually has a high chemical oxygen demand, TDS, inorganic ions such as chloride and sulfide, and precious metals such as chromium. At a relatively low pressure (12 bar) and the most commonly used DS (NaCl), a pilot-scale NF-FO system demonstrated a great potential to treat wastewater continuously from the leather industry along with DS regeneration [13]. At a high feed crossflow rate (≥400 LPH), the NF module provided a high water flux (60 LMH) with over 80% salt rejection. Another study indicated that the NF-FO system could remove over 95% TDS, 88% COD, 100% chromium, along with other ions [11].

The hybrid NF-FO system shows an exceptional ability to remove over 97% of COD in the pharmaceutical industry [14]. In a real coke-oven plant, NF-FO removed over 96% of cyanide, phenols, etc. [7]. For such applications, the NF-FO system not only offered benefits with environmental aspects but also required less energy to operate. Compared to a conventional RO, a hybrid NF-FO system consumed at least 13.6% less energy [12]. In order to further benefit both the environmental and economic aspects, process optimization to improve the NF recovery rate is recommended.

Other than removing organic waste compounds, NF-FO is also good at removing some specific ions. As mentioned in Section 8.3.2, a novel modification method has been applied to the NF membrane. This helps the surface to generate a positive charge, especially in acidic conditions, and further repel and concentrate ammonia ions [25]. The NF membrane normally shows a negative charge under neutral and alkaline conditions. Abdullah et al. successfully doped the NF membrane in an FO system to remove the cationic copper ion based on this concept [25].

The NF-FO system has also shown a great potential to treat and recycle municipal wastewater [22]. The municipal wastewater usually includes a large amount of glucose, ammonium chloride, potassium salt, calcium chloride, and magnesium sulfate. Installing the flat-sheet/hollow fiber NF membrane in an FO-membrane bioreactor with various novel high-molecular-weight DSs enhanced the bioreactor's water flux and water treatment efficiency. Thus, this study demonstrated the advantages of an NF-FO-MBR hybrid; it allowed high flux rates, while the increase in reverse solute transport was negligible and did not contribute to toxicity for even modest molecular weights. On the other hand, the potential issue of long-term draw solute accumulation in the bioreactor tank was highlighted. Following draw solute recovery by MD, the permeate water quality was also high enough for the purposes of recycling [22], although some further polishing via RO might be necessary prior to potable uses.

8.4.3 CHEMICAL AND FOOD PROCESSING

The superior characteristics of the NF-FO system also benefit the fine chemical and food industries. In 2017, Shibuya et al. developed a novel hybrid system to generate bioethanol from rice straw (Figure 8.12). In this system, the NF pre-treatment removed inhibitors of the saccharification process, and the FO concentrated the sugars.

The fermentation inhibitor was removed in the NF step, with water continuously supplied in order to hydrolyze polysaccharides to monomeric sugars. This step was followed by the FO module

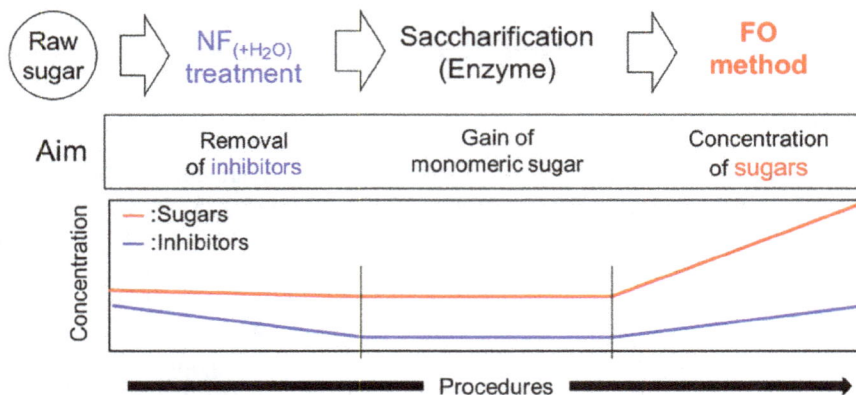

FIGURE 8.12 Block flow diagram with concentration change of NF-FO system to concentrate sugar production. (Adapted from Ref. [20].)

to further increase the sugar concentration. The hybrid NF-FO process provides a higher sugar concentration than stand-alone NF (107 vs 67.6 g/L) and a higher ethanol yield (grams of ethanol produced by consuming 1 g of total monomeric sugars) than stand-alone FO (0.3 vs 0.01) [20].

NF also shows some potential in food processing, such as in concentrating the bioactive components of fruit and vegetable juices [66]. Using a sequence of NF and FO processes, the concentration of the total soluble solid in black carrot juice such as sucrose rose from 10.4 to 43.4 Brix (grams of soluble solid per 100 g of solution) [21]. To a certain extent, the hybrid NF-FO systems offer an alternative technology to thermal evaporation in the food processing area.

8.5 CONCLUSION

Along with the ongoing development of NF membranes, a hybrid NF-FO system shows a great potential to solve many real-world challenges in water stream processing. The NF module can act as a pre-treatment or post-treatment for the FO unit with respect to the target feed or product solution, or it can recover the diluted DS within the FO unit. Divalent draw solutes such as Na_2SO_4 and $MgCl_2$ are usually considered as candidate DSs for an NF membrane that provide high water flux and low reverse solute transport.

The NF membrane can also be modified for application. While NF membranes are usually negatively charged under alkaline conditions and positively charged under acidic conditions, the hollow fiber configuration and surface modifications such as doping provide a highly hydrophilic surface and an adjustable surface charge. These exceptional characteristics make the NF membrane operationally useful in many different situations, along with an enhanced antifouling ability.

Because of the previously mentioned powerful features and a lower energy consumption compared to the conventional and equivalent RO and FO processes, it is apparent that the NF-FO system is now being applied in many diverse areas, in terms of brackish water desalination, wastewater treatment, and chemical processing. With some optimization in terms of novel NF membrane module design, suitable operating parameters, and innovative surface modification, the hybrid NF-FO system offers a promising technology in future sustainable water treatment.

REFERENCES

1. G. L. Stephens et al., "Earth's water reservoirs in a changing climate," *Proc Royal Soc A: Math Phys Eng Sci*, vol. 476, no. 2236, p. 20190458, 2020, doi: 10.1098/rspa.2019.0458.
2. V. Gitis and N. Hankins, "Water treatment chemicals: Trends and challenges," *J Water Process Eng*, vol. 25, pp. 34–38, Oct. 2018, doi: 10.1016/J.JWPE.2018.06.003.
3. A. W. Mohammad, Y. H. Teow, W. L. Ang, Y. T. Chung, D. L. Oatley-Radcliffe, and N. Hilal, "Nanofiltration membranes review: Recent advances and future prospects," *Desalination*, vol. 356, pp. 226–254, Jan. 2015, doi: 10.1016/J.DESAL.2014.10.043.
4. B. van der Bruggen, M. Mänttäri, and M. Nyström, "Drawbacks of applying nanofiltration and how to avoid them: A review," *Sep Purif Technol*, vol. 63, no. 2, pp. 251–263, Oct. 2008, doi: 10.1016/J.SEPPUR.2008.05.010.
5. H. Wang et al., "Comprehensive analysis of a hybrid FO-NF-RO process for seawater desalination: With an NF-like FO membrane," *Desalination*, vol. 515, p. 115203, doi: 10.1016/j.desal.2021.115203.
6. M. Giagnorio, F. Ricceri, M. Tagliabue, L. Zaninetta, and A. Tiraferri, "Hybrid forward osmosis–nanofiltration for wastewater reuse: System design," *Membranes (Basel)*, vol. 9, no. 5, p. 61, doi: 10.3390/membranes9050061.
7. R. Kumar and P. Pal, "A novel forward osmosis-nano filtration integrated system for coke-oven wastewater reclamation," *Chem Eng Res Des*, vol. 100, pp. 542–553, doi: 10.1016/j.cherd.2015.05.012.
8. K. Y. Wang, T.-S. Chung, and J.-J. Qin, "Polybenzimidazole (PBI) nanofiltration hollow fiber membranes applied in forward osmosis process," *J Memb Sci*, vol. 300, no. 1, pp. 6–12, 2007, doi: 10.1016/j.memsci.2007.05.035.
9. S. Phuntsho et al., "A closed-loop forward osmosis-nanofiltration hybrid system: Understanding process implications through full-scale simulation," *Desalination*, vol. 421, pp. 169–178, doi: 10.1016/j.desal.2016.12.010.

10. B. Corzo, T. de la Torre, C. Sans, R. Escorihuela, S. Navea, and J. J. Malfeito, "Long-term evaluation of a forward osmosis-nanofiltration demonstration plant for wastewater reuse in agriculture," *Chem Eng J*, vol. 338, pp. 383–391, doi: 10.1016/j.cej.2018.01.042.

11. P. Pal, M. Sardar, M. Pal, S. Chakrabortty, and J. Nayak, "Modelling forward osmosis-nanofiltration integrated process for treatment and recirculation of leather industry wastewater," *Comput Chem Eng*, vol. 127, pp. 99–110, doi: 10.1016/j.compchemeng.2019.05.018.

12. J. E. Kim et al., "Environmental and economic impacts of fertilizer drawn forward osmosis and nanofiltration hybrid system," *Desalination*, vol. 416, pp. 76–85, doi: 10.1016/j.desal.2017.05.001.

13. P. Pal, S. Chakrabortty, J. Nayak, and S. Senapati, "A flux-enhancing forward osmosis–nanofiltration integrated treatment system for the tannery wastewater reclamation," *Environ Sci Pollut Res Int*, vol. 24, no. 18, pp. 15768–15780, doi: 10.1007/s11356-017-9206-z.

14. P. Pal, P. Das, S. Chakrabortty, and R. Thakura, "Dynamic modelling of a forward osmosis-nanofiltration integrated process for treating hazardous wastewater," *Environ Sci Pollut Res Int*, vol. 23, no. 21, pp. 21604–21618, doi: 10.1007/s11356-016-7392-8.

15. S. Phuntsho, S. Hong, M. Elimelech, and H. K. Shon, "Forward osmosis desalination of brackish groundwater: Meeting water quality requirements for fertigation by integrating nanofiltration," *J Memb Sci*, vol. 436, pp. 1–15, Jun. 2013, doi: 10.1016/J.MEMSCI.2013.02.022.

16. M. Giagnorio, F. Ricceri, and A. Tiraferri, "Desalination of brackish groundwater and reuse of wastewater by forward osmosis coupled with nanofiltration for draw solution recovery," *Water Res (Oxford)*, vol. 153, pp. 134–143, doi: 10.1016/j.watres.2019.01.014.

17. B.-M. Jun, T. P. N. Nguyen, S.-H. Ahn, I.-C. Kim, and Y.-N. Kwon, "The application of polyethyleneimine draw solution in a combined forward osmosis/nanofiltration system," *J Appl Polym Sci*, vol. 132, no. 27, doi: 10.1002/app.42198.

18. S. Zhao, L. Zou, and D. Mulcahy, "Brackish water desalination by a hybrid forward osmosis–nanofiltration system using divalent draw solute," *Desalination*, vol. 284, pp. 175–181, 2012, doi: 10.1016/j.desal.2011.08.053.

19. H. T. Nguyen, N. C. Nguyen, S.-S. Chen, and S.-Y. Wu, "Concentrate of surfactant-based draw solutions in forward osmosis by ultrafiltration and nanofiltration," *Water Sci Technol Water Suppl*, vol. 15, no. 5, pp. 1133–1139, doi: 10.2166/ws.2015.060.

20. M. Shibuya et al., "Development of combined nanofiltration and forward osmosis process for production of ethanol from pretreated rice straw," *Bioresour Technol*, vol. 235, pp. 405–410, doi: 10.1016/j.biortech.2017.03.158.

21. M. C. Roda-Serrat et al., "Processing of black carrot juice by nanofiltration and forward osmosis," *Chem Eng Trans*, vol. 87, doi: 10.3303/CET2187092.

22. F. Parveen and N. Hankins, "Comparative performance of nanofiltration and forward osmosis membranes in a lab-scale forward osmosis membrane bioreactor," *J Water Procs Eng*, vol. 28, pp. 1–9, doi: 10.1016/j.jwpe.2018.12.003.

23. S. Dutta, P. Dave, and K. Nath, "Performance of low pressure nanofiltration membrane in forward osmosis using magnesium chloride as draw solute," *J Water Process Eng*, vol. 33, p. 101092, doi: 10.1016/j.jwpe.2019.101092.

24. K. Y. Wang, Q. Yang, T.-S. Chung, and R. Rajagopalan, "Enhanced forward osmosis from chemically modified polybenzimidazole (PBI) nanofiltration hollow fiber membranes with a thin wall," *Chem Eng Sci*, vol. 64, no. 7, pp. 1577–1584, 2009, doi: 10.1016/j.ces.2008.12.032.

25. W. N. A. S. Abdullah, S. Tiandee, W. Lau, F. Aziz, and A. F. Ismail, "Potential use of nanofiltration like-forward osmosis membranes for copper ion removal," *Chin J Chem Eng*, vol. 28, no. 2, pp. 420–428, doi: 10.1016/j.cjche.2019.05.016.

26. S. Jafarinejad, H. Park, H. Mayton, S. L. Walker, and S. C. Jiang, "Concentrating ammonium in wastewater by forward osmosis using a surface modified nanofiltration membrane," *Env Sci Water Res Technol*, vol. 5, no. 2, pp. 246–255, doi: 10.1039/c8ew00690c.

27. M. B. M. Y. Ang, Y.-T. Lu, S.-H. Huang, J. C. Millare, H.-A. Tsai, and K.-R. Lee, "Surfactant-assisted interfacial polymerization for improving the performance of nanofiltration-like forward osmosis membranes," *J Polym Res*, vol. 29, no. 3, doi: 10.1007/s10965-022-02942-6.

28. L. Setiawan, R. Wang, K. Li, and A. G. Fane, "Fabrication of novel poly(amide–imide) forward osmosis hollow fiber membranes with a positively charged nanofiltration-like selective layer," *J Memb Sci*, vol. 369, no. 1, pp. 196–205, 2011, doi: 10.1016/j.memsci.2010.11.067.

29. J. Su, Q. Yang, J. F. Teo, and T.-S. Chung, "Cellulose acetate nanofiltration hollow fiber membranes for forward osmosis processes," *J Memb Sci*, vol. 355, no. 1, pp. 36–44, 2010, doi: 10.1016/j.memsci.2010.03.003.

30. X. Liu, S. Qi, Y. Li, L. Yang, B. Cao, and C. Y. Tang, "Synthesis and characterization of novel antibacterial silver nanocomposite nanofiltration and forward osmosis membranes based on layer-by-layer assembly," *Water Res (Oxford)*, vol. 47, no. 9, pp. 3081–3092, doi: 10.1016/j.watres.2013.03.018.

31. S. O. Alaswad, S. Al-aibi, E. Alpay, and A. Sharif, "Efficiency of organic draw solutions in a forward osmosis process using nano-filtration flat sheet membrane," *J Chem Eng Process Technol*, vol. 9, no. 1, p. 370, 2018.

32. E. Nagy, "Pressure-retarded osmosis (PRO) process," Basic Equations of Mass Transport Through a Membrane Layer, pp. 505–531, Jan. 2019, doi: 10.1016/B978-0-12-813722-2.00021-2.

33. T. Y. Cath, A. E. Childress, and M. Elimelech, "Forward osmosis: Principles, applications, and recent developments," *J Memb Sci*, vol. 281, no. 1, pp. 70–87, 2006, doi: 10.1016/j.memsci.2006.05.048.

34. L. Chekli, S. Phuntsho, H. K. Shon, S. Vigneswaran, J. Kandasamy, and A. Chanan, "A review of draw solutes in forward osmosis process and their use in modern applications," *Desal Water Treat*, vol. 43, no. 1–3, pp. 167–184, 2012, doi: 10.1080/19443994.2012.672168.

35. T.-S. Chung, S. Zhang, K. Y. Wang, J. Su, and M. M. Ling, "Forward osmosis processes: Yesterday, today and tomorrow," *Desalination*, vol. 287, pp. 78–81, doi: 10.1016/j.desal.2010.12.019.

36. S. Zhao, L. Zou, C. Y. Tang, and D. Mulcahy, "Recent developments in forward osmosis: Opportunities and challenges," *J Memb Sci*, vol. 396, pp. 1–21, doi: 10.1016/j.memsci.2011.12.023.

37. M. M. Ling and T. S. Chung, "Desalination process using super hydrophilic nanoparticles via forward osmosis integrated with ultrafiltration regeneration," *Desalination*, vol. 278, no. 1–3, pp. 194–202, Sep. 2011, doi: 10.1016/J.DESAL.2011.05.019.

38. I. Ban et al., "Synthesis of poly-sodium-acrylate (PSA)-coated magnetic nanoparticles for use in forward osmosis draw solutions," 2019, doi: 10.3390/nano9091238.

39. D. Li, X. Zhang, J. Yao, G. P. Simon, and H. Wang, "Stimuli-responsive polymer hydrogels as a new class of draw agent for forward osmosis desalination," *Chem Commun (Camb)*, vol. 47, no. 6, pp. 1710–1712, doi: 10.1039/c0cc04701e.

40. R. W. Holloway, R. Maltos, J. Vanneste, and T. Y. Cath, "Mixed draw solutions for improved forward osmosis performance," *J Memb Sci*, vol. 491, pp. 121–131, Oct. 2015, doi: 10.1016/J.MEMSCI.2015.05.016.

41. A. Achilli, T. Y. Cath, E. A. Marchand, and A. E. Childress, "The forward osmosis membrane bioreactor: A low fouling alternative to MBR processes," *Desalination*, vol. 239, no. 1–3, pp. 10–21, Apr. 2009, doi: 10.1016/J.DESAL.2008.02.022.

42. F. Parveen and N. Hankins, "Integration of forward osmosis membrane bioreactor (FO-MBR) and membrane distillation (MD) units for water reclamation and regeneration of draw solutions," *J Water Process Eng*, vol. 41, p. 102045, doi: 10.1016/j.jwpe.2021.102045.

43. P. Zhao et al., "Polyelectrolyte-promoted forward osmosis process for dye wastewater treatment – Exploring the feasibility of using polyacrylamide as draw solute," *Chem Eng J*, vol. 264, pp. 32–38, Mar. 2015, doi: 10.1016/J.CEJ.2014.11.064.

44. C. Ju and H. Kang, "Zwitterionic polymers showing upper critical solution temperature behavior as draw solutes for forward osmosis," *RSC Adv*, vol. 7, no. 89, pp. 56426–56432, doi: 10.1039/c7ra10831a.

45. M. S. Nawaz, F. Parveen, G. Gadelha, S. J. Khan, R. Wang, and N. P. Hankins, "Reverse solute transport, microbial toxicity, membrane cleaning and flux of regenerated draw in the FO-MBR using a micellar draw solution," *Desalination*, vol. 391, pp. 105–111, Aug. 2016, doi: 10.1016/J.DESAL.2016.02.023.

46. M. S. Nawaz, G. Gadelha, S. J. Khan, and N. Hankins, "Microbial toxicity effects of reverse transported draw solute in the forward osmosis membrane bioreactor (FO-MBR)," *J Memb Sci*, vol. 429, pp. 323–329, Feb. 2013, doi: 10.1016/J.MEMSCI.2012.11.057.

47. R. W. Holloway, A. E. Childress, K. E. Dennett, and T. Y. Cath, "Forward osmosis for concentration of anaerobic digester centrate," *Water Res (Oxford)*, vol. 41, no. 17, pp. 4005–4014, 2007, doi: 10.1016/j.watres.2007.05.054.

48. J. R. McCutcheon, R. L. McGinnis, and M. Elimelech, "Desalination by ammonia–carbon dioxide forward osmosis: Influence of draw and feed solution concentrations on process performance," *J Memb Sci*, vol. 278, no. 1–2, pp. 114–123, Jul. 2006, doi: 10.1016/J.MEMSCI.2005.10.048.

49. L. P. Raman, M. Cheryan, and N. Rajagopalan, "Consider nanofiltration for membrane separations," *Chem Eng Prog*, vol. 90, no. 3, pp. 68–74, 1994.

50. Gary Miner, "Nanofiltration: Principles and applications," *Am Water Works Assoc J*, vol. 97, no. 11, p. 121.

51. A. P. Duarte, M. T. Cidade, and J. C. Bordado, "Cellulose acetate reverse osmosis membranes: Optimization of the composition," *J Appl Polym Sci*, vol. 100, no. 5, pp. 4052–4058, doi: 10.1002/app.23237.

52. H. Strathmann, "Membrane separation processes: Current relevance and future opportunities," *AIChE J*, vol. 47, no. 5, pp. 1077–1087, doi: 10.1002/aic.690470514.

53. J.-H. Choi, K. Fukushi, and K. Yamamoto, "A submerged nanofiltration membrane bioreactor for domestic wastewater treatment: The performance of cellulose acetate nanofiltration membranes for long-term operation," *Sep Purif Technol*, vol. 52, no. 3, pp. 470–477, 2007, doi: 10.1016/j.seppur.2006.05.027.

54. R. H. Lajimi, A. ben Abdallah, E. Ferjani, M. S. Roudesli, and A. Deratani, "Change of the performance properties of nanofiltration cellulose acetate membranes by surface adsorption of polyelectrolyte multilayers," *Desalination*, vol. 163, no. 1–3, pp. 193–202, doi: 10.1016/S0011–9164(04)90189–0.

55. R. Haddada, E. Ferjani, M. S. Roudesli, and A. Deratani, "Properties of cellulose acetate nanofiltration membranes. Application to brackish water desalination," *Desalination*, vol. 167, no. 1–3, pp. 403–409, 2004, doi: 10.1016/j.desal.2004.06.154.

56. R. E. Kravath and J. A. Davis, "Desalination of sea water by direct osmosis," *Desalination*, vol. 16, no. 2, pp. 151–155, Apr. 1975, doi: 10.1016/S0011–9164(00)82089–5.

57. M. R. Kosuri and W. J. Koros, "Defect-free asymmetric hollow fiber membranes from Torlon®, a polyamide–imide polymer, for high-pressure CO2 separations," *J Memb Sci*, vol. 320, no. 1–2, pp. 65–72, Jul. 2008, doi: 10.1016/J.MEMSCI.2008.03.062.

58. K. Y. Wang and T.-S. Chung, "Polybenzimidazole nanofiltration hollow fiber for cephalexin separation," *AIChE J*, vol. 52, no. 4, pp. 1363–1377, doi: 10.1002/aic.10741.

59. H. Jain and M. C. Garg, "Fabrication of polymeric nanocomposite forward osmosis membranes for water desalination—A review," *Env Technol Innov*, vol. 23, p. 101561, Aug. 2021, doi: 10.1016/J.ETI.2021.101561.

60. J. Wang and X. Liu, "Forward osmosis technology for water treatment: Recent advances and future perspectives," *J Clean Prod*, vol. 280, p. 124354, Jan. 2021, doi: 10.1016/J.JCLEPRO.2020.124354.

61. A. F. H. Abdul Rahman and M. N. Abu Seman, "Modification of commercial ultrafiltration and nanofiltration membranes by UV-photografting technique for forward osmosis application," *Mater Today : Proc*, vol. 17, pp. 590–598, 2019, doi: 10.1016/j.matpr.2019.06.339.

62. S. Zou, Y. Gu, D. Xiao, and C. Y. Tang, "The role of physical and chemical parameters on forward osmosis membrane fouling during algae separation," *J Memb Sci*, vol. 366, no. 1, pp. 356–362, 2011, doi: 10.1016/j.memsci.2010.10.030.

63. D. Emadzadeh, W. J. Lau, and A. F. Ismail, "Synthesis of thin film nanocomposite forward osmosis membrane with enhancement in water flux without sacrificing salt rejection," *Desalination*, vol. 330, pp. 90–99, doi: 10.1016/j.desal.2013.10.003.

64. A. Seidel and M. Elimelech, "Coupling between chemical and physical interactions in natural organic matter (NOM) fouling of nanofiltration membranes: Implications for fouling control," *J Memb Sci*, vol. 203, no. 1, pp. 245–255, 2002, doi: 10.1016/S0376–7388(02)00013–3.

65. V. I. Rich and R. M. Maier, "Aquatic environments," Environmental Microbiology: Third Edition, pp. 111–138, Jan. 2015, doi: 10.1016/B978–0–12–394626–3.00006–5.

66. N. A. Arriola, G. D. dos Santos, E. S. Prudêncio, L. Vitali, J. C. C. Petrus, and R. D. M. Castanho Amboni, "Potential of nanofiltration for the concentration of bioactive compounds from watermelon juice," *Int J Food Sci Technol*, vol. 49, no. 9, pp. 2052–2060, doi: 10.1111/ijfs.12513.

9 Application of NF for Agricultural and Food Industry Wastewater

Kah Chun Ho and Aida Isma M. I.
SEGi University

CONTENTS

9.1 INTRODUCTION

9.1.1 AGRICULTURAL WASTEWATER

Agricultural wastewater is produced from various farm activities such as animal feeding processes and the handling of agricultural products. Figure 9.1 demonstrates the water consumption and wastewater generation by agricultural, municipal, and industrial sectors [1]. As shown in the figure, the agriculture sector has the highest consumption of water (38%) and generation of wastewater (32%) compared to the municipal and industrial sectors. Hence, recycling agricultural wastewater help to conserve resources and reduce operating cost.

In general, agricultural wastewater is mainly contributed by the edible oil industry and agricultural runoff. Wastewater from edible oil industries is mainly produced during vegetable oil production and processing through physical refining such as degumming, deacidification, deodorization, and neutralization. In addition, chemical refining of vegetable oils produces massive amount of wastewater approximately $10-25\,m^3$ per metric ton of refined oil [2]. Specifically, edible oil wastewater is generated from acidification of soapstock, filter backwash, and equipment cleaning during vegetable oil refining. Table 9.1 indicates the characteristics of wastewater released from various edible oil industries. As shown, the palm oil mill discharges wastewater is acidic with relatively high chemical oxygen demand (COD), biological oxygen demand (BOD), total dissolved solids, and total suspended solids [3]. Discharge of untreated palm oil mill wastewater into the water bodies can endanger aquatic life by depleting dissolved oxygen. Besides, high turbidity (65,590–69,410 NTU) of palm oil mill wastewater

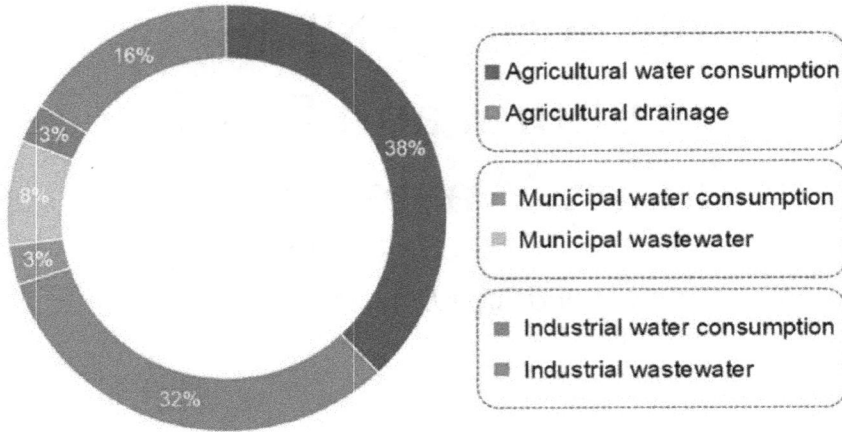

FIGURE 9.1   Water consumption and wastewater generation by main sectors [1].

TABLE 9.1
Characteristics of Wastewater Released from Various Edible Oil Industries

	Palm Oil Mill [5]	Corn Oil [6]	Sunflower Oil [7]	Soya Edible Oil [8]	Olive Oil [9]
pH	3.9	2.79	8.4	4.2	4.43
COD (mg/L)	114,800	12,880	946	17,000	130
BOD (mg/L)	25,000	1,932	N/A	4,340	41
TDS (mg/L)	32,000	N/A	2,326	6,250	31
TSS (mg/L)	44,000	N/A	N/A	450	26

TDS = Total dissolved solids, N/A= Not available

restricts the sunlight availability by aquatic plants while appreciable oil and grease contents (2,234–27,166 ppm) can harm water-resistant parts of seabirds and aquatic mammals [4].

Secondly, agricultural wastewater is contributed by agricultural runoff from farming activities caused by the discharge of fertilizer and pesticides in water bodies, which subsequently leads to its accumulation in the riverbank. Chemical fertilizers with high-level of nitrogen, phosphorus, potassium, and the necessary minor and trace elements are used frequently in croplands during the harvesting period as well as in the pre-harvest cycle. It has been reported that farmers apply 5–10 times of the agriculture phosphorus requirements onto soils to meet the nitrogen requirements due to the imbalance weight ratios of nitrogen to phosphorus (3:1) in the fertilizer [10]. This eventually led to the fertilizer wastages as a result of agricultural runoff. The runoff of fertilizer can also contribute to air pollution, soil acidification, biodiversity damage, and water eutrophication [11]. In addition, pesticides are mainly used to protect plants against harmful organisms and pests. The concentration and types of pesticides in the agriculture wastewater can vary with the processed products and the season. Organochlorine pesticides are the most hazardous pesticides available in the market [12]. Pesticides can cause harmful effects to human beings such as respiratory inflammation, skin itchiness, drowsiness, and even loss of life [13].

9.1.2   FOOD INDUSTRY WASTEWATER

Food processing is a crucial process in the food supply chain. However, the water footprint of food processing is drawing much concern nowadays due to the considerable water consumption in the

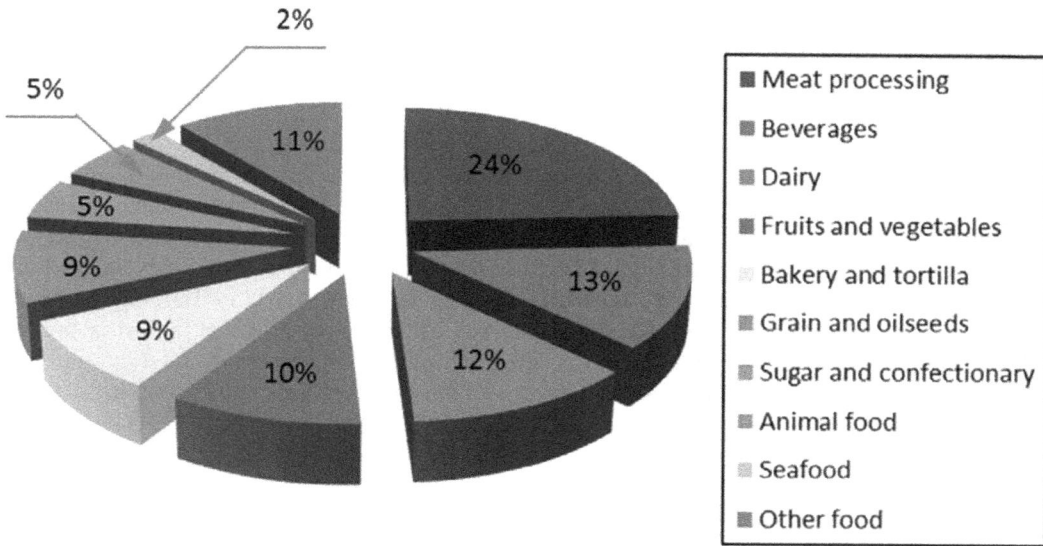

FIGURE 9.2 Distribution of water consumption in the food industries [14].

TABLE 9.2
Quality of Typical Food Industry Wastewater

	Dairy [17]	Brewery [18]	Potato Processing [19]	Winery [20]	Meat Processing [21]
pH	7.8	3–11	4.5	2.5–12.9	3.5–6.6
COD (mg/L)	2,000–6,000	2,000–32,500	37,000	320–49,105	800–135,000
TOC (mg/L)	120	N/A	N/A	41–7,363	7,884–8,759
TN (mg/L)	300–400	25–450	620	10–415	50–300
TP (mg/L)	20–70	0.5–216	560	2–280	107.7–120.2

TP = Total phosphorus, TN = total nitrogen, N/A = Not available

production and the large amounts of wastewater generation (as shown in Figure 9.2) [14]. Generally, the food industry wastewater is produced during food processing, cooling, heating, and cleaning. It is reported that the meat-based diet has a higher water footprint compared to that of vegetables by 36%. For instance, roughly 29, 31, and 112 L of water are needed to obtain 1 g of animal protein from egg, milk, and meat, respectively, while around 21 L of water are used to obtain 1 g of cereal protein [15].

Table 9.2 shows the characteristics of typical food industry wastewater. As can be seen from the table, meat processing industry contributes the most polluted wastewater through slaughtering, washing, processing, and packing process. It is reported that the slaughterhouse industry in Europe generates an average of 145 million m³ of wastewater per annum [57]. Meat processing wastewater contains lots of organic matter content, BOD, and COD because of blood, mucosa, and tallow. Besides, blood from meat increases the nitrogen and phosphorus content in the wastewater which are the nutrients for bacteria. Furthermore, the huge number of organic and inorganic materials for instance carbohydrates (starch for coating), proteins and fat, oil, and grease present in the wastewater discharged from fast-food processing industry. Disposal of such waste is a significant concern for this industry and its detrimental environmental risk. More recently, it has been reported that microplastic (MP) is found in the wastewater discharged from livestock farms (8–40 n/L) and fish ponds (13–27 n/L). This is because MPs are easily accessible for digestion by the aquatic life, which may cause many harmful impacts on organisms such as feeding reduction, growth inhibition, breeding, and sometimes death [16].

9.2   NF TREATMENT

Nanofiltration (NF) membrane separation process is considered an effective method for treatment for agricultural and food industry wastewater to remove dissolved organic carbon (DOC), hardness, and several organic micropollutants. NF membranes are effective in recovering clean water for recycling from the agricultural wastewater [22,23], conventional irrigation system [24], aquaculture farm [25], livestock [26,27], and food industries, including dairy industry [28–31], food processing industry [22,32–34], and beverage industry [35], as well as recover by-products like whey protein [36,37] and antioxidant compounds [38]. This section discusses the treatment of agricultural and food industry wastewater as a single NF and integrated NF process focusing on reuse, recycle, and resource recovery approaches.

9.2.1   Single NF Process

In general, single NF process is used to treat low-strength wastewater discharged from the fruit juice, beverage, dairy, seafood, and vegetable oils industries [39]. NF can decrease the organic content and at the same time recover valuable food components from wastewater. For instance, fruits and vegetable processing wastewater are commonly studied to recover various antioxidants and dietary fibers. In addition, animal processing wastewater contains high amount of proteins. For example, proteins and saccharides can be extracted from the cheese whey processing wastewater in dairy industry.

Table 9.3 shows some of the recent applications of single NF process for agricultural and food industry wastewater treatment. Ochando-Pulido and Martinez-Ferez [22] have treated two-phase olive mill wastewater using three types of membranes. The membranes include microfiltration (MF), ultrafiltration (UF), and loose NF. Their result proved that loose NF membrane with molecular weight cut-off (MWCO) of 3,000 Da improved the effluent quality by transferring the phenolic fraction to the concentrate stream. The membrane also reduced the COD by 60% in the treated effluent compliant to the indications given by the Food and Agriculture Organization for irrigation

TABLE 9.3
Recent Application of Single NF for Agricultural and Food Industry Wastewater Treatment

NF Membrane	Wastewater	Performance	Reference
TFC poly(piperazine amide) NF membrane	Diazinon (pesticide)	Water flux = 41.56 L/m^{2}h Rejection of diazinon = 98.8%	[41]
NF270, NF200, and NF90 (Dow Filmtec)	Industrial lupin beans wastewater	Water flux = 33 L/h m^2 Rejection of lupanine = 99.5% Rejection of oil = 99.8%	[42]
PES-based nanocomposite NF membranes modified with β-CD/MWCNTs	Powder milk solution	Water flux = 21.5 kg/m^2 FRR = 89% Irreversible resistance = 11.1%	[43]
TA-BM/PES blend membranes	Licorice solution	Rejection of licorice dye = 97.3% Flux reduction = 20.8%	[44]
TiO$_2$/3-CPTES/metformin/ PES NF membrane	Licorice solution	Permeation flux = 25 kg/m^{2}h Rejection of COD = 88% Rejection of dye = 98%	[45]
Yttria-stabilized ZrO$_2$ (8YSZ) NF membrane	Carbofuran (pesticide)	Water flux = 3.9–4.2 L/m^{2}h Rejection of carbofuran = 89%	[46]

reuse. Later, their research team further conducted the further treatment of olives and olive oil washing wastewaters using commercial NF (DK 2540F1072) with lower MWCO of 150–300 Da from GE Water & Process Tech [40]. Similar quality of effluent was obtained and authors further reported that acid (citric acid monohydrate) cleaning of 15 minutes and alkaline/detergent (sodium hydroxide) cleaning of 15 minutes could recover the permeability of the membrane satisfactorily.

In addition, Esteves et al. [42] employed commercial NF membranes (NF270, NF200, and NF90) from Dow Filmtec to treat wastewaters from lupin beans processing. It is reported that the NF270 membrane demonstrated the highest rejection of lupanine (99.5%) and the organic matter (94%). NF270 is typically produced as polypiperazine thin-film composite (TFC) with high salt rejection (>97%) via interfacial polymerization. Similar result was obtained by Tan et al. [47] using NF270 treating aerobically treated palm oil mill wastewater. Later, Karimi et al. [41] studied the effect of triethylamine (TEA) as an accelerator in the aqueous phase during the interfacial polymerization of polypiperazine TFC NF membrane. The NF membranes were then used to remove two types of pesticides: atrazine and diazinon, which are the most common contaminants found in agricultural wastewater. In general, the modified NF membranes produced with TEA demonstrated higher rejection and water flux than the pristine membranes. The modified membrane with 2% (w/v) TEA showed higher water flux (41.56 L/m²h) and diazinon rejection (98.8%) compared to that of pristine membrane of 22 L/m²h and 95.2%, respectively. This is because of the higher roughness, density, and hydrophilicity by the addition of TEA.

More recently, many studies have reported that incorporating different nanomaterials into NF membranes has elevated its performances and imparts advantageous characteristics possessed by the nanomaterials [48]. These nanomaterials have large surface area, high chemical and physical stability, active surface, structural stability, surface functionality, and high regular/ordered channel-like structure that significantly improve water transport and rejection of membrane. Such properties have improved membrane permeation and rejection properties [49]. Rahimi et al. [43] have incorporated β-cyclodextrin functionalized with multiwalled carbon nanotubes (β-CD/MWCNTs) into polyethersulfone (PES) NF membranes. The research examined the fouling propensity of membranes using powder milk solution which is commonly present as contaminant in dairy industry wastewater. It was reported that the NF membrane incorporated by β-CD/MWCNTs with concentration of 0.5 wt% exhibited high flux recovery ratio (FRR) value (76%–84%) of the powder milk solution over a three conservative-cycle process. High FRR signified lower occurrence of irreversible membrane fouling caused by the powder milk due to the addition of β-CD/MWCNTs.

In addition, modified NF process was employed for the treatment of wastewater from the licorice processing industry that contains high COD, high turbidity, and strong color. Oulad et al. [44] synthesized PES NF membrane embedded with tannic acid-coated boehmite nanomaterials (TA-BM) for color rejection from licorice solution (1 g/L). The PES NF membrane with 0.5 wt% TA-BM demonstrated a high color rejection of 97.3% after 8 hours because of the hydrophilic hydroxyl groups on the membrane surface. Their studies were also supported by Barahimi et al. [45] who added titanium oxide (TiO₂)-based nanomaterials functionalized with metformin using silane coupling agent 3-cyanopropyltriethoxysilane (CPTES). The modified NF membrane with 1 wt% of nanomaterials exhibited high permeation flux (25 kg/m²h), COD rejection (88%), and color rejection (98%) during membrane treatment of licorine wastewater at 5 bar after 150 minutes. This was mainly due to the hydrophilicity and ultrafine properties of the modified PES membrane evidenced by the topography images in Figure 9.3 [45].

9.2.2 Integrated NF Process

In recent years, several NF integration processes were reported to provide extraordinary efficiencies and greater control over the process when treating specific pollutants in the wastewater. Typically, the NF process can be integrated with other existing treatment techniques such as biological degradation, adsorption, Fenton, electrodialysis, ozonation, coagulation, and flocculation (Table 9.4).

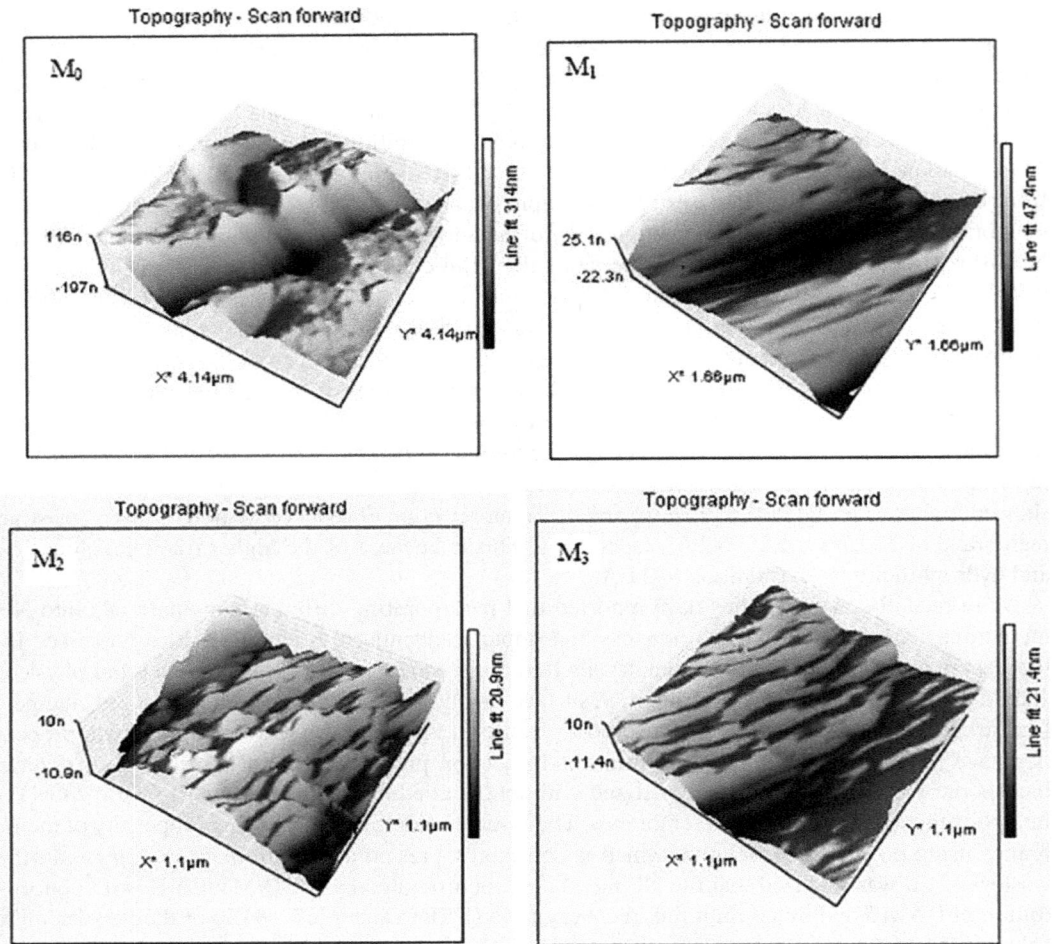

FIGURE 9.3 Topography of membrane M0, M1, M2, and M3 with concentration of TiO_2/CPTES/metformin nanomaterials of 0, 0.1, 0.5, and 1 wt%, respectively [45].

Several integrations of the pressurized membrane processes have been applied for agricultural and food wastewater treatment. For instance, Bortoluzzi et al. [28] have compared two integrated processes: (i) MF and NF, and (ii) MF and reverse osmosis (RO) under various operating pressures to treat dairy wastewater. It was reported that MF and RO integrated system is more efficient compared to the MF and NF integrated system in removing turbidity (100%), color (100%), total kjeldahl nitrogen (TKN) (94%), and total organic carbon (84%). However, the operating pressure (30 bar) of RO was significantly higher than that of NF (20 bar) which can be translated to the higher operating cost. Later, Kim et al. [50] performed the environmental and economic analysis of an integrated fertilizer-drawn forward osmosis (FDFO)-NF system. The FDFO utilized fertilizers solution as its draw solution where it contains fertilizer nutrients and potential for crops irrigation after the FO process. It was shown that the FDFO-NF system had lower environmental effect compared to that of conventional RO hybrid systems because of lesser energy utilization and cleaning chemicals. The energy utilization for wastewater treatment by the FDFO-NF hybrid system was 1.08 kWh/m^3, which was lesser than required for MF-RO and UF-RO systems by 13.6% and 21%, respectively.

Recently, the amounts of wastewater released by olive oil industries have drawn much attention due to the escalating production of olive oil to meet the surge in worldwide demand. As mentioned

TABLE 9.4

Recent Application of Integrated NF for Agricultural and Food Industry Wastewater Treatment

Integrated NF Process	Wastewater	Performance	Reference
SBR, sand filtration, UF, NF, RO	Meat rendering	Rejection of conductivity > 85% Rejection of phosphate > 97% Rejection of TKN > 67% Rejection of nitrate > 79% Rejection of ammonia > 76%	[27]
MF, NF	Dairy	Rejection of turbidity = 100% Rejection of color = 100% Rejection of TKN = 94% Rejection of TOC = 84%	[28]
FO, NF	Synthetic fertilizer	Recovery rate of 84%	[50]
Fenton, NF	Olive oil washing	Rejection of COD = 84.9% Rejection of conductivity = 42.3%	[40]
MBR, NF/electrodialysis	Brewery	Rejection of COD = 81.2% Rejection of conductivity = 86.9%	[51]
Ozonation, NF	Baker's yeast	Rejection of COD = 95.19% Rejection of total hardness = 99.41% Rejection of chloride = 95.74% Rejection of conductivity = 89.57%	[34]
Cationic exchanger, activated carbon, NF, RO	Fermentation	Purity of succinic acid crystal = 96.7% Yield of succinic acid = 65%	[52]
Coagulation and flocculation, UF, NF, RO, adsorption	Olive mill	Purity of phenol = 90%	[53]
MF, coagulation-flocculation, biodegradation and solar photocatalysis, NF, RO	Table olive	Rejection of COD = 82%–93% Rejection of total phenols = 81%–100% Rejection of EC = 60%–98%	[54]

in Section 9.2.1, single NF process can produce high-quality treated olive oil wastewater. However, direct filtration of the raw wastewater usually caused the rapid formation of membrane fouling. Javier Miguel Ochando-Pulido and Martinez-Ferez [40] have integrated Fenton reaction with NF process to treat two-phase olives and olive oil washing wastewaters. The Fenton aims to remove the high organic matter content in the olive oil wastewater through advanced oxidation processes. It was reported that membrane fouling is still present because of the extremely high organic acids in the wastewater. Recently, Aldana et al. [54] further evaluated the effects of several integrated NF processes including MF, coagulation-flocculation, biodegradation, and solar photocatalysis for table olive processing wastewater treatment. It was reported that NF followed by either aerobic biodegradation or solar TiO_2 photocatalysis decreased the pollutants amount in wastewater satisfactorily to be discharged into the municipal sewage system. The retentate from NF membrane filtration was extremely rich in polyphenolic compounds which can be recovered as valuable by-products. This was then verified by Njimou et al. [53] where they have integrated few processes including coagulation-flocculation, UF, NF, RO, and adsorption for treating olive mill wastewater. The adsorption was added to recover phenols from the portion from the membrane process. The adsorbent was synthesized from orange-peel powder as strong support mixed with aluminum oxide (Al_2O_3). The NF concentrate contained 1,691.496 mg/L of polyphenol and the synthesized adsorbent could separate phenol and hydroxytyrosol from the NF concentrate reaching 90% concentration. In addition, integration of the sequential batch reactor (SBR), sand filtration, UF, NF, and RO was employed for the treatment of rendering plant wastewater (Figure 9.4) [27]. The pretreatment of the secondary effluent with sand filtration had

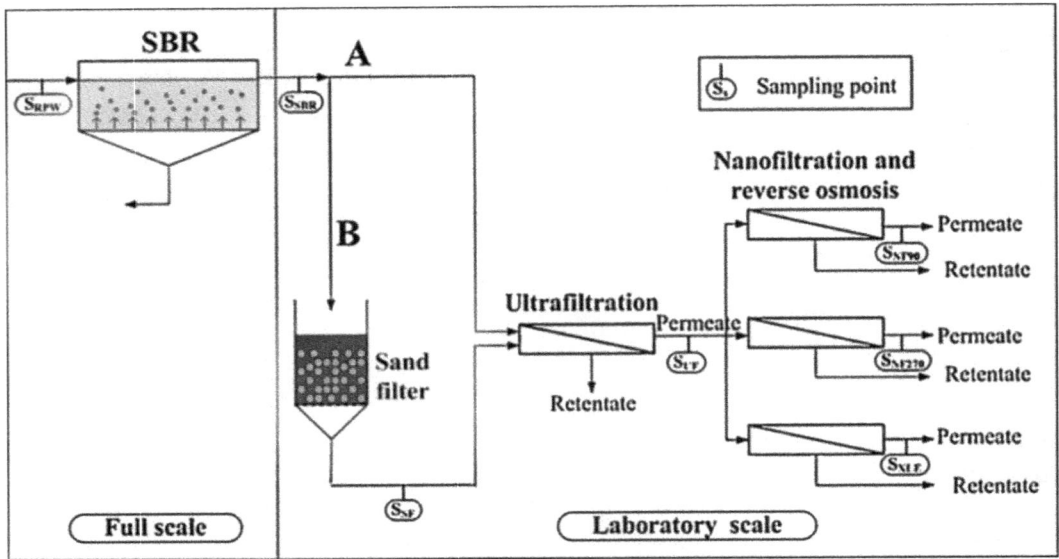

FIGURE 9.4 Schematic representation of the rendering plant wastewater treatment [27].

enhanced membranes performances by decreasing the turbidity, DOC, and COD. The treated wastewater could be reused for washing factory floors and steam generation in rendering plants.

9.3 CHALLENGES AND RECENT STRATEGIES IN NF OPERATION

NF membrane separations are a promising technology to be incorporated as a resource recovery strategies and wastewater treatment, due to the fact that permeated water can be reclaimed in the production processes. NF separation processes can also be utilized as an option in contrast to traditional treatment methods, allowing for more cost-effective production and greater product quality in terms of technological functioning and nutritional value, as well as incorporating social, environmental, and economic sustainability.

Water reconditioning and reuse allows the agricultural and food industries to extend the water cycle within a processing operation. It has the ability to extract and recover high value-added compounds [55,56]. Water is required at various phases of food production chain, including irrigation, processing, cooling, heating, and cleaning [14]. NF has been successfully used in variety of applications in the fruit juice, beverage, dairy, sugar, lactic acid, and vegetable oils processing sectors. NF is also utilized for wastewater treatment, value-added components and solvents extraction from food waste, and food products deacidification/enrichment. Despite all the positive aspects of NF membranes, there are still several unsolved issues and challenges which include membrane fouling and remediation options, recovery of valuable by-products, as well as additional treatment of concentrates. Therefore, to secure the efficiency of NF membrane operation, significant efforts are being made to design membranes that overcome those limitations.

9.3.1 MEMBRANE FOULING AND REMEDIATION OPTIONS

NF membranes are very selective, removing a large percentage of total organic carbon and other organic molecules with a greater molecular weight. The charge groups, polarity (such as hydrophilicity or hydrophobicity), and the molecular weight distribution of the chemicals in the solution have a significant impact on NF separation performance [39]. Membrane fouling can occur as a result of organic matter adsorption due to steric exclusion. The membrane materials, transport mechanism, and operating parameters such as temperature, pressure, pH, and feed concentration all have an impact on the optimum efficiency of this NF membrane [57]. The mass transport of NF membrane is also affected by the crossflow velocity, diffusivity, solubility, and the presence of salts [58].

Bortoluzzi et al. [28] evaluated the dairy wastewater treatment using MF, NF, and RO in double-stage integrated filtration systems, whereby protein, amino acids, and polysaccharides played major role as blocking agents causing membrane fouling. Results revealed that the combined MF and NF system reduced 100% turbidity, 96% color, 84% total organic carbon (TOC), and 58% TKN. Furthermore, solids deposition and pressure on the membrane surface affected the permeate flux, as higher pressure produced higher flux. Chen et al. [31] correlated the separation performance and physiochemical properties of NF membrane for dairy wastewater treatment and observed that contact angle and higher roughness resulted in higher irreversible fouling. This implied that foulants adsorption at the membrane surface was the major fouling mechanism rather than the pore blocking (affected by pore size) or cake formation (its compressibility was influenced by trans membrane pressure (TMP)). Moreover, the retentions of salts and lactose by the NF membranes have been shown to be controlled by steric hindrance.

One of the significant challenges was the decline in permeate flux due to membrane fouling when using NF membrane for direct treatment of wastewater. The success of NF depends on the interactions between the membrane and the feed solution, causing biofouling, inorganic fouling, or scaling [59]. Balcıoğlu and Gönder [34] observed relatively lower flux decline during treatment of wastewater from baker's yeast with 18% of membrane fouling and the optimum permeate qualities achieved with the used of the combined membranes (NF and RO) treatment.

9.3.2   Recovery of Valuable By-products

Food processing wastewater contains useful by-products that can be collected to reduce pollutant levels in effluent streams and improve quality of water for reuse or disposal. Integration of NF membranes separation with agricultural and food industry wastewater containing value-added products (such as pigment, polysaccharide, and protein) could increase the feasibility of techno-economic in food wastewater treatment. However, apart from the desired final product obtained, separation or recovery of intermediate and/or secondary metabolites with high added value compounds is a concern [60].

One of the earliest industries to adopt NF technology was the dairy industry [61]. Salty waste streams such as NF whey demineralization waste, ion exchange, electrodialysis, chromatography brines, and clean-in-place (CIP) wastewater, are produced in dairy processing plants, i.e., originating from dairy processes [62]. Every 1 kg of produced cheese generates 9 L of whey which will be further processed into various products, as well as salts from the milk, and the cheese-salting processes will be concentrated in the waste streams [62].

To recover phenolic compounds from olive mill wastewater, Kontos et al. [32] used a combination of crystallization and membrane filtration approach. NF treatment was used by Esteves et al. [42] to reclaimed lupanine from lupin beans processing wastewater. The NF membrane, which came from a lupanine-rich retentate, had the maximum lupanine rejection of 99.5%. Omwene et al. [52] used ion exchange resins in combination with NF and RO to produce bio-based succinic acid. At pH less than 2, the NF membrane rejected 53.1%, 51.8%, 46.6%, and 39.8%, respectively, of succinic acid, lactic acid, formic acid, and acetic acid.

NF is used widely in sugar refinery industry because it is feasible to separate smaller molecules such as fructose, glucose, amino acids, and peptides producing higher flux which results in higher rejection due to inhibitors such as acetic acid [63]. Sjölin et al. [64] found that the NF membrane maintained 40%–90% of total sugar. To reduce sugar loss, energy input, and evaporator size, NF was also used to filter-diluted juice. To reduce the possibility of microbial development, the NF procedure for galacto-oligosaccharide (GOS) purification was carried out at 5°C and 60°C [65]. Using an NF membrane at 45 bar, 85% of product purity (monosaccharide content) and recovery yield of

82% oligosaccharide were obtained. Inulin-type fractions can also be separated from sugar solution. Organic acids separated from saccharides using NF before purification for lactic acid separation from fermentation broth appeared promising with 93.28% of glucose rejection utilizing a low-flux NF membrane [66]. Córdova et al. [67] established a purification approach for GOS, a potent prebiotic, employing three stages of commercial NF at various temperatures and total suspended solids (TSS) (Brix). Enzymatic step was added to eliminate lactose, which improved the flux because there was less lactose present, but the gain in GOS purified was minimal.

Food and nutrient-rich fertilizer consumption has increased, implying a greater need for resource recovery from urban wastewater which can contribute significantly to the circular economy [68]. Urban wastewater may be converted into important sources of clean water, energy, and nutrients [69,70]. Membrane capabilities involve recovering valuable components from by-products, diluted effluents, and wastewater.

NF membrane techniques can be used to generate products and components with desirable properties that conventional techniques cannot offer. As a result, additional research should be conducted to improve the effectiveness of technologies recovering resources by constructing a pilot-scale model of various wastewater treatment solutions. NF membrane separation techniques are categorized as green technology since they are more energy-efficient and avoid the use of additives and chemicals, which is beneficial for the environment and human health. It is recommended that green technologies and renewable natural resources to be used to reduce the industrial carbon footprint and achieve development of sustainability.

9.3.3   Additional Treatment of Concentrates

The NF membrane is employed in the food industries to concentrate food and beverages naturally without degrading the products. Membranes can be used to produce new food items or upgrade current ones as creative approaches for producing tailored-functionality ingredients. The formation of a concentrate flow is an issue inherent with NF membrane processes. The concentrate is often an unwanted by-product of the purification process and must be further managed before being discharged to water bodies. NF has been used to ultrafiltrate whey in the dairy industry to recover lactose from retentate and reduce the wastewater load [71]. The phenolic fraction was concentrated and recovered from olive oil washing wastewater resulting in a significant TSS abatement (78.02%) and supernatant recovery of 85.7% with no phenolic fraction loss [22].

Brião et al. [29] employed NF and RO to recover water from dairy washing water, allowing it to be used as cooling water on-site as well as a white milk-based stream for fermented milk beverages. The recovered water failed to fulfill the drinking water quality criteria for cleaning equipment, pipelines and tanks in the CIP system, as well as for feed boilers because the COD in the permeate is greater than 40 mg/L. The results showed that NF membranes produced high permeate flux but low electrical conductivity (EC) and lactose rejection, resulting in COD in the permeate. Higher TMP caused more irreversible fouling, which was linked to flux degradation and the formation of a gel layer. Several factors influenced the yield of value-added compounds, and the number of stages was determined by the purity requirement [71]. As a result, cost-effective extraction procedures that generate high yields of various value-added components must be developed. To gain additional value, other extraction methods must verify that the extracted compounds are safe enough to be used in practical foods, pharmaceuticals, or animal food. Efforts to improve the separation effectiveness and clarity of value-added compounds from NF membrane separation are certainly needed.

9.3.4   Recent Strategies

New techniques to control/mitigate fouling and avoid future equipment and operating issues are crucial. A high-priority research topic is integrating membrane systems with improved cleaning performance, and new module designs and materials could give attractive alternatives.

Membrane fouling is influenced by variety of factors, including membrane physical characteristics, feed solution chemistry, properties of natural organic matter, hydrodynamics, and membrane operating conditions. In the operation of NF separation, including processing trains, concentration polarization is unavoidable and more critical. Fouling can be controlled by improving hydrodynamic conditions on the NF membrane. Improved shear stress, crossflow velocity, flow pattern at the membrane surface, and feed flowrate are some of the fouling mitigation characteristics that could be used to lower concentration polarization on membranes and increase mass transfer coefficient and turbulence.

Mitigation of membrane fouling can also be done by pretreatment of the feed solution. This procedure can be modified according to the application, the membrane performance, feed solution quality, and permeate demand. Pretreatment is an essential and cost-effective step in reducing cleaning time and extending membrane life. Monitoring the pretreatment system and being aware of how trace pollutants, such as trace oils and solvents, could damage the membrane is critical. Parameters such as feed pH, permeate and concentrate flowrates, and the feed of temperature need to be controlled and could help reduce membrane fouling.

Membranes are made from polymers and inorganic materials like metals and ceramics. Polyesters, PES, polyvinylidene fluoride, polyacrylonitrile, and polypropylene copolymers are the most prevalent polymers used in membrane manufacturing [72]. Figure 9.5 shows the common methods of membrane fabrication [73]. Surface modification techniques such as coating, mixing, grafting, ultraviolet irradiation, membrane functionalization, and additive crosslinking have been developed to improve the membrane hydrophilicity and antifouling capabilities [74,75]. In the production of PES membranes, the use of 5% silicon dioxide (SiO_2) nanoparticles and 1.7% Pluronic F127 as additives was compared to the NF of xylitol mixed solution [76]. Results showed that the PES/Pluf127 membrane had the highest xylitol permeate flux and higher fouling resistance than the pure PES and PES/SiO_2 membranes. This was owing to the pore-forming agent, which caused the membrane pore to disintegrate.

Several researchers have attempted to reduce fouling and improve the hydrophilicity and performance of polymeric membranes by incorporating various nanoparticles additives such as MWCNT,

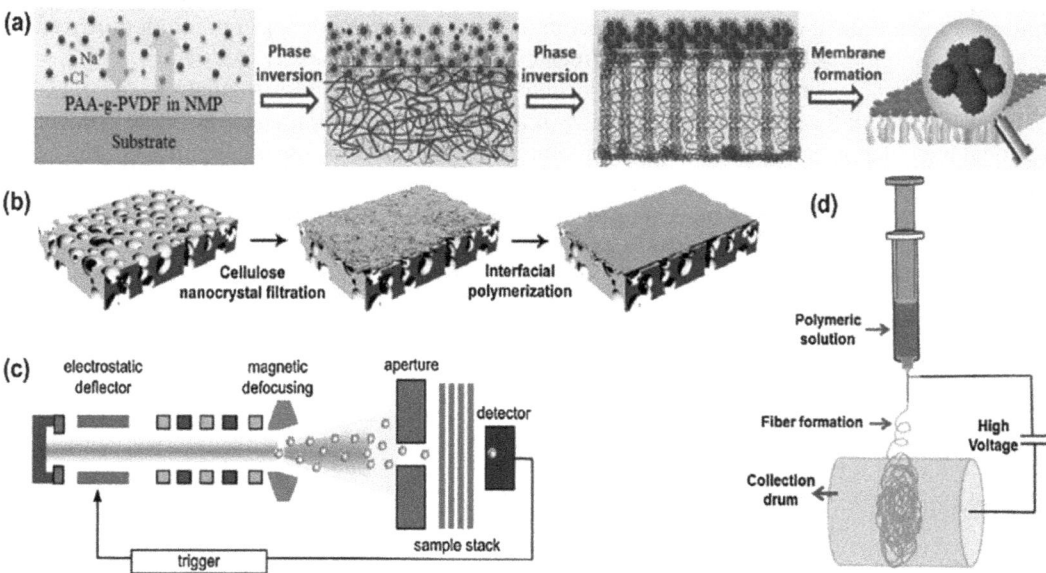

FIGURE 9.5 Methods of membrane fabrication: (a) phase inversion process; (b) interfacial polymerization methodology; (c) track-etching method; and (d) electrospinning process [73].

titanium dioxide (TiO$_2$), magnesium oxide (MgO), graphene oxide, and zinc oxide (ZnO), which allows greater control over the ability to produce the required structure and reduce membrane fouling [77].

Colloidal fouling can be mitigated by membrane surface patterning. Physical modification of surfaces of membrane reduces fouling by altering the hydrodynamics at the solid–liquid boundary [78]. Within the nanometer to micrometer scale, Ref. [79] studied many line-and-groove patterns, circular pillars, rectangular, and pyramids. Reference [80] suggests a nanoscale line-and-groove design on NF membranes to raise the threshold flux by 20%–30% during colloidal suspension membrane filtration. Biofouling resistance has also been researched using irregular shapes such as shark skin-like patterns [81].

Nanohybrid membranes are polymeric membranes that have been functionalized with discrete nanomaterials. In a mixed-matrix membrane, inorganic elements such as zeolites and silicate are incorporated into the polymer membrane matrix to increase selectivity and other practical properties. These membranes are well known as polymer nanocomposites, which are a subset of nanohybrids [82]. For continued advancement of membrane technology, more research investigations on fouling reduction, optimization of permeate flowrate, and membrane lifetime extension are needed.

9.4 CONCLUSION AND FUTURE PERSPECTIVES

The use of NF for agricultural and food industry wastewater has been discussed, as well as the challenges, causes, and barriers of using NF for water reuse, recycling, and resource recovery. Due to increase in water demand and rigorous wastewater disposal standards, the industry will be obliged to evaluate treatment options and viability. Treatment at the processing plant may become more interesting, particularly if valuable by-products can be reclaimed for additional profits. Recent developments in single and integrated NF membranes have been reviewed, and combined treatment technologies have been proposed due to completely different effluent quality produced by various food processing stages.

Membrane fouling and remediation options, as well as the recovery of beneficial by-products and subsequent treatment of concentrates are all the major concerns and challenges in the NF membrane applications. Pretreatment of feed solution, modification of membrane surfaces, enhancing membrane hydrophilicity, and integrating various additives to achieve the appropriate structure could minimize membrane fouling. To advance membrane technology, more research into fouling reduction, membrane lifecycle extension, and permeate flowrate optimization is required.

Moving forward, communicating on NF technology to manufacturers, regulators, and consumers, as well as the information on water quality improvements that NF membrane could produce, is a critical issue that must be addressed. Collaboration among stakeholders is essential for improving knowledge in this area in the interest of all parties involved. Academia must have access to food processing procedures in order to create a personalized solution for the production, and the findings must be made plain and concisely available to the people in order to connect not only scientists but all investors.

ABBREVIATIONS

BOD	Biological Oxygen Demand
CIP	Clean-in-place
COD	Chemical Oxygen Demand
CPTES	Cyanopropyltriethoxysilane
DOC	Dissolved Organic Carbon
EC	Electrical Conductivity
FDFO	Fertilizer-Drawn Forward Osmosis
FRR	Flux Recovery Ratio
GOS	Galacto-oligosaccharides

MP	Microplastic
MWCNT	Multiwalled Carbon Nanotubes
MWCO	Molecular Weight Cut-Off
NF	Nanofiltration
PES	Polyethersulfone
RO	Reverse Osmosis
SBR	Sequential Batch Reactor
TA-BM	Tannic Acid-Coated Boehmite Nanomaterials
TEA	Triethylamine
TFC	Thin-Film Composite
TKN	Total Kjeldahl Nitrogen
TMP	Trans Membrane Pressure
TOC	Total Organic Carbon
TSS	Total Suspended Solids
β-CD/MWCNTs	B-Cyclodextrin-Functionalized Multiwalled Carbon Nanotubes

REFERENCES

1. M. Khodakarami, M. Bagheri, Recent advances in synthesis and application of polymer nanocomposites for water and wastewater treatment, *J. Clean. Prod.* 296 (2021). https://doi.org/10.1016/j.jclepro.2021.126404.
2. E.A. Sharghi, A. Shorgashti, B. Bonakdarpour, The study of organic removal efficiency and membrane fouling in a submerged membrane bioreactor treating vegetable oil wastewater, *Int. J. Eng.* 29 (2016). https://doi.org/10.5829/idosi.ije.2016.29.12c.02.
3. K.C. Ho, Y.X. Teoh, Y.H. Teow, A.W. Mohammad, Life cycle assessment (LCA) of electrically-enhanced POME filtration: Environmental impacts of conductive-membrane formulation and process operating parameters, *J. Environ. Manage.* 277 (2021). https://doi.org/10.1016/j.jenvman.2020.111434.
4. Y.W. Cheng, C.C. Chong, M.K. Lam, W.H. Leong, L.F. Chuah, S. Yusup, H.D. Setiabudi, Y. Tang, J.W. Lim, Identification of microbial inhibitions and mitigation strategies towards cleaner bioconversions of palm oil mill effluent (POME): A review, *J. Clean. Prod.* 280 (2021). https://doi.org/10.1016/j.jclepro.2020.124346.
5. J.O. Iwuagwu, J.O. Ugwuanyi, Treatment and valorization of palm oil mill effluent through production of food grade yeast biomass, *J. Waste Manag.* 2014 (2014) 1–9. https://doi.org/10.1155/2014/439071.
6. S. Aslan, B. Alyüz, Z. Bozkurt, M. Bakaoğlu, Characterization and biological treatability of edible oil wastewaters, *Polish J. Environ. Stud.* 18 (2009) 533–538.
7. P. Ghasemian, E. Abdollahzadeh Sharghi, L. Davarpanah, The influence of short values of hydraulic and sludge retention time on performance of a membrane bioreactor treating sunflower oil refinery wastewater, *Int. J. Eng. Trans. A Basics.* 30 (2017) 1417–1424. https://doi.org/10.5829/ije.2017.30.10a.01.
8. K. Rajkumar, M. Muthukumar, R. Sivakumar, Novel approach for the treatment and recycle of wastewater from soya edible oil refinery industry – An economic perspective, *Resour. Conserv. Recycl.* 54 (2010) 752–758. https://doi.org/10.1016/j.resconrec.2009.12.005.
9. S. Babić, O. Malev, M. Pflieger, A.T. Lebedev, D.M. Mazur, A. Kužić, R. Čož-Rakovac, P. Trebše, Toxicity evaluation of olive oil mill wastewater and its polar fraction using multiple whole-organism bioassays, *Sci. Total Environ.* 686 (2019) 903–914. https://doi.org/10.1016/j.scitotenv.2019.06.046.
10. A. Rosemarin, B. Macura, J. Carolus, K. Barquet, F. Ek, L. Järnberg, D. Lorick, S. Johannesdottir, S.M. Pedersen, J. Koskiaho, N.R. Haddaway, T. Okruszko, Circular nutrient solutions for agriculture and wastewater – A review of technologies and practices, *Curr. Opin. Environ. Sustain.* 45 (2020) 78–91. https://doi.org/10.1016/j.cosust.2020.09.007.
11. M. Colella, M. Ripa, A. Cocozza, C. Panfilo, S. Ulgiati, Challenges and opportunities for more efficient water use and circular wastewater management. The case of Campania Region, Italy, *J. Environ. Manage.* 297 (2021). https://doi.org/10.1016/j.jenvman.2021.113171.
12. K. Rani, G. Dhania, Bioremediation and biodegradation of pesticide from contaminated soil and water – A noval approach, 3 (2014) 23–33.
13. J. Nie, Y. Sun, Y. Zhou, M. Kumar, M. Usman, J. Li, J. Shao, L. Wang, D.C.W. Tsang, Bioremediation of water containing pesticides by microalgae: Mechanisms, methods, and prospects for future research, *Sci. Total Environ.* 707 (2020). https://doi.org/10.1016/j.scitotenv.2019.136080.

14. S. Li, S. Zhao, S. Yan, Y. Qiu, C. Song, Y. Li, Y. Kitamura, Food processing wastewater purification by microalgae cultivation associated with high value-added compounds production – A review, *Chinese J. Chem. Eng.* 27 (2019) 2845–2856. https://doi.org/10.1016/j.cjche.2019.03.028.

15. a K. Chapagain, a Y. Hoekstra, The green, blue and grey water footprint of farm animals and animal products, *Unesco.* 1 (2010) 80. http://www.waterfootprintnetwork.org/Reports/Report47-WaterFootprintCrops-Vol1.pdf%5Cnhttp://wfn.project-platforms.com/Reports/Report-48-WaterFootprint-AnimalProducts-Vol1.pdf.

16. F. Wang, B. Wang, L. Duan, Y. Zhang, Y. Zhou, Q. Sui, D. Xu, H. Qu, G. Yu, Occurrence and distribution of microplastics in domestic, industrial, agricultural and aquacultural wastewater sources: A case study in Changzhou, China, *Water Res.* 182 (2020). https://doi.org/10.1016/j.watres.2020.115956.

17. D. Karadag, O.E. Köroılu, B. Ozkaya, M. Cakmakci, A review on anaerobic biofilm reactors for the treatment of dairy industry wastewater, *Process Biochem.* 50 (2015) 262–271. https://doi.org/10.1016/j.procbio.2014.11.005.

18. S. Tejedor-Sanz, J.M. Ortiz, A. Esteve-Núñez, Merging microbial electrochemical systems with electrocoagulation pretreatment for achieving a complete treatment of brewery wastewater, *Chem. Eng. J.* 330 (2017) 1068–1074. https://doi.org/10.1016/j.cej.2017.08.049.

19. Y. Zhang, H. Su, Y. Zhong, C. Zhang, Z. Shen, W. Sang, G. Yan, X. Zhou, The effect of bacterial contamination on the heterotrophic cultivation of *Chlorella pyrenoidosa* in wastewater from the production of soybean products, *Water Res.* 46 (2012) 5509–5516. https://doi.org/10.1016/j.watres.2012.07.025.

20. L.A. Ioannou, G.L. Puma, D. Fatta-Kassinos, Treatment of winery wastewater by physicochemical, biological and advanced processes: A review, *J. Hazard. Mater.* 286 (2015) 343–368. https://doi.org/10.1016/j.jhazmat.2014.12.043.

21. M. Galib, E. Elbeshbishy, R. Reid, A. Hussain, H.S. Lee, Energy-positive food wastewater treatment using an anaerobic membrane bioreactor (AnMBR), *J. Environ. Manage.* 182 (2016) 477–485. https://doi.org/10.1016/j.jenvman.2016.07.098.

22. J.M. Ochando-Pulido, A. Martinez-Ferez, Novel micro/ultra/nanocentrifugation membrane process assessment for revalorization and reclamation of agricultural wastewater, *J. Environ. Manage.* 222 (2018) 447–453. https://doi.org/10.1016/j.jenvman.2018.05.092.

23. M.P. Zacharof, S.J. Mandale, P.M. Williams, R.W. Lovitt, Nanofiltration of treated digested agricultural wastewater for recovery of carboxylic acids, *J. Clean. Prod.* 112 (2016) 4749–4761. https://doi.org/10.1016/j.jclepro.2015.07.004.

24. A. Chávez, Organic micropollutant removal by a nanofiltration pilot plant used to treat spring water from a wastewater-irrigated valley, *Int. J. Membr. Sci. Technol.* 4 (2018) 64–74. https://doi.org/10.15379/2410-1869.2017.04.02.04.

25. L.Y. Ng, C.Y. Ng, E. Mahmoudi, C.B. Ong, A.W. Mohammad, A review of the management of inflow water, wastewater and water reuse by membrane technology for a sustainable production in shrimp farming, *J. Water Process Eng.* 23 (2018) 27–44. https://doi.org/10.1016/j.jwpe.2018.02.020.

26. L. Lan, X. Kong, H. Sun, C. Li, D. Liu, High removal efficiency of antibiotic resistance genes in swine wastewater via nanofiltration and reverse osmosis processes, *J. Environ. Manage.* 231 (2019) 439–445. https://doi.org/10.1016/j.jenvman.2018.10.073.

27. M. Racar, D. Dolar, A. Špehar, K. Košutić, Application of UF/NF/RO membranes for treatment and reuse of rendering plant wastewater, *Process Saf. Environ. Prot.* 105 (2017) 386–392. https://doi.org/10.1016/j.psep.2016.11.015.

28. A.C. Bortoluzzi, J.A. Faitão, M. Di Luccio, R.M. Dallago, J. Steffens, G.L. Zabot, M. V. Tres, Dairy wastewater treatment using integrated membrane systems, *J. Environ. Chem. Eng.* 5 (2017) 4819–4827. https://doi.org/10.1016/j.jece.2017.09.018.

29. V.B. Brião, A.C. Vieira Salla, T. Miorando, M. Hemkemeier, D.P. Cadore Favaretto, Water recovery from dairy rinse water by reverse osmosis: Giving value to water and milk solids, *Resour. Conserv. Recycl.* 140 (2019) 313–323. https://doi.org/10.1016/j.resconrec.2018.10.007.

30. I.M. Stoica, E. Vitzilaiou, H. Lyng Røder, M. Burmølle, D. Thaysen, S. Knøchel, F. van den Berg, Biofouling on RO-membranes used for water recovery in the dairy industry, *J. Water Process Eng.* 24 (2018) 1–10. https://doi.org/10.1016/j.jwpe.2018.05.004.

31. Z. Chen, J. Luo, X. Hang, Y. Wan, Physicochemical characterization of tight nanofiltration membranes for dairy wastewater treatment, *J. Memb. Sci.* 547 (2018) 51–63. https://doi.org/10.1016/j.memsci.2017.10.037.

32. S.S. Kontos, F.K. Katrivesis, T.C. Constantinou, C.A. Zoga, I.S. Ioannou, P.G. Koutsoukos, C.A. Paraskeva, Implementation of membrane filtration and melt crystallization for the effective treatment and valorization of olive mill wastewaters, *Sep. Purif. Technol.* 193 (2018) 103–111. https://doi.org/10.1016/j.seppur.2017.11.005.

33. B. Ozbey-Unal, C. Balcik-Canbolat, N. Dizge, B. Keskinler, Treatability studies on optimizing coagulant type and dosage in combined coagulation/membrane processes for table olive processing wastewater, *J. Water Process Eng.* 26 (2018) 301–307. https://doi.org/10.1016/j.jwpe.2018.10.023.

34. G. Balcıoğlu, Z.B. Gönder, Baker's yeast wastewater advanced treatment using ozonation and membrane process for irrigation reuse, *Process Saf. Environ. Prot.* 117 (2018) 43–50. https://doi.org/10.1016/j.psep.2018.04.006.

35. N. Dizge, C. Akarsu, Y. Ozay, H.E. Gulsen, S.K. Adiguzel, M.A. Mazmanci, Sono-assisted electrocoagulation and cross-flow membrane processes for brewery wastewater treatment, *J. Water Process Eng.* 21 (2018) 52–60. https://doi.org/10.1016/j.jwpe.2017.11.016.

36. A. Macedo, D. Azedo, E. Duarte, C. Pereira, Valorization of goat cheese whey through an integrated process of ultrafiltration and nanofiltration, *Membranes (Basel).* 11 (2021). https://doi.org/10.3390/membranes11070477.

37. Y.E. Meneses, R.A. Flores, Feasibility, safety, and economic implications of whey-recovered water in cleaning-in-place systems: A case study on water conservation for the dairy industry, *J. Dairy Sci.* 99 (2016) 3396–3407. https://doi.org/10.3168/jds.2015–10306.

38. A. Giacobbo, A. Meneguzzi, A.M. Bernardes, M.N. de Pinho, Pressure-driven membrane processes for the recovery of antioxidant compounds from winery effluents, *J. Clean. Prod.* 155 (2017) 172–178. https://doi.org/10.1016/j.jclepro.2016.07.033.

39. K. Nath, H.K. Dave, T.M. Patel, Revisiting the recent applications of nanofiltration in food processing industries: Progress and prognosis, *Trends Food Sci. Technol.* 73 (2018) 12–24. https://doi.org/10.1016/j.tifs.2018.01.001.

40. J.M. Ochando-Pulido, A. Martinez-Ferez, Operation setup of a nanofiltration membrane unit for purification of two-phase olives and olive oil washing wastewaters, *Sci. Total Environ.* 612 (2018) 758–766. https://doi.org/10.1016/j.scitotenv.2017.08.287.

41. H. Karimi, A. Rahimpour, M.R. Shirzad Kebria, Pesticides removal from water using modified piperazine-based nanofiltration (NF) membranes, *Desalin. Water Treat.* 57 (2016) 24844–24854. https://doi.org/10.1080/19443994.2016.1156580.

42. T. Esteves, A.T. Mota, C. Barbeitos, K. Andrade, C.A.M. Afonso, F.C. Ferreira, A study on lupin beans process wastewater nanofiltration treatment and lupanine recovery, *J. Clean. Prod.* 277 (2020). https://doi.org/10.1016/j.jclepro.2020.123349.

43. Z. Rahimi, A.A. Zinatizadeh, S. Zinadini, M.C.M. van Loosdrecht, β-cyclodextrin functionalized MWCNTs as a promising antifouling agent in fabrication of composite nanofiltration membranes, *Sep. Purif. Technol.* 247 (2020). https://doi.org/10.1016/j.seppur.2020.116979.

44. F. Oulad, S. Zinadini, A.A. Zinatizadeh, A.A. Derakhshan, Fabrication and characterization of a novel tannic acid coated boehmite/PES high performance antifouling NF membrane and application for licorice dye removal, *Chem. Eng. J.* 397 (2020). https://doi.org/10.1016/j.cej.2020.125105.

45. V. Barahimi, R.A. Taheri, A. Mazaheri, H. Moghimi, Fabrication of a novel antifouling TiO2/CPTES/metformin-PES nanocomposite membrane for removal of various organic pollutants and heavy metal ions from wastewater, *Chem. Pap.* 74 (2020) 3545–3556. https://doi.org/10.1007/s11696-020-01178–2.

46. H. Qin, W. Guo, X. Huang, P. Gao, H. Xiao, Preparation of yttria-stabilized ZrO2 nanofiltration membrane by reverse micelles-mediated sol-gel process and its application in pesticide wastewater treatment, *J. Eur. Ceram. Soc.* 40 (2020) 145–154. https://doi.org/10.1016/j.jeurceramsoc.2019.09.023.

47. Y.H. Tan, W.J. Lau, P.S. Goh, N. Yusof, A.F. Ismail, Nanofiltration of aerobically-treated palm oil mill effluent: Characterization of the size of colour compounds using synthetic dyes and polyethylene glycols, *J. Eng. Sci. Technol.* 13 (2018) 1–10.

48. M.N. Subramaniam, P.S. Goh, W.J. Lau, A.F. Ismail, Exploring the potential of photocatalytic dual layered hollow fiber membranes incorporated with hybrid titania nanotube-boron for agricultural wastewater reclamation, *Sep. Purif. Technol.* 275 (2021). https://doi.org/10.1016/j.seppur.2021.119136.

49. F. Mehrjo, A. Pourkhabbaz, A. Shahbazi, PMO synthesized and functionalized by p-phenylenediamine as new nanofiller in PES-nanofiltration membrane matrix for efficient treatment of organic dye, heavy metal, and salts from wastewater, *Chemosphere.* 263 (2021). https://doi.org/10.1016/j.chemosphere.2020.128088.

50. J.E. Kim, S. Phuntsho, L. Chekli, S. Hong, N. Ghaffour, T.O. Leiknes, J.Y. Choi, H.K. Shon, Environmental and economic impacts of fertilizer drawn forward osmosis and nanofiltration hybrid system, *Desalination.* 416 (2017) 76–85. https://doi.org/10.1016/j.desal.2017.05.001.

51. B. Sawadogo, Y. Konaté, G. Lesage, F. Zaviska, M. Monnot, M. Heran, H. Karambiri, Brewery wastewater treatment using MBR coupled with nanofiltration or electrodialysis: Biomass acclimation and treatment efficiency, *Water Sci. Technol.* 77 (2018) 2624–2634. https://doi.org/10.2166/wst.2018.232.

52. P.I. Omwene, Z.B.O. Sarihan, A. Karagunduz, B. Keskinler, Bio-based succinic acid recovery by ion exchange resins integrated with nanofiltration/reverse osmosis preceded crystallization, *Food Bioprod. Process.* 129 (2021) 1–9. https://doi.org/10.1016/j.fbp.2021.06.006.

53. J.R. Njimou, J. Godwin, H. Pahimi, S.A. Maicaneanu, F. Kouatchie-Njeutcha, B.C. Tripathy, A. Talla, T. Watanabe, N.G. Elambo, Biocomposite spheres based on aluminum oxide dispersed with orange-peel powder for adsorption of phenol from batch membrane fraction of olive mill wastewater, *Colloids Interface Sci. Commun.* 42 (2021). https://doi.org/10.1016/j.colcom.2021.100402.

54. J.C. Aldana, J.L. Acero, P.M. Álvarez, Membrane filtration, activated sludge and solar photocatalytic technologies for the effective treatment of table olive processing wastewater, *J. Environ. Chem. Eng.* 9 (2021). https://doi.org/10.1016/j.jece.2021.105743.

55. C.M. Galanakis, E. Agrafioti, Sustainable water and wastewater processing, *Sustain. Water Wastewater Process.* (2019) 1–377. https://doi.org/10.1016/C2017–0–02118–3.

56. Y.E. Meneses, B. Martinez, X. Hu, Water reconditioning in the food industry, *Sustain. Water Wastewater Process.* (2019) 329–365. https://doi.org/10.1016/B978–0–12–816170–8.00010–7.

57. N.F. Ghazali, N.D.A. Razak, Recovery of saccharides from lignocellulosic hydrolysates using nano-filtration membranes: A review, *Food Bioprod. Process.* 126 (2021) 215–233. https://doi.org/10.1016/j.fbp.2021.01.006.

58. Y. Roy, D.M. Warsinger, J.H. Lienhard, Effect of temperature on ion transport in nanofiltration membranes: Diffusion, convection and electromigration, *Desalination.* 420 (2017) 241–257. https://doi.org/10.1016/j.desal.2017.07.020.

59. W.M. Samhaber, Uses and problems of nanofiltration in the food industry, *Chemie Ing. Tech.* 77 (2005) 583–588.

60. M. Reig, X. Vecino, J.L. Cortina, Use of membrane technologies in dairy industry: An overview, *Foods.* 10 (2021). https://doi.org/10.3390/foods10112768.

61. B. Van der Bruggen, M. Mänttäri, M. Nyström, Drawbacks of applying nanofiltration and how to avoid them: A review, *Sep. Purif. Technol.* 63 (2008) 251–263. https://doi.org/10.1016/j.seppur.2008.05.010.

62. G.Q. Chen, S. Talebi, S.L. Gras, M. Weeks, S.E. Kentish, A review of salty waste stream management in the Australian dairy industry, *J. Environ. Manage.* 224 (2018) 406–413. https://doi.org/10.1016/j.jenvman.2018.07.056.

63. L. Zou, S. Zhang, J. Liu, Y. Cao, G. Qian, Y.Y. Li, Z.P. Xu, Nitrate removal from groundwater using negatively charged nanofiltration membrane, *Environ. Sci. Pollut. Res.* 26 (2019) 34197–34204. https://doi.org/10.1007/s11356-018-3829–6.

64. M. Sjölin, J. Thuvander, O. Wallberg, F. Lipnizki, Purification of sucrose in sugar beet molasses by uti-lizing ceramic nanofiltration and ultrafiltration membranes, *Membranes (Basel).* 10 (2020). https://doi.org/10.3390/membranes10010005.

65. S. Pruksasri, T.H. Nguyen, D. Haltrich, S. Novalin, Fractionation of a galacto-oligosaccharides solution at low and high temperature using nanofiltration, *Sep. Purif. Technol.* 151 (2015) 124–130. https://doi.org/10.1016/j.seppur.2015.07.015.

66. B. Oonkhanond, W. Jonglertjunya, N. Srimarut, P. Bunpachart, S. Tantinukul, N. Nasongkla, C. Sakdaronnarong, Lactic acid production from sugarcane bagasse by an integrated system of lignocel-lulose fractionation, saccharification, fermentation, and ex-situ nanofiltration, *J. Environ. Chem. Eng.* 5 (2017) 2533–2541. https://doi.org/10.1016/j.jece.2017.05.004.

67. A. Córdova, C. Astudillo, L. Santibañez, A. Cassano, R. Ruby-Figueroa, A. Illanes, Purification of galacto-oligosaccharides (GOS) by three-stage serial nanofiltration units under critical transmem-brane pressure conditions, *Chem. Eng. Res. Des.* 117 (2017) 488–499. https://doi.org/10.1016/j.cherd.2016.11.006.

68. E. Neczaj, A. Grosser, Circular economy in wastewater treatment plant – Challenges and barriers, *Proceedings.* 2 (2018) 614. https://doi.org/10.3390/proceedings2110614.

69. A. Grobelak, A. Grosser, M. Kacprzak, T. Kamizela, Sewage sludge processing and management in small and medium-sized municipal wastewater treatment plant-new technical solution, *J. Environ. Manage.* 234 (2019) 90–96. https://doi.org/10.1016/j.jenvman.2018.12.111.

70. B. Maryam, H. Büyükgüngör, Wastewater reclamation and reuse trends in Turkey: Opportunities and challenges, *J. Water Process Eng.* 30 (2019). https://doi.org/10.1016/j.jwpe.2017.10.001.

71. R.D. Dhineshkumar, Review on membrane technology applications in food and dairy processing, *J. Appl. Biotechnol. Bioeng.* 3 (2017) 399–407.

72. S.T. Yang, Bioprocessing for value-added products from renewable resources, *Bioprocess. Value-Added Prod. from Renew. Resour.* (2007). https://doi.org/10.1016/B978-0-444-52114-9.X5000–2.

73. N. García Doménech, F. Purcell-Milton, Y.K. Gun'ko, Recent progress and future prospects in development of advanced materials for nanofiltration, *Mater. Today Commun.* 23 (2020). https://doi.org/10.1016/j.mtcomm.2019.100888.

74. J. Ayyavoo, T.P.N. Nguyen, B.M. Jun, I.C. Kim, Y.N. Kwon, Protection of polymeric membranes with antifouling surfacing via surface modifications, *Colloids Surfaces A Physicochem. Eng. Asp.* 506 (2016) 190–201. https://doi.org/10.1016/j.colsurfa.2016.06.026.

75. C. Hoffmann, H. Silau, M. Pinelo, J.M. Woodley, A.E. Daugaard, Surface modification of polysulfone membranes applied for a membrane reactor with immobilized alcohol dehydrogenase, *Mater. Today Commun.* 14 (2018) 160–168. https://doi.org/10.1016/j.mtcomm.2017.12.019.

76. K.A. Faneer, R. Rohani, A.W. Mohammad, Nanofiltration of xylitol mixed solution using polyethersulfone membrane in presence of silicon dioxide nanoparticles or Pluronic F127 additives : A comparative study, *Int. J. Biomass Renew.* 5 (2016) 21–26.

77. A. Jalal Sadiq, K.M. Shabeeb, B.I. Khalil, Q.F. Alsalhy, Effect of embedding MWCNT-g-GO with PVC on the performance of PVC membranes for oily wastewater treatment, *Chem. Eng. Commun.* 207 (2020) 733–750. https://doi.org/10.1080/00986445.2019.1618845.

78. S.T. Weinman, E.M. Fierce, S.M. Husson, Nanopatterning commercial nanofiltration and reverse osmosis membranes, *Sep. Purif. Technol.* 209 (2019) 646–657. https://doi.org/10.1016/j.seppur.2018.09.012.

79. Z. Zhou, B. Ling, I. Battiato, S.M. Husson, D.A. Ladner, Concentration polarization over reverse osmosis membranes with engineered surface features, *J. Memb. Sci.* 617 (2021). https://doi.org/10.1016/j.memsci.2020.118199.

80. A. Malakian, S.M. Husson, Understanding the roles of patterning and foulant chemistry on nanofiltration threshold flux, *J. Memb. Sci.* 597 (2020). https://doi.org/10.1016/j.memsci.2019.117746.

81. W. Choi, C. Lee, C.H. Yoo, M.G. Shin, G.W. Lee, T.S. Kim, H.W. Jung, J.S. Lee, J.H. Lee, Structural tailoring of sharkskin-mimetic patterned reverse osmosis membranes for optimizing biofouling resistance, *J. Memb. Sci.* 595 (2020). https://doi.org/10.1016/j.memsci.2019.117602.

82. S.B. Mohamed Khalith, R. Ramalingam, S.K. Karuppannan, M.J.H. Dowlath, R. Kumar, S. Vijayalakshmi, R. Uma Maheshwari, K.D. Arunachalam, Synthesis and characterization of polyphenols functionalized graphitic hematite nanocomposite adsorbent from an agro waste and its application for removal of Cs from aqueous solution, *Chemosphere.* 286 (2022). https://doi.org/10.1016/j.chemosphere.2021.131493.

10 Palm Oil Mill Secondary Effluent Treatment Via Nanofiltration Membrane Photocatalytic Reactor (MPR)

Ummi Kalsum Hasanah Mohd Nadzim, Nur Hanis Hayati Hairom, Chin Yin Ying, Rais Hanizam Madon, Dilaeleyana Abu Bakar Sidik, Hazlini Dzinun, and Zawati Harun
Universiti Tun Hussein Onn Malaysia

Sofiah Hamzah and Alyza Azzura Abd Rahman Azmi
Universiti Malaysia Terengganu

CONTENTS

DOI: 10.1201/9781003261827-10

10.1   PALM OIL MILL EFFLUENT AND ITS ENIGMA

As the second-largest palm oil producer after Indonesia, the palm oil sector in Malaysia is a prominent industry that contributes significantly to the country's economic growth and development.
With the increase in palm oil production, there is a large amount of lignocellulosic biomass generated during the process, namely oil palm fronds, oil palm trunks, palm-pressed fiber, empty fruit
bunches, palm shells and palm oil mill effluent (POME). POME has been identified as the most
expensive and difficult to dispose of due to its high generation volume [1]. Furthermore, POME is
a high-strength pollutant rich in organic content in the form of total suspended solids, total solids,
oil and grease, and volatile suspended solids that will increase biochemical oxygen demand (BOD)
and chemical oxygen demand (COD) [2]. It also contains lignocellulosic residues with a mixture of
carbohydrates and oil. Therefore, it needs proper treatment before being discharged into the river.
The untreated POME may pose adverse impacts on the environment and human health.

Several conventional methods have been developed for treating POME, namely ponding system, open tank digester and aeration system, and land application system. However, those methods
have several limitations, particularly to comply with the standards regulations of the Environmental
Quality Act 1974 (Act 127). The ponding system, which acts as waste stabilization pond, was developed for POME treatment since 1982. More than 85% of palm oil mills use ponding system due to
its low cost and simple operating system [3].

Owing to the limitations of conventional methods, membrane photocatalytic reactor (MPR) can
be considered as one of the promising methods due to its advantages. The MPR has been developed
by combining photocatalysis and a membrane filtration system. Compared to conventional photo-
reactors, MPR allows operation with high catalyst amounts due to the presence of the membrane.
Simultaneously, MPR provides the advantage of controlling the molecule residence time in the reactor.
It realizes a continuous process that simultaneously separates product from the reaction environment.
Other essential advantages are simple configuration, small installation size, efficient use of catalyst
and UV light, and high possibility of reusing photocatalyst for subsequent experimental runs [4].

This chapter focuses on the potential of MPR as an alternative treatment method for post-treated
POME. The system integrates nanoparticles (e.g., zinc oxide) as photocatalyst, and several parameters also influence the effectiveness of the treatment. This chapter also elucidates the mechanism of
membrane fouling that occurs during the filtration process in MPR. This integrated approach can be
implemented to treat POME in compliance with the environmental regulations in order to maintain
a green environment for future generations.

10.2   PALM OIL MILL SECONDARY EFFLUENT

In general, the biological systems for treating POME in Malaysia are ponding system, open tank
digester and extended aeration system or closed anaerobic digester, and land application system
[5]. After the biological treatment of POME, palm oil mill secondary effluent (POMSE) or post-
treated POME is generated. POMSE is characterized as a thick brownish liquid with bad odor and
has higher pH value but lower BOD and COD than POME [1]. The POMSE has very high content
of degradable organic substances due to the presence of residual oil in it [6]. Consequently, the
BOD and COD values of POMSE do not meet the discharge limit as required by the environmental
regulations. High COD and BOD values can cause serious pollution issue [2]. Tables 10.1 and 10.2
summarize the characteristics of POMSE and the standard discharge limit for POME outlined in
Environment Quality Act 1974 as stipulated by Department of Environment, Malaysia.

TABLE 10.1

Characteristics of Palm Oil Mill Effluent (POME) and Palm Oil Mill Secondary Effluent (POMSE)

Parameter	POME Mohammad et al. [7]	POMSE Fadzil et al. [8]	POMSE Shairah et al. [1]
BOD (mg/L)	10,250–43,750	160	200–300
COD (mg/L)	15,000–100,000	1,600	3,000–5,000
TSS (mg/L)	5,000–54,000	14,787	1,500–2,000
pH	3.4–5.2	9.0	7.8–8.5
Color (ADMI)	N/A	N/A	3,000–6,000
Total iron (mg/L)	N/A	N/A	0–1.7
Turbidity (NTU)	N/A	N/A	2,000–3,000
Oil and grease (mg/L)	130–18,000	N/A	N/A

BOD, biochemical oxygen demand; COD, chemical oxygen demand; TSS, total suspended solids; N/A, not available.

TABLE 10.2

Standard Discharge Limit of Environmental Quality Act 1974 for POME [1]

Parameters	DOE Discharge Limit (1986 Onwards)	Environmental Quality Act
BOD (mg/L)	50	100
COD (mg/L)	1,000	1,000
Total solids (mg/L)	1,500	1,500
Suspended solids (mg/L)	400	400
Oil and grease (mg/L)	50	50
Ammoniacal nitrogen (mg/L)	100	150
Total nitrogen (mg/L)	200	200
pH	5	5–9
Temperature (°C)	45	45

BOD, biochemical oxygen demand; COD, chemical oxygen demand.

The discharge of POMSE with a high content of organic matter into the waterways will lead to some adverse effects. Organic contamination may occur when there are large amounts of organic compounds acting as substrates for microorganisms in the waterbody. The amount of dissolved oxygen (DO) in the receiving water can be consumed during the decomposition process and lead to oxygen depletion [9]. If there are large amounts of suspended solids in the organic effluent, the light source will be obstructed from reaching the photosynthetic organisms [10]. In addition, organic pollutants found in drinking water sources will also affect human health. Therefore, the end product of each process should be treated well before being released into the environment as it can bring chain effects to the ecosystem. Thus, POMSE treatment using MPRs is a promising alternative to ensure this effluent will achieve the standard regulations before being discharged through public waterways.

10.3 OVERVIEW OF MEMBRANE PHOTOCATALYTIC REACTOR

In recent years, MPR has been used to treat POMSE, as reported by Sidik et al. [11–13]. MPR is a hybrid reactor in which a photocatalytic degradation system is coupled with a membrane filtration process. In MPR, photocatalytic degradation acts as a pre-treatment process where it can minimize flux decline and membrane fouling, while the membrane plays its role as a simple barrier for photocatalysts as well as a selective barrier for the molecules to degrade [14]. In general, MPR combines photocatalysis with pressure-driven membrane processes, such as microfiltration (MF), nanofiltration (NF), and ultrafiltration (UF). When suspended catalyst is employed, membrane fouling will tend to occur.

As a more advanced alternative than conventional methods, MPR has several advantages. The advantages include confining the photocatalyst in the reaction environment through membrane, realizing a continuous process by separating the catalyst and products simultaneously and controlling the residence time of molecules in the reactor [15]. Basically, a conventional photoreactor needs to operate together with some additional operations, such as flocculation, coagulation, and sedimentation, to remove the catalyst from the treated product. By using MPR, the additional operations can be eliminated and at the same time can achieve energy savings. Moreover, in MPR, it is possible to reuse the photocatalyst in subsequent runs but it is nearly impossible to use in conventional separation system [16].

10.3.1 CONFIGURATION OF MEMBRANE PHOTOCATALYTIC REACTOR

The MPR has two configurations in which the catalyst used can be suspended in the mixture or immobilized on the substrate material and act as photocatalytic membrane [17]. For MPR with a suspended photocatalyst, the photocatalyst will be dispersed in solution and the membrane is used to recover the photocatalyst from reaction solution. Furthermore, the main advantage of this system is that the photocatalyst can have higher contact surface area with the pollutants compared to that of immobilized [18], which enhances the photocatalytic efficiency.

Based on the photocatalytic degradation system coupled with membrane process, the suspended MPR can be divided into integrative type and split type. In a split-type MPR, the photocatalytic reactor and membrane module are completely split. The configuration can be clearly structured and easy to install. The split-type MPR is depicted in Figure 10.1. An ultraviolet (UV) lamp with a peak wavelength of 253.7 nm is placed in the photocatalytic feed tank. This hybrid system consists of a photocatalytic reactor and a cross-flow MF membrane module. The filtration membrane is placed between transparent quartz cases. To homogenize the solution, a motorized stirrer is included in the feed tank [16].

Before commencing the photocatalytic reaction, the UV lamp is turned off while the POMSE mixture with the photocatalyst, such as zinc oxide (ZnO), is stirred well to achieve adsorption–desorption equilibrium. The water chiller serves to maintain the operating temperature at 25°C. Then, the mixture is transferred into the photocatalytic reactor to perform the photocatalysis reaction and proceed with membrane separation process. The permeate obtained at the end of the process is collected for analysis, while the retentate is returned to the feed tank [18]. When using this type of MPR, UV irradiation or reactive oxygen species will not cause damage to the membrane. The issue of membrane damage can occur in immobilized MPR, as there is direct contact between the membrane surface and UV light. The application of split-type MPR involves the transfer of photocatalyst to the membrane module for separation. This will lead to the accumulation of photocatalyst at the corner of the flow line and thus affect the photocatalytic performance [18].

10.3.2 PHOTOCATALYSIS PROCESS

As mentioned above, photocatalysis process is involved in the MPR operation. The photocatalysis process is categorized as an emerging technology in wastewater treatment, as it can undergo rapid oxidation, no polycyclic product formation, and pollutant oxidation up to parts per billion (ppb) level [19]. During the photocatalysis process, UV light irradiates the photocatalyst and drives it to the photoexcitation

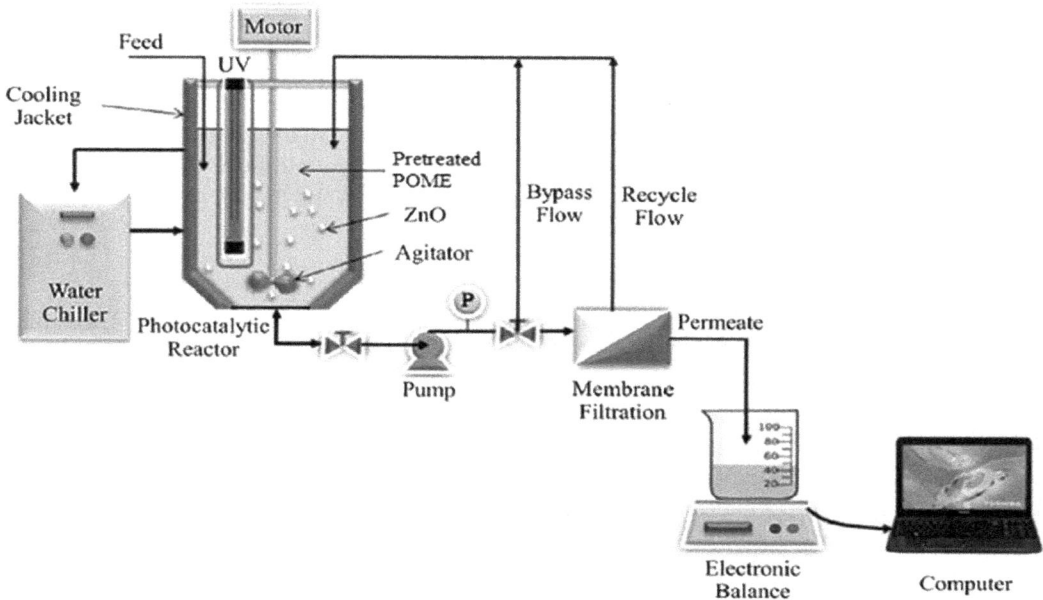

FIGURE 10.1 Schematic diagram of a split-type MPR with suspended photocatalyst. (Adapted from Hairom et al. [14].)

stage. The photoexcitation of the photocatalyst will then be followed by the creation of an electron–hole pair on the surface. As a result, reactive hydroxyl radicals are formed through reactions between holes and hydroxide ions. The hydroxyl radicals then aid the degradation process and reduce organic compounds to carbon dioxide, water, and inorganic constituents [4]. In this photocatalysis process, various types of nanoparticles can be employed such as zinc oxide (ZnO), titanium dioxide (TiO_2), tin dioxide (SnO_2), cadmium sulfide (CdS), zinc sulfide (ZnS), and iron oxide (Fe_2O_3). Nevertheless, ZnO has been claimed as the best photocatalyst that can improve wastewater treatment due to its advantages in terms of its stability under severe processing environments [20], antibacterial/antimicrobial activity [21], high photosensitivity, low cost, non-toxic, and environmentally friendly [22].

10.3.3 Nanofiltration Membranes

In the previous section, it has been mentioned that MPR is a hybrid system that involves a combination of photocatalytic degradation system with a membrane filtration process. Due to the presence of membrane, there is a high possibility that the photocatalyst can be reused for further experiments [4]. Theoretically, membrane is referred to as a selectively permeable barrier to ions and organic molecules [23]. The pore size of membrane provides an indication of the average pore size on the membrane surface. In addition, it also describes the particle size that tends to be rejected by the membrane [24].

Normally, four main types of membranes are available, namely MF, NF, UF, and reverse osmosis (RO). Each membrane type has a specific range of pore size. For example, the pore size of MF ranges from 0.1 to 5 μm, which is the largest pore size among the other membrane types [24]. For UF, the pore size ranges from 0.1 to 0.01 μm. Due to the reduction in pore size, the required osmotic pressure will be higher than MF. Furthermore, NF membrane has the pore size of 0.01–0.001 μm, which is significantly smaller than MF and UF. NF has a major advantage that it is able to filter particles such as salts, synthetic dyes, and sugars. Generally, it is confined to specialized uses [24]. On the other hand, RO membrane has pore size range from 0.001 to 0.0001 μm, which is the finest separation membrane available. It is more ideal for salt or metallic ion filtration than water [24].

In a study conducted by Hairom et al. [4], UF and NF membranes were employed in the treatment of industrial dye wastewater via MPR. The purpose was to demonstrate their performance in dye wastewater treatment under the same transmembrane pressure applied. At the end of the experiment, the NF membrane was able to exhibit more prominent performance compared to the UF membrane. There was a substantial volume of wastewater treated when NF membrane was used. This is because small-sized size particles or ions are able to easily pass through the UF membrane; thus NF can be considered an appropriate technique for dye wastewater treatment in MPR [4].

10.4   ZINC OXIDE NANOPARTICLES

There are various types of nanoparticles that can be used in the photocatalytic degradation which include zinc oxide (ZnO), titanium dioxide (TiO$_2$), tin dioxide (SnO$_2$), cadmium sulfide (CdS), zinc sulfide (ZnS), and iron oxide (Fe$_2$O$_3$). Nevertheless, ZnO has been claimed as the best photocatalyst that may enhance wastewater treatment due to its advantages. ZnO has a very high potential to be used as a catalyst in MPR due to its stable wurtzite crystal structure. In addition, ZnO has a large exciton binding energy (60 meV) which can provide efficient UV light at room temperature. Due to these properties, ZnO is an interesting photocatalyst in wastewater treatment through photodegradation mechanisms. Moreover, its chemical stability, lower cost, non-toxic properties, and photosensitivity allow it to be among the most suitable catalyst for use in MPR [4].

Previously, Hairom et al. [4] explained that the addition of ZnO nanoparticles has outstanding performance for industrial dye wastewater treatment via MPR. However, the efficacy of ZnO nanoparticles in MPR for POMSE treatment instead of dye wastewater is still limited. Furthermore, the best conditions of POMSE treatment through MPR have not been verified. With reference to relevant studies, MPR is one of the promising methods owing to its advantages. Therefore, the subsequent section presents the factors affecting the performance of MPR for POMSE treatment.

10.5   FACTORS AFFECTING THE PHOTOCATALYTIC PROCESS

10.5.1   pH

pH value plays an important role in controlling the photocatalytic efficiency and fouling behavior in the MPR system [4]. The charge of the nanoparticles and the membrane surface will vary depending on the pH of the solution causing changes in the molecular charge of the treated POMSE, ZnO photocatalyst and the membrane surface as well. As a result, it affects the photocatalytic process as well as the performance of the membrane during the process. To elucidate the effect of pH on MPR performance for POMSE treatment, the pH value of the treated sample can be varied. It can be adjusted using hydrochloric acid (HCl) or sodium hydroxide (NaOH).

Before starting the experiment, the isoelectric point (IEP) of photocatalyst, POMSE molecules, and membrane surface should be identified beforehand. IEP is a pH value where the zeta potential value is zero, representing no electrical charge on the surface of the particle. Zeta potential is a measure of the charge on the surface of the particles [25]. For NF membranes and ZnO photocatalysts, the identified IEPs are around 5.5 and 9 [4]. POMSE contains heavy metals; thus, there will be positive charges existing on the particle surfaces. Theoretically, if the pH value exceeds the IEP value, the membrane surface will have negative charges and vice versa [4]. Similar conditions are applied to POMSE particles and ZnO surfaces. They will carry positive charges when the pH value is lower than its IEP value. If the pH value is relatively close to the IEP value, there may be weak repulsion force between the particle and the surface involved.

Based on this condition, the particles will be easily deposited on the membrane surface inducing a cake layer formation [4]. Consequently, the permeate flux rate will be affected. For example, when a solution with pH 2 is prepared, the surface of the NF membrane carries positive charge as its IEP value exceeds the pH value. Similar to the surface of photocatalyst and POMSE particles, they also exist as positive charges. Therefore, there is a repulsion force generated between the surfaces.

This may prevent cake formation on the membrane surface and thus enhance the process efficiency and minimize the fouling phenomenon. Thus, it can be inferred that a proper pH application can minimize the membrane fouling.

10.5.2 LIGHT INTENSITY

On the other hand, light intensity also has an effect on the degradation process [4]. There are two types of light sources involved, namely UV light and visible light. Different light sources provide different light intensities that determine the performance of photoactivation activity. Light intensity is referred as the sources to initiate photocatalytic activity in POMSE treatment. The use of photocatalyst helps absorb the light present and bring it to a higher energy level. The energy provided is capable of inducing a chemical reaction to occur [26].

The photocatalytic degradation process begins with the photoexcitation of ZnO and leads to the generation of electron–hole pair on the surface of the nanoparticles. The hole on ZnO has high oxidative potential. As a result, direct oxidation of POMSE particles may occur. Moreover, the reaction between the hole and the hydroxide ion has produced highly reactive hydroxyl radicals. Under extremely unstable state, hydroxyl radicals will then lead to organic chemical degradation [4]. This can explain the importance of light sources and photocatalysts in the photocatalytic degradation process.

Furthermore, light can be divided into three stages, namely low, middle, and high light intensities. At low light intensity, reaction with electron hole formation is predominant, whereas the electron–hole recombination phase is negligible. At this stage, the reaction rate will increase in response to an increase in light intensity. Although middle light intensity is involved, the electron–hole pair separation and the recombinant process are considered complete with each other. This will result in a relatively low reaction rate being achieved. When high light intensity is applied, the reaction rate is not compromised by the light intensity [18]. In other words, the photocatalytic reaction rate will increase when the light intensity increases within standard levels. From previous study, Hairom et al. [4] revealed that the light source with higher light intensity performed better than others.

10.6 ANALYSIS OF WATER QUALITY

The permeate is collected for analysis at the end of the process [18]. Water quality is analyzed in terms of color, BOD, COD, and turbidity.

10.6.1 COLOR

Scientifically, color is categorized as true color and apparent color. True color is referred to as the color present after the removal of particulate matter through a filtration process. Typically, the filtration process is performed using a filter with a pore size of 0.45 μm. Apparent color is the presence of color that can be seen. The combination of true color and particulate matter will give an apparent color [27]. For example, the presence of contaminants such as iron will cause oxygen to be present which changes the color of the waterbody from clear to yellow or red sediments. Under this condition, the color formed by the iron is known as the apparent color rather than the true color. The true color of the water can be determined by filtering the sample [28].

There are several sources that can lead to different apparent colors. In natural waters, the presence of soluble organic matter such as humic acid originated from decaying crops will result in brownish-colored waterbody [29]. The metal dissolution in pipelines can also discolor the drinking water. Serious corrosion of iron pipes will generate a brownish color, while corrosion of copper pipes will give blue-green color on sanitary [30]. Thus, the condition of household pipes can definitely affect the color of the water.

Furthermore, color is not a toxic feature, but according to the Environmental Protection Agency, color is known as a secondary parameter that affects the palatability and appearance of water [28]. Occasionally, color can cause the formation of organic matter; for example, trihalomethanes will form from the chlorination of chlorinated organic compounds. Moreover, color can be determined spectrophotometrically or using a visual comparator. The standard measurement unit is in Hazen unit (HU), which is defined in terms of platinum-cobalt standard. For application in true color, it is often referred to as True Color Unit (TCU). This standard is developed for water analysis with a yellow appearance and is considered not applicable for waters with various colors [27]. In addition, color testing can also be performed by comparing visually to a set of platinum-cobalt standards in Nessler tubes. Samples will be collected in bottles that have been rinsed clean and then refrigerated [28]. Colors can be removed using granulated activated carbon (GAC) filters which are relatively inexpensive and simple to use. Contaminants attached to activated coal or charcoal will be removed. The GAC filters will be replaced after a period of time to maintain process efficiency [31].

10.6.2 BIOCHEMICAL OXYGEN DEMAND

BOD is known as a bioassay test that measures the oxygen consumed by microorganisms while stabilizing biologically decomposable organic matter under aerobic conditions. Organic matter acts as food for bacteria and simultaneously cells receive energy from organic matter during oxidation [32]. BOD testing operates based on the principle that if adequate oxygen is provided, aerobic biological decomposition by microorganisms will proceed until all waste products are consumed [33]. It can be measured directly and the dilution procedure is applied. The BOD test sometimes can be referred to as BOD_5 as it is a standard test that provides information about the wastewater organic strength. It indicates the amount of sample consuming oxygen over a 5-day period which is measured under standard conditions.

The calculation of BOD_5 is mainly based on the difference between the initial DO reading and the final DO reading. Theoretically, the initial DO reading will be higher than the final DO reading. Throughout the 5 days of incubation, the amount of oxygen that has been consumed by the microorganisms in the water leads to oxygen depletion at the end of the process [9]. In general, a period of 5 days is not sufficient to perform complete oxidation, but it provides sufficient time for microbial acclimation and substantial oxidation [34]. Changes in the concentration of DO represent the oxygen demand for respiration of biological microorganisms. High BOD_5 reading indicates high oxygen demand for the biological microorganism respiration.

Before starting the experiment, the bottles for blank and samples are prepared and labeled. Then, the dilution water with specific amount of sample cell is added into the bottles. The sample cell added is calculated as follows:

$$\text{Sample added (mL)} =$$

$$\left[\left(\text{minimum allowable depletion, mg/L}\right)\middle/\text{estimated BOD, mg/L}\right] \times \text{volume of BOD bottle used, mL} \tag{10.1}$$

The initial DO concentration is determined using DO meter (model HI9146) before placing the bottles in the incubator. After 5 days, the final DO value was measured again and the BOD concentration can be identified using Equation (10.2) [35].

$$BOD_5 =$$

$$\left[\left(\text{Initial DO reading} - \text{final DO reading}\right)\middle/\text{sample size added, mL}\right] \times \text{volume of BOD bottle used, mL} \tag{10.2}$$

In addition, BOD testing can be divided into three measurable categories, namely ultimate BOD (UBOD), nitrogenous BOD (nBOD), and carbonaceous BOD (cBOD) [32]. For the UBOD, it measures all the biodegradable materials in the sample. The UBOD values and oxygen consumption rates are frequently used in mathematical models to predict the effects of effluents on waterbodies. Therefore, the oxygen consumption rate is always determined together with the UBOD value in the testing [34]. Furthermore, cBOD describes the oxygen consumed when carbonaceous compounds are oxidized to carbon dioxide and other by-products. Typically, this involves the oxidation of inorganic compounds such as ferrous iron and sulphite [36]. In fact, the oxidation of organic carbon involves a series of bio-chemical reactions mediated by various microorganisms consuming specific substrates present in the process. The total cBOD can be determined through experiment until all the organic carbon involved is oxidized. However, the period can take around 20–50 days or longer depending on the condition [34].

On the other hand, nBOD describes the oxygen consumed during the oxidation of nitrogenous compounds to nitrate and nitrite. The cBOD and nBOD differ in terms of the types of bacteria involved in oxidation of reduced nitrogen [37]. Bacteria such as nitrifiers are related to suspended solids and are usually found in water with low concentrations. The nitrifiers typically have slower growth rates. It prefers to grow on the surface of the sample bottle which affects the bottle and enhances the nitrification process. Incorrect results may be obtained. Therefore, a short-term measurement is suggested to identify the nBOD value. By detecting the concentration of ammonia within 1–3 days, accurate results can be obtained [34].

10.6.3 Chemical Oxygen Demand

COD is an alternative test to BOD to determine the organic matter concentration in wastewater samples. The most significant difference between COD and BOD is the duration of the process completion. BOD takes about 5 days to complete, but COD only requires a few hours to complete. The COD testing uses typical chemical substances such as potassium dichromate in 50% sulphuric acid solution to oxidize both organic and inorganic matter in wastewater samples [38]. The determination of COD is important in identifying the amount of waste in the water. Waste containing high amounts of organic matter will require further treatment to minimize the organic waste content before being discharged into public waterways. Untreated waste discharged into nearby rivers can lead to the consumption of organic matter by microbes in the water. Consequently, the microbes consume oxygen to break down the organic waste. This nutrient-rich but lack of oxygen condition has led to the occurrence of eutrophication and aquatic life is also compromised [38].

10.6.4 Turbidity

Turbidity refers to the cloudiness of water which typically stems from the particles suspended in the water that may be algae, suspended oils, minerals, oils, or bacteria. Turbidity is an optical measurement that indicates the presence of suspended particles [39]. The more particles present in water, the higher the turbidity. To determine the substances present in the wastewater, jar test analysis can be considered as one of the options. The principle of jar testing can be explained in two basic terms, namely coagulation and flocculation. Coagulation represents the process of chemical addition to destabilize stable charged particles, whereas flocculation means a slow mixing technique that promotes agglomeration and deposition of particles.

The jar testing is a method that stimulates a full-scale water treatment process that provides a reasonable idea of the chemical treatment methods to the system operator. Since jar testing imitates full-scale operation, system operators can use this test to determine which chemical treatments can show better performance with the raw water of the system. Jar testing can be conducted by adjusting the amount of chemical for treatment and the sequence of additions into the sample jars. The mixture is then stirred so that the floc formation, development, and deposition can be observed during the process.

However, jar testing is not considered a priority parameter for the final treatment of POMSE. There are several disadvantages of using jar testing to optimize coagulation. One of the criteria involved is the frequency of testing. When jar testing is conducted in batch mode, it is only able to provide a snapshot of influent water quality and unable to represent the continuous dynamic changes of the complete system. On the other hand, the optimal dosing level in different scales of jar test will differ from each other, as they differ in hydrodynamics [40]. The simplest method of measuring turbidity is to use a turbidity meter. Turbidity meter is usually fast and inexpensive. It uses nephelometry (90° scattering) or other optical scattering detection techniques for fast and accurate turbidity measurements on the water sample involved [41].

10.7 MEMBRANE PERFORMANCE

MPR is known as a hybrid system that combines photocatalytic degradation process with membrane filtration process [14]. Both processes are crucial as the system performance may affect overall treatment efficiency. In membrane filtration process, there is a membrane placed between the transparent quartz cases [16]. In this section, the membrane performance is discussed further in the aspects of flux decline and membrane fouling mechanisms.

10.7.1 Flux Decline

During the membrane filtration process, the phenomenon of flux decline usually occurs. Typically, the initial permeate flux in the membrane separation process is relatively high. However, after certain period of operation, the flux value decreases gradually to a much lower amount than the initial permeate flux. To investigate the behavior of membrane fouling in flux decline analysis, the normalized flux of membrane can be calculated using the following equation.

$$\text{Normalized flux} = \text{Solution flux}, J/\text{Pure water flux}, J_0 \qquad (10.3)$$

A comparison study of fouling behaviors can be made by plotting a graph of normalized flux against operating time. The percentage of flux decline can also be obtained according to the simplified equation as follows [4]:

$$\text{Percentage of flux decline} = \left(1 - J/J_0\right) \times 100\%. \qquad (10.4)$$

The causes that lead to flux decline are the concentration polarization and fouling phenomenon. Concentration polarization occurs when there is accumulation of retained solutes near the membrane surface, while the fouling phenomenon involves adsorption, pore blocking, and decomposition of solidified solutes. The rejection of solute by the membrane has resulted in the accumulation of solute on the membrane and elevated the concentration of solute near the membrane surface. As a result, flux decline occurs due to an increase in the osmotic pressure of the solution close to the membrane surface. The formation of osmotic pressure has led to loss of driving force and thus caused flux decline [42]. In addition, membrane fouling is initiated by the accumulation of inorganic, organic, and colloidal substances on the membrane surface or in its pore. The occurrence of membrane fouling results in flux decline, rapid increase in pressure on transmembrane, and deterioration of mechanical properties [43].

10.7.2 Mechanisms of Membrane Fouling

When studying the flux decline behavior, the blocking filtration laws is also applied. The laws are modified, and there are three main assumptions considered, namely constant pressure drop, constant membrane-resistant involvement, and a specific resistance of cake that is invariant with time [44]. There are four blocking filtration laws which include complete blocking model, intermediate blocking model, standard blocking model, and cake formation model. Each type of blocking filtration law has a different description and linearized equation as tabulated in Table 10.3.

For a complete blocking model, each particle is presumed to reach the membrane surface and completely block the membrane pores without any overlap. The particles involved are greater than the membrane pores. For intermediate blocking, each particle is considered to partially block the membrane pores or settle on other particles resulting in partial overlap. In terms of standard blocking model, the directly involved particles block the pore channels on the inside rather than the membrane surface due to the small-sized solute being absorbed. This indicates that the volume of the membrane pores decreases with respect to permeate volume. Cake formation occurs when the solute is larger than the pore size of the membrane. Thus, the particles block the pores of the membrane surface and also deposit on other particles resulting in a layer of substance forming on the surface [14]. The schematic diagram of the membrane fouling mechanisms is depicted in Figure 10.2.

TABLE 10.3
Blocking Filtration Laws

Blocking Filtration Laws	n	Linearized Equation
Complete blocking model	2	$-\ln (J_0/J)-1=K_t$
Intermediate blocking model	1	$J_0/J-1=K_t$
Standard blocking model	1.5	$\sqrt{J_0/J}-1=K_t$
Cake formation model	0	$(J_0/J)^2-1=K_t$

n is the constant parameter depending on the fouling mechanisms.
Adapted from Hairom et al. [44].

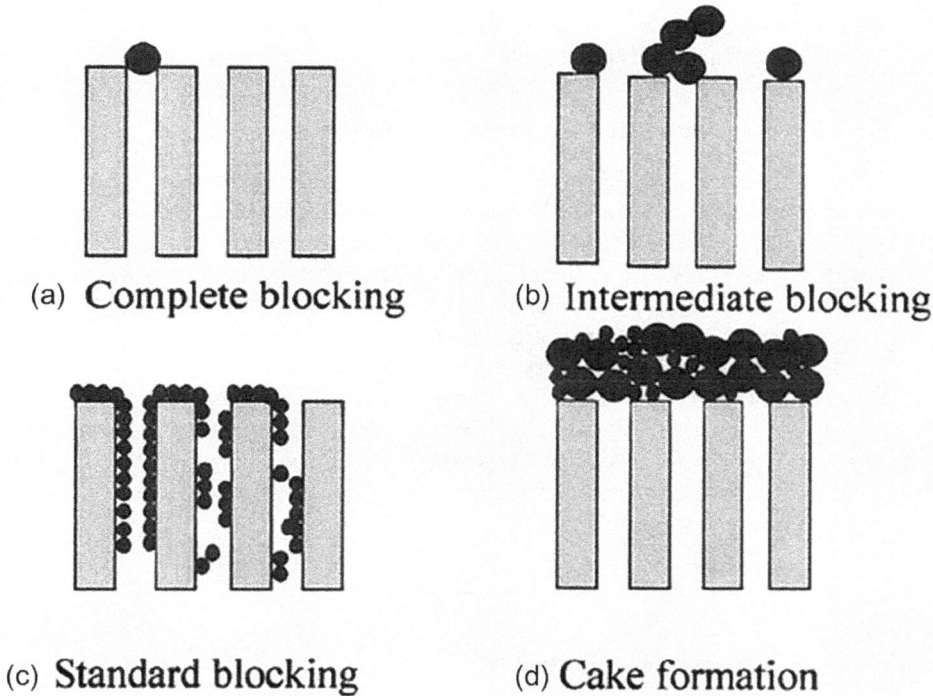

(a) Complete blocking   (b) Intermediate blocking

(c) Standard blocking   (d) Cake formation

FIGURE 10.2 Schematic diagram of membrane fouling mechanisms: (a) complete blocking model, (b) intermediate blocking model, (c) standard blocking model, and (d) cake formation model. (Adapted from Aghdam et al. [45].)

10.8 EFFECT OF pH AND LIGHT INTENSITY ON POMSE TREATMENT VIA NANOFILTRATION MEMBRANE PHOTOCATALYTIC REACTOR

Experiments were conducted to evaluate the performance of NF membrane for treating POMSE via MPR under different pH and light intensities. The water quality after treatment was also analyzed in terms of color, BOD, COD, and turbidity. Prior to conducting the experiments, POMSE samples obtained from a selected palm oil mill were analyzed. Table 10.4 summarizes the initial characteristics of POMSE before being treated with NF.

10.8.1 PERFORMANCE OF NF MEMBRANE

To determine the performance of NF membrane, the flux decline behavior was analyzed against different pH (2, 4, 7, 10, and 11), and light intensities (UV and visible lights) as depicted in Figure 10.3. Based on Figure 10.3a, it can be seen that there was a gradual flux decline formed at the beginning of 0.5 hour (30 minutes). After that, the flux constantly declined for the entire period of time. For pH values of 2, 4, and 7, the fouling that occurred can be explained by the presence of weak repulsion forces between the contact surfaces. If the pH of the solution differs from the membrane IEP, a great repulsion force is generated. In other words, treatment under pH 10 was the optimum condition, as it led to a lower flux decline compared to other pH values. At pH 10, the amount of water that can pass through the membrane was higher, resulting in a lower flux decline.

Based on Figure 10.3b, UV light demonstrated lower flux decline compared to visible light. At zero hours, the permeability of NF membrane was close to 1. As the processing time increased, the flux passing through the membrane decreased accordingly. After 3 hours, treatment under UV light recorded normalized flux of 0.39 passing through the NF membrane, while only 0.17 normalized flux passing through membrane under visible light treatment. The reason may be due to blockage of solutes on the membrane pores or rapid cake formation leading to relatively gradual flux decline [8]. Thus, UV light provided optimum condition for POMSE treatment via MPR.

10.8.2 EFFECT OF pH ON POST-TREATED POME QUALITY

Figure 10.4 depicts the effect of different pH values on water quality after treatment. Generally, the results showed a positive trend for most pH values for POMSE treatment via MPR (Figure 10.4a). At pH 2, 4, 7, and 10, the percentage of color removal was almost 100%, while pH 11 recorded 99% color removal. This is because at higher pH (alkaline), the functional group such as carboxylic group will become anionic in nature forming carboxylate. When the functional group exists as anion, they are not suitable for the adsorption of negatively charged color causative molecules as interionic repulsion forces are generated from each other [46].

On the other hand, Figure 10.4b shows the turbidity removal percentage of post-treated POME after treatment under different pH conditions. For most pH values, the turbidity removal percentages achieved almost 100%, while samples treated at pH 2 achieved 99% removal. This reason is because higher pH has a tendency toward sedimentation and alkaline condition is more suitable for sedimentation or turbidity removal [47].

TABLE 10.4
Characteristics of POMSE before Treatment

Substance	Color (HU)	Turbidity (NTU)	Chemical Oxygen Demand (mg/L)	Biochemical Oxygen Demand (mg/L)	pH
Palm oil mill secondary effluent	2,540.67	541.67	721.67	147.42–375.90	8.51

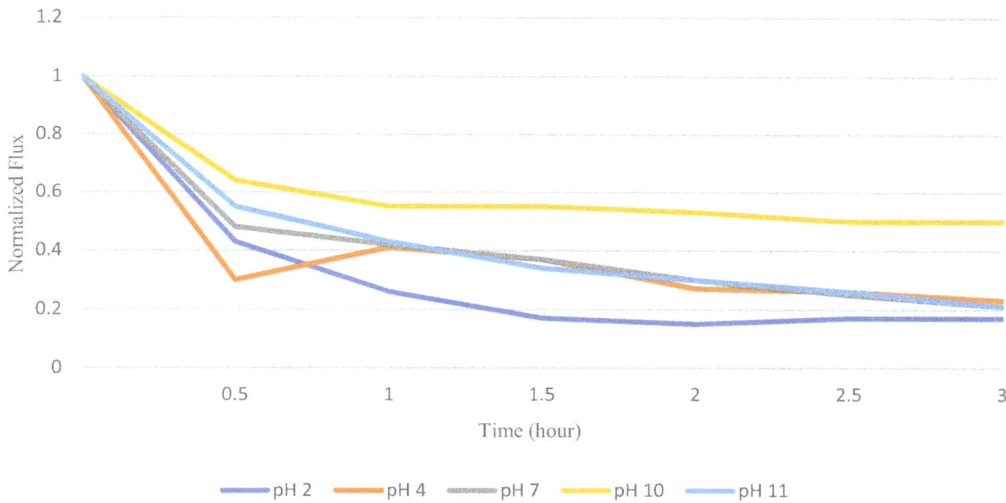

(a)

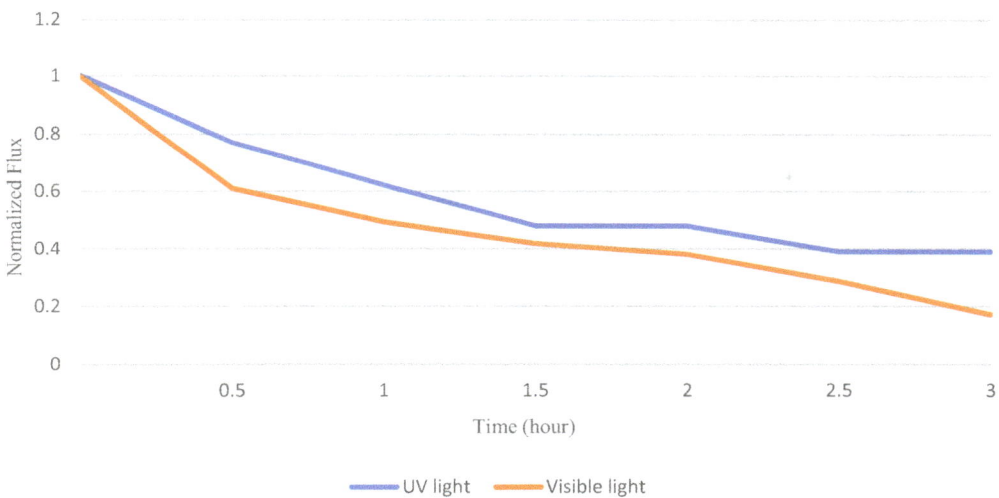

(b)

FIGURE 10.3 Membrane flux decline against time under different (a) pH and (b) light intensities.

Furthermore, Figure 10.5a exhibits that all COD values after treatment complied with the dis-charge limit standard. It was also found that the COD values at high pH were higher than others, especially at pH 10. This can be explained by the fact that ozone initially converts suspended sol-ids to dissolved solids, and subsequent ozonation destructs high suspended solids into very small molecules. As a consequence, COD levels in wastewater are enhanced, thus leading to higher COD readings [48]. In contrast, pH 7 had the highest COD removal with lower COD value. The high COD removal efficiency achieved at pH 7 may be due to the generation of hydroxyl (OH) radicals [49]. These radicals are strong, non-selective chemical oxidants that can react rapidly with most organic compounds. This enables degradation of organic compounds leading to low COD readings [50].

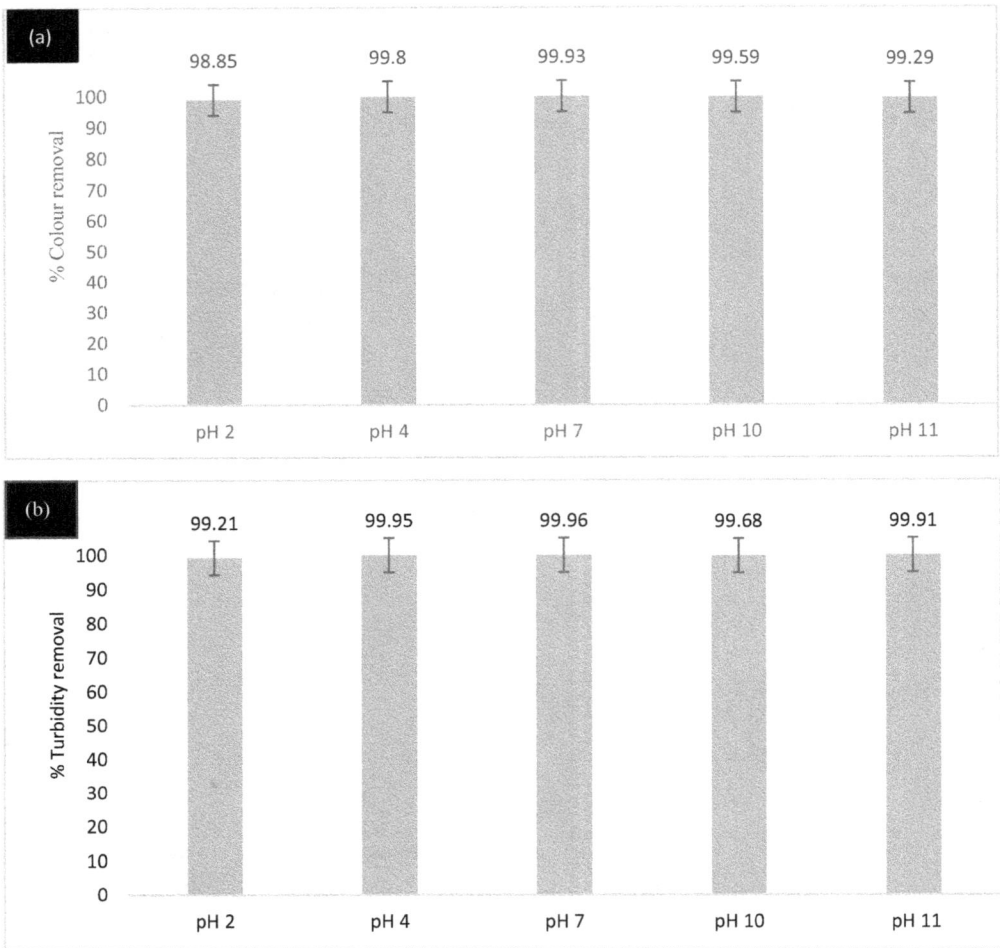

FIGURE 10.4   Removal percentages of (a) color and (b) turbidity of post-treated POME samples after treatment under different pH.

Meanwhile, results in Figure 10.5b showed that the optimum condition for post-treated POME after treatment was at pH 10 as it exhibited a small standard deviation value of BOD reading. In fact, it can be considered that pH 10 was not a favorable growth environment for microorganisms in post-treated POME. Generally, the microorganisms that can be found in POME are bacteria, algae, and fungi with different species [51]. For bacteria, the favorable growth environment is in the pH range of 6.5–7.0, which is a neutral condition [52]. Moreover, the optimum pH condition that promotes algae growth is from pH 8.2 to 8.7 [53]. Fungi are able to thrive at pH 5–6, which prefer slightly acidic conditions [5].

10.8.3   EFFECT OF LIGHT INTENSITY ON POST-TREATED POME QUALITY

Figure 10.6 presents the color and turbidity removal percentages of post-treated POME samples after treatment under different light intensities. Based on the results obtained in Figure 10.6a, it can be concluded that the post-treated POME treated under UV light recorded optimum color removal due to the small standard deviation range. Algae and suspended sediment particles are common

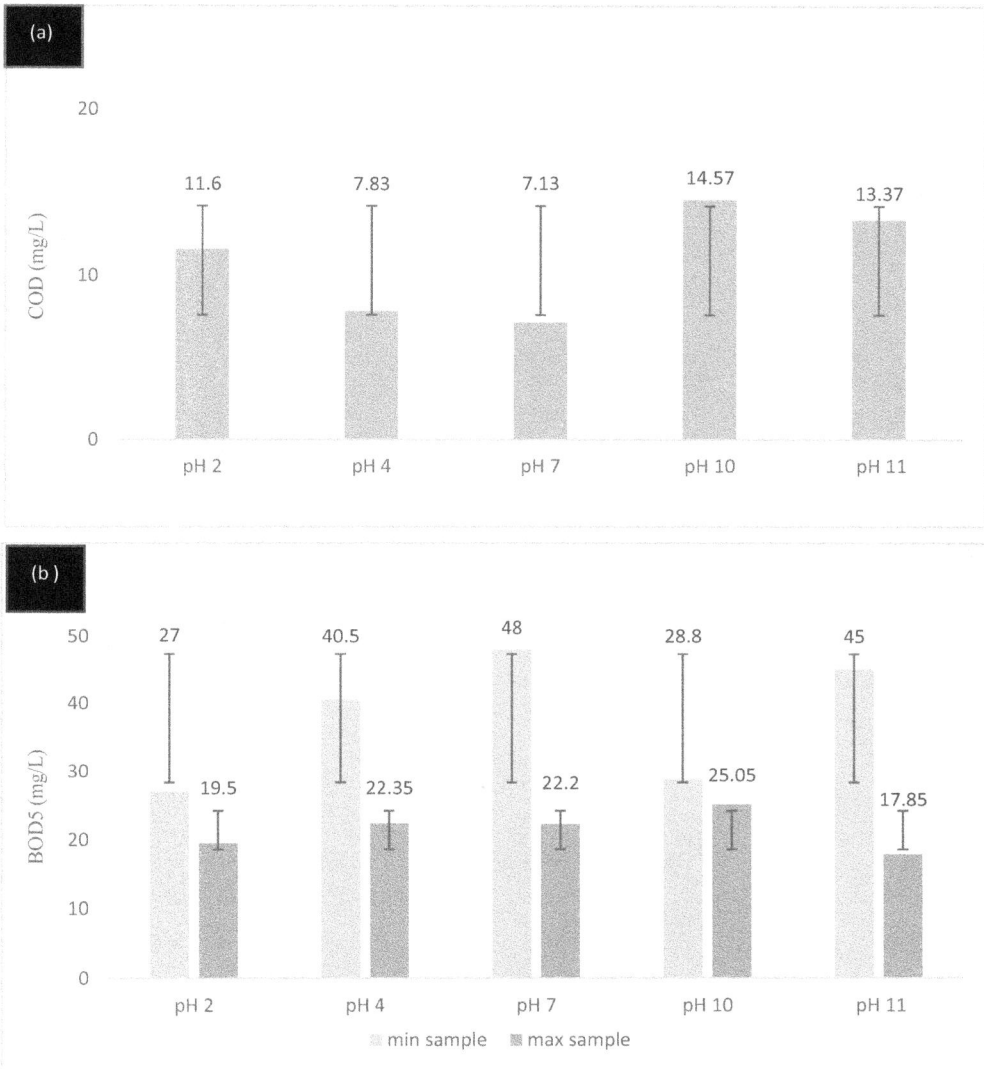

FIGURE 10.5 (a) COD and (b) BOD values of post-treated POME samples after treatment under different pH.

particulate matter that cause naturally colored waters [54]. UV light and visible light are both categorized as low light intensity ranges. Thus, the formation of electron–hole pairs is dominant. The pollutant degradation rate increases linearly with light intensity [19].

Moreover, the turbidity removal percentage under visible light was slightly higher than that under UV light. Based on the terms of turbidity removal, it can be referred to as removing existing particles mechanically or chemically and reducing the turbidity in water [55]. In this study, fluorescent lamp was used as the visible light source. The working principle involved is through the mercury vapor ionization in a glass tube. One of the advantages of fluorescent lamp is that it can provide diffused light that can reduce harsh shadows [56]. Diffused light is known as soft light and is scattered. It may come from all directions that it seems to wrap around the objects [57]. Thus, it can be concluded that visible light can perform slightly better than UV light, as it provided diffused light which could enhance the turbidity removal.

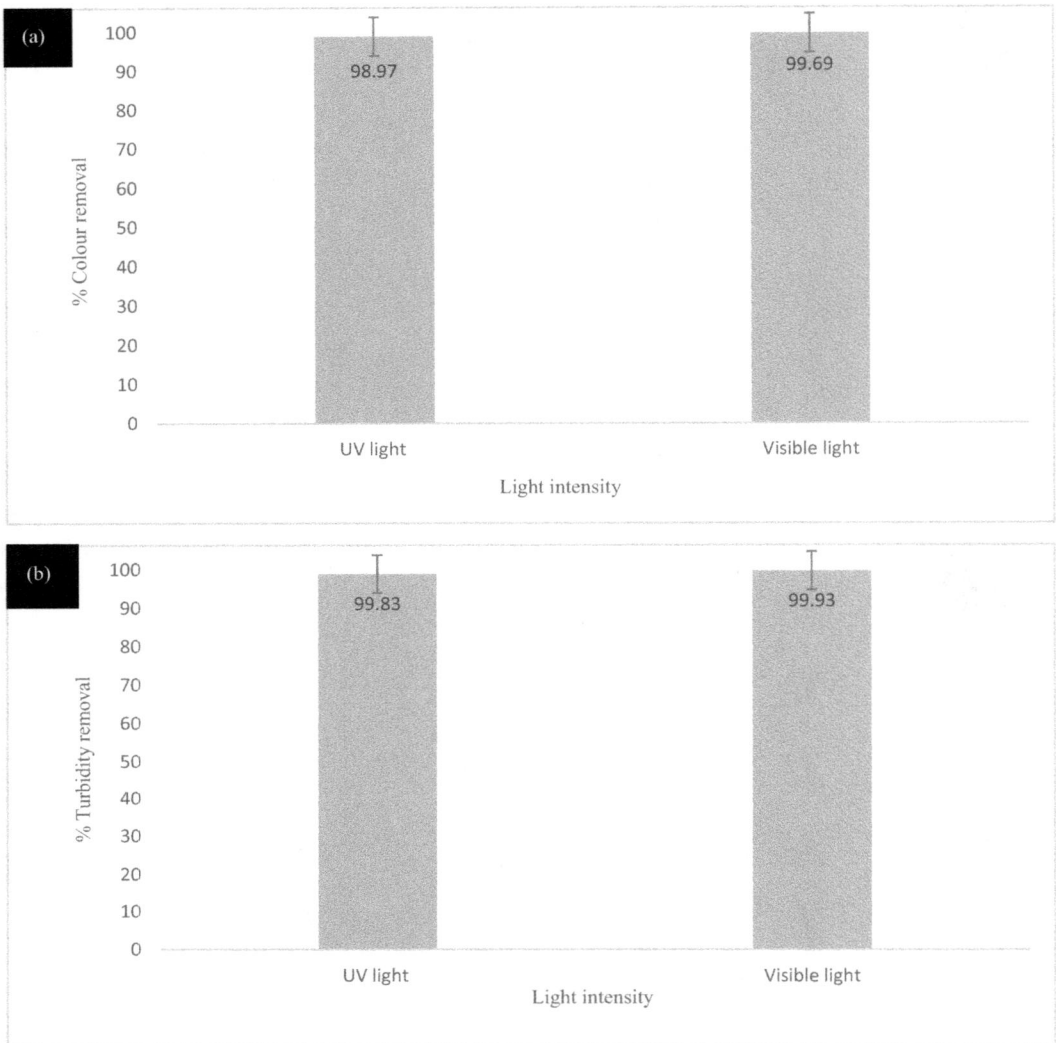

FIGURE 10.6 Removal percentages of (a) color and (b) turbidity of post-treated POME samples after treatment under different light intensities.

Meanwhile, the COD values obtained by the samples treated under visible light were slightly lower than that of UV light, as presented in Figure 10.7a. No standard deviation was observed in the COD values under visible light, indicating the result accuracy. If the COD value is high, it indicates that a high amount of oxidizable substances present in the water [58]. Similarly, the BOD values obtained from UV light and visible light were not much different, but visible light has a smaller standard deviation than UV light. Lower BOD value indicates that lesser living organisms can be found in the water treated under visible light.

10.8.4 Nanofiltration Membrane Fouling

In this study, Matlab R2016b was used to analyze the membrane fouling mechanisms occurring while treating the pretreated POME via MPR under UV light and visible light. Table 10.5 summarizes the rate constants of NF membrane for post-treated POME treatment under UV and visible lights.

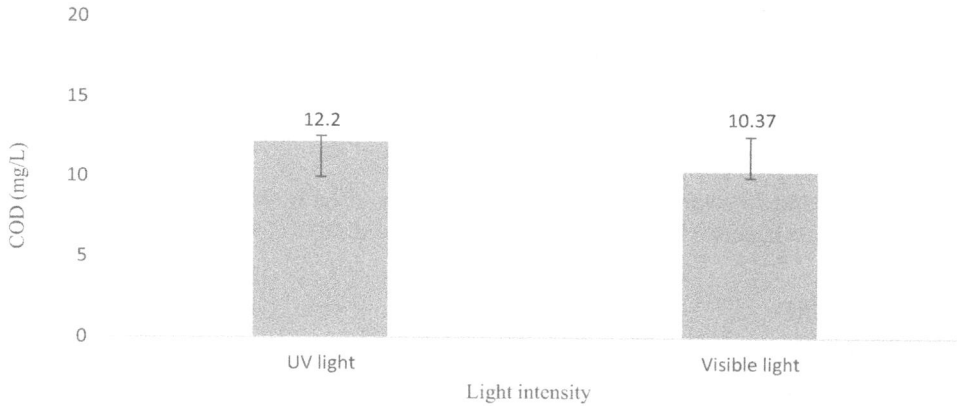

FIGURE 10.7 COD values of post-treated POME samples after treatment under different light intensities.

TABLE 10.5
Rate Constants of Nanofiltration Membrane after Treatment under Different Light Intensities

Light Intensity	Blocking Filtration	Rate Constant	
	Laws	R^2	K (s^{-1})
UV light	Complete blocking	<0.1	$K_c = 2.4 \times 10^{-4}$
	Intermediate blocking	0.9042	$K_i = 1.6 \times 10^{-4}$
	Standard blocking	0.8800	$K_s = 4.4 \times 10^{-2}$
	Cake filtration	0.8810	$K_{cf} = 5.5 \times 10^{-4}$
Visible light	Complete blocking	<0.1	$K_c < 0.1$
	Intermediate blocking	0.8186	$K_i = 3.1 \times 10^{-4}$
	Standard blocking	0.8891	$K_s = 1.1 \times 10^{-4}$
	Cake filtration	0.5966	$K_{cf} = 1.5 \times 10^{-3}$

Based on Table 10.5, the intermediate blocking, standard blocking, and cake filtration mechanisms were well fitted with the experimental results. However, the complete blocking model was poorly fitted with experimental results as the $R^2 < 0.1$. This is mostly likely due to the involvement of various solute sizes in the post-treated POME which do not allow each particle to completely block the membrane pore entrance [14]. The intermediate blocking mechanisms occurred when NF was incorporated in treating post-treated POME samples. During the treatment process, the sticky surface condition of the membrane promotes the attachment of particles or flocs and forms a gel layer. Thus, this may lead to the intermediate blocking mechanism to occur [59]. Similar results were also found for visible light. However, for visible light, standard blocking occurred, which could be due to small-sized solutes being absorbed into the membrane pores [60].

10.9 CONCLUSION

Since it is considered a pollutant rich in organic content, POMSE cannot be discharged into waterbodies without proper treatment. To date, conventional methods that are commonly used in palm oil mills around Malaysia still do not comply with the standard regulations for effluent discharge. As an alternative, MPR can be considered as one of the promising methods due to its advantages. This chapter provides elucidation on the advantages of MPR for POMSE treatment. This system combines photocatalysis and membrane separation processes. For photocatalysis, ZnO is more prominent to be used as catalyst in MPR compared to other metal oxides as it has superior features. This approach is seen as a promising method to assist palm oil mills in complying with the environmental regulations as well as reducing contamination to the waterbodies.

ACKNOWLEDGMENT

The authors would like to acknowledge the financial support given by Ministry of Higher Education Malaysia through Fundamental Research Grant Scheme (FRGS/1/2021/TK0/UTHM/02/14).

NOMENCLATURE

J Solution flux

J_0 Pure water flux

n Fouling mechanism-dependent constant parameter

K_t Coefficient (acquired from slopes of liner curves)

K_c Blocking constant for complete blocking law

K_i Blocking constant for intermediate blocking law

K_s Blocking constant for standard blocking law

K_{cf} Blocking constant for cake filtration blocking law

R^2 Degree of model fitness

SYMBOLS DEFINITION

BOD Biochemical oxygen demand

DO   Dissolved oxygen

REFERENCES

1. Shairah, N., Shahrifun, A., Nazilah, N., Aris, A., Omar, Q., & Ahmad, N. (2015). Characterization of palm oil mill secondary effluent (POMSE). *Malaysian Journal of Civil Engineering, 27*(1), 144–151.
2. Bala, J.D., Lalung, J., & Ismail, N. (2015). Studies on the reduction of organic load from palm oil mill effluent (POME) by bacterial strains. *International Journal of Recycling of Organic Waste in Agriculture, 4*(1), 1–10. https://doi.org/10.1007/s40093-014-0079-6.
3. Rupani, P.F., Singh, R.P., Ibrahim, M.H., & Esa, N. (2010). Review of current palm oil mill effluent (POME) treatment methods: Vermicomposting as a sustainable practice. *World Applied Sciences Journal, 10*(10), 1190–1202.
4. Hairom, N.H.H., Mohammad, A.W., Ng, L.Y., & Kadhum, A.A.H. (2015). Utilization of self-synthesized ZnO nanoparticles in MPR for industrial dye wastewater treatment using NF and UF membrane. *Desalination and Water Treatment, 54*(4–5), 944–955.
5. Loan, L.W., Kassim, M.A., Muda, K., Kheang, L.S., & Abdullah, N. (2014). Performance evaluation of Industrial Effluent Treatment Systems (IETSs) – An insight for biotechnology advances in agro-based wastewater treatment. *World Applied Sciences Journal, 30*(30A), 316–325. https://doi.org/10.5829/idosi.wasj.2014.30.icmrp.45.
6. Ahmad, A.L., Ismail, S., & Bhatia, S. (2003). Water recycling from palm oil mill effluent (POME) using membrane technology. *Desalination, 157*(1–3), 87–95. https://doi.org/10.1016/S0011-9164(03)00387-4.
7. Mohammad, S., Baidurah, S., Kobayashi, T., Ismail, N., Leh, C.P. (2021). Palm oil mill effluent treatment processes – A review. *Processes, 9*(5), 739. https://doi.org/10.3390/pr9050739.
8. Fadzil, N.A.M., Zainal, Z., & Abdullah, A.Z. (2013). COD removal for palm oil mill secondary effluent by using UV/ferrioxalate/TiO$_2$/O$_3$ system. *International Journal of Emerging Technology and Advanced Engineering, 3*, 237–243.
9. Environmental Protection Agency. (2012, March 6). *5.2 dissolved oxygen and biochemical oxygen demand.* https://archive.epa.gov/water/archive/web/html/vms52.html.
10. Lenntech. (2018). *Organic compounds in freshwater.* https://www.lenntech.com/aquatic/organic-pollution.htm.

11. Sidik, D.A.B., Hairom, N.H.H., & Mohammad, A.W. (2019). Performance and fouling assessment of different membrane types in a hybrid photocatalytic membrane reactor (PMR) for palm oil mill secondary effluent (POMSE) treatment. *Process Safety and Environmental Protection, 130*, 265–274.
12. Sidik, D.A.B., Hairom, N.H.H., Mohammad, A.W., Abdul Halim, N., Ahmad, M.K., Hamzah, S., & Sulaiman, N. (2020). The potential control strategies of membrane fouling and performance in membrane photocatalytic rector (MPR) for treating palm oil mill secondary effluent (POMSE). *Chemical Engineering Research and Design, 162*.
13. Sidik, D.A.B., Hairom, N.H.H., Adnan, R.M., Afendi, A.S., Rahman, F.A.A., & Hisam, N.F. (2021). Performance of membrane photocatalytic reactor on palm oil mill secondary effluent using zinc oxide – *Cymbopogon citratus* (ZnO-CC). *Multidisciplinary Applied Research and Innovation, 2*(1), 111–117.
14. Hairom, N.H.H., Mohammad, A.W., & Kadhum, A.A.H. (2014). Effect of various zinc oxide nanoparticles in membrane photocatalytic reactor for Congo red dye treatment. *Separation and Purification Technology, 137*, 74–81.
15. Hairom, N.H.H., Mohammad, A.W., & Kadhum, A.A.H. (2015). Influence of zinc oxide nanoparticles in the nanofiltration of hazardous Congo red dyes. *Chemical Engineering Journal, 260*, 907–915.
16. Mozia, S. (2010). Photocatalytic membrane reactors (PMRs) in water and wastewater treatment: A review. *Separation and Purification Technology, 73*(2), 71–91. https://doi.org/10.1016/j.seppur.2010.03.021.
17. Molinari, R., Lavorato, C., & Argurio, P. (2017). Recent progress of photocatalytic membrane reactors in water treatment and in synthesis of organic compounds. A review. *Catalysis Today, 281*, 144–164. https://doi.org/10.1016/j.cattod.2016.06.047.
18. Zheng, X., Shen, Z.-P., Shi, L., Cheng, R., & Yuan, D.-H. (2017). Photocatalytic membrane reactors (PMRs) in water treatment: Configurations and influencing factors. *Catalysts, 7*(8), 224. https://doi.org/10.3390/catal7080224.
19. Zhang, W., Ding, L., Luo, J., Jaffrin, M.Y., & Tang, B. (2016). Membrane fouling in photocatalytic membrane reactors (PMRs) for water and wastewater treatment: A critical review. *Chemical Engineering Journal, 302*, 446–458. https://doi.org/10.1016/j.cej.2016.05.071.
20. Dimapilis, E.A.S., Hsu, C.-S., Mendoza, R.M.O., & Lu, M.-C. Zinc oxide nanoparticles for water disinfection. *Sustainable Environment Research, 28*, 47–56.
21. Shah, A. (2022). ZnO photocatalysts for the treatment of wastewater. *Encyclopedia*. https://encyclopedia.pub/entry/19526.
22. Malhotra, S.P.K. (2020). ZnO nanostructures and nanocomposites as promising photocatalysts for the remediation of wastewater pollution. In Kumar, V., Singh, J., & Kumar, P. (Eds.), *Environmental Degradation: Causes and Remediation Strategies*, Volume 1 (pp. 120–132). https://doi.org/10.26832/aesa-2020-edcrs-09.
23. Candela, L. (2018). *Components and Structure*. https://courses.lumenlearning.com/boundless-biology/chapter/components-and-structure/.
24. Promo, E. (2009). *Pore size*. http://membranes.edu.au/wiki/index.php/Pore_Size.
25. Solution, E.W. (2016). *Isoelectric point (IEP) test*. http://pssnicomp.com/applications/isoelectric-point-iep/isoelectric-point-test/.
26. Oshida, Y. (2013). Chapter 4: Oxidation and oxides. In Oshida Y. (Ed.), *Bioscience and Bioengineering of Titanium Materials* (pp. 87–115). https://doi.org/10.1016/B978-008045142-8/50004–9.
27. Sheets, F. (2013). *Colour (true)*, (December), 548–551.
28. Oram, M.B. (2014). *Water color odor appearance and taste*. https://www.water-research.net/index.php/water-testing/water-color-appearance-and-taste.
29. Cooper, L., & Abi-ghanem, R. (2016). *The value of humic substances in the carbon lifecycle of crops: Humic acids, fulvic acids, and beyond*. https://humagro.com/wp-content/uploads/2017/01/The-Value-of-Humic-Substances-in-the-Carbon-Lifecycle-of-Crops-AgroPages-Article-HG.pdf.
30. Washington State Department of Health. (2018). *Color, taste and odor problems in drinking water*. https://doh.wa.gov/sites/default/files/legacy/Documents/Pubs//331-286.pdf
31. Water.com, F. (2018). *Colour in drinking water*. http://www.filterwater.com/t-color-in-water.aspx.
32. Biowizard.com. (2018). *Running the biochemical oxygen demand (BOD) test*. https://www.ebsbiowizard.com/running-the-biochemical-oxygen-demand-bod-test-3563/.
33. Nagwekar, P.R. (2014). Removal of organic matter from wastewater by activated sludge process – Review. *International Journal of Science, Engineering and Technology Research, 3*(5), 1260–1263.
34. Penn, M.R., Pauer, J.J., & Mihelcic, J.R. (2009). *Biochemical oxygen demand*. http://www.eolss.net/.
35. Choudhary, A. (2016). *Determination of biochemical oxygen demand (BOD) in waste water*. https://www.pharmaguideline.com/2013/06/determination-of-biological-oxygen.html.

36. Hach.com. (2017, May 27). *What is CBOD?* https://support.hach.com/app/answers/answer_view/a_id/1010292/~/what-is-cbod%3F.

37. Ajayi, A.A., Peter-Albert, C.F., Ajojesu, T.P., Bishop, S.A., Olasehinde, G.I., & Siyanbola, T.O. (2016). Biochemical oxygen demand and carbonaceous oxygen demand of the Covenant University sewage oxidation pond. *Covenant Journal of Physical and Life Sciences (CJPL)*, 4(1), 11–19.

38. Hanna. (2017a, May 10). *Guide to chemical oxygen demand (COD) testing.* http://blog.hannainst.com/cod-testing.

39. Hanna. (2017b, July 27). *The complete guide to measuring turbidity in water.* http://blog.hannainst.com/turbidity-guide.

40. Stanley, S.J. (2000). *Process Modelling and Control of Enhanced Coagulation.* American Water Works Association.

41. Environmental, F. (2014, September 5). *Measuring turbidity, TSS, and water clarity.* https://www.fondriest.com/environmental-measurements/equipment/measuring-water-quality/turbidity-sensors-meters-and-methods/.

42. Bhattacharjee, S. (2017). *Concentration polarization smart membrane products.* http://www.water-planet.com/wpcontent/uploads/2017/07/ConcentrationPolarization_FINAL_7-11-17.pdf.

43. Wang, K., Abdala, A.A., Hilal, N., & Khraisheh, M.K. (2017). Chapter 13 – Mechanical characterization of membranes. In Hilal N., Ismail, A.F., Matsuura, T., & Oatley-Radcliffe, D. (Eds.), *Membrane Characterization* (pp. 259–306). https://doi.org/10.1016/B978-0-444-63776-5.00013–9.

44. Hairom, N.H.H., Mohammad, A.W., & Kadhum, A.A.H. (2014b). Nanofiltration of hazardous Congo red dye: Performance and flux decline analysis. *Journal of Water Process Engineering*, 4(C), 99–106.

45. Aghdam, M.A., Mirsaeedghazi, H., Aboonajmi, M., & Kianmehr, M.H. (2015). Effect of ultrasound on different mechanisms of fouling during membrane clarification of pomegranate juice. *Innovative Food Science and Emerging Technologies*, 30(June), 127–131. https://doi.org/10.1016/j.ifset.2015.05.008.

46. Mohammed, R.R. (2013). Decolorisation of biologically treated palm oil mill effluent (POME) using adsorption technique. *International Refereed Journal of Engineering and Science (IRJES)*, 2(10), 1–11.

47. Koohestanian, A., Hosseini, M., & Abbasian, Z. (2008). The separation method for removing of colloidal particles from raw water. *American-Eurasian Journal of Agricultural & Environmental Sciences (JAES)*, 4(2), 266–273.

48. Yasar, A., Ahmad, N., Chaudhry, M.N., Rehman, M.S.U., & Khan, A.A.A. (2007). Ozone for color and COD removal of raw and anaerobically biotreated combined industrial wastewater. *Polish Journal of Environmental Studies*, 16(2), 289–294.

49. Sheikh, A., Reza, A., & Sardar, M. (2013). Chemical oxygen demand removal from synthetic wastewater containing non-beta Lactam antibiotics using advanced oxidation processes: A comparative study. *Archives of Hygiene Sciences*, 2(1), 23–30.

50. Brienza, M., & Katsoyiannis, I.A. (2017). Sulfate radical technologies as tertiary treatment for the removal of emerging contaminants from wastewater. *Sustainability*, 9(9), 1604.

51. Tan, K.M., Liew, W.L., Muda, K., & Kassim, M.A. (2015). Microbiological characteristics of palm oil mill effluent. *International Congress on Chemical, Biological, and Environmental Sciences (ICCBES)*, 186–200.

52. Microbiology (n.d). *The effects of pH on microbial growth.* https://www.coursehero.com/study-guides/microbiology/the-effects-of-ph-on-microbial-growth/.

53. Cerulli, P. (n.d). *In what conditions of water does algae grow?* https://www.cuteness.com/article/conditions-water-algae-grow.

54. Water Science School. (2018). *Water color.* https://www.usgs.gov/special-topics/water-science-school/science/water-color.

55. IWA Publishing. (n.d). *Simple options to remove turbidity.* https://www.iwapublishing.com/news/simple-options-remove-turbidity#:~:text=Settling%20and%20decanting%20is%20a, top%20into%20a%20second%20container.

56. Whelan, M. (2013). *The fluorescent lamp.* http://edisontechcenter.org/Fluorescent.html.

57. Amit, A. (2018). *What is diffused light?* https://sciencing.com/diffused-light-5470956.html.

58. Realtech Inc. (n.d). *Chemical oxygen demand (COD).* https://realtechwater.com/parameters/chemical-oxygen-demand/.

59. Abdulsalam, M., Che Man, H., Idris, A.I., Yunos, K.F., & Abidin, Z.Z (2018). Treatment of palm oil mill effluent using membrane bioreactor: Novel processes and their major drawbacks. *Water*, 10(9), 1165.

60. Bowen, W.R., Calvo, J.I., & Hernandez, A. (1995). Steps of membrane blocking in flux decline during protein microfiltration. *Journal of Membrane Science*, 101(1–2), 153–165.

11 Organic Solvent Recovery by Nanofiltration Membrane

Thomas McKean and Ranil Wickramasinghe
University of Arkansas

CONTENTS

11.1 INTRODUCTION

11.1.1 SUSTAINABILITY OF ORGANIC SOLVENT RECOVERY

Organic solvent recovery has emerged as a potentially lucrative opportunity to implement more sustainable practices. Organic solvents are a key component of a variety of industrial processes, including many chemical reactions, purifications, cleaning processes, and analytical processes [1]. The use of organic solvents is so prevalent that it is estimated that it can account for as much as 80%–90% of a total manufacturing cost when including necessary water, 60% of energy use including recovery and disposal via incineration, and potentially 70% of the process's combined capital and operating cost [1–3]. Thus, improving the ability of a process to more efficiently utilize organic solvents may dramatically improve the mass, energy, and monetary considerations. Focusing on solvent recovery also offers a route to decreasing the carbon footprint of associated processes since unrecovered solvent is typically disposed of via incineration that releases greenhouse gases. It has

DOI: 10.1201/9781003261827-11

been reported that industrial-scale incineration can create as much as 6.7 kg CO_2 per kg of organic solvent treated [1]. Further solvent recovery also improves the environmental aspects of different processes by reducing solvent inventory and waste, more solid waste rather than liquid, stronger environmental compliance, lower energy use, and process length [1,2]. These efforts also result in highly desirable cost decreases and are of great interest in a wide variety of industries.

Distillation is the most common process employed for organic solvent recovery. As a thermal separation technique, it is notorious for poor efficiency and for being energy intensive. Recovery efficiencies as low as 50% are not uncommon in older systems [2]. Furthermore, processes in which organic solvents are important often contain thermally sensitive components, particularly in the pharmaceutical industry.

High-value products must be preserved during separation, which results in any thermal-based separation technique being a naturally poor fit for these products. In fact, due to these limitations and the poor efficiency of solvent recovery processes, it is often more economically feasible to incinerate the waste solvent, in turn polluting the atmosphere and wasting a potentially valuable solvent [4]. To address this, membrane processes are quickly beginning to replace these thermal separations due to their inherent simplicity and strong selectivity. Pressure-driven membrane processes reject contaminants by size exclusion. The simple nature of this process allows for operation without troublesome phase changes and at shorter batch times than distillation [5]. Additionally, as a non-thermal technique, product is preserved during recovery and energy efficiency is greatly decreased compared to distillation. Figure 11.1 compares the two processes. For example, Geens et al. demonstrated a dramatically lower energy requirement for a membrane recovery system of nearly 200 times lower than conventional distillation, notably at a low concentration [6]. Another common problem using distillation for solvent recovery that is avoided by membrane technology is azeotrope formation [1]. A final important advantage of membrane technologies for solvent recovery is its modular nature and its ability to work in combination with other processes such as distillation, adsorption, and chromatography [1,5]. Rundquist et al. demonstrated a successful combined membrane and distillation process that consumed nine times less energy than conventional distillation [2]. These performance and operating cost advantages will be key to increasing the economic feasibility of commercial solvent recovery.

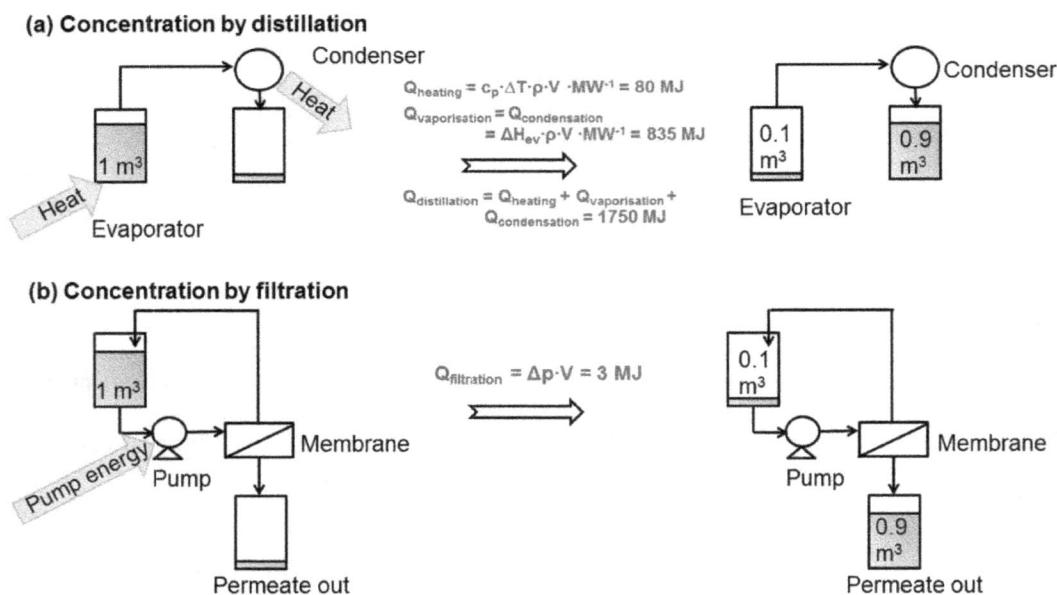

(a) Concentration by distillation

$Q_{heating} = c_p \cdot \Delta T \cdot \rho \cdot V \cdot MW^{-1} = 80$ MJ

$Q_{vaporisation} = Q_{condensation} = \Delta H_{ev} \cdot \rho \cdot V \cdot MW^{-1} = 835$ MJ

$Q_{distillation} = Q_{heating} + Q_{vaporisation} + Q_{condensation} = 1750$ MJ

(b) Concentration by filtration

$Q_{filtration} = \Delta p \cdot V = 3$ MJ

FIGURE 11.1 Comparison between concentrating 1 m³ of feed using distillation and membrane processes. (Reprinted with permission from Ref. [7]. Copyright American Chemical Society 2014.)

11.1.2 Organic Solvent Nanofiltration (NF)

NF for the recovery of organic solvents, commonly referred to as organic solvent NF or organic solvent nanofiltration (OSN), is a common membrane process used to recover low-molecular-weight high-value solvents. NF encompasses membranes with pores sufficiently small to reject molecules in the range of 200–1,000 Da [8]. NF utilizes a combination of steric effects determined by the membrane pores and architecture, electrostatic interactions that arise from the slight charge present on the membrane surface in aqueous media, and dielectric exclusion for the rejection of contaminants or products of interest [9]. OSN largely relies on steric exclusion considering the lack of charge and thermal sensitivity of most organic solvents and products. This leads NF to be a particularly effective solution for organic solvent recovery given the tightness of the membranes compared to other filtration processes. NF also shows operational advantages over other filtration technologies. Similar to reverse osmosis (RO), NF is a pressure-driven process that employs a membrane strongly permeable to the solvent and impermeable to the solute. Given the larger pore size compared to RO, NF also shows improved water permeation at a lower operating pressure [9,10]. Typical NF pressures range from 345–1550 Kpa (50 to 225 PSI). RO is commonly run as high as 5515 kPa (800 PSI) for comparison [12].

The most common problem exhibited by NF membranes is fouling. Fouling involves foreign substances adsorbing to the membrane surface and inhibiting permeate flux. An inherent trade-off exists between fouling and separation ability. Reducing the pore size increases the rejection by the membrane but often increases its susceptibility to fouling [13]. Significant research has been undertaken to address membrane fouling and improve other material properties of the membrane. In addition to fouling, reduction of the chemical stability and mechanical integrity of the membranes can also be responsible for a loss of performance over time. Chemical stability in the presence of harsh organic solvents and highly basic solutions found in OSN processes is of key concern. Stability for high-temperature and high-pressure operation is also important [14]. Significant time and funding are required to study the use of the materials in application and the manufacturing of these materials. Current manufacturing can be slow and produce membranes with inconsistent pore sizes, and manufacturing processes for novel materials still need to be developed [15,16].

This chapter will discuss recent trends in different areas of OSN with the focus on progress toward more sustainable practices. Different types of membranes will be surveyed, discussing the state-of-the-art polyamide (PA) thin-film composite (TFC) membranes and their formation. Membrane stability in different solvents will also be discussed. It is recognized that ceramic membranes and ceramic nanomaterial additives will be necessary for sustainable processes moving forward. Different types of hybrid and composite membranes will be discussed. Pore flow and solution diffusion transport models will be presented to discuss solvent transport through the membrane. Finally, prevalent applications of the technology will be highlighted, including recent uses of OSN in the pharmaceutical, food, and petrochemical industries.

11.2 OSN MEMBRANE MATERIALS

11.2.1 Polymer Membranes

11.2.1.1 PA Active Layer

Polymer materials are particularly suited for membrane applications due to their low production costs, robust chemical modification capabilities, favorable combination of mechanical integrity and elasticity, and simple scalability [14]. State-of-the-art NF membranes belong to a class of membranes known as PA TFC membranes. These membranes rely on a thin active layer of PA formed using interfacial polymerization (IP) on the surface of a support layer [17]. The thin layer grown on

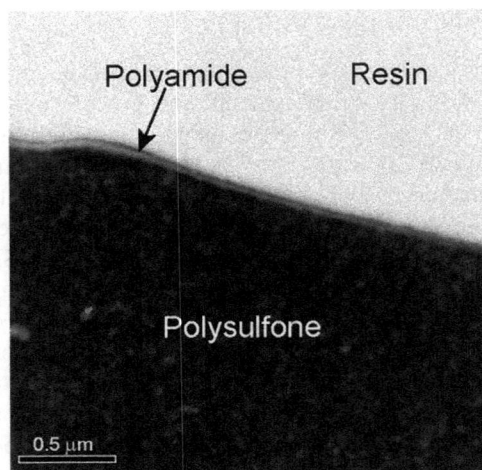

FIGURE 11.2 TEM image of a cross section of a NF270 Polysulfone NF membrane showing the PA layer. (Reprinted with permission from Ref. [25]. Copyright 2010 Elsevier.)

a polysulfone (PS) support is shown in Figure 11.2. This change represented a significant improvement over the previous state-of-the-art integrally skinned asymmetric membranes, particularly in aprotic polar solvents such as tetrahydrofuran (THF) and acetone [18]. The resulting aromatic PA forms a dense top layer capable of efficiently separating small molecules and ions from the carrier solvents. Extremely tight pores only a few nanometers in diameter form a highly uniform network of solvent transport pathways, leading to a relatively high water flux among competitive membranes [19]. PA NF membranes of excellent homogeneity were reported to display efficient rejection of solutes as small as 0.5 Å with a layer thickness as low as 30–40 nm [20]. These membranes can be susceptible to swelling that disrupts the transport pathways of solvent molecules through the membrane [21]. Current work on PA membranes focuses on incorporating different stabilizing and cross-linking agents into the membrane to preserve its strong transport and selectivity properties during operation. PA can be naturally cross-linked by varying reaction conditions [22], introducing another polymer [23], or various inorganic nanomaterials during formation [24].

Ongoing efforts have focused on understanding the IP process that forms the PA layer in parallel to improving its stability. PA is frequently synthesized from m-phenylenediamine (MPD) and 1,3,5-benzenetricarbonyl (TMC) at an organic/aqueous-phase interface of aqueous-phase MPD and organic-phase TMC. TMC is frequently soaked in a porous PS or polyethersulfone (PES) support membrane that is exposed to a solution containing a high concentration of MPD. The high porosity of PS and PES provides high TMC loading and ultimately strong water flux during operation. Initial characterizations of this process show the film grows very quickly at first before stopping abruptly [26]. As the PA layer forms, contact between the MPD and TMC is limited by the ability of the MPD to diffuse across the PA layer. Different studies have sought to determine the effects of changing the MPD and TMC solutions on the resulting PA layer. For example, Khare et al. studied the cross-linking of the resulting PA layer when varying the TMC concentration present in the PS pores. Over a range of 0.005–0.8 wt%, PA cross-linking increased with TMC concentration until TMC concentration reached 0.1 wt%. It was observed that branching began to dominate over cross-linking, and this branching had a negative effect on water permeation [27]. However, it is typically observed that the thickness of the resulting PA layer is largely dependent on the MPD supply during the reaction [28]. Grzebyk et al. reported a system capable of continuously supplying MPD to the reaction and produced a TFC membrane with a 1,282 nm thick layer of PA after 15 minutes of reaction time [29].

11.2.1.2 Materials Stable for OSN

Material and solvent considerations are particularly important for OSN considering the harsh nature of the employed solvents. Previously discussed PA TFC membranes are widely used in aqueous applications, but the PS layer is known to be unstable in harsh organic solvents such as ketones, so different materials must be employed as the support layer in OSN applications [30]. Kim et al. report an OSN membrane with a PA active layer formed through IP on a drop-casted polyacrylonitrile (PAN) support membrane. The membrane was stable in a variety of organic solvents but showed higher flux in polar solvents (i.e., methanol, ethanol, acetone) compared to nonpolar solvents (i.e., isopropanol, methyl ether ketone, hexane) [30]. PA layers were also formed on other solvent-stable polymers such as polyetherketone (PEEK) and polyimide (PI). PI membranes demonstrated a higher flux than PEEK membranes in both polar and nonpolar organic solvents [31]. In addition to PI membranes with a PA active layer, PI membranes have been used for OSN applications with active layers formed by coating with PDMS or polypyrrole (PPy), and thermal cross-linking [32–34]. PI also is popular in commercial applications and is sold commercially by Lenzing under the trade name P84. P84 is known for its stability in polar and nonpolar solvents [35]. Other polymers known for their robustness and success in other application, such as polybenzimidazole (PBI) and polyvinylidene fluoride (PVDF), are also being explored for potential applications in OSN [36,37]. Table 11.1 summarizes the polymer materials discussed.

11.2.2 CERAMIC MEMBRANES

Recent developments and future directions for OSN membrane materials include the use of ceramic materials. Ceramic materials are extremely stable alternatives to polymer membranes and are capable of high-temperature operation in a wide range of solvents. These membranes are made with the aid of a polymer precursor that allows for manipulation of the morphology of the final ceramic membrane. The polymer/ceramic solution is cast into the desired shape and the polymer is removed through sintering to leave the final membrane. Hollow fiber membranes that are particularly important to industry are made through this process [38]. Production of ceramic membranes is typically more complex and energy intensive than the manufacturing of polymer membranes, so polymers are generally favored for low-temperature or mild solvent processes. Similarly to polymer membranes, ceramic membranes consist of an active layer responsible for solute rejection in combination with a support layer. Different metal oxide materials, namely Al, Ti, Zr, and Si, have been studied for both the active and support layers [39]. Pure ceramic membranes with ceramic support and active layers are typically formed by coating a powder layer on the surface of the support. This method easily produces membranes of approximately 30 nm pore size, which is considered ultrafiltration (UF) rather than NF based on the molecular weight cut-off (MWCO) [7]. Two strategies to reduce the pore size to the NF range include using zeolite and silica particles as the active layer [40].

Zeolite materials are a family of ceramics that show a highly defined and rigid network of pores. The small pore size of 0.3–1.3 nm coupled with the inherent stability of their ceramic makeup has led to extensive study in different harsh environments for separations and catalytic processes [41]. This combination of strong resistance and uniform pore network leads to effective uses in OSN, and efficient performance has been reported for a variety of organic solvents [42]. These materials are formed from a silicate backbone made up by $[SiO_4]^{4-}$ tetrahedral units. Substituting Si^{4+} with Al^{3+} provides sites for transition metal cations to intercalate into the structure. These transition metal cations are often responsible for the adsorption of contaminants onto the membrane. Four classes of zeolite materials have been explored extensively in academic and industrial applications: Linde type A (LTA), faujisite (FAU), mordenite (MOR), and mobile five (MFI). LTA was discovered first, typically containing a Na^+ cation in a configuration commonly referred to as NaA. FAU zeolites were then developed using Cu^{2+} and Co^{2+}, followed by the industrial development of MOR and MFI zeolites for stronger and more selective oxidation of heavy petrochemical compounds [43]. These

TABLE 11.1
Structures of Polymer Materials used in OSN

Polymer	Structure
PAN	
PI	
PEEK	
PBI	
PVDF	
PS	

Source: Adapted with permission from Ref. [7]. Copyright American Chemical Society 2014.

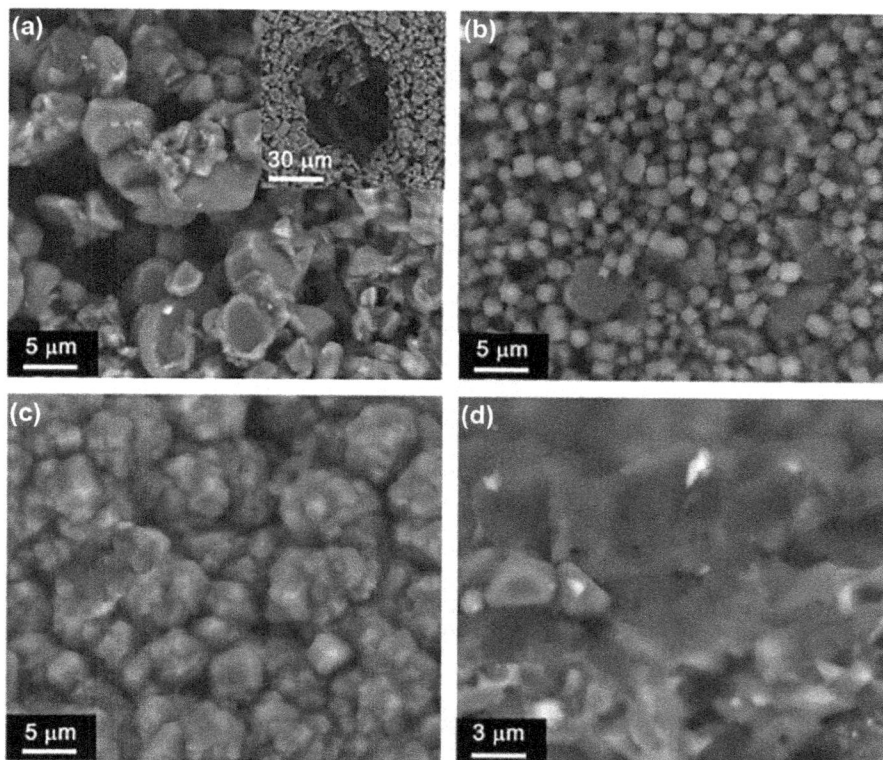

FIGURE 11.3 SEM images depicting the process of growing a zeolite on an alumina support layer. Panel (a) shows the bare support membrane, (b) shows the seeded membrane, (c) shows a top view of the zeolite layer, and (d) shows a cross section. (Reprinted from Ref. [44] with permission. Copyright WILEY-VCH Verlag GmbH & Co. KGaA, Weinheim 2011.)

zeolite structures make up the active layer of OSN membranes and are deposited on the surface of the support layer using a dip or vacuum coating process [44]. Figure 11.3 contains Scanning electron microscopy (SEM) images of the coating process at different stages. This simple coating process allows for the formation of different membrane configurations, namely hollow fiber membranes particularly relevant to industry. Asghari et al. studied FAU and MOR active layers for the dehydration of ethanol. MOR demonstrated better separation, but FAU showed higher flux, reaching 1.56 kg/m²h. The flux also increased as the temperature increased [45]. Cao et al. used a NaA layer to dehydrate ethanol at a much higher permeate flux of 19.7 kg/m²h at 348 K [46].

Another method of reducing the pore size of ceramic membranes involves using a silica active layer. The pore size of silica can be reduced to single nanometer range using surfactants such as cetyltrimethylammonium bromide (CTAB) or sodium dodecyl sulfate. Silica membranes are often functionalized by grafting a silane molecule with a long-chain carbon tail that increases the hydrophobicity of the surface. Kujawa et al. study the addition of the silane molecule $C_6F_{13}C_2H_4Si(OEt)_3$ (C6) to the surface of alumina and titania membranes for the removal of methyl tert-butyl ether (MTBE), and ethyl acetate (EtAc) from water. The C6 chain was dip-coated onto either the alumina or titania substrate, both with 5 nm pore size. The titania membranes achieved approximately 4 kg/m²h for EtAc compared to 2 for alumina. This trend reversed for MTBE, where alumina reached 2.75 kg/m²h as opposed to 1.75 kg/m²h for titania. The selectivity was much stronger for MTBE using titania, and the authors claim the titania membrane was superior to the alumina [47]. Van Gestel et al. further confirmed that these long-chain silane molecules are effective for organic solvents by comparing the *n*-hexane permeability among ceramic membranes modified by different chain-length silanes. They found that the *n*-hexane permeability significantly increased as the silane chain length increased, reaching a maximum of 6.1 L/m²h bar for C8 membranes. This modification also eliminated the water flux through the membrane [48].

Ceramic membranes, both zeolites and silica, show great promise for high temperature, very harsh solvent, or extreme pH applications since polymer membranes cannot withstand these harsh operating conditions. The high cost of these ceramic membranes leads polymer membranes to be a more efficient alternative at more moderate conditions. However, combining the two in the form of hybrid membranes and composite membranes has allowed researchers to design membrane material configurations such that the weaknesses of one material are compensated for by the strengths of the other. The remainder of this section will discuss the future direction of OSN materials that focuses on the combination of polymer and ceramic materials in hybrid and composite OSN membranes.

11.2.3 Hybrid and Composite Membranes

Hybrid and composite OSN membranes contain both an organic polymer component and an inorganic ceramic component. Hybrid membranes designate those that contain a pure support layer and a pure active layer, and composite membranes refer to polymer membranes with small ceramic particles dispersed throughout. Composite membranes are at the forefront of nearly all fields of polymer science, where hybrid membranes were developed specifically for OSN applications. Both ceramic-based hybrid membranes with polymer active layers and polymer-based membranes with ceramic active layers have been studied with promising results (Figure 11.4).

11.2.4 Hybrid Ceramic Membranes

Beginning with ceramic-based hybrid membranes with polymer active layers, this configuration was a natural extension to increase the performance of ceramic membranes. The terminating hydroxyl groups on the surface of ceramic membranes are inherently hydrophilic which is not desirable for OSN application. A variety of chemistries have been developed to perform modifications to materials containing these terminating hydroxyl groups. Doing so introduces the necessary

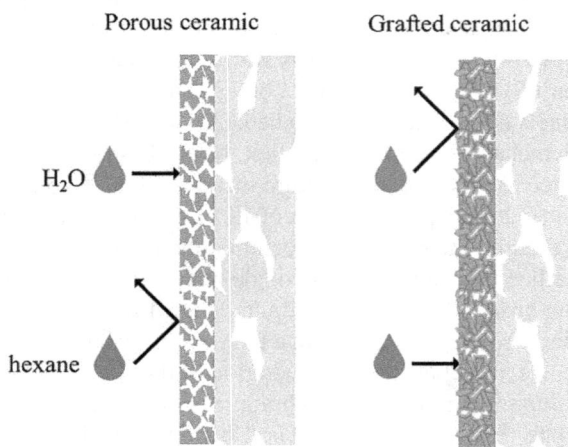

FIGURE 11.4 Schematic of hybrid ceramic membranes modified for OSN. A polymer grafted onto a ceramic support is depicted. Polymer membranes are modified to perform identically, swapping the ceramic support for a polymer and the polymer active layer for a ceramic. (Reprinted with permission from Ref. [49]. Copyright Elsevier 2020.)

hydrophobicity while also reducing the pore size to the single nanometer range necessary for OSN [50]. A hydrophobic PDMS layer coated on the surface of an α-alumina membrane through an automated dip coating process displayed extremely high stability in a toluene environment, maintaining a permeance of 1.6 L/m^{2}h bar for as long as 40 hours [51]. Introducing a compatible functionalization can increase the ability of the layer to coat the substrate. By first depositing an aminosilane monolayer linker through a vapor phase deposition, a membrane with a notably smaller pore size can be formed [52]. Tanardi et al. extended this concept to investigate mercaptopropyltriethoxysilane (MPTES) as a linker on a mesoporous γ-alumina membrane. First the silane was deposited on the ceramic membrane in a vapor phase reaction, and then the modified membranes were soaked in a PDMS/toluene solution for 24 hours to complete the reaction. This refined process produced the most uniform membrane from the group to date [53].

Besides PDMS, polymer chains of different materials can be grafted to the surface of ceramic membranes to create a structure called a "polymer brush." Polymer brushes of different materials have been shown to modify macroporous ceramic substrates and successfully reduce the pore size [49]. These results demonstrated the effectiveness of polymer chains grafted to the surface of ceramic membranes, but the technique to graft the brush to the microporous substrate is not capable of modifying mesoporous ceramic membranes used in OSN applications. Instead, more precise methods were investigated to grow polymer chains on the surface of ceramics. Atom Transfer Radical Polymerization (ATRP) is one such process. It is estimated the ideal polymer chain length for OSN membranes is 4–20 monomers, a level ATRP can easily reach [49,54]. An initiator is first attached to the hydroxyl groups on the membrane surface, followed by the controlled growth of the desired polymer from the initiator. Merlet et al. used ATRP to grow polystyrene chains on the surface of an γ-alumina ceramic membrane using an organosilane initiator. The resulting membrane showed stable attachment with a final pore size below 2 nm and a permeability of 1.6 L/m^{2}h bar for toluene. The authors also report controlling the resulting chain length of the polystyrene by changing ATRP conditions such as reaction time and monomer:initiator ratio [55]. Kruk et al. report growing PAN chains from the surface of mesoporous silica particles. The resulting layer was highly uniform with a tunable thickness. Cu(II) halide is used as a deactivator to obtain precise control of the reaction. While this modification was not tested directly in OSN applications, the principle of the results could be extended with promise to an OSN membrane [56].

11.2.5 Hybrid Polymer Membranes

Polymer membranes also greatly benefit from ceramics added as a particle to fill the space between the polymer chains. Membrane properties are optimized when the ceramic is homogeneously dispersed throughout the polymer. Hybrid membranes only integrate the ceramic into the active layer rather than being dispersed throughout the entire membrane in a composite. Ceramic nanoparticles consistently show the best performance as an additive in both configurations, widely attributed to their ability to better disperse throughout the polymer matrix as their diameter shrinks. First focusing on hybrid membranes, the most effective configuration is to incorporate the ceramic into an interfacially polymerized PA layer. In fact, the addition of the ceramic to the PA layer improves its applicability in OSN applications. Such membranes are often referred to as thin-film nanocomposite (TFN) membranes. SiO$_2$ and TiO$_2$ are very common and inexpensive ceramic nanoparticles that demonstrate the performance of TFN membranes. SiO$_2$ has an abundance of surface hydroxyl groups that lend it as a natural TFN alternative. As such, SiO$_2$ was identified as a filler that could increase the flux of polar organic solvents. A polyethyleneimine-based active layer formed through IP on the surface of a PS membrane produced a flux of approximately 1.3 L/m^{2}h bar for isopropyl alcohol (IPA) at an optimal SiO$_2$ content of 0.1 wt% [57]. TiO$_2$ offers similar benefits to SiO$_2$ but requires a surface modification to be effectively dispersed in a PA layer of a TFN membrane. Amine and chloride compounds can be used to improve the ability of TiO$_2$ particles to be integrated into a PA film during IP. The presence of these nanoparticles also can greatly reduce the swelling of the membrane and in turn improve its stability in organic solvents. The addition of these nanoparticles was shown to improve the methanol flux and the selectivity against different organic dyes [58]. Zhang et al. compared SiO_2 to TiO_2 as nanoparticle filler for a PEI active layer on the surface of a PAN support. SiO_2 showed a stronger permeability in IPA, reaching 1.46 L/m^2h bar compared to just 0.61 L/m^2h bar for TiO_2 membranes [59].

While SiO_2 and TiO_2 nanoparticles do show some utility in TFC membranes at an extremely low cost, researchers have consistently found that employing different ceramic nanomaterials with greater surface area leads to superior OSN performance. Current research focuses on zeolite, metal organic frameworks (MOFs), graphene and graphene oxide (GO), and other two-dimensional (2D) materials. Zeolites are a strong candidate given their previously reported promising results as the active layer of ceramic membranes. IP can be used to form a PA layer containing the zeolite nanoparticle UZM-5 on the surface of a PEI support membrane. The membranes effectively rejected oil from a solvent mixture of toluene and methyl ether ketone (MEK) with a flux of 13.85 L/m^2h [60]. MOF structures resemble zeolites where metal and organic components are highly ordered and tunable to form a consistent network of pores that are particularly effective for filtration applications. MOFs are largely considered the next generation of zeolite materials given their resemblance but superior tunability that addresses some of the limitations of traditional zeolites. The organic component of the structure may be changed to manipulate pore sizes, flexibility of the material, and affinity for host polymer chains to ultimately optimize the removal of desired contaminants [61]. In particular, tuning the structure to the affinity of the host polymer can minimize the formation of voids between the polymer and filler that led to a decline in rejection compared to the pure polymer [62]. Sorribas et al. study the incorporation of zeolitic imidazolate framework (ZIF) and Materials Institute Lavoisier (MIL) MOFs into a PA layer synthesized on a PI support. The membranes were tested for the rejection of styrene oligomers in THF and methanol solutions and found that while the incorporated MOFs did not significantly change the rejection, they did dramatically increase the permeability (160% to 3.9 61 L/m^2h bar). They attribute this increase to an increase in pore size provided by the MOFs, supported by an increase in permeance as MOF size increased [61]. The incorporation of the MOFs into the active layer is shown in Figure 11.5.

GO and carbon-based nanomaterials are another class of rapidly emerging materials that are drawing aggressive attention in filtration applications given their excellent chemical and mechanical

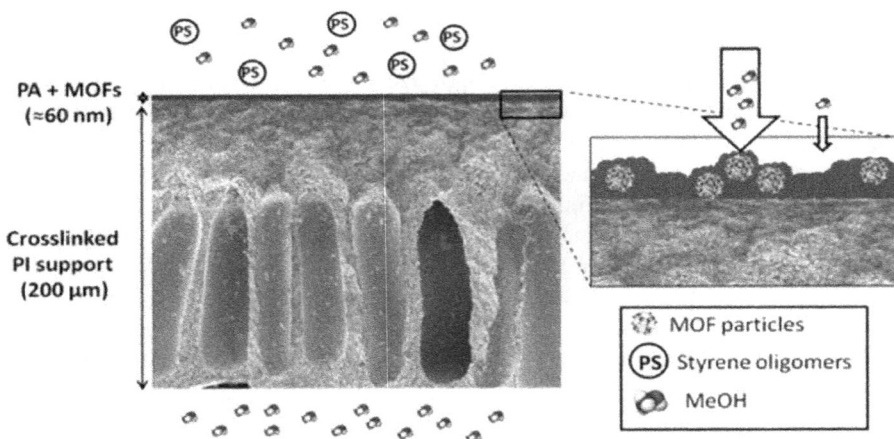

FIGURE 11.5 Composite PA layer filled with MOF particles. (Reprinted with permission from Ref. [61]. Copyright American Chemical Society 2013.)

stabilities. These materials can exist as 2D materials that possess a relatively large X and Y dimension and a near atomically thick Z dimension or a rolled configuration of these 2D nanosheets called a carbon nanotube (CNT). An unexpected early result of 4–5x expected water permeation through CNT cores sparked this interest [63]. The most common of these materials currently is GO for its ease of production and ability to be coated on the surface of a polymer membrane [64]. GO was found to effectively disperse throughout the PA layer during IP on a PI support. The addition increased the ethanol permeance to 41.47 L/m^2h MPa [65]. Multi-walled CNTs (MWCNTs) can also be integrated into the active PA layer. These MWCNTs provided efficient pathways for solvent permeation without sacrificing rejection, improving the permeation by an order of magnitude [66]. Further development of carbon-based nanomaterials continues to give performance increases. Graphene quantum dots (GQD) are the smallest form of graphene reported to date. Shen et al. report the integration of GQDs into a PA active layer for OSN. By studying the IP process, they were able to determine that the GQDs were able to provide space between the PA polymer chains that led to a more uniform network of interconnected pores without sacrificing polymer-polymer interactions that preserve stability. The membrane was found to show reliably high flux in methanol (11.1 L/m^2h bar) and the harsher DMF (5.9 L/m^2h bar) [67]. Other similarly structured 2D materials such as MoS_2 have also been studied as an additive to the PA active layer with similar strong performance [68,69].

The final configuration of hybrid polymer membranes employs a pure nanomaterial as the active layer on a microporous or UF support. Inspired by pure nanomaterial membranes in other applications, advances in 2D material fabrication technology to produce larger and more uniform layers of 2D materials have unlocked the possibility of eliminating the PA layer all together with excellent results. This direction is also attractive from a sustainability standpoint, as the fabrication of these types of membranes dramatically reduces the use of many hazardous chemicals associated with polymer membrane synthesis [70]. Reduced GO (rGO) was studied as the active layer of a microfiltration support layer by using vacuum filtration to coat the flakes onto the support. Permeability as high as 77.2 L/m^2h bar was achieved for methanol permeation in this configuration [71]. MoS_2 similarly functions as an active layer. MoS_2 nanoflakes are synthesized in an aqueous environment which leaves residual water molecules between nanoflakes. Without a supporting liquid that persists during water removal for OSN the flakes would aggregate as the water was removed. Glycerol is one such liquid that was used to support the drying of the nanosheets into a desired interlayer spacing of approximately 1.0 nm. Without this buffer, the solvent transport pathways between the nanoflakes would be eliminated. 3.2 L/m^2h bar IPA permeance was recorded over a 7-day period following this process [72]. Recent work by Zhang et al. reports achieving a massive methanol permeance of 670 L/m^2h bar by creating an active membrane layer inspired by the structure of leaf veins from an rGO-nickel phosphate nanotube composite nanomaterial. The nickel phosphate nanotubes enhance the performance of the rGO layer both by increasing the interlayer spacing between rGO nanosheets and providing additional transport pathways for solvent molecules [3]. The membrane is shown in Figure 11.6.

11.2.6 Composite Membranes

Composite membranes are a more common configuration of membranes that contain both polymer and ceramic components. This can be attributed both to the ease of distributing a ceramic nanomaterial into the precursor polymer membrane solution and to the widespread success, composite membranes have shown in a variety of interdisciplinary applications. The polymer and nanomaterial components must be compatible to ensure proper dispersion of the nanomaterial into the resulting polymer matrix. In OSN applications, the presence of the nanomaterial is very effective at stabilizing the polymer membrane against swelling and often providing additional pathways for solvent transport. Gevers et al. were first to investigate the use of composite membrane for OSN, incorporating

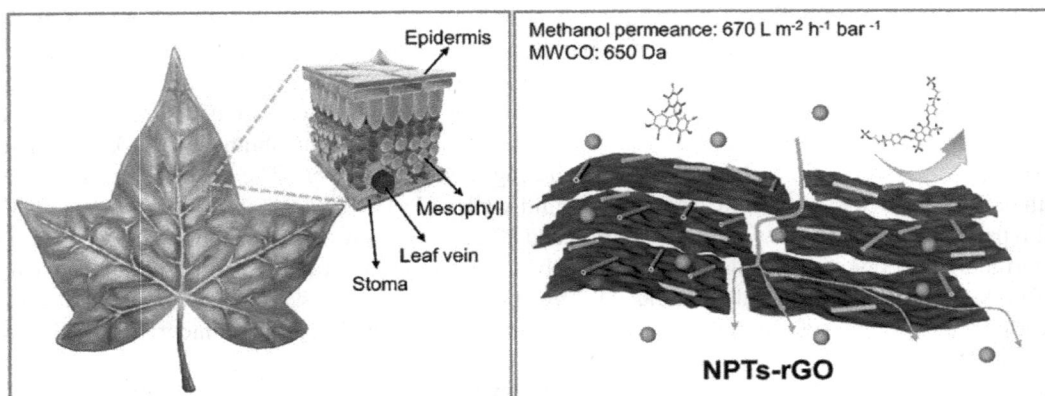

FIGURE 11.6 Schematic of leaf vein-inspired pure nanomaterial membrane for outstanding OSN performance. (Reprinted with permission from Ref. [3]. Copyright Elsevier 2022.)

silica, carbon fillers, and zeolites into a PDMS membrane to reduce its swelling in organic solvents. The permeance did decrease for all solvents tested (toluene, EtAc, dichloromethane [DCM], and THF), though the zeolite-filled membranes did show a relatively acceptable range of permeance from 0.55 L/m^{2}h bar for EtAc to 0.71 L/m^{2}h bar for DCM [73]. As with previously discussed TFN membranes, MOFs tend to show better permeation and have thus attracted more research attention for polymer composites than more traditional zeolite or SiO$_2$ nanoparticles. MOFs have been incorporated into stable polymers such as PVDF to study the effect of filling the space between chains with ZIF-8 particles. The flexible and inherently hydrophobic ZIF-8 particles provided diffusion pathways for organic solvents (ethanol and IPA) and led to a strong permeability of 14.7 L/m^{2}h bar of IPA and 37.7 L/m^{2}h bar for ethanol. It was noted that a high concentration of filler led to agglomeration and ultimately a decrease in mechanical properties. 10 wt% ZIF-8 was the optimal concentration for mechanical properties [74]. Sani et al. also came to this same conclusion that a moderate concentration of MOF led to better performance. They found polyphenylsulfone membranes filled with copper-1,3,5-benzenetricarboxylate (Cu-BTC) nanoparticles at 0.5–1 wt% showed better MWCO and rejection than 3 wt% [75]. This can be attributed to aggregation of MOFs at high loading. Different strategies have been explored to maximize MOF dispersion in polymer membranes, including in-situ growth of nanoparticles within the polymer membrane [76], using a resin microsphere to block the MOFs from interacting with each other [77], and compositing the MOF with a polymer or 2D material [78,79].

Carbon-based nanomaterials, specifically GO and CNTs, are also growing in popularity for composite OSN membranes. 2D materials effectively orient themselves perpendicular to the direction of solvent flow during membrane synthesis, optimizing the barrier layer and its thickness. The abundance of oxo-groups on the surface of GO allows it to disperse within a variety of polymers and perform well as a stabilizer [80]. Incorporating GO into a PBI membrane produced a comparable flux to membranes with a pure GO active layer in multiple tested solvents, producing a peak permeability of 54 L/m^{2}h bar [81]. CNTs possess similar advantages to GO but are more difficult to incorporate into polymer membranes given their lack of surface functionality. A polar carboxyl group added to the surface of MWCNTs allowed for uniform dispersion into a PI membrane. The membrane was tested in ethanol, achieving 1.92 L/m^{2}h bar [82]. The same group also tested an amine functionalization of the MWCNTs in an effort to improve the interactions between the MWCNTs and the PI. The result was an amidation reaction between the polymer and amine functional group on the MWCNT that improved both the permeance and rejection of the previously tested membranes [83]. Grosso et al. repeated this result for aminated MWCNTs and also discussed their inherently strong antifouling properties [84]. Table 11.2 summarizes the results of the previously discussed membranes.

TABLE 11.2

Summary of Membranes Discussed

Membrane Class	Membrane Material	Solvent	Solute	Flux or Permeability	Rejection	Ref.
Polymer	PAN	Methanol	Oleic acid	1.55 ton/m^{2}day	>90%	30
	PEEK	THF	Polystyrene	27 L/m^{2}h (30 bar)	99%	31
		Methanol	Polystyrene	60 L/m^{2}h (30 bar)	99%	31
	PI	THF	Polystyrene	45 L/m^{2}h (30 bar)	99%	31
		Methanol	Polystyrene	51 L/m^{2}h (30 bar)	99%	31
	PBI	Hexane	L-α-lecithin	80.8 L/m^{2}h (10 bar)	92%	36
	PVDF	Hexane	Soybean oil	4 kg/m^{2}h	95%	37
Ceramic	FAU (Zeolite)	Ethanol		1.56 kg/m^{2}h		45
	MOR (Zeolite)	Ethanol		1.15 kg/m^{2}h		45
	NaA (Zeolite)	Ethanol		19.7 kg/m^{2}h		46
	C6 (Silane) on Alumina	Water	EtAc	2 kg/m^{2}h		47
			MTBE	2.75 kg/m^{2}h		47
	C6 (Silane) on Titania	Water	EtAc	4 kg/m^{2}h		47
			MTBE	1.75 kg/m^{2}h	99%	47
	(C8) Silane	n-hexane	Polyethylene Glycol (PEG)	5 L/m^{2}h bar	99%	48
Hybrid Ceramic	Alumina coated with PDMS	Toluene	Polystyrene	1.6 L/m^{2}h bar	99%	51
	Alumina coated with MPTES	Toluene		2.1 L/m^{2}h bar		53
	Alumina coated with polystyrene	Toluene	Diphenylanthracene	1.6 L/m^{2}h bar	90%	55
Hybrid Polymer	PS coated with SiO$_2$ composite active layer	IPA	Crystal Violet	1.3 L/m^{2}h bar	99%	57
	PAN coated with SiO$_2$ composite active layer	IPA	PEG	1.46 L/m^{2}h bar	99%	59
	PAN coated with TiO$_2$ composite active layer	IPA	PEG	0.61 L/m^{2}h bar	97%	59
	PEI coated with UZM-5 (zeolite) composite layer	Toluene/MEK mixture	Oil	13.85 L/m^{2}h (15 bar)	96%	60
	PI coated with ZIF (MOF) composite layer	Methanol	Polystyrene	2.1 L/m^{2}h bar	99%	61
	PI coated with MIL (MOF) composite layer	THF	Polystyrene	11 L/m^{2}h bar	99%	61
	PI coated with MIL (MOF) composite layer	Methanol	Polystyrene	4.2 L/m^{2}h bar	99%	61
	PI coated with GO composite layer	Ethanol	Rhodamine B	41.47 L/m^{2}h MPa	98%	65
	Polyketone coated with GQD composite layer	Methanol	Rose Bengal	11.1 L/m^{2}h bar	98%	67
	Polyketone coated with GQD composite layer	THF	Rose Bengal	5.9 L/m^{2}h bar	98%	67
	rGO	Methanol	Brilliant Yellow	77.2 L/m^{2}h bar	86.2%	71
	MoS$_2$	IPA	Rose Bengal	3.2 L/m^{2}h bar	~90%	72
	rGO-Nickel Phosphate Nanotube	Methanol	Reactive Black	670 L/m^{2}h bar	98.2%	3

(Continued)

TABLE 11.2 (*Continued*)
Summary of Membranes Discussed

Membrane Class	Membrane Material	Solvent	Solute	Flux or Permeability	Rejection	Ref.
Composite	PDMS filled with zeolite	EtAc	Wilkinson Catalyst	0.55 L/m²h bar	98.5%	73
		DCM	Wilkinson Catalyst	0.71 L/m²h bar	97%	73
	PVDF filled with ZIF-8	IPA	Rose Bengal	14.7 L/m²h bar	99.5%	74
		Ethanol	Rose Bengal	37.7 L/m²h bar	99.2%	74
	PBI filled with GO	Acetone	Pharmaceutical (MW 420 g/mol)	54 L/m²h bar	93%	81
	PI filled with MWCNTs	Ethanol	Tetracycline	1.92 L/m²h bar	92.1%	82

11.2.7 COMMERCIAL OSN MEMBRANES

Commercial OSN membranes focus on stability in a wide range of solvents rather than maximizing permeability and rejection. All membranes discussed in this section report stability in many organic solvents, including the solvents discussed to this point. Evonik-MET Ltd. offers some of the most well-known commercial OSN membranes based on PI. Duramem employed cross-linked PI, Puramem used P84 PI, and Puramem S600 added a silicone layer to the surface of the P84 [85,86]. Many other commercial membranes report a silicone active layer. Early Koch MPF membranes consisted of a PDMS layer on a PAN support, SolSep, another industry leader, is rumored to also use a silicone layer as the active layer in a proprietary configuration, and GMT-GmbH has also introduced a TFC membrane utilizing a silicone active layer. Outside of silicone and PI, PoroGen produces PEEK hollow fiber membrane, and PolyAn produces a proprietary composite membrane [7]. Commercial ceramic membranes are less common. Inopor TiO₂-based membranes are the most well-known. The company has a wide range of offerings of membranes with different pore sizes and surface configurations (hydrophobic or hydrophilic). Newer ceramic membrane products consist of a combination of α-alumina support with a γ-alumina interlayer. Energy Research Centre of the Netherlands (ECN) offers this alumina combination with active layers of 1,2-bis(triethoxysilyl) ethane (BTESE) and cetyltrimethylammonium bromide (CTAB), Polydimethylsiloxane (PDMS), and polyimide (PI) [7,87].

11.3 TRANSPORT MODELING IN OSN PROCESSES

OSN processes are often described using the solution diffusion model and pore flow model. These two models develop mathematical descriptions of flow through the membrane as a function of membrane, solvent, and solute properties. Denser membranes are best described using the solution diffusion model, whereas looser more porous membranes are more accurately described with the pore flow model. Figure 11.7 compares these models.

11.3.1 SOLUTION DIFFUSION MODEL

The solution diffusion model was developed to describe transport in a variety of fluid flow processes, including NF, RO, dialysis, gas separations, etc. Pores are often treated as fixed free-volume elements responsible for transport. In more dense membranes, this description is not always accurate. Rather, natural motion of polymer chains and motion fueled by solvent flux causes these free-volume elements to exist dynamically rather than statically. This model assumes solvent flux is

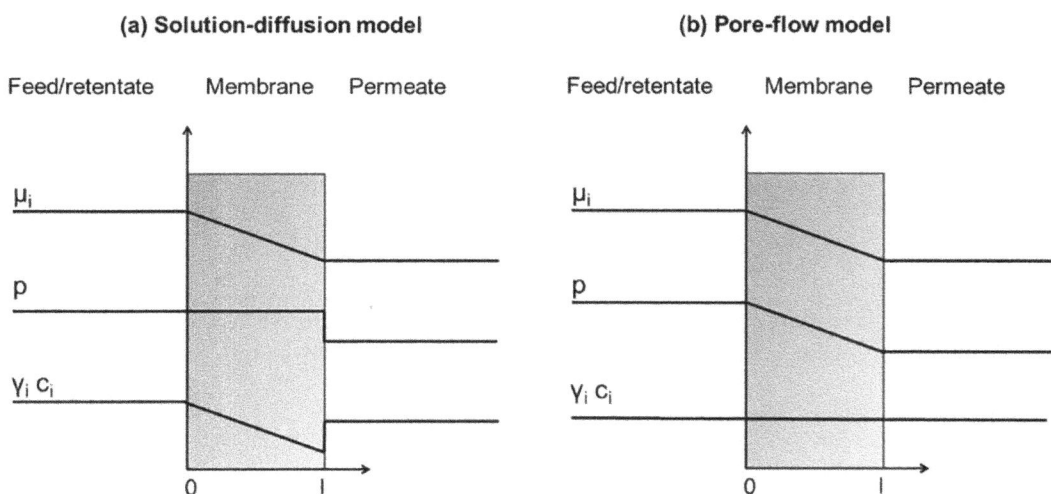

FIGURE 11.7 Comparison between solution diffusion and pore flow models of OSN transport. (Reprinted with permission from Ref. [7]. Copyright American Chemical Society 2014.)

related only to diffusive transport and that a pressure gradient does not exist across the membrane. Under these assumptions, solute flux can be described through Fick's Law:

$$J_i = -\frac{RTL_i}{c_i}\frac{dc_i}{dx} = -D_i\frac{dc_i}{dx} \qquad (11.1)$$

where D_i replaces RTL_i/c_i and represents the diffusion coefficient of the solute. In the case of OSN applications, these assumptions must be altered given the pressure gradient present. The equation is expanded and integrated over the membrane thickness to give an expression for both the flux of the solute and the solvent. In a dilute solution, this is further simplified to give:

$$J_V = \frac{D_V K_V c_{V,0}}{l}\left(1 - e^{-\frac{v_V(\Delta p - \Delta \pi)}{RT}}\right) \qquad (11.2)$$

where V for solvent has replaced i for solute. K_v is the sorption coefficient for the solvent, $\Delta\pi$ represents the differential osmotic pressure and is calculated through the van't Hoff equation. Lastly, in the case where $v_V(\Delta p - \Delta\pi)/RT$ is small, the simplified solution diffusion model can be developed:

$$J_V = \frac{D_V K_V c_{V,0}}{l}\left(\frac{v_V(\Delta p - \Delta\pi)}{RT}\right) = L_V(\Delta p - \Delta\pi) \qquad (11.3)$$

$$J_i = \frac{D_i K_i}{l}(c_{i,0} - c_{i,l}) = P_i \Delta c_i \qquad (11.4)$$

where L_V and P_i are the solvent permeability and solute permeability coefficients of the simple solution diffusion model [7,88]. However, modeling OSN processes requires further development of this simple model to account for the interactions between the solute and the solvent throughout the process. High pressure experiments demonstrated that flux does not increase linearly with pressure as described in Equation (11.3). Deviations from the simplified model are explained by its failure to consider the effect of pressure on solute transport. The Maxwell-Stefan equation was used to improve the simplified model and describe the interactions between solute and solvent such as friction or convective coupling. This equation was developed by performing a force balance on different components in the system under isothermal conditions. The driving force is given by the equation:

$$df_i = \frac{x_i}{RT} \nabla \mu_i^m \tag{11.5}$$

where $\nabla \mu_i^m$ is the chemical potential in the membrane that depends on the activity and ultimately the mixture composition. Incorporating the mutual friction force between the solute and solvent produces:

$$df_i = -\sum_{i \neq j} f_{ij} x_i x_j \left(u_i - u_j \right) \tag{11.6}$$

where u_i is the relative velocity and f_{ij} is the drag coefficient ($f_{ij} = f_{ji}$). Defining the Maxwell-Stefan $D_{ij} = \dfrac{1}{f_{ij}}$ results in the following expression:

$$\sum_{i \neq j} \frac{x_j J_i - x_i J_j}{c_{\text{tot}} D_{ij}} = -\frac{x_i}{RT} \nabla \mu_i^m \tag{11.7}$$

Finally, introducing the mass flux $n_i = w_i \rho J_i$ and expressing Equation (11.7) in terms of mass fractions, the equations that describe a binary mixture are:

$$n_1 + \left(\frac{w_2 n_1 - w_1 n_2}{w_m} \right) \varepsilon_1 = -\frac{\rho D_{1m}}{w_m} \nabla w_1 \tag{11.8}$$

$$n_2 + \left(\frac{w_1 n_2 - w_2 n_1}{w_m} \right) \varepsilon_2 = -\frac{\rho D_{2m}}{w_m} \nabla w_2 \tag{11.9}$$

where ρ is the mass density of the different components of the system (solute, solvent, and membrane), w_m is the membrane mass fraction [7,88,89]:

$$w_m = 1 - w_1 - w_2 \tag{11.10}$$

$$\varepsilon_1 = \frac{D_{1m}}{D_{12}} \tag{11.11}$$

$$\varepsilon_2 = \frac{MW_2}{MW_1} \frac{D_{2m}}{D_{12}} \tag{11.12}$$

Marchetti and Livingston found this model most accurate describing the performance of PI and PDMS-coated TFC membranes [88].

11.3.2 PORE FLOW MODEL

While the solution diffusion model considers the membrane as a platform for mobile free-volume elements, the pore flow model revolves around the idea that fixed pores within the network of polymer chains are the pathways for solvent transport through the membrane. Further, this model assumes the concentration of solvent and solute passing through the pores is uniform, therefore assuming the chemical potential gradient is only a function of pressure. Under these conditions, Darcy's law can be used to model the flux through the pores of a membrane:

$$J_i = k \frac{p_0 - p_l}{l} \tag{11.13}$$

where l is the thickness of membrane and k is the permeability coefficient. Pore geometry and surface properties influence k. This expression is used to develop the well-known Hagen-Poiseuille model to calculate the velocity of the fluid passing through the membrane (V):

$$V = \frac{\varepsilon r_p^2}{8 l \tau} \frac{\Delta p - \Delta \pi}{\eta} = K_{HP} \left(\Delta p - \Delta \pi \right) \tag{11.14}$$

The membrane is described in this equation through the pore size r_P, the tortuosity τ, and the porosity ε, and the solvent is described by the viscosity, η. This model was shown to be particularly effective for describing PDMS membranes prone to swelling by Robinson et al. [7,88,90].

11.4 RECENT APPLICATIONS OF OSN

Commercial OSN aims to separate a desired chemical compound from an organic solvent or isolate a purified organic solvent for future use. This process is an ideal replacement for conventional thermal separation techniques both for its energy efficiency and its nondestructive nature. Thermal effects are particularly pronounced for organic solvents and compatible chemical compounds, leading OSN processes to be a particularly good fit for associated industrial processes. Membrane-based technologies like OSN allow for separation processes to be performed at room temperature, eliminating the risk of degrading the desired product. In fact, OSN has been identified as a candidate for the best available technology for sustainable organic solvent recovery given its energy efficiency, mild operating conditions, stability in harsh environments, and low solid waste generation [1]. The most common industries to employ OSN are the pharmaceuticals, petrochemicals, and food and beverage industries. Numerous processes for solvent recovery to reduce stored volume of solvent and by association reducing the risk of exposure and solvent exchange of a high boiling point solvent for a low boiling point solvent have been reported using OSN [2,91]. Recent research focuses on using OSN for the purification of various feed streams to isolate a desired product or solvent. Active pharmaceutical ingredients (API) are targeted in the pharmaceutical industry, lubricants are dewaxed in the petrochemical industry, and vegetable oils are refined in the food and beverage industry.

11.4.1 API ISOLATION

A major revenue source for many pharmaceutical companies is the sale of APIs. APIs are extremely precise compounds that must be refined until pure by separating them from their carrier solvent. In OSN processes, this is accomplished by using the membrane to reject them from the solvent that passes through the membrane. Higher rejection efficiency of OSN membranes indicates a more efficient recovery of the API. In their sustainability analysis, Szekely et al. demonstrate OSN as a substantially more sustainable technology than chromatography and recrystallization for the recovery

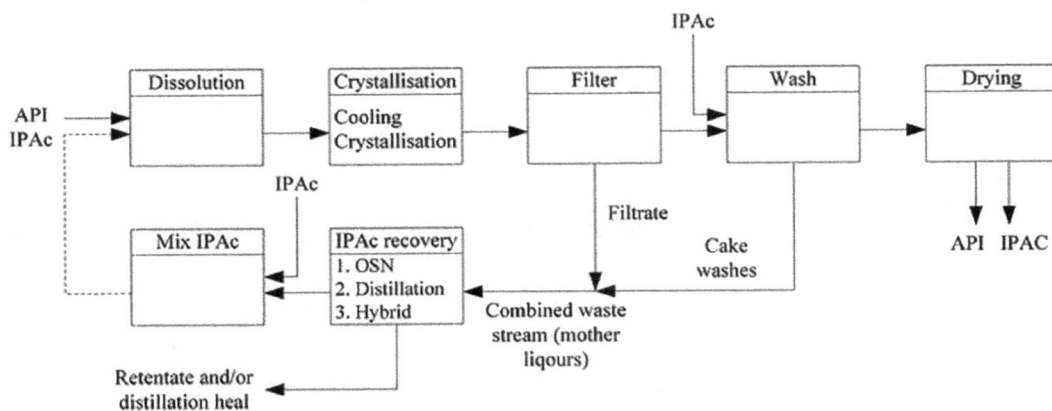

FIGURE 11.8   Process flow diagram of API recovery. (Reprinted with permission from Ref. [1]. Copyright Royal Society of Chemistry 2014.)

of APIs, both in terms of simplicity and energy efficiency [1]. Figure 11.8 describes the process. Geens et al. report an extremely robust study of the removal of APIs of varying molecular weights from three different organic solvents. A commercially available PI membrane (StarMem, Evonik LTD, Austria) was chosen to isolate APIs of 189, 313, 435, 531, and 721 Da from toluene, methylene chloride, and methanol. It was found that the rejection was dependent more on the solvent than the molecular weight of the API. Methanol consistently displayed better rejections of >90% for all APIs, whereas toluene and methylene chloride did not give as good as promising results. Toluene showed low rejection for all species but the largest, which the authors attribute to the small size of the APIs and claim that 600 Da is required for efficient separation in toluene based on rejection curve modeling. Methylene chloride partially dissolved the top layer of the membrane and thus was found to be ineffective in combination with StarMem membranes [6]. More recent work explores perfluoropolymers and perfluoropolymers with hydrophilic modifications for API isolation. The authors identified perfluoropolymers to be particularly attractive for pharmaceutical applications since the only solvent they are particularly unstable in is perfluorosolvents, which are not used in pharmaceutical applications. They study the rejection of four APIs (molecular weights between 432 and 809 Da) in a variety of industrially relevant solvents, including some mixtures. All membrane configurations produced at least 92% rejection for each of the four studied APIs, and the hydrophilic modification gave the best results of 95% rejection of an API with 809 Da molecular weight at an elevated temperature of 75°C [92].

11.4.2   Lubricant Dewaxing

Industrial lubricants are constantly in high demand worldwide and are a major commodity of the petrochemical industry. The process of producing these lubricants creates a waxy intermediate that must be purified using volatile organic compounds to remove the wax and achieve the desired final properties of the oil. Incorporating a NF unit to traditional thermal-based solvent recovery dramatically improves the solvent recovery efficiency by adding additional solvent recycling streams to the process. First, a thermal separation is used to remove the wax from a feed containing oil, wax, and solvent. OSN is then used to further separate the solvent from the oil before another thermal recovery step is performed. ExxonMobil and W.R. Grace developed a massive OSN plant for the dewaxing of lubrication oil referred to as MAX-DEWAX™. MAX-DEWAX™ is the largest OSN operation in existence, recovering enough solvent to save approximately 2 million barrels of crude oil annually. This translated to a 36,000 less barrels of fuel oil consumed annually, 20,000 less tons of greenhouse gas emissions annually, 4 million less gallons of cooling water used annually, and as much as 200 tons less dewaxing solvents released annually. The companies report a flux as high as 12.9 L/m²h at 4.1 MPa using a PI membrane and a solvent mixture containing MEK, methyl isobutyl ketone, and toluene [93]. This process has been replaced by catalytic cracking processes that avoid the waxy intermediate on the highest scale, but similar processes are still employed and studied in other applications. Modified polymer membranes, hybrid polymer membranes, and polymer composite membranes have been reported to reach >10 L/m²h and >90% oil rejection [60, 94–96]. More recent work attempts to use modeling to understand the results of experiments. Xin et al. use molecular modeling of the differences in rubbery PDMS-based membranes and glassy P84 PI membranes. They found the greater mobility of the polymer chains in the PDMS-based membranes should be more strongly influenced by the operating temperature, pressure, and oil concentration than the PI membranes, and these results were confirmed experimentally. They also note the rigid pores present in the PI membrane lead to a combination of pore flow and solution diffusion transport, where the more mobile PDMS-based membrane behaves according to solution diffusion [97].

11.4.3 Vegetable Oil Refining

Naturally occurring oils and fats represent some of the most promising natural feedstocks for future energy-intensive processes. The free fatty acids (FAAs), fatty acid esters, and fatty alcohols can be manipulated into a variety of products through well-understood chemistry. Oils are extracted from seeds using an extraction solvent (commonly *n*-hexane) that is then recovered through a steam distillation and evaporation process. Using waste oils can reduce the feed cost by 2–3x, but these oils contain a variable content of FAAs and thus variable fluid properties, making processing difficult. OSN offers the potential to increase the viability of waste oils when used as a pretreatment step [98,99]. Commercial membranes such as StarMem, SolSep, and Desal-DK were shown to reject oil from *n*-hexane with reasonable effectiveness, reaching a maximum of 78% at 4.8 L/m²h using a SolSep membrane [100]. Ceramic membranes are often used in conjunction with the nonpolar *n*-hexane. Tres et al. report up to complete rejection of different soybean oils using a zirconia-based membrane. More current research seeks to understand the fouling and long-term stability of OSN membranes using different feeds of different oils and concentrations. Shi et al. observe a near total decline in flux and a small drop in rejection when increasing the oil concentration in the feed from 5 to 50 wt%. They found oil to accumulate on the membrane's surface in layers and the high viscosity of the concentrated feed to be the major influences on performance decline [101]. Their process is shown in Figure 11.9.

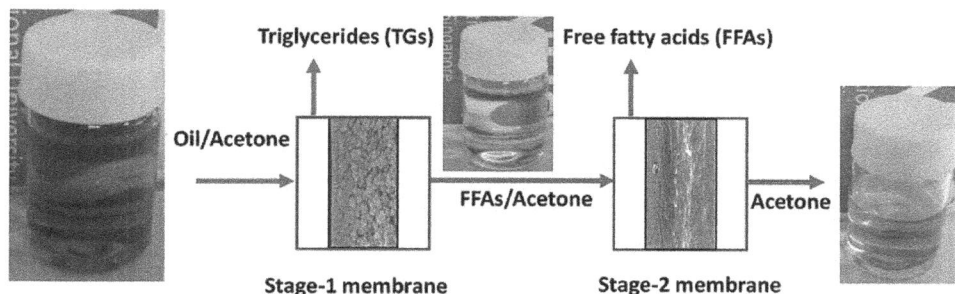

FIGURE 11.9 Flow of the OSN processes to separate triglycerides and FAAs from vegetable oil. (Reprinted with permission from Ref. [101]. Copyright Elsevier 2019.)

11.5 CONCLUSION

Organic solvent recovery will be critical as more sustainable manufacturing processes are developed in the future. Though there is a significant effort to replace environmentally unfriendly organic solvent by more environmentally friendly solvent e.g. in the manufacture of membranes, solvent recovery will be essential if sustainable manufacturing processes that promote a circular economy are to be developed. Further, it is highly unlikely that all environmentally unfriendly solvents can be replaced. Thus there will be a continuing effort to develop new solvent-resistant membranes.

Future membranes will not only focus on maximizing flux while delivering a highly specific separation, ease of manufacture including development of sustainable manufacturing processes will be essential. While numerous composite membranes as well as membrane based on advanced nanomaterials are likely to be reported, development of commercially viable membranes will be challenging. Commercially viable membranes need to be easily manufactured in a highly consistent manner while remaining economically viable. It is likely many new innovations in the area of solvent-resistant membranes will be seen in the future.

ACKNOWLEDGMENTS

This research was partially supported by BARD, The United States – Israel Binational Agricultural Research and Development Fund, Senior Research Fellow Award No. FR-40–2020 and National Science Foundation Industry/University Cooperative Research Center for Membrane Science, Engineering and Technology, the National Science Foundation (IIP 1822101), and the University of Arkansas.

REFERENCES

1. G. Szekely, M. F. Jimenez-Solomon, P. Marchetti, J. F. Kim, and A. G. Livingston, "Sustainability assessment of organic solvent nanofiltration: from fabrication to application," *Green Chemistry*, vol. 16, no. 10, pp. 4440–4473, 2014, doi: 10.1039/C4GC00701H.
2. E. M. Rundquist, C. J. Pink, and A. G. Livingston, "Organic solvent nanofiltration: a potential alternative to distillation for solvent recovery from crystallisation mother liquors," *Green Chemistry*, vol. 14, no. 8, pp. 2197–2205, 2012, doi: 10.1039/C2GC35216H.
3. L. Zhang et al., "Leaf-veins-inspired nickel phosphate nanotubes-reduced graphene oxide composite membranes for ultrafast organic solvent nanofiltration," *Journal of Membrane Science*, vol. 649, p. 120401, 2022, doi: 10.1016/J.MEMSCI.2022.120401.
4. Y. Cui and T. S. Chung, "Pharmaceutical concentration using organic solvent forward osmosis for solvent recovery," *Nature Communications*, vol. 9, no. 1, pp. 1–9, 2018, doi: 10.1038/s41467-018-03612-2.
5. J. F. Kim, G. Szekely, M. Schaepertoens, I. B. Valtcheva, M. F. Jimenez-Solomon, and A. G. Livingston, "In situ solvent recovery by organic solvent nanofiltration," *ACS Sustainable Chemistry and Engineering*, vol. 2, no. 10, pp. 2371–2379, 2014, doi: 10.1021/SC5004083.
6. J. Geens, B. de Witte, and B. van der Bruggen, "Removal of API's (active pharmaceutical ingredients) from organic solvents by nanofiltration," vol. 42, no. 11, pp. 2435–2449, 2007, doi: 10.1080/01496390701477063.
7. P. Marchetti, M. F. J. Solomon, G. Szekely, and A. G. Livingston, "Molecular separation with organic solvent nanofiltration: a critical review," 2014, doi: 10.1021/cr500006j.
8. K. Werth, P. Kaupenjohann, M. Knierbein, and M. Skiborowski, "Solvent recovery and deacidification by organic solvent nanofiltration: experimental investigation and mass transfer modeling," *Journal of Membrane Science*, vol. 528, pp. 369–380, 2017, doi: 10.1016/J.MEMSCI.2017.01.021.
9. O. Labban, C. Liu, T. H. Chong, and J. H. Lienhard, "Fundamentals of low-pressure nanofiltration: membrane characterization, modeling, and understanding the multi-ionic interactions in water softening," *Journal of Membrane Science*, vol. 521, pp. 18–32, 2017, doi: 10.1016/J.MEMSCI.2016.08.062.
10. A. W. Mohammad, Y. H. Teow, W. L. Ang, Y. T. Chung, D. L. Oatley-Radcliffe, and N. Hilal, "Nanofiltration membranes review: recent advances and future prospects," *Desalination*, vol. 356, pp. 226–254, 2015, doi: 10.1016/J.DESAL.2014.10.043.

11. "Nanofiltration (NF)." https://www.dupont.com/water/technologies/nanofiltration-nf.html (accessed Mar. 30, 2022).

12. J. C. Schrotter, S. Rapenne, J. Leparc, P. J. Remize, and S. Casas, "Current and emerging developments in desalination with reverse osmosis membrane systems," *Comprehensive Membrane Science and Engineering*, vol. 2, pp. 35–65, 2010, doi: 10.1016/B978-0-08-093250-7.00047-5.

13. Y. H. Chiao et al., "Zwitterion co-polymer PEI-SBMA nanofiltration membrane modified by fast second interfacial polymerization," *Polymers*, vol. 12, no. 2, p. 269, 2020, doi: 10.3390/POLYM12020269.

14. C. Wang, M. J. Park, D. H. Seo, E. Drioli, H. Matsuyama, and H. Shon, "Recent advances in nanomaterial-incorporated nanocomposite membranes for organic solvent nanofiltration," *Separation and Purification Technology*, vol. 268, p. 118657, 2021, doi: 10.1016/J.SEPPUR.2021.118657.

15. H. Guo et al., "Nanofiltration for drinking water treatment: a review," 2021, doi: 10.1007/s11705-021-2103-5.

16. T. A. Siddique, N. K. Dutta, and N. R. Choudhury, "Nanofiltration for arsenic removal: challenges, recent developments, and perspectives," *Nanomaterials*, vol. 10, no. 7, p. 1323, 2020, doi: 10.3390/NANO10071323.

17. H. Mokarizadeh, S. Moayedfard, M. S. Maleh, S. I. G. P. Mohamed, S. Nejati, and M. R. Esfahani, "The role of support layer properties on the fabrication and performance of thin-film composite membranes: the significance of selective layer-support layer connectivity," *Separation and Purification Technology*, vol. 278, p. 119451, 2021, doi: 10.1016/J.SEPPUR.2021.119451.

18. M. F. Jimenez Solomon, Y. Bhole, and A. G. Livingston, "High flux membranes for organic solvent nanofiltration (OSN)—Interfacial polymerization with solvent activation," *Journal of Membrane Science*, vol. 423–424, pp. 371–382, 2012, doi: 10.1016/J.MEMSCI.2012.08.030.

19. L. Lin, R. Lopez, G. Z. Ramon, and O. Coronell, "Investigating the void structure of the polyamide active layers of thin-film composite membranes," *Journal of Membrane Science*, vol. 497, pp. 365–376, 2016, doi: 10.1016/J.MEMSCI.2015.09.020.

20. Y. Liang et al., "Polyamide nanofiltration membrane with highly uniform sub-nanometre pores for sub-1 Å precision separation," *Nature Communications*, vol. 11, no. 1, pp. 1–9, 2020, doi: 10.1038/s41467-020-15771-2.

21. M. Fathizadeh, A. Aroujalian, A. Raisi, and M. Fotouhi, "Preparation and characterization of thin film nanocomposite membrane for pervaporative dehydration of aqueous alcohol solutions," *Desalination*, vol. 314, pp. 20–27, 2013, doi: 10.1016/J.DESAL.2013.01.001.

22. S. M. Aharoni and S. F. Edwards, "Gels of rigid polyamide networks," *Macromolecules*, vol. 22, no. 2, p. 169, 1989, Accessed: May 19, 2022. [Online]. Available: https://pubs.acs.org/sharingguidelines.

23. S. Peng Sun, T. A. Hatton, and T.-S. Chung, "Hyperbranched polyethyleneimine induced cross-linking of polyamideàimide nanofiltration hollow fiber membranes for effective removal of ciprofloxacin," *Environmental Science and Technology*, vol. 45, pp. 4003–4009, 2011, doi: 10.1021/es200345q.

24. K. Zarshenas, H. Dou, S. Habibpour, A. Yu, and Z. Chen, "Thin film polyamide nanocomposite membrane decorated by polyphenol-assisted Ti3C2TxMXene nanosheets for reverse osmosis," *ACS Applied Materials and Interfaces*, vol. 14, no. 1, pp. 1838–1849, 2022, doi: 10.1021/ACSAMI.1C16229/ASSET/IMAGES/LARGE/AM1C16229_0005.JPEG.

25. F. A. Pacheco, I. Pinnau, M. Reinhard, and J. O. Leckie, "Characterization of isolated polyamide thin films of RO and NF membranes using novel TEM techniques," *Journal of Membrane Science*, vol. 358, no. 1–2, pp. 51–59, 2010, doi: 10.1016/J.MEMSCI.2010.04.032.

26. R. Nadler and S. Srebnik, "Molecular simulation of polyamide synthesis by interfacial polymerization," *Journal of Membrane Science*, vol. 315, no. 1–2, pp. 100–105, 2008, doi: 10.1016/J.MEMSCI.2008.02.023.

27. V. P. Khare, A. R. Greenberg, and W. B. Krantz, "Investigation of the viscoelastic and transport properties of interfacially polymerized barrier layers using pendant drop mechanical analysis," 2004, doi: 10.1002/app.20964.

28. T. D. Matthews, H. Yan, D. G. Cahill, O. Coronell, and B. J. Mariñas, "Growth dynamics of interfacially polymerized polyamide layers by diffuse reflectance spectroscopy and Rutherford backscattering spectrometry," *Journal of Membrane Science*, vol. 429, pp. 71–80, 2013, doi: 10.1016/J.MEMSCI.2012.11.040.

29. K. Grzebyk, M. D. Armstrong, and O. Coronell, "Accessing greater thickness and new morphology features in polyamide active layers of thin-film composite membranes by reducing restrictions in amine monomer supply," *Journal of Membrane Science*, vol. 644, p. 120112, 2022, doi: 10.1016/J.MEMSCI.2021.120112.

30. I. C. Kim, J. Jegal, and K. H. Lee, "Effect of aqueous and organic solutions on the performance of polyamide thin-film-composite nanofiltration membranes," *Journal of Polymer Science Part B: Polymer Physics*, vol. 40, no. 19, pp. 2151–2163, 2002, doi: 10.1002/POLB.10265.

31. M. F. Jimenez-Solomon, P. Gorgojo, M. Munoz-Ibanez, and A. G. Livingston, "Beneath the surface: influence of supports on thin film composite membranes by interfacial polymerization for organic solvent nanofiltration," *Journal of Membrane Science*, vol. 448, pp. 102–113, 2013, doi: 10.1016/J. MEMSCI.2013.06.030.

32. K. Vanherck, "Hollow filler based mixed matrix membranes," *Chemical Communications*, no. 46, pp. 2492–2494, 2010, doi: 10.1039/b924086a.

33. X. Li, P. Vandezande, and I. F. J. Vankelecom, "Polypyrrole modified solvent resistant nanofiltration membranes," *Journal of Membrane Science*, vol. 320, no. 1–2, pp. 143–150, 2008, doi: 10.1016/J. MEMSCI.2008.03.061.

34. W. Feng, J. Li, C. Fang, L. Zhang, and L. Zhu, "Controllable thermal annealing of polyimide membranes for highly-precise organic solvent nanofiltration," *Journal of Membrane Science*, vol. 643, p. 120013, 2022, doi: 10.1016/J.MEMSCI.2021.120013.

35. Y. H. See Toh, F. W. Lim, and A. G. Livingston, "Polymeric membranes for nanofiltration in polar aprotic solvents," *Journal of Membrane Science*, vol. 301, no. 1–2, pp. 3–10, 2007, doi: 10.1016/J. MEMSCI.2007.06.034.

36. M. H. Davood Abadi Farahani and T. S. Chung, "A novel crosslinking technique towards the fabrication of high-flux polybenzimidazole (PBI) membranes for organic solvent nanofiltration (OSN)," *Separation and Purification Technology*, vol. 209, pp. 182–192, 2019, doi: 10.1016/J.SEPPUR.2018.07.026.

37. X. Li, B. Chen, W. Cai, T. Wang, Z. Wu, and J. Li, "Highly stable PDMS–PTFPMS/PVDF OSN membranes for hexane recovery during vegetable oil production," *RSC Advances*, vol. 7, no. 19, pp. 11381–11388, 2017, doi: 10.1039/C6RA28866A.

38. B. F. K. Kingsbury and K. Li, "A morphological study of ceramic hollow fibre membranes," *Journal of Membrane Science*, vol. 328, no. 1–2, pp. 134–140, 2009, doi: 10.1016/J.MEMSCI.2008.11.050.

39. M. Amirilargani, M. Sadrzadeh, E. J. R. Sudhölter, and L. C. P. M. de Smet, "Surface modification methods of organic solvent nanofiltration membranes," *Chemical Engineering Journal*, vol. 289, pp. 562–582, 2016, doi: 10.1016/J.CEJ.2015.12.062.

40. S. Ruthusree, S. Sundarrajan, and S. Ramakrishna, "Progress and perspectives on ceramic membranes for solvent recovery," *Membranes (Basel)*, vol. 9, no. 10, 2019, doi: 10.3390/MEMBRANES9100128.

41. Y. Li and J. Yu, "Emerging applications of zeolites in catalysis, separation and host–guest assembly," *Nature Reviews Materials*, vol. 6, no. 12, pp. 1156–1174, 2021, doi: 10.1038/s41578-021-00347-3.

42. K.-I. Okamoto, H. Kita, K. Horii, K. Tanaka, and M. Kondo, "Zeolite NaA membrane: preparation, single-gas permeation, and pervaporation and vapor permeation of water/organic liquid mixtures," 2001, doi: 10.1021/ie0006007.

43. P. J. Smeets, J. S. Woertink, B. F. Sels, E. I. Solomon, and R. A. Schoonheydt, "Transition-metal ions in zeolites: coordination and activation of oxygen," *Inorganic Chemistry*, vol. 49, no. 8, pp. 3573–3583, 2010, doi: 10.1021/IC901814F/ASSET/IMAGES/MEDIUM/IC-2009-01814F_0001.GIF.

44. Z. Wang, Q. Ge, J. Gao, J. Shao, C. Liu, and Y. Yan, "High-performance zeolite membranes on inexpensive large-pore supports: highly reproducible synthesis using a seed paste," *ChemSusChem*, vol. 4, no. 11, pp. 1570–1573, 2011, doi: 10.1002/CSSC.201100252.

45. M. Asghari, S. R. Mousavi, and T. Mohammadi, "A comprehensive comparative study on morphology and pervaporative performance of porous-supported mesoporous zeolitic membranes," *Microporous and Mesoporous Materials*, vol. 280, pp. 174–186, 2019, doi: 10.1016/J.MICROMESO.2019.01.044.

46. Y. Cao et al., "High-flux NaA zeolite pervaporation membranes dynamically synthesized on the alumina hollow fiber inner-surface in a continuous flow system," *Journal of Membrane Science*, vol. 570–571, pp. 445–454, 2019, doi: 10.1016/J.MEMSCI.2018.10.043.

47. J. Kujawa, S. Cerneaux, and W. Kujawski, "Removal of hazardous volatile organic compounds from water by vacuum pervaporation with hydrophobic ceramic membranes," *Journal of Membrane Science*, vol. 474, pp. 11–19, 2015, doi: 10.1016/J.MEMSCI.2014.08.054.

48. T. van Gestel et al., "Surface modification of γ-Al$_2$O$_3$/TiO$_2$ multilayer membranes for applications in non-polar organic solvents," *Journal of Membrane Science*, vol. 224, no. 1–2, pp. 3–10, 2003, doi: 10.1016/S0376-7388(03)00132-7.

49. R. B. Merlet, M. A. Pizzoccaro-Zilamy, A. Nijmeijer, and L. Winnubst, "Hybrid ceramic membranes for organic solvent nanofiltration: state-of-the-art and challenges," *Journal of Membrane Science*, vol. 599, p. 117839, 2020, doi: 10.1016/J.MEMSCI.2020.117839.

50. M. Amirilargani, M. Sadrzadeh, E. J. R. Sudhölter, and L. C. P. M. de Smet, "Surface modification methods of organic solvent nanofiltration membranes," *Chemical Engineering Journal*, vol. 289, pp. 562–582, 2016, doi: 10.1016/J.CEJ.2015.12.062.

51. S. M. Dutczak et al., "Composite capillary membrane for solvent resistant nanofiltration," *Journal of Membrane Science*, vol. 372, no. 1–2, pp. 182–190, 2011, doi: 10.1016/J.MEMSCI.2011.01.058.

52. A. F. M. Pinheiro, D. Hoogendoorn, A. Nijmeijer, and L. Winnubst, "Development of a PDMS-grafted alumina membrane and its evaluation as solvent resistant nanofiltration membrane," *Journal of Membrane Science*, vol. 463, pp. 24–32, 2014, doi: 10.1016/J.MEMSCI.2014.03.050.

53. C. R. Tanardi, A. F. M. Pinheiro, A. Nijmeijer, and L. Winnubst, "PDMS grafting of mesoporous γ-alumina membranes for nanofiltration of organic solvents," *Journal of Membrane Science*, vol. 469, pp. 471–477, 2014, doi: 10.1016/J.MEMSCI.2014.07.010.

54. J. Pyun, T. Kowalewski, and K. Matyjaszewski, "Synthesis of polymer brushes using atom transfer radical polymerization," *Macromolecular Rapid Communications*, vol. 24, no. 18, pp. 1043–1059, 2003, doi: 10.1002/MARC.200300078.

55. R. B. Merlet, M. Amirilargani, L. C. P. M. de Smet, E. J. R. Sudhölter, A. Nijmeijer, and L. Winnubst, "Growing to shrink: nano-tunable polystyrene brushes inside 5 nm mesopores," *Journal of Membrane Science*, vol. 572, pp. 632–640, 2019, doi: 10.1016/J.MEMSCI.2018.11.058.

56. M. Kruk, B. Dufour, E. B. Celer, T. Kowalewski, M. Jaroniec, and K. Matyjaszewski, "Grafting monodisperse polymer chains from concave surfaces of ordered mesoporous silicas," *Macromolecules*, vol. 41, no. 22, pp. 8584–8591, 2008, doi: 10.1021/MA801643R/SUPPL_FILE/MA801643R_SI_001.PDF.

57. M. R. S. Kebria, M. Jahanshahi, and A. Rahimpour, "SiO2 modified polyethyleneimine-based nanofiltration membranes for dye removal from aqueous and organic solutions," *Desalination*, vol. 367, pp. 255–264, 2015, doi: 10.1016/J.DESAL.2015.04.017.

58. M. Peyravi, M. Jahanshahi, A. Rahimpour, A. Javadi, and S. Hajavi, "Novel thin film nanocomposite membranes incorporated with functionalized TiO2 nanoparticles for organic solvent nanofiltration," *Chemical Engineering Journal*, vol. 241, pp. 155–166, 2014, doi: 10.1016/J.CEJ.2013.12.024.

59. H. Zhang et al., "Mineralization-inspired preparation of composite membranes with polyethyleneimine–nanoparticle hybrid active layer for solvent resistant nanofiltration," *Journal of Membrane Science*, vol. 470, pp. 70–79, 2014, doi: 10.1016/J.MEMSCI.2014.07.019.

60. M. Namvar-Mahboub, M. Pakizeh, and S. Davari, "Preparation and characterization of UZM-5/polyamide thin film nanocomposite membrane for dewaxing solvent recovery," *Journal of Membrane Science*, vol. 459, pp. 22–32, 2014, doi: 10.1016/J.MEMSCI.2014.02.014.

61. S. Sorribas, P. Gorgojo, C. Téllez, J. Coronas, and A. G. Livingston, "High flux thin film nanocomposite membranes based on metal-organic frameworks for organic solvent nanofiltration," *Journal of American Chemical Society*, vol. 135, no. 40, pp. 15201–15208, 2013, doi: 10.1021/JA407665W/SUPPL_FILE/JA407665W_SI_001.PDF.

62. B. Zornoza, C. Tellez, J. Coronas, J. Gascon, and F. Kapteijn, "Metal organic framework based mixed matrix membranes: an increasingly important field of research with a large application potential," *Microporous and Mesoporous Materials*, vol. 166, pp. 67–78, 2013, doi: 10.1016/J.MICROMESO.2012.03.012.

63. M. Majumder, N. Chopra, R. Andrews, and B. J. Hinds, "Enhanced flow in carbon nanotubes," *Nature*, vol. 438, no. 7064, pp. 44–44, 2005, doi: 10.1038/438044a.

64. M. Hu and B. Mi, "Enabling graphene oxide nanosheets as water separation membranes," *Environmental Science and Technology*, vol. 47, p. 49, 2013, doi: 10.1021/es400571g.

65. Y. Li, C. Li, S. Li, B. Su, L. Han, and B. Mandal, "Graphene oxide (GO)-interlayered thin-film nanocomposite (TFN) membranes with high solvent resistance for organic solvent nanofiltration (OSN)," *Journal of Materials Chemistry A*, vol. 7, no. 21, pp. 13315–13330, 2019, doi: 10.1039/C9TA01915D.

66. S. Roy, S. A. Ntim, S. Mitra, and K. K. Sirkar, "Facile fabrication of superior nanofiltration membranes from interfacially polymerized CNT-polymer composites," *Journal of Membrane Science*, vol. 375, no. 1–2, pp. 81–87, 2011, doi: 10.1016/J.MEMSCI.2011.03.012.

67. Q. Shen et al., "Development of ultrathin polyamide nanofilm with enhanced inner-pore interconnectivity via graphene quantum dots-assembly intercalation for high-performance organic solvent nanofiltration," *Journal of Membrane Science*, vol. 635, p. 119498, 2021, doi: 10.1016/J.MEMSCI.2021.119498.

68. S. Li, S. Du, S. Liu, B. Su, and L. Han, "Ultra-smooth and ultra-thin polyamide thin film nanocomposite membranes incorporated with functionalized MoS2 nanosheets for high performance organic solvent nanofiltration," *Separation and Purification Technology*, vol. 291, p. 120937, 2022, doi: 10.1016/J.SEPPUR.2022.120937.

69. X. Wang, Q. Xiao, C. Wu, P. Li, and S. Xia, "Fabrication of nanofiltration membrane on MoS2 modified PVDF substrate for excellent permeability, salt rejection, and structural stability," *Chemical Engineering Journal*, vol. 416, p. 129154, 2021, doi: 10.1016/J.CEJ.2021.129154.

70. M. Kamali, D. P. Suhas, M. E. Costa, I. Capela, and T. M. Aminabhavi, "Sustainability considerations in membrane-based technologies for industrial effluents treatment," *Chemical Engineering Journal*, vol. 368, pp. 474–494, 2019, doi: 10.1016/J.CEJ.2019.02.075.

71. L. Huang et al., "Reduced graphene oxide membranes for ultrafast organic solvent nanofiltration," *Advanced Materials*, vol. 28, no. 39, pp. 8669–8674, 2016, doi: 10.1002/ADMA.201601606.

72. B. Y. Guo et al., "MoS2 membranes for organic solvent nanofiltration: stability and structural control," *Journal of Physical Chemistry Letters*, vol. 10, no. 16, pp. 4609–4617, 2019, doi: 10.1021/ACS. JPCLETT.9B01780/ASSET/IMAGES/LARGE/JZ9B01780_0006.JPEG.

73. L. E. M. Gevers, I. F. J. Vankelecom, and P. A. Jacobs, "Solvent-resistant nanofiltration with filled polydimethylsiloxane (PDMS) membranes," *Journal of Membrane Science*, vol. 278, no. 1–2, pp. 199–204, 2006, doi: 10.1016/J.MEMSCI.2005.10.056.

74. A. Karimi, A. Khataee, M. Safarpour, and V. Vatanpour, "Development of mixed matrix ZIF-8/polyvinylidene fluoride membrane with improved performance in solvent resistant nanofiltration," *Separation and Purification Technology*, vol. 237, p. 116358, 2020, doi: 10.1016/J.SEPPUR.2019.116358.

75. N. A. A. Sani, W. J. Lau, and A. F. Ismail, "Polyphenylsulfone-based solvent resistant nanofiltration (SRNF) membrane incorporated with copper-1,3,5-benzenetricarboxylate (Cu-BTC) nanoparticles for methanol separation," *RSC Advances*, vol. 5, no. 17, pp. 13000–13010, 2015, doi: 10.1039/C4RA14284E.

76. J. Campbell, G. Székely, R. P. Davies, D. C. Braddock, and A. G. Livingston, "Fabrication of hybrid polymer/metal organic framework membranes: mixed matrix membranes versus in situ growth," *Journal of Materials Chemistry A*, vol. 2, no. 24, pp. 9260–9271, 2014, doi: 10.1039/C4TA00628C.

77. J. Dai et al., "Fabrication and characterization of a defect-free mixed matrix membrane by facile mixing PPSU with ZIF-8 core–shell microspheres for solvent-resistant nanofiltration," *Journal of Membrane Science*, vol. 589, p. 117261, 2019, doi: 10.1016/J.MEMSCI.2019.117261.

78. Y. Li, J. Li, R. B. Soria, A. Volodine, and B. van der Bruggen, "Aramid nanofiber and modified ZIF-8 constructed porous nanocomposite membrane for organic solvent nanofiltration," *Journal of Membrane Science*, vol. 603, p. 118002, 2020, doi: 10.1016/J.MEMSCI.2020.118002.

79. H. Yang, N. Wang, L. Wang, H. X. Liu, Q. F. An, and S. Ji, "Vacuum-assisted assembly of ZIF-8@GO composite membranes on ceramic tube with enhanced organic solvent nanofiltration performance," *Journal of Membrane Science*, vol. 545, pp. 158–166, 2018, doi: 10.1016/J.MEMSCI.2017.09.074.

80. R. Ding et al., "Graphene oxide-embedded nanocomposite membrane for solvent resistant nanofiltration with enhanced rejection ability," *Chemical Engineering Science*, vol. 138, pp. 227–238, 2015, doi: 10.1016/J.CES.2015.08.019.

81. F. Fei, L. Cseri, G. Szekely, and C. F. Blanford, "Robust covalently cross-linked polybenzimidazole/ graphene oxide membranes for high-flux organic solvent nanofiltration," *ACS Applied Materials and Interfaces*, vol. 10, no. 18, pp. 16140–16147, 2018, doi: 10.1021/ACSAMI.8B03591/ASSET/IMAGES/ LARGE/AM-2018–03591J_0006.JPEG.

82. M. H. Davood Abadi Farahani, D. Hua, and T. S. Chung, "Cross-linked mixed matrix membranes consisting of carboxyl-functionalized multi-walled carbon nanotubes and P84 polyimide for organic solvent nanofiltration (OSN)," *Separation and Purification Technology*, vol. 186, pp. 243–254, 2017, doi: 10.1016/J.SEPPUR.2017.06.021.

83. M. H. Davood Abadi Farahani, D. Hua, and T. S. Chung, "Cross-linked mixed matrix membranes (MMMs) consisting of amine-functionalized multi-walled carbon nanotubes and P84 polyimide for organic solvent nanofiltration (OSN) with enhanced flux," *Journal of Membrane Science*, vol. 548, pp. 319–331, 2018, doi: 10.1016/J.MEMSCI.2017.11.037.

84. V. Grosso et al., "Polymeric and mixed matrix polyimide membranes," *Separation and Purification Technology*, vol. 132, pp. 684–696, 2014, doi: 10.1016/J.SEPPUR.2014.06.023.

85. E. Resource Efficiency GmbH, "Technical information – DuraMem® Membrane Products," 2017, Accessed: May 29, 2022. [Online]. Available: www.duramem.com.

86. "Puramem Products Technical Information." https://corporate.evonik.com/Downloads/membrane-separation/Flyer-PuraMem-Technical-Information.pdf (accessed May 29, 2022).

87. R. Merlet et al., "Comparing the performance of organic solvent nanofiltration membranes in non-polar solvents," *Chemie Ingenieur Technik*, vol. 93, no. 9, pp. 1389–1395, 2021, doi: 10.1002/CITE.202100032.

88. P. Marchetti and A. G. Livingston, "Predictive membrane transport models for organic solvent nanofiltration: how complex do we need to be?," *Journal of Membrane Science*, vol. 476, pp. 530–553, 2015, doi: 10.1016/J.MEMSCI.2014.10.030.

89. D. R. Paul, "Reformulation of the solution-diffusion theory of reverse osmosis," *Journal of Membrane Science*, vol. 241, no. 2, pp. 371–386, 2004, doi: 10.1016/J.MEMSCI.2004.05.026.

90. J. P. Robinson, E. S. Tarleton, C. R. Millington, and A. Nijmeijer, "Solvent flux through dense polymeric nanofiltration membranes," *Journal of Membrane Science*, vol. 230, no. 1–2, pp. 29–37, 2004, doi: 10.1016/J.MEMSCI.2003.10.027.

91. E. M. Rundquist, "Application and evaluation of organic solvent nanofiltration in pharmaceutical processing," 2013.

92. J. Chau, K. K. Sirkar, K. J. Pennisi, G. Vaseghi, L. Derdour, and B. Cohen, "Novel perfluorinated nanofiltration membranes for isolation of pharmaceutical compounds," *Separation and Purification Technology*, vol. 258, p. 117944, 2021, doi: 10.1016/J.SEPPUR.2020.117944.

93. R. M. Gould, L. S. White, and C. R. Wildemuth, "Membrane separation in solvent lube dewaxing," *Environmental Progress*, vol. 20, no. 1, pp. 12–16, 2001, doi: 10.1002/EP.670200110.

94. Y. Kong, D. Shi, H. Yu, Y. Wang, J. Yang, and Y. Zhang, "Separation performance of polyimide nanofiltration membranes for solvent recovery from dewaxed lube oil filtrates," *Desalination*, vol. 191, no. 1–3, pp. 254–261, 2006, doi: 10.1016/J.DESAL.2005.09.014.

95. S. Monjezi, M. Soltanieh, A. C. Sanford, and J. Park, "Polyaniline membranes for nanofiltration of solvent from dewaxed lube oil," vol. 54, no. 5, pp. 795–802, 2018, doi: 10.1080/01496395.2018.1512617.

96. M. Namvar-Mahboub and M. Pakizeh, "Development of a novel thin film composite membrane by interfacial polymerization on polyetherimide/modified SiO2 support for organic solvent nanofiltration," *Separation and Purification Technology*, vol. 119, pp. 35–45, 2013, doi: 10.1016/J.SEPPUR.2013.09.003.

97. Y. Xin and F. Yin, "A combined experimental and molecular simulation study of lube oil dewaxing solvent recovery using membrane," *Separation and Purification Technology*, vol. 261, p. 118278, 2021, doi: 10.1016/J.SEPPUR.2020.118278.

98. K. Werth, P. Kaupenjohann, M. Knierbein, and M. Skiborowski, "Solvent recovery and deacidification by organic solvent nanofiltration: experimental investigation and mass transfer modeling," *Journal of Membrane Science*, vol. 528, pp. 369–380, 2017, doi: 10.1016/J.MEMSCI.2017.01.021.

99. A. N. Phan and T. M. Phan, "Biodiesel production from waste cooking oils," *Fuel*, vol. 87, no. 17–18, pp. 3490–3496, 2008, doi: 10.1016/J.FUEL.2008.07.008.

100. S. Darvishmanesh, T. Robberecht, P. Luis, J. Degrève, and B. van der Bruggen, "Performance of nanofiltration membranes for solvent purification in the oil industry," *JAOCS, Journal of the American Oil Chemists' Society*, vol. 88, no. 8, pp. 1255–1261, 2011, doi: 10.1007/S11746-011-1779-Y.

101. G. M. Shi, M. H. Davood Abadi Farahani, J. Y. Liu, and T. S. Chung, "Separation of vegetable oil compounds and solvent recovery using commercial organic solvent nanofiltration membranes," *Journal of Membrane Science*, vol. 588, p. 117202, 2019, doi: 10.1016/J.MEMSCI.2019.117202.

12 Recovery of Xylose from Oil Palm Frond (OPF) Bagasse Hydrolysate Using Commercial Spiral-Wound Nanofiltration Membrane

N.F.M Roli, H.W. Yussof, S.M. Saufi, and M.N. Abu Seman
Universiti Malaysia Pahang

A.W. Mohammad
University of Sharjah
Universiti Kebangsaan Malaysia

CONTENTS

12.1 INTRODUCTION: RESEARCH BACKGROUND

There are many types of agricultural waste biomass in Malaysia, especially from palm oil industry and plantation. Malaysia as the second largest palm oil producer in the world has generated more than 80 million tonnes of biomass every year [1] and expected to increase up to more than 100 million tonnes by 2020 [2]. The wastes generated from this palm oil industry include empty fruit bunch, fronds, trunks, fiber, shell, palm kernel and palm oil mill effluent [1]. Data reported in 2009 showed oil palm fronds (OPFs) could be considered the most abundant agriculture waste among these wastes in Malaysia with the amount of 44.84 million tonnes [3]. Normally, these OPF wastes are not properly disposed and mostly left to rot on the fields, which leads to environmental pollution [4,5]. However, the lignocellulosic materials in this huge amount of agriculture waste have the potential to be extracted and used to produce value-added materials such as biosugar [6]. Therefore, many works have been done to recover biosugar from oil palm biomass using various separation processes or as a substitute for non-renewable energy sources.

Lignocellulosic biomass (LCB) can be found in OPF which is generated from the OPF bagasse. Due to low cost, renewable and widespread availability in nature of the LCB, bioconversion of LCB into bioproduct and chemical are receiving interest recently. The LCB of palm oil normally contains high amount of cellulose in the range of 37.3%–46.5%, followed by lignin (27.6%–32.5%) and hemicellulose (25.3%–33.8%). The LCB of palm oil contains major carbohydrates such as glucan, xylan, and arabinan, with 31%, 17.3%, and 0.5%, respectively [7]. These cellulose and hemicellulose can be hydrolyzed into biosugars such as glucose and xylose, respectively. Biosugars such as xylose from hemicellulose in oil palm biomass is a high-value product that is derived from several saccharides. Xylose is a pentose sugar and a transitional output in xylitol production [8]. In the food industry, Xylitol has been used as an alternative sweetener that has equal sweetness as sucrose [9]. Meantime, another potential monosaccharide of interest which can be extracted from this LCB of palm oil is glucose due to the high cellulose content of LCB in palm oil [7]. Therefore, these two biosugars of xylose and glucose can be easily produced from the same OPF waste through bioconversion process such as fermentation.

Effective separation is required in order to separate monosaccharide sugar such xylose and glucose after formation of desired product through microbial fermentation. From previous study, the estimated cost of production of xylitol using chemical process (i.e. hydrogenation of xylose to xylitol) was about $350/ton, which is less than 20% of the total cost of xylitol production ($2,300–2,500/ton), while xylose crystal production contributes the most with more than 80% of the total cost of xylitol production [10]. This high cost of xylose crystal production may come from different factors. First, in the hemicellulose hydrolysates, beside sugars, the composition of other components including non-sugar components is very complex where the purification steps are not easy and very tedious. Second, xylose crystallization process is inhibited by the sugar impurities which have almost similar physicochemical properties of the xylose [11].

The various separation processes in the biorefinery industry include, for example, desugarization of molasses, glucose-fructose separation and xylose recovery from spent sulphite liquor for production of xylitol. Thus, chromatographic separation would be the most suitable technique in sugar biorefinery [12]. Currently, in the biorefining industry, chromatography has become one of the vital separation methods. However, this technique requires higher cost than other separation methods, and selection of the stationary phase is usually difficult [13]. Warner and Nochumson [14] have reported on cost of biopharmaceutical production using chromatographic method. The estimated cost for traditional ion-exchange removal of DNA and endotoxins was $3,800 for the membrane cartridge and $3,700 for the column. As an alternative separation method for separation of xylose from glucose, nanofiltration (NF) is more cost-effective and offers easy maintenance [15,16]. The cost of NF membrane is 83% cheaper than chromatography. In this present study, a pilot-scale spiral-wound membrane system was designed with commercial spiral-wound NF membrane to evaluate the separation of biomass hydrolysate from OPF bagasse. This spiral-wound module was designed in compact mode, thus offering optimum membrane surface as possible in a given volume [17].

12.1.1 Issues in Separation of Sugars from Oil Palm Frond Biomass

The number of oil palm plantations keeps increasing from year to year, and these contributed to a large amount of oil palm residues that can be categorized as waste. Currently, OPF is classified as one of the biomasses obtained from the palm oil industry which is under-utilized [18]. A previous study by Zahari et al. [19] showed that OPF can be converted into biosugar by only pressing the fresh OPF to get the juice. Meanwhile, the solid waste from the extracted OPF known as OPF bagasse was normally disposed and not utilized. A proper utilization of palm oil residues needs to be developed in order to reduce environmental problem and at the same time produce beneficial products from OPF bagasse. Thus, Sabiha-Hanim et al. [18] suggested to produce renewable sugar by hydrolyzing the solid waste. By 2025, it is predicted that the world sugar production will grow by 2.1% per annum to reach 210 metric tonnes with developing countries contributing 79% of global sugar production [20]. Moreover, these sugars have found applications in agriculture, pharmaceuticals, cosmetics and fine chemicals [21]. However, these sugars can be replaced by using renewable sugar from the hydrolyzed solid-waste biomass.

LCB from OPF bagasse can be hydrolyzed into biosugar and consists of xylose and glucose [22]. Xylose and glucose mainly come from hydrolysis of hemicellulose and cellulose of agriculture waste, contributing roughly 55% and 25% of total sugar, respectively [23]. However more than 80% of the production cost comes solely from separation of these renewable sugars, where the recovery method relied on chromatographic separation [10]. In addition, present separation methods proved to be expensive for concentration of xylose reaction liquor for subsequent production of pure xylitol at yields of 50%–60% using the chemical reduction method [24]. Hence, separation of monosaccharides using alternative method which is more cost-effective and can obtain the pure fraction of a specific monosaccharide is needed.

The possibility of sugar separation appears most popular using commercial membrane, especially NF membrane. NF is a favourable and cost-competitive membrane separation technology. It has a molecular weight cut-off (MWCO) in between 150 and 1,000 g/mol, which allows high retention of compounds with molecular weight up to 150–250 g/mol [25]. NF has many applications in separation of fermentation broth, sugar fractionation, and sugar concentration [26]. In food industry, membrane characterization and modelling of NF are also important in purification process of saccharides in order to understand the process behaviour [27]. The effects of NF separation of saccharides have been studied in the past [16,25,26,28,29]. The main operating parameters that will affect the NF membrane on sugar separation such as feed concentration, permeation flux, temperature and pressure have been investigated.

In recent years, most of the researchers are using model solution in sugar separation. Sjoman et al. [16] have done separation of xylose from glucose with plate-and-frame filtration system using flat-sheet NF commercial membranes which are Desal-5 DK, Desal-5 DL and NF270. The separation was made of xylose and glucose model solution in different mass ratios and total monosaccharide concentration. Separation of sugar from molasses by ultrafiltration and NF using a pilot-scale system has been carried out by Mousavi and Moghadam [29] to determine sugar concentration, permeate flux, rejection, density and adsorption. Screening study using design of experiments has been conducted by previous researchers for separation of different molecular weights of carbohydrates such as inulin, sucrose, glucose and fructose. The model solution of carbohydrates was separated using NF membrane with MWCO of 600 Da with pilot crossflow unit. The operational parameters such as transmembrane pressure, total feed concentration (C_0) and retentate flow (Q_R) were screened [28]. However, no report presently is available on separation of sugar from OPF bagasse hydrolysate using pilot-scale commercial spiral-wound NF membrane separation system (NMSS), particularly on separating xylose from glucose. Therefore, this present work aims to investigate the performance of pilot-scale commercial spiral-wound NMSS by using Desal-5 DK, Desal-5 DL and NF90 for separation of OPF bagasse hydrolysate from OPF bagasse.

12.1.2 Commercial Spiral-Wound Nanofiltration Membrane and Pilot-Scale Set-up

The flow diagram and design of NF system for separation of xylose from glucose from biomass hydrolysate are shown in Figures 12.1 and 12.2, respectively. The pilot-scale membrane system was equipped with a 25 L feed tank. The feed liquid (pure water/synthetic sugar solution/OPF bagasse hydrolysate) was pumped with a Grundfos pump from the tank to the membrane housing. The switch control for system was placed inside the control box. The pilot-scale crossflow NF system was fabricated by Bumificient Sdn Bhd, Kuala Lumpur. The membranes were loaded inside the membrane housing and covered three different commercial spiral-wound membranes which are Desal-5 DK,

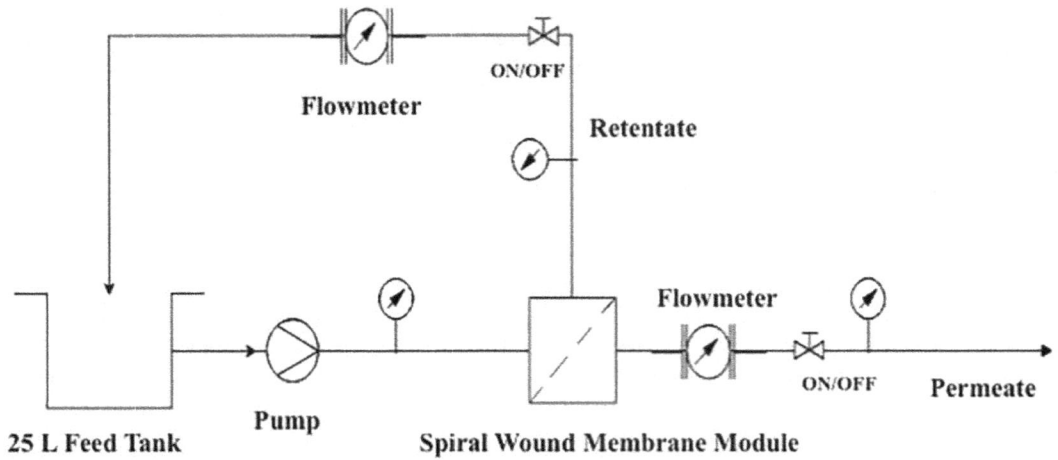

FIGURE 12.1 Schematic set-up for NF spiral-wound module system.

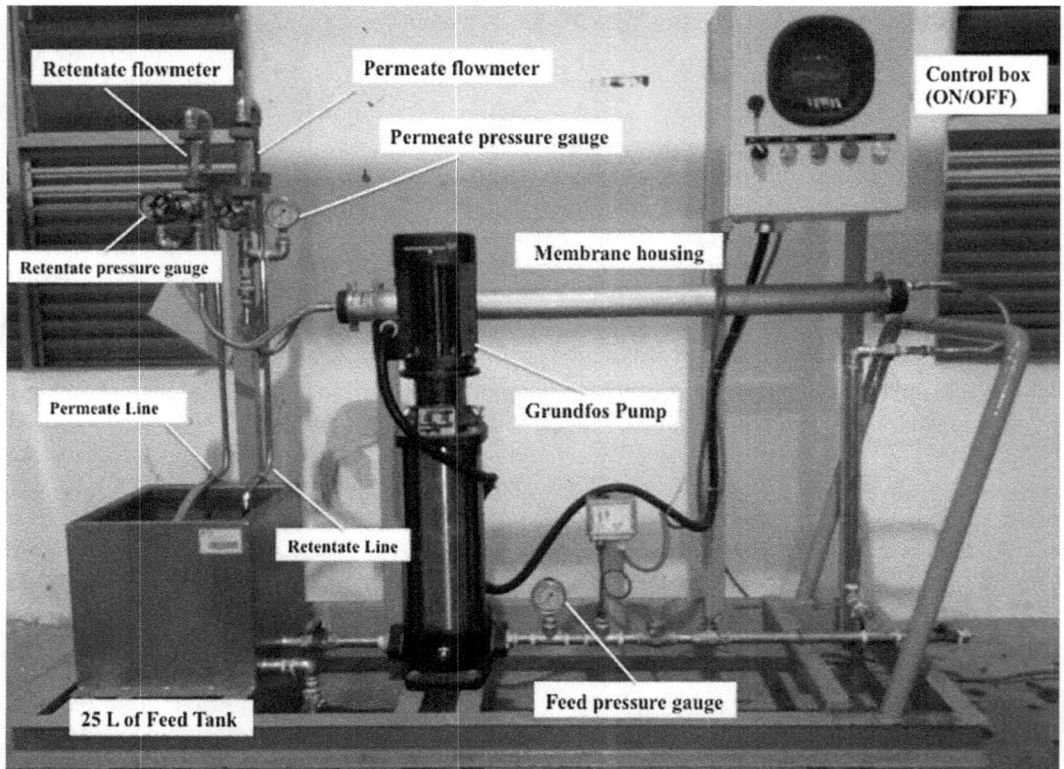

FIGURE 12.2 Pilot-scale NF spiral-wound membrane system.

TABLE 12.1

Properties of Commercial Spiral-Wound NF Membranes Provided by the Manufacturer

Properties	NF90	Desal-5 DL	Desal-5 DK
Manufacturer	Dow/Filmtec	GE Osmonics	GE Osmonics
Support material	Polysulfone	Polysulfone	Polysulfone
Surface material	Polyamide	Polyamide	Polyamide
Average pore diameter (nm)	0.68 [25]	0.90 [30]	0.84 [30]
Maximum temperature (°C)	45	50–90	50
Maximum pressure (bar)	41	2–30	4.6–26.6
Molecular weight cut-off (g/mol)	100–400	150–300	150–300

TABLE 12.2

Pure Water Permeability (PWP) of Commercial NF Membranes

Membrane	PWP (L/m²h·bar)
NF90	1.33 ± 0.05
Desal-5 DL	1.28 ± 0.24
Desal-5 DK	6.78 ± 0.06

Desal-5 DL (GE Water & Process Technologies, USA) and NF90 (Dow Filmtec Membranes, USA). Some chemical and physical characteristics of the membranes are given in Table 12.1. The MWCO of the membranes are in the range of 150–300 g/mol. The membranes have same support and surface material, which are polysulfone and polyamide, respectively. The maximum pressures reported for these commercial NF membranes are 4.6–26.6 bar for the Desal-5 DK, 2–30 bar for the Desal-5 DL and 41 bar for the NF90. However, due to maximum pressure of the Grundfos pump (15 bar), the filtration was run at a maximum operating pressure of 10 bar, which is below the membrane's operating pressure limit.

12.2   PURE WATER PERMEABILITY OF COMMERCIAL SPIRAL-WOUND NF MEMBRANES

Pure water permeability (PWP) values for all commercial NF membranes are tabulated in Table 12.2. The data show Desal-5 DK membrane has the highest PWP among the tested commercial NF membranes. It was observed that NF90 and Desal-5 DL membranes exhibited almost similar PWP values even when both membranes have different average pore size. It was clear that PWP was not only affected by the pore size but also by other properties including membrane thickness, porosity and hydrophilicity that may contribute to higher water flux.

12.3   PERFORMANCE OF PILOT-SCALE SPIRAL-WOUND NANOFILTRATION MEMBRANE

The performance of pilot-scale spiral-wound NF membrane was evaluated using three different commercial NF spiral-wound membranes, which are Desal-5 DK, Desal-5 DL and NF90, for sugar separation. Firstly, the NF separation was tested using synthetic sugar solution model (xylose and glucose) for the selection of best membrane and then followed by the real sample which was OPF bagasse hydrolysate. In this work, the performance of NF membrane was measured based on the separation factor of xylose.

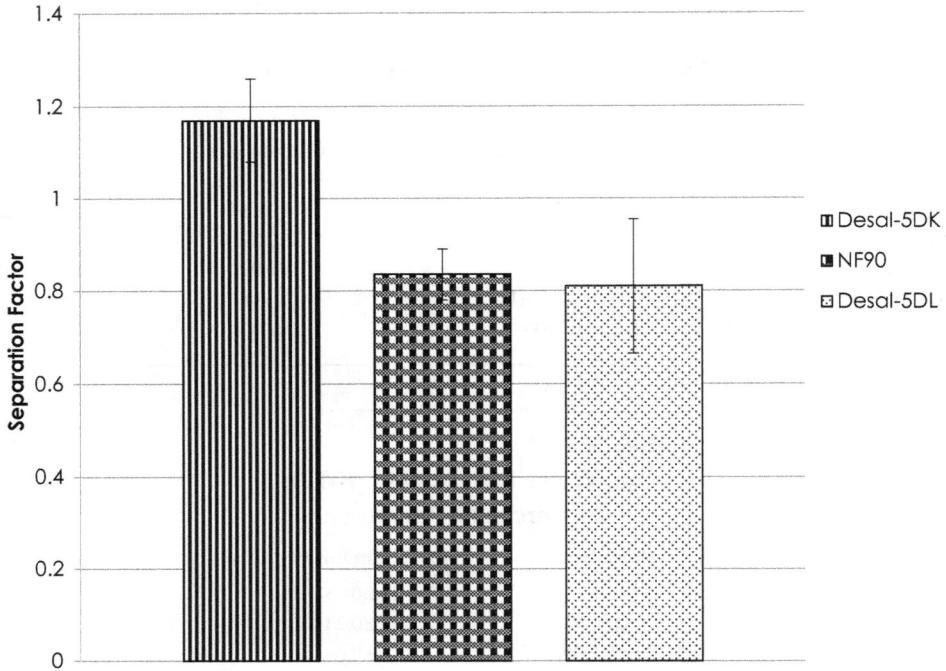

FIGURE 12.3 Separation factor of xylose for synthetic sugar model solution at 10 bar of operation pressure.

12.3.1 SEPARATION OF SUGARS FROM SYNTHETIC SOLUTION MODEL

Figure 12.3 shows that Desal-5 DK performs very well with the highest xylose separation factor of 1.17, while the other two membranes (NF90 and Desal-5 DL) exhibited xylose separation factor less than 1 (0.84 and 0.81, respectively) indicating that these two membranes failed to separate xylose from glucose. Similar results were also obtained by Sjoman et al. [8], where they reported that Desal-5 DK could achieve the highest xylose separation factor up to 3.8 which was higher than Desal-5 DL membrane (3.5). The higher value of xylose separation factor reported by Sjoman et al. [8] compared to our study was due to the higher operating pressure of 40 bar applied in their work while only 10 bar was applied in this current study. It was expected that the higher xylose separation factor could also be achieved if the higher pressure pump was used. The details for effect of pressure, sugar concentrations and type of synthetic solution (single and binary) on the sugars separation can be found elsewhere [31,32].

12.3.2 SEPARATION OF SUGARS FROM OIL PALM FROND BAGASSE HYDROLYSATE

12.3.2.1 Preparation of Oil Palm Frond Bagasse Hydrolysate

OPF bagasse was obtained from local oil palm plantation in Felda Lepar Hilir, Pahang, Malaysia. The OPF bagasse were washed and cleansed from stains and dirt and then were dried under sunlight and stored at room temperature. The samples were chipped, grinded and sieved into size of less than 1.0–2.0 mm according to Nanda et al. [33]. After that, the samples were sealed in plastic bag before used for further analysis.

Analyses of OPF bagasse composition (xylose, glucose, acetic acid, furfural and 5-hydroxy-methyl-furfural (HMF)) were carried out after a quantitative acid hydrolysis under optimized conditions [34]. OPF bagasse sample with dry weight of 250 g was placed in 5,000 mL of Schott bottle.

FIGURE 12.4   OPF bagasse treated in media containing 6% HNO_3 and autoclaved at 122°C for 9 minutes.

Then, fresh sample was soaked in diluted nitric acid solution. Treatments were performed in media containing 6 wt% HNO_3 and autoclaved at 122°C for 9 minutes as shown in Figure 12.4.

12.3.2.2   Pre-filtration of Oil Palm Frond Bagasse Hydrolysate

Due to the presence of suspended solids in the 15 L of OPF bagasse hydrolysate that has been prepared in Section 12.3.2.1, pre-treatment of the hydrolysate solution was necessary to prevent the NF membrane fouling, plugging and deterioration. This may lead to decline in flux and rejection or even membrane failure [35]. Two stages were selected for initial pre-treatment of OPF bagasse hydrolysate: cloth filtration and cartridge depth micro-filtration. The OPF bagasse hydrolysate was filtered first with the cloth filter to eliminate the large size of fiber for three times until only small particle left in the hydrolysate solution. The method was repeated if larger size of fiber was still present in the hydrolysate solution. The cartridge depth filter was a polypropylene porous membrane with a pore size of 5–10 µm. The hydrolysate was filtered at 1–2 bar until the hydrolysate solution became clear solution and no precipitate of fiber appeared. After these two steps, the majority of large and medium particles were removed from the OPF bagasse hydrolysate. This was very important to prevent the NF membrane from damage. The pre-filtration schematic system for second stage is shown in Figure 12.5.

12.3.2.3   Composition of Dilute Acid Hydrolysis

In this work, the concentrations of xylose, glucose, acetic acid, furfural and HMF released from OPF bagasse were obtained under optimum operational condition of dilute nitric acid (HNO_3) hydrolysis based on the work by Rodríguez-Chong et al. [34]. The optimum conditions used for

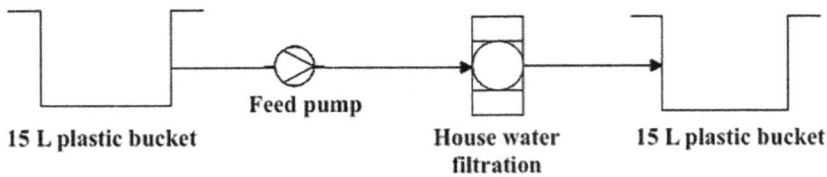

FIGURE 12.5 Schematic diagram for pre-filtration system for the second stage.

TABLE 12.3
Composition from Dilute Nitric Acid Hydrolysis of Biomass

Composition (g/L)	This Study	[34]	[36]
Xylose	16.57	18.6	22.01
Glucose	2.36	2.87	1.91
Acetic acid	2.07	0.90	2.42
Furfural	1.22	1.32	0.21
HMF	1.03	–	–
Raw material	OPF bagasse	Sugarcane bagasse	Corn stover

hydrolysis were 6 wt% of HNO_3, temperature of 122°C and duration for 9 minutes. The composition of dilute nitric acid hydrolysis of biomass was compared in Table 12.3. Zhang et al. [36] applied the condition of 0.6% of HNO_3, 150°C and 1 minute to hydrolyze corn stover. According to Table 12.3, the highest xylose concentration obtained by Zhang et al. [36] is 22.01 g/L, while from this present work, the xylose concentration was slightly lower compared to the other two studies. The glucose concentration obtained from this present study was slightly lower than Rodríguez-Chong et al. [34] with 2.87 g/L compared to 2.36 g/L. The lowest glucose concentration of 1.91 g/L was reported by Zhang et al. [36]. Acetic acid was generated in the hydrolysis of acetyl groups of hemicelluloses. The acetic acid obtained from this study was in between the values of Rodríguez-Chong et al. [34] and Zhang et al. [36]. Both researchers did not find HMF in their works, while small amount of HMF which is 1.03 g/L has been generated from this work.

12.3.2.4 Performance of Desal-5 DK NF Membrane on Sugars Separation

A pilot-scale crossflow system using spiral-wound NF membrane was used to separate xylose from OPF bagasse hydrolysate. Xylose was produced by dilute acid hydrolysis of OPF bagasse and the composition of hydrolysate can be found in Table 12.4. However, some inhibitory compounds were also generated. In this work, Desal-5 DK membrane has been chosen (based on its performance in Section 12.3.1) to separate the xylose from inhibitors such as furfural, HMF and acetic acid. The influence of operating parameter such as pressure on the separation performance of the membrane was investigated.

Table 12.4 shows the comparison on pilot-scale crossflow NF membrane with previous studies. The highest xylose rejection was reported by Weng et al. [37] with 85.7% of xylose rejection, whereas xylose rejection from this study was the lowest at 10%–30%. This was because applied pressure on the membrane system used in this study was lower compared to the other two studies. The operating pressure gave high impact on membrane performance since NF is a pressure-driven process. According to Weng et al. [37], an increase in pressure increases the retentions of monosaccharides.

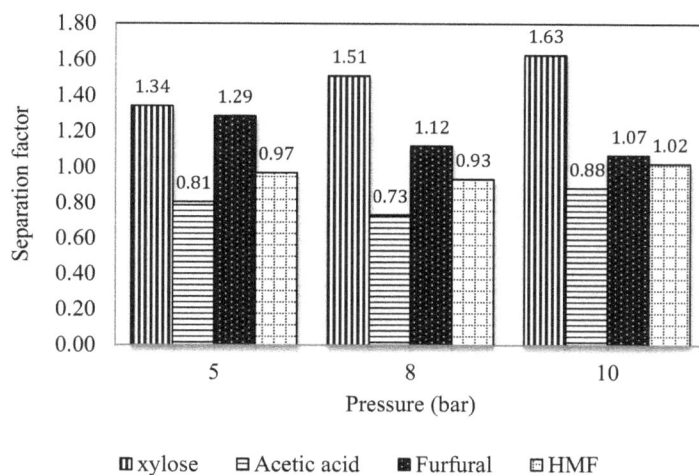

FIGURE 12.6 Effect of pressure on xylose separation factor of biosugar.

TABLE 12.4
Comparison on Pilot-Scale Crossflow Nanofiltration Membrane of Biosugar from Different Biomass Sources

	This Study	[37]	[40]
Raw material	OPF bagasse	Rice husk	Red grape
Membrane module	Spiral wound	Spiral wound	Spiral wound
Molecular weight cut-off (MWCO) (Da)	150–300	300	200
Applied pressure (bar)	5–10	24.5–34.3	41
Biosugar rejection (%)	10–30	85.7	99

The xylose separation factor is shown in Figure 12.6. At high pressure of 10 bar, Desal-5 DK membrane gave the highest separation factor up to 1.63. According to the results, all inhibitors present in OPF bagasse hydrolysate were retained more due its lower separation factor. By comparing stokes diameter of the solutes, the molecular sizes of inhibitor and sugar were in the following order (glucose = 0.726 nm; xylose = 0.638 nm) > furans (HMF = 0.463 nm; furfural = 0.438 nm) > carboxylic acid (acetic acid = 0.412 nm) [37]. While the average pore diameter of the Desal-5 DK membrane ranged from 0.84 to 1 nm [26,30], the inhibitors should easily pass through the membrane. However, as the NF membrane separation mechanism involve sieve and charge effects, these smaller sizes of inhibitors were most probably repelled by the membrane due to the charge effect. Weng et al. [37] reported that the retention of inhibitor (i.e. acetic acid) could be higher than xylose with the same membrane when the charge exclusion dominates the separation mechanism.

Although the average values of the membrane pore size were slightly larger than the size of xylose, Sjöman et al. [8] has observed separation factor of xylose up to 3.8 due to high osmotic pressure that pushed the xylose molecules passed through the membrane. Patil et al. [38] obtained 96% of xylose purity from inulin mixture-cascaded NF using spiral-wound module, whereas Feng et al. [39] in their study obtained 54.5% of xylose purity using NF-3 spiral-wound membrane at pressure of 6 bar. This suggested that the separation of xylose from inhibitors could be achieved.

12.4   CONCLUSIONS

In this work, three commercial NFs Desal-5 DK, Desal-5 DL and NF90 membranes were evaluated for separation of xylose from glucose using pilot-scale system. Results from synthetic sugar solution revealed that the Desal-5 DK exhibited a very good separation of xylose from glucose with separation factor of 1.17, while the other two membranes were unable to separate the sugars. When applying to the real biosugars extracted from the OPF bagasse hydrolysate, the Desal-5 DK would separate xylose from glucose with up to 1.63 separation factor higher than synthetic sugar solution. In conclusion, the spiral-wound NF Desal-5 DK membrane has a great potential in separation and recovery of xylose from glucose from the biomass waste, especially OPF. It is believed the higher separation factor could be achieved with pilot plant system working at pressure higher than 10 bar. Therefore, further investigation is required for biosugars separation using higher pressure NF pilot plant system.

ACKNOWLEDGEMENTS

The authors wish to thank the Ministry of Education and Universiti Malaysia Pahang for the financial aid through research grant LRGS/2013/UKM-UKM/PT/03 and RDU190171. The authors also wish to acknowledge the Ministry of Education Malaysia for sponsoring Fatihah postgraduate's study via MyBrain15.

REFERENCES

1. Ahmad Rizal, N.F.A., Ibrahim, M.F., Zakaria, M.R., Abd-Aziz, S., Yee, P.L. and Ali Hassan, M. 2018. Pre-treatment of Oil Palm Biomass for Fermentable Sugars Production. *Molecules* 23: 1381.
2. Agensi Inovasi Malaysia. 2013. National Biomass Strategy 2020: New Wealth Creation for Malaysia's Biomass Industry Version 2.0, 2013. https://www.cmtevents.com/MediaLibrary/BStgy2013RptAIM.pdf.
3. Yusoff, Z. 2012. Review of Research Activities on Malaysian Palm Oil-Based Green Technology. *Res. Develop. Depart. Mech. Eng.* 1(2): 89–106.
4. Lim, K.O., Zainal, Z.A., Quadir, G.A. and Abdullah, M.A. 2000. Plant Based Energy Potential and Biomass Potential in Malaysia. *Int. Energy J.* 1(2): 77–88.
5. Mohd Nor, N. 2008. The Effects of Hydrothermal Treatment on the Physico – Chemical Properties of Oil Palm Frond (OPF) Derived Hemicellulose. MSc Thesis. Universiti Sains Malaysia, Penang.
6. Siti-Normah, M., Sabiha-Hanim, S. and Noraishah, A. 2012. Effects of pH, Temperature, Enzyme and Substrate Concentration on Xylooligosaccharides Production. *Int. Chem. Mol. Eng.* 6(12): 1391–1395.
7. Sudiyani, Y., Styarini, D., Triwahyuni, E., Sembiring, K.C., Aristiawan, Y., Abimanyu, H. and Han, M.H. 2013. Utilization of Biomass Waste Empty Fruit Bunch Fiber of Palm Oil for Bioethanol Production Using Pilot–Scale Unit. *Energy Procedia* 32: 31–38.
8. Sjöman, E., Mänttäri, M., Nyström, M., Koivikko, H. and Heikkilä, H. 2008. Xylose Recovery by Nanofiltration from Different Hemicellulose Hydrolyzate Feeds. *J. Membr. Sci.* 310(1–2): 268–277.
9. Rangaswamy S. 2003. Xylitol Production from D-Xylose by Facultative Anaerobic Bacteria. PhD Thesis. Virginia Polytechnic Institute and State University.
10. da Silva, S. S. and Chandel, A. K. 2012. *D-Xylitol: Fermentative Production, Application and Commercialization.* (S. S. da Silva & A. K. Chandel, Eds.). Springer.
11. Mah, K. H., Yussof, H. W., Jalanni, N. a., Seman, M. N. A. and Zainol, N. 2014. Separation of Xylose from Glucose Using Thin Film Composite (TFC) Nanofiltration Membrane: Effect of Pressure, Total Sugar Concentration and Xylose/Glucose Ratio. *Jurnal Teknologi.* 70(1): 93–98.
12. Hellstén, S. 2013. Recovery of Biomass-derived Valuable Compounds Using Chromatographic and Membrane Separations. PhD Thesis. Lappeenranta University of Technology, Finland.
13. Liu, H., Zhao, L., Fan, L., Jiang, L., Qiu, Y. and Xia, Q. 2016. Establishment of a Nanofiltration Rejection Sequence and Calculated Rejections of Available Monosaccharides. *Sep. Purif. Tech.* 163: 319–330.
14. Warner, T.N. and Nochumson, S. 2003. Rethinking the Economics of Chromatography: New Technologies and Hidden Costs. *BioPharm Int.* 16(1): 58–60.

15. Grandison, S.A., Goulas, A.K. and Rastall, R.A. 2002. The Use of Dead-End and Cross-Flow Nanofiltration to Purify Prebiotic Oligosaccharides from Reaction Mixtures. *Songklanakarin J. Sci. Technol.* 24: 915–928.

16. Sjoman, E., Manttari, M., Nystrom, M., Koivikko, H. and Heikkila, H. 2007. Separation of Xylose from Glucose by Nanofiltration from Concentrated Monosaccharide Solutions. *J. Membr. Sci.* 292(1–2): 106–115.

17. Senthilmurugan, S., Ahluwalia, A. and Gupta, S.K. 2005. Modeling of a Spiral-Wound Module and Estimation of Model Parameters Using Numerical Techniques. *Desalination* 173(3): 269–286.

18. Sabiha-Hanim, S., Noor, M.A.M. and Rosma, A. 2011. Effect of Autohydrolysis and Enzymatic Treatment on Oil Palm (*Elaeis guineensis* Jacq.) Frond Fibres for Xylose and Xylooligosaccharides Production. *Bioresource Tech.* 102(2): 1234–1239.

19. Zahari, M. H. 2012. Production and Separation of Glucose from Cellulose Hydrolysate Using Membrane Reactor: Effect of Transmembrane Pressure and Cross Flow Velocity. Undergraduate Thesis. Universiti Malaysia Pahang.

20. OECD-FAO Agricultural Outlook 2016–2025. 2016. OECD/Food and Agriculture Organization of the United Nations "Sugar", in *OECD-FAO Agricultural Outlook 2016–2025*. OECD Publishing, Paris. https://doi.org/10.1787/agr_outlook-2016-9-en.

21. Howard, R.L., Abotsi, E., Jansen van Rensburg, E.L. and Howard, S. 2003. Lignocellulose Biotechnology: Issues of Bioconversion and Enzyme Production. *African J. Biotech.* 2(12): 602–619.

22. Hashim, F. S. 2017. Production of Renewable Glucose from Oil Palm Frond Bagasse by Using Sacchariseb C6 through Enzymatic Hydrolysis. MSc Thesis. Universiti Malaysia Pahang.

23. Flickinger, M.C. and Drew, S.W. 1999. *Encyclopedia of Bioprocess Technology: Fermentation, Biocatalysis, and Bioseparation*. John Wiley & Sons.

24. Murthy, G.S., Sridhar, S., Shyam Sunder, M., Shankaraiah, B. and Ramakrishna, M. 2005. Concentration of Xylose Reaction Liquor by Nanofiltration for the Production of Xylitol Sugar Alcohol. *Sep. Purif. Tech.* 44(3): 221–228.

25. Qi, B., Luo, J., Chen, X., Hang, X. and Wan, Y. 2011. Separation of Furfural from Monosaccharides by Nanofiltration. *Bioresource Tech.* 102(14): 7111–7118.

26. Weng, Y.H., Wei, H.J., Tsai, T.Y., Chen, W.H., Wei, T.Y., Hwang, W.S. and Huang, C.P. 2009. Separation of Acetic Acid from Xylose by Nanofiltration. *Sep. Purif. Tech.* 67(1): 95–102.

27. Almazán, J.E., Romero-Dondiz, E.M., Rajal, V.B. and Castro-Vidaurre, E.F. 2015. Nanofiltration of Glucose: Analysis of Parameters and Membrane Characterization. *Chem. Eng. Res. and Des.* 94: 485–493.

28. Moreno-Vilet, L., Bonnin-Paris, J., Bostyn, S., Ruiz-Cabrera, M.A. and Moscosa-Santillán, M. 2014. Assessment of Sugars Separation from a Model Carbohydrates Solution by Nanofiltration Using a Design of Experiments (DOE) Methodology. *Sep. Purif. Tech.* 131: 84–93.

29. Mousavi, S.M. and Moghadam, M.T. 2009. Separation of Sugar from Molasses by Ultrafiltration and Nanofiltration. *World Appl. Sci. J.* 7(5): 632–636.

30. Bargeman, G., Vollenbroek, J.M., Straatsmac, J., Schroënd, C.G.P.H. and Boomd, R.M. 2005. Nanofiltration of Multi-component Feeds. Interactions between Neutral and Charged Components and Their Effect on Retention. *J. Membr. Sci.* 247: 11–20.

31. Roli, N.F.M., Yussof, H. W., Saufi, S.M., Abu Seman, M.N. and Mohammad, A.W. 2016. Separation of Xylose From Glucose Using Pilot Scale Spiral Wound Commercial Membrane. *J. Teknologi* 78(12):1–5.

32. Roli, N.F.M., Yussof, H.W., Abu Seman, M.N., Saufi, S.M. and Mohammad, A.W. 2017. Separating Xylose from Glucose Using Spiral Wound Nanofiltration Membrane: Effect of Cross-Flow Parameters on Sugar Rejection. 2nd International Conference on Chemical Engineering (ICCE). *IOP Conf. Series: Mater. Sci. Eng.* 162: 012035. (Scopus).

33. Nanda, S., Dalai, A.K. and Kozinski, J.A. 2014. Butanol and Ethanol Production from Lignocellulosic Feedstock: Biomass Pretreatment and Bioconversion. *Energy Sci. and Eng.* 2(3): 138–148.

34. Rodríguez-Chong, A., Ramírez, J.A., Garrote, G. and Vázquez, M. 2004. Hydrolysis of Sugar Cane Bagasse Using Nitric Acid: A Kinetic Assessment. *J. Food Eng.* 61(2): 143–152.

35. Md Zain, M. and Mohammad, A.W. 2016. Clarification of Glucose from Cellulose Hydrolysate by Ultrafiltration with Polyethersulfone Membrane. *Int. J. Biomass Renew.* 5(1), 14–18.

36. Zhang, R., Lu, X.B., Sun, Y.S., Wang, X.Y. and Zhang, S.T. 2011. Modeling and Optimization of Dilute Nitric Acid Hydrolysis on Corn Stover. *J. Chem. Tech. Biotech.* 86: 306–314.

37. Weng, Y.H., Wei, H.J., Tsai, T.Y., Lin, T.H., Wei, T.Y., Guo, G.L. and Huang, C.P. 2010. Separation of Furans and Carboxylic Acids from Sugars in Dilute Acid Rice Straw Hydrolyzates by Nanofiltration. *Bioresource Tech.* 101(13): 4889–4894.

38. Patil, N.V., Feng, X., Sewalt, J.J.W., Boom, R.M. and Janssen, A.E.M. 2015. Separation of an Inulin Mixture Using Cascaded Nanofiltration. *Sep. Purif. Tech.* 146: 261–267.
39. Feng, Y.M., Chang, X.L., Wang, W.H. and Ma, R.Y. 2009. Separation of Galacto-Oligosaccharides Mixture by Nanofiltration. *J. Taiwan Ins. Chem. Eng.* 40(3): 326–332.
40. Salgado, C.M., Palacio, L., Prádanos, P., Hernández, A., González-Huerta, C. and Pérez-Magariño, S. 2015. Comparative Study of Red Grape Must Nanofiltration: Laboratory and Pilot Plant Scales. *Food Bioprod. Process.* 94: 610–620.

13 Adsorptive Nanomembranes for Treatment of Wastewater Containing Ammonia

Rosiah Rohani, Dharshini Mohanadas,
and Nagarajan R. Periasamy
Department of Chemical and Process Engineering
Faculty of Engineering & Built Environment
Universiti Kebangsaan Malaysia

CONTENTS

13.1 INTRODUCTION: BACKGROUND AND DRIVING FORCES

Water is a vital part of our everyday life and is essential to the global living system. Water is required by humans for various purposes such as agriculture, industry, electricity, drinking, cooking, and transportation. Water is severely polluted due to the wastewater discharges from the industrial, domestic, or agricultural sectors. The wastewater discharged contains various types of pollutants, and these can affect the natural biological community and even humans. Water pollution is defined by the Environmental Quality Act of 1974 [1] as any direct or indirect difference to the physical, thermal, biological, or radioactive characteristics of any part of the environment that discharges, emits, or places this waste up to affect its use and cause a condition dangerous and detrimental to the health, safety, and welfare of the public, or other life such as birds, wildlife, fish and aquatic life and aquatic plants.

Wastewater contains organic pollutants such as detergents, pesticides, processed food, and pharmaceutical products, and inorganic pollutant such as salts and heavy metals. Ammonia is also

DOI: 10.1201/9781003261827-13

247

TABLE 13.1
Ammonia Level in Wastewater That Arises from Various Industries

Wastewater	Ammonia Level (mg/L)
Municipalities	0–200
Fertilizer manufacturing	500–2,000
Food and pharmaceuticals	2,000–7,000
Petroleum refinery	5–1,000
Landfill leachate	1,000–4,000

Source: [7].

considered as one of the major pollutants that could cause water pollution. Ammonia is one of the important elements that has become the basic need for plant as a nutrient source such as fertilizer [2] and has a wide potential in plantation and coolant industries. In general, there are two forms of ammonia present, known as non-ionic (NH_3) and ionic (NH_4^+). Non-ionic ammonia is the most toxic compound for aquatic life and needs to be removed from the aquaculture waters [3,4] because it does not has charge and has the ability to dissolve in lipid. The presence of excessive ammonia in water body is possibly due to human activity such as rapid industrialization and increasing living standards. Ammonia potentially causes difficulty in breathing problems including trachea-bronchitis, bronchiolitis, laryngitis, bronchopneumonia, and pulmonary edema [5]. Ammonia needs to be removed from wastewater in order to reduce environmental problems and the impact toward human health.

Ammonia can be very harmful if the content in the water exceeds its minimum limit. Ammonia that exists in the environment of aquatic life is harmful to all vertebrates, which can potentially cause convulsions, coma, and death [6]. Manufacturing industries (Table 13.1) such as chemical fertilizers, pharmaceuticals, and petroleum refining are the major sectors that cause increase of ammonia content in wastewater [7]. The Environmental Quality (Industrial Effluent) Regulations 2009 was introduced by the Department of Environment under the Environmental Quality Act 1974 to specify the parameters that need to be followed before treated wastewater is discharged into water bodies.

Researchers are looking for inexpensive and suitable technologies for wastewater treatment. There are three effective water treatment methods that are being practiced to remove the ammonia content in water, which are physical treatment, chemical treatment, and biological treatment [8]. The advantages and disadvantages of these treatments are clearly mentioned in Table 13.2. Figure 13.1 depicts the overview of a sludge management flow chart, utilizing the pretreatment approach. Currently, the unit operations and processes are integrated, and it is referred to as the primary, secondary, and tertiary treatment. Primary treatment is a preliminary purification of the physical and chemical characteristics of wastewater, whereas the secondary treatment indicates the biological treatment of wastewater. The wastewater that undergoes primary and secondary processes can be converted to safe and high-quality water via tertiary treatment processes [9].

Physical treatments such as filtration using filter press, centrifugal separation, and sedimentation are processes that are carried out in wastewater treatment. Conventional methods such as ion separation using resins, chlorination, air stripping, and biological treatment are also utilized for an effective ammonia removal process [10]. Rohani et al. [11] mentioned that membrane extraction technique is a modern physical method, which can be implemented to remove ammonia content in wastewater. The use of liquid membranes (liquid membrane) has been given attention by the researchers in removing any heavy metals from the solution. The advantages that can be obtained through the liquid membranes are low capital and operating costs, minimal energy consumption as well as solvent, and high concentration factor [12]. In addition to the use of liquid membranes, other

TABLE 13.2

Comparison of Advantages/Disadvantages of Water Treatment Methods

Water Treatment Method	Advantages	Disadvantages
Physical treatment	– High efficiency – Simple process – Easy to maintain	– High equipment costs – High maintenance costs
Chemical treatment	– High efficiency – Save time	– High cost of chemicals – High sediment yield – Proper chemical storage methods – High energy costs
Biological treatment	– Environmentally friendly – Easy Operation – Low cost – Microbes are readily available	– Slow process – Microbes are easily deactivated

Source: [9].

techniques of utilizing membranes such as membrane distillation (MD) [13] and membrane contactor [14] are also promising in removing ammonia.

Landfill leachate, coke plant wastewater, digested swine wastewater, wastewater from the fertilizer industry, petroleum refinery wastewater, metallurgical wastewater, and household wastewater are the main factors of ammonia contamination [15,16]. China has disclosed the release of 2.5 million tons of ammonia nitrogen into water per year [17]. Caustic soda solvent is being used in the oil refining industry, to discard sulfur within the hydrocarbon streams that possibly contain ammonia. Sulfidic caustic solution or wasted caustic waste streams are produced as a result of compound absorption during caustic scrubbing process [18–20]. Ammonia affects aquatic life both acutely and chronically, and it contributes to environmental eutrophication. The harmful effects of ammonia on aquatic animals include gill destruction, reduction of blood oxygen-carrying capacity, ATP inhibition, and liver and kidney damage.

The removal of dissolved ammonia in wastewater is very important to ensure that the ammonia content is at a safe level for release to the environment. A large amount of treated wastewater from the industrial sector is released into open areas such as rivers to be used as a source of raw water for drinking water. Hasan et al. [21] stated that the ammonia content in raw water far exceeds the allowable levels and thus complicates the water treatment process for water treatment plants using conventional methods. This high concentration level can lead to shut down of the water treatment plant, thereby disrupting the water supply.

There are various conventional methods, namely nitrification-denitrification biological treatment, air stripping method, and breakpoint chlorination that are performed for the removal of dissolved ammonia [22]. Biological treatment is also a conventional method for the removal of ammonia and organic matter, but it is considered less effective due to the low bio-conversion as well as the unpleasant environmental factors [23]. Mandowara and Bhattacharya [22] mentioned that controlling the pH is difficult and high chemical cost is needed to treat high load of water during the chlorination process. The ion-exchange process is an alternative method, which requires expensive organic resins, as well as the need for large quantity of regeneration makes this process less relevant to use. Chlorination method is usually performed for industrial wastewater, as it contains high levels of ammonia content. However, this is a high-cost and a high-maintenance technique because of the usage of strong chemicals and also the oxidation process [23].

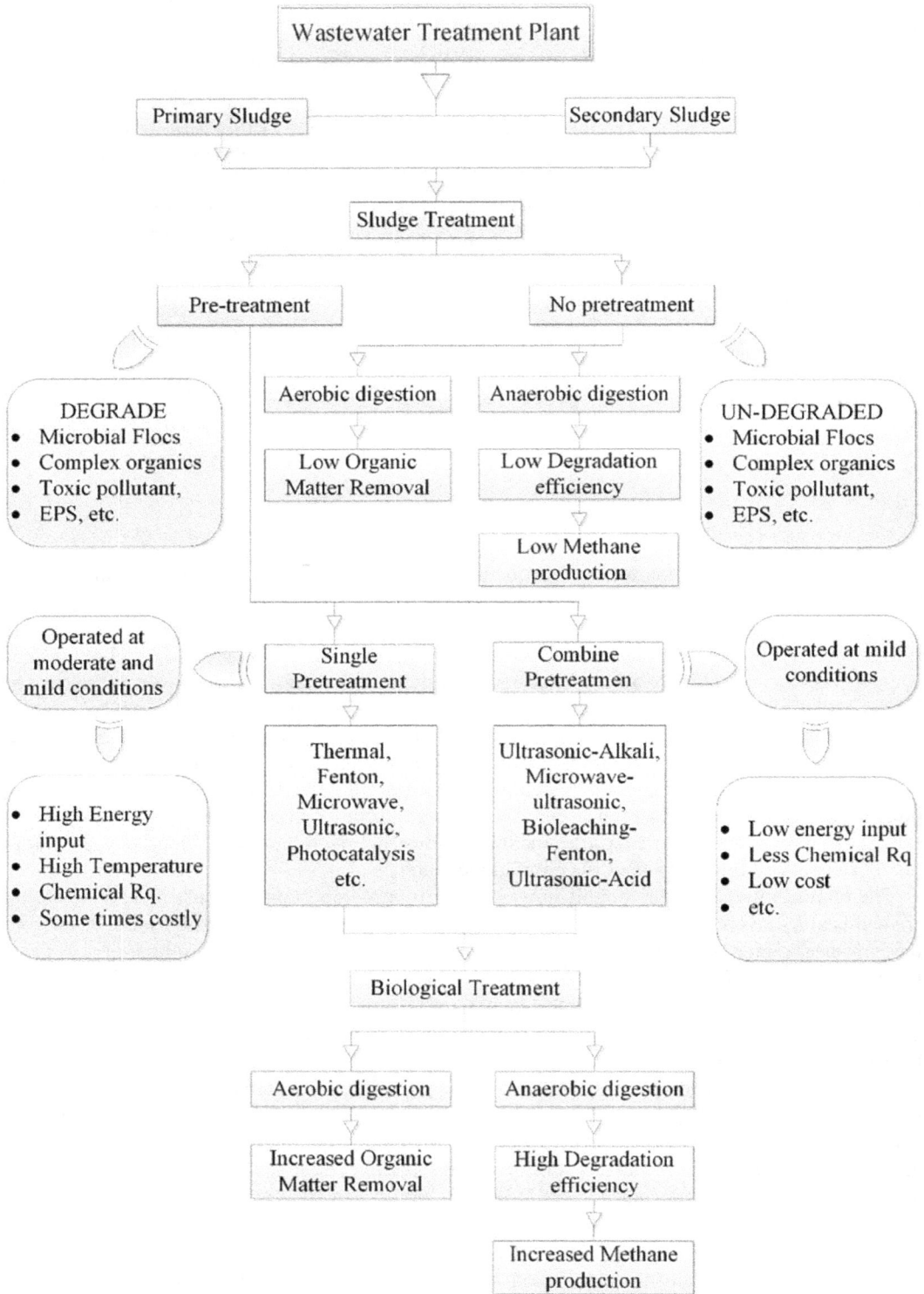

FIGURE 13.1   Flow chart of wastewater treatment plant involving various routes. (Source: [21].)

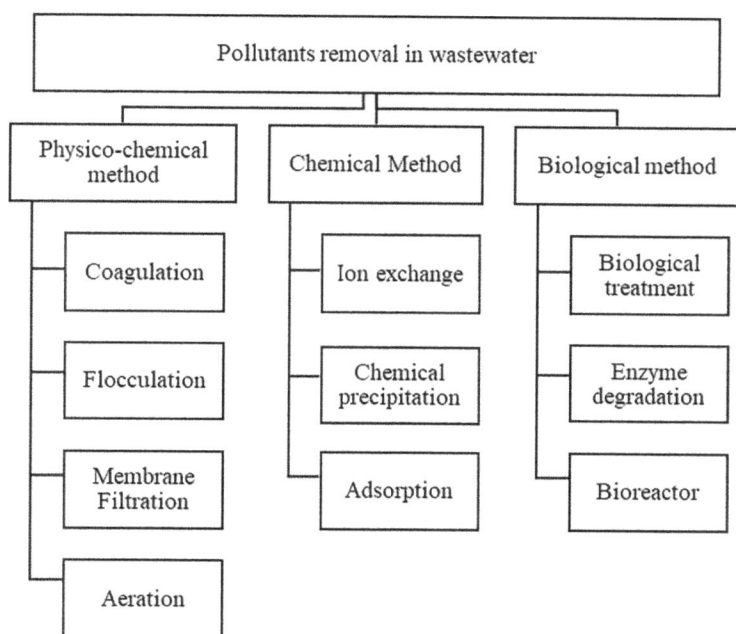

FIGURE 13.2   Techniques for removal of pollutants and wastewater treatment.

13.2   COMMON METHODS OF TREATMENT OF AMMONIA WASTEWATER

There are several physico-chemical, chemical, and biological processes that have been reported as the methods to treat wastewater via aeration, coagulation, flocculation, precipitation, absorption, oxidation, solvent extraction, evaporation, carbon adsorption, ion exchange, membrane filtration, electrochemical, biodegradation, and phytoremediation technology. Every treatment process has its own advantages and disadvantages, not only due to the cost but also in term of its effectiveness, marketability, practicality, reliability, environmental impact, sludge formation, difficult operation, pretreatment requirement, and potential in producing toxic by-products. Figure 13.2 shows the techniques that can be used to remove pollutants in wastewater.

Among these approaches, biological process is one of the most demanding techniques for wastewater treatment in removing ammonia [24], as it transforms ammonium ion (NH_4) into its elemental nitrogen gas (N_2) to establish a circular economy [25]. Air stripping, oxidation, precipitation, photocatalysis, and chemical coagulation are also well-known techniques for water treatment [26]. These methods necessitate complex configuration, tolerate high concentration NH_4 levels (>2 g/L NH_4-N), and possess high maintenance [25]. Table 13.3 presents the summary of various water treatment technologies for the removal of ammonia.

13.2.1   MEMBRANE FILTRATION

Membrane functions as a semi-permeable barrier to allow some particles to pass through it. Membrane filtration is among the promising method used in treatment of raw water [27]. Separation using membrane is a technology that leads to many advantages. It gives high separation efficiency, has no phase change, and is energy saving, easy to scale up, and environment-friendly. The common membranes used for ammonia rejection are ultrafiltration (UF), nanofiltration (NF), and reverse osmosis (RO). Water permeable flux through the membrane increases proportionally with the applied pressure of the feed that indicates as pressure-driven filtration [28].

TABLE 13.3

Summary of Various Water Treatment Methods for Ammonia Removal

Technology	Working Conditions	Advantages	Challenges	Outlet Concentration (mg/L NH_4-N)	Removal Efficiency (%)	Reference
Adsorption	Wide ranges of temperature and pH	• Simple operation • Efficient NH_4 removal • Operate in low NH_4 concentrations	• Various adsorbents depict different removal efficiencies	1	43–100	[27]
Air stripping	pH range: 10.8–11.5 High NH_4 concentration >15°C	• Widely used process for Wastewater pretreatment • Simple equipment • Not sensitive to toxic substances	• Great chemical demands • Hindered by low temperature • Time-consuming • Enormous energy-consuming • Scaling and fouling	500–1,000	50–99	[27,28]
Membrane filtration	Liquid–liquid reverse osmosis membrane is needed	• Low field and space • Simple operation	• Produces dilute concentrate stream • Only concentrating max 1 order of magnitude • Membrane fouling • Membrane cleaning	1	60–99	[29,30]
Struvite precipitation	Requires certain pH and temperature	• A valuable slow-release fertilizer • Medium cost	• Needs phosphate to proceed • Many competitors for phosphate (calcium, magnesium phosphates) • Introduce new pollutants	29–100	20–99	[31–33]
Biological treatment	Growth of heterotrophic or photosynthetic (or phototrophic) algae or bacteria Sensitive to temperature	• Free of chemical reagents • No need for complicated configuration	• Produces biomass • Only operates at low concentrations • Long start-up time	<5	70–99.5	[27,34]
Photocatalysis	In both liquid and gas phase	• Solar energy is free • No secondary pollution	• Efficient photocatalysts are still under development • Efficiency depends on light source		35–100	[35,36]

i. Pressure Filtration Membranes

The use of membrane filtration under pressure driving force in wastewater treatment allows recovery and reuse of water. This practice leads to more environmental friendly processes by reducing waste generation and is cost-effective, due to the recovery and reuse of high-value components [29]. Membrane filtration process involves several membrane types including microfiltration (MF), UF, NF, and RO. These membranes are considered as the most environmental friendly method available to date [30]. The main difference between these membranes is the pore size. Under different filtration processes, various elements in the water can be permeated or rejected. RO is a filtration procedure that generally removes all elements and allows only pure water to pass through. UF generally rejects large molecules such as protein and viruses, while MF rejects suspended solids and other larger compounds. Meanwhile, NF is a process that commonly removes 50%–90% monovalent ions such as chloride and sodium. Figure 13.3a and Table 13.4 show the general differences of membranes based on its pore size.

Membrane system can be operated based on two configurations: dead end and cross-flow. Dead end allows the flow of solution to deposit on the membrane surface without circulation of the concentrate, while cross-flow is the feed that flows tangential to the membrane surface with the concentrate to be recycled back. This causes the dead end to have a higher fouling and require frequent cleaning compared to cross-flow method. Figure 13.3b and c shows the mechanism of the flow pattern.

UF membranes have more than 90% efficiency in removing heavy metals, and they have been widely used in water treatment to treat colloidal particles, heavy metals, and natural organic matters (NOM). However, UF alone does not remove NOM and heavy metals effectively due to its relatively larger pore size [31]. RO membrane is suitable for desalination process, while other membranes are more applicable for other water and wastewater treatment [32–34]. RO has shown good removal efficiency of nitrogen ammonia (NH_3-N) in water. It removes various dissolved materials via size exclusion and charge rejection. Thin-film composite membrane was able to reject 98% iron, 93% manganese, and 45% ammonia, as well as suspended solid, chemical oxygen demand, biochemical oxygen demand, and total organic carbon completely [35]. Molar ratio, ionic strength, and pH are few factors that could influence ammonia permeability across the RO membrane.

In the application of NF for heavy metals and ammonia removal, applied pressure needs to be increased for effective separation. pH of the aqueous solution is one of the important factors because this process combines both the size exclusion and electrical interaction between ions in feed solution and the charged NF membranes [36]. Transmembrane pressure (TMP) for polyamide NF membrane was reported at 16 bar, while the ammonium removal in the investigated NF membranes (NF90, NF200, and NF270) has the average of 95.3% at pH 7. Thus, it can be concluded that the retention mechanism of ammonium is based on electrochemical mechanism, where the pH and TMP critically influenced the retention value of ammonium [37].

ii. Liquid Membranes

A liquid membrane is formed from a thin organic phase layer between two aqueous solutions of different compositions. There are several types of liquid membranes in removing metal ions, namely emulsion liquid membrane, bulk liquid membrane, and supported liquid membrane. Emulsified and bulky liquid membranes are unsupported liquid membranes, while flat sheet membranes and hollow fiber membranes are supporting liquid membranes [12]. Figure 13.4a demonstrates a schematic diagram of a flat sheet support fluid membrane in a membrane separation process.

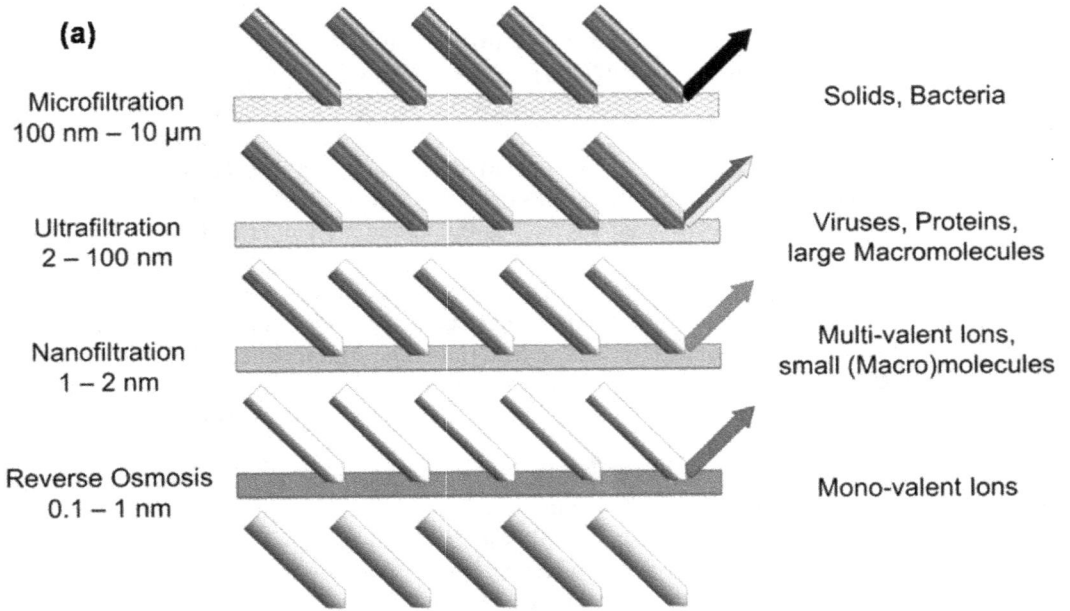

(a)

Microfiltration
100 nm – 10 μm

Solids, Bacteria

Ultrafiltration
2 – 100 nm

Viruses, Proteins,
large Macromolecules

Nanofiltration
1 – 2 nm

Multi-valent Ions,
small (Macro)molecules

Reverse Osmosis
0.1 – 1 nm

Mono-valent Ions

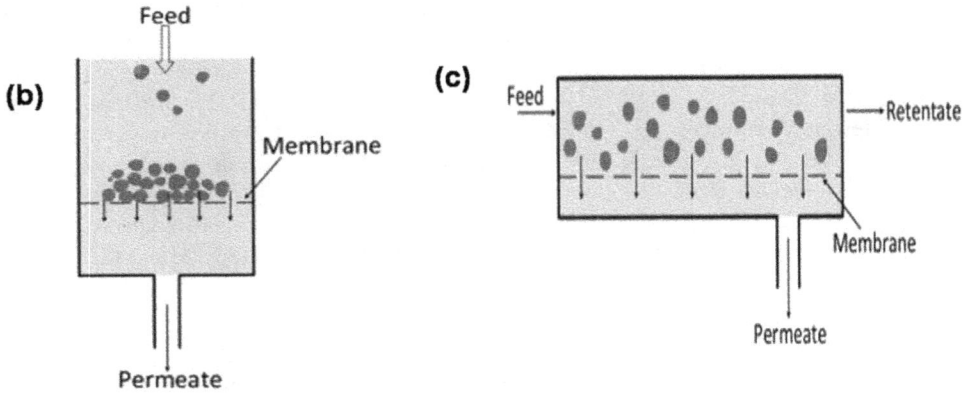

(b)

Feed

Membrane

Permeate

(c)

Feed

Retentate

Membrane

Permeate

FIGURE 13.3 (a) Membrane filtration process based on the membrane pore size (Source: [31]) and the mechanism of (b) dead end and (c) cross-flow filtration. (Source: [32].)

TABLE 13.4
Pore Size and Operational Pressure for Different Membrane Types

Membrane Type	Pore Size (μm)	Operational Pressure (bar)
Microfiltration	0.1–10	1–6.2
Ultrafiltration	0.01–0.1	1–10
Nanofiltration	0.001	20–40
Reverse osmosis	<0.001	30–100

Source: [30].

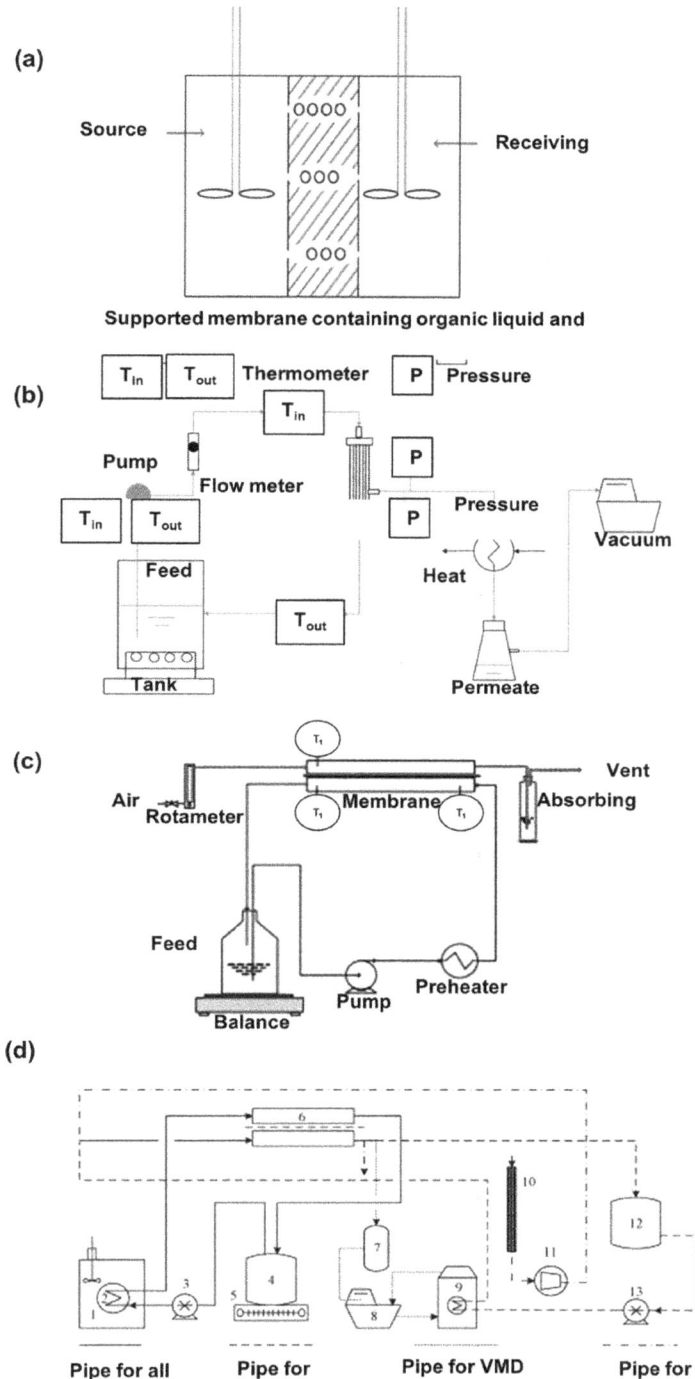

FIGURE 13.4 (a) Schematic diagram of a flat sheet support fluid membrane (Source: [12]) and the experimental preparation using (b) VMD (Source: [45]) and (c) SGMD (Source: [41]). (d) Schematic of MD apparatus consisting (1) thermostatic bath, (2) coil for heating, (3) gear pump for feed circulation, (4) feed tank, (5) electric balance, (6) flat sheet membrane module, (7) buffer tank, (8) vacuum pump, (9) chiller, (10) pipe for air drying, (11) air fan, (12) permeate tank for DCMD, and (13) gear pump for permeate circulation in DCMD. (Source: [44].)

Supporting fluid membrane (SLM) is one of the three-phase fluid membrane systems that is supported by capillary forces within the pores of microbial polymer as well as inorganic thin films [38]. The liquid SLM process is related to the membrane-based extraction process due to the factors of mass transfer rate as well as high selectivity. Wastewater treatment has been studied by combining the advantages of solvent extraction and the use of membranes through SLM process. SLM is capable of separating two different phase solution phases if the organic liquid cannot mix with the aqua feed and stripping flow [39].

Liquid membrane systems require supports, and they play a significant part in transfer as well as efficiency in the separation process. Three important components in the liquid membrane are the supporting membrane, the organic solvent, and the carrier. The permeability rate and separation efficiency depend on the type of fluid used as well as the support in the manufacture of SLM. A study found that most industries apply microbial polymer membranes, and in many cases, the selection for polymeric supports influences the level of SLM stability [38]. The observed characteristics of microlithic polymer membranes are high hydrophobic and porosity, small pore size, and perfect tortuosity.

13.2.2 Membrane Distillation

Apart from the liquid membrane method, the use of a MD system as well as a contact membrane for ammonia removal is greatly increasing among the researchers. Separation by the MD process is subject to the level of volatility of different materials as well as the difference of vapor pressure in both parts of the membrane which is the driving force for this process [40].

MD eliminates the volatile compounds such as ammonia, and due to their low energy requirements, focus on MD has been increasing [41]. Membrane for MD is made up of superhydrophobic-type membranes. Theoretically, only vapor penetrates the membrane, while the liquid will not wet the membrane or penetrate through the membrane pores [13]. MD refers to the separation process of a mixture across a hydrophobic membrane through a thermal-driven process method [42]. The differences in the temperature and the concentration at the inlet port produce a vapor pressure gradient, which is the driving force for any vapor molecules of the higher volatile compounds to move from the feed to the membrane.

According to previous studies, there are several types of MD processes, namely vacuum membrane distillation (VMD) [13], modified direct contact membrane distillation (MDCMD) and direct contact membrane distillation (DCMD) [43], gas membrane distillation sweep (gas sweep) (SGMD) [41] and water gap membrane distillation (air gap) [42]. Schematic diagrams showing the experimental preparation for VMD and SGMD are clearly illustrated in Figure 13.4b and c. The percentage rate of ammonia removal using MD method can be seen in Table 13.5 based on the studies that have been conducted. Percentage of ammonia removal, η, is determined using Equation (13.1), where C_o and C_t are the ammonia concentration in the feed flow at the module inlet and outlet, respectively:

$$\eta\% = (C_o - Ct)/C_o \times 100\%. \tag{13.1}$$

TABLE 13.5
Percentage Rates of Ammonia Removal Using Different Distillation Membranes

Distillation Membrane Type	Percentage Rate (%)	Reference
Sweep gas	97	[41]
MDCMD	99.5	[43]
DCMD	52	[43]
VMD	78–99	[13]

TABLE 13.6

Configuration Differences of the Three MDs

MD Type	Overall Mass Transfer Coefficient of Ammonia, K_a	Selectivity, β
VMD	Very high	Very low
DCMD	Moderate	Very high
SGMD	Very low	Moderate

Source: [44].

MD configuration differences were studied in the ammonia removal process from water. Factors influencing the separation process are feed temperature, feed velocity, diffuser, pH, and initial concentration of the feed. Moreover, membrane properties influence the separation performance of water via MD [44]. Ding et al. [44] used flat sheet membranes in studies conducted with different pore sizes as well as thicknesses. Figure 13.4d shows a schematic diagram for the MD-based separation process for the three types of configurations.

The total/overall mass transfer coefficient (K_a) of the ammonia increases, as the pore size of the membrane increases and the thickness is low, but the selective force, β occurs otherwise. The configuration differences between the three types of MD, namely VMD, DCMD, and SGMD, are summarized in Table 13.6. The total mass transfer coefficient is determined by the following Equation (13.2), where F_f is the feed flow rate, A is the total membrane area, V is the feed volume in the tank, and C_o and C_t are the ammonia concentration in the feed flow at the module inlet and outlet, respectively.

$$K_a = V/At \ln F_f \left(C_o/C_t\right). \tag{13.2}$$

High feed temperatures indicate high K_a values but low selectivity, β. The feed velocity affects the value of K_a for VMD, and there is no effect for the value of β on all three types of MD. The values of β and K_a increase for the SGMD configuration, as the sweep gas velocity increases, but no effect occurs due to the diffuser velocity in the DCMD. Feed pH is important in determining the performance of the separation process for all three configurations, and this factor can increase the values of β and K_a.

13.2.3 MEMBRANE CONTACTOR (MC)

MC is a membrane technology that has gained the attention of researchers for ammonia removal. The MC is good for surface ammonia removal because the large surface area can help the separation of ammonia more quickly in the wastewater [10]. MC allows direct contact and mass transfer between the gas and liquid phases without scattering among each other [45]. Figure 13.5 illustrates the MC of hollow fiber liquids.

The liquid MC functions in the presence of a concentration gradient between the fluid on the inside (lumen) and also the fluid flowing outside the fiber (shell), where molecular transfer occurs through the hollow membrane wall [46]. Hollow fiber membranes in MC systems are often used in ammonia removal processes, such as the use of polypropylene (PP) membranes [23,46,47], polyvinylidene fluoride (PVDF) membranes [48], polytetrafluoroethylene (PTFE) membranes [49], and also PTFE flat sheets [10]. The MC requires a stripping solution to accelerate the process of molecular transfer in the hollow membrane. The most commonly used strippers are sulfuric acid [10,46,50] and phosphoric acid [48]. Ammonia from the wastewater can be recovered in the form of ammonium salts, using contact membranes of liquids [49]. Ammonia is capable to permeate across the gas-filled pores in the membrane, and it reduces the concentration of ammonia over the time.

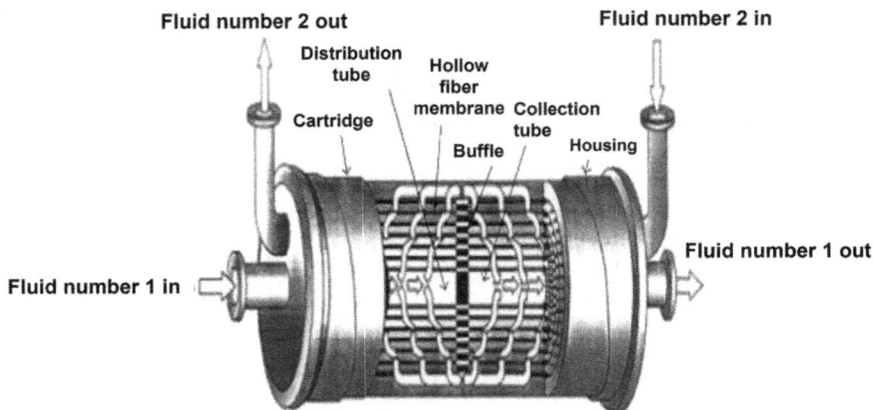

FIGURE 13.5   Membrane contactor of hollow fiber liquids. (Source: [12].)

Darestani et al. [14] described that the most commonly used stripping solution is sulfuric acid in order to produce ammonium sulfate. Researchers have proven the effectiveness of MCs in removing ammonia from wastewater with promising removal percentage as depicted in Table 13.7.

The removal percentage significantly increases as the pH of the ammonia feed solution increases up to pH 10. Report shows that when the pH exceeds 10, the ammonia removal process drops drastically. In addition, increasing the velocity of the ammonia feed affects the ammonia removal rate [23]. The stripping velocity as well as the initial concentration of ammonia had no effect on the ammonia removal. This summary indicates that MC can efficiently remove ammonia from wastewater under different set conditions, but it would require high-cost maintenance and operating conditions to keep the performance at the optimum level.

Based on the findings reported from various literatures on the advantages and disadvantages of different membrane filtration types in removing ammonia in comparison to adsorption, it can be said that the common methods using membrane and adsorption are highly promising to solve the ammonia presence issue in water/wastewater. Several attempts are made to combine two techniques, namely adsorption and membrane, to enhance the separation of ammonia from water, and these have been reported in open literature in recent years. Adsorptive membrane could overcome the drawbacks of each individual technique by using a hybrid process, in a single step only. Adsorptive membrane is further discussed in the next section.

13.3   ADSORPTIVE MEMBRANES

Adsorptive membranes offer functionality of both adsorption and filtration. The adsorptive membrane is primarily based on the adsorption process, which is a mass transfer process in which chemicals are chemically and physically bonded to the solid surfaces. Adsorption is a simple and practical method that allows for design flexibility and excellent resistance to harmful chemicals. Interestingly, it is a reversible process, because the adsorbents can be regenerated through desorption process that is considered cost-effective. The characteristics of the adsorbents used determine their performance [51]. Ammonia removal process can be improved when the adsorbent material is combined with the polymer to produce an adsorptive membrane. Adsorption is a process performed, to treat ammonia using an adsorbent. Polymer membrane with the combination of two elements leads to an effective process of ammonia removal with promising results. The performance of the modified polymeric membrane with adsorption capability is highly impacted by the presence of convenient and cost-effective adsorption-membrane filtration system.

TABLE 13.7

Percentage of Ammonia Removal Using Different Polymer Membranes in Membrane Contactor System

Type of Polymer Membrane	Ammonia Removal Percentage (%)	Reference
Polyethylene	43–73	[7]
PVDF	85	[48]
PP	99	[50]
PP	99	[23]
PP	99.83	[10]

Ammonia removal from wastewater is supported by adsorptive membrane due to its specific adsorption groups and unique morphological features. In most adsorptive membranes, the presence of an affinity complex that forms in the adsorbent would cause them to slow down the rate-limiting mass transport process. Adsorptive membranes with large surface areas and intra-particle diffusion have short residence durations, low back pressures, and a great volumetric capacity at a large scale. These properties influence the investigation of the morphological structure of the adsorptive membranes which are to be explored in understanding the pollutants removal efficiency during the adsorption process [52].

Adsorption is based on the physicochemical concept of adsorbate mass transfer to the adsorbent surface (solid) in the form of gas or liquid [53]. It is reported that the existence of additional ions and elements in the adsorbent might leak into the solvent, impacting the removal of ammonia. This limits the ability of the adsorbent to do direct adsorption of ammonia. Therefore, addition of adsorbent in the membrane structure could aid the filtration process. The natural properties of ammonia that does not has strong bonding with the adsorbent, unlike the metal ions, could lead to the attachment of the ammonia onto the adsorbent for a short while only. This advantage allows the membrane to be used for a few times without fail, and also no treatment process for regeneration is required, since there is no occurrence of chemical binding (Rohani et al. [11]). It is also reported that polysulfone (PSF)-zeolite mixed matrix membrane could remove 99% and 95% of ammonia from water, which are comparable to MC, but can be run at a lower cost operating conditions [54].

13.3.1 TYPES OF POLYMER AND ADDITIVES FOR ADSORPTIVE MEMBRANES IN REMOVAL OF AMMONIA

Polymeric membranes are in the group of organic membranes that are made from a source of cellulose or a modified organic polymer. Polymeric membranes have low resistance to high temperatures compared to inorganic membranes such as ceramics and metals. The flexibility in terms of material selection for module construction makes these polymeric membranes a great candidate for water treatment [55]. Table 13.8 shows several types of membrane that have been used for ammonia removal. These polymers are among the few that have been repeatedly reported in the literature, which were incorporated with composites/additives/fillers during the membrane synthesis.

Meanwhile, there are various adsorptive materials such as biopolymer, activated carbon, chelating agent, nanocomposites, and nanomaterials that have been used for adsorptive membrane technology. The advantage of using nanomaterials as fillers has been mentioned widely due to their unique ability to increase membrane properties in term of hydrophilicity, anti-fouling resistance, permeability or rejection, adsorptive potential, high surface area, free surface energy, small size, active atomic property, and reactivity. Ratio of the high surface area to the volume of the nanomaterial has been found to increase the adsorption efficiency [56].

TABLE 13.8
**Several Types of Membrane Used in the
Process of Ammonia Removal**

Type of Polymeric Membrane	Reference
Polypropylene (PP)	[23,49]
PP and polytetrafluoroethylene (PTFE)	[10]
PTFE	[49]
PVDF	[11,48]
Polyethersulfone (PES)	[56]
Polysulfone (PSF)	[3,57]
Polyethylene (PE)	[7]

Apart from that, nanomaterials such as carbon nanotubes (CNTs), graphene oxide, titanium oxide, silicone dioxide, metal-organic framework, silver and zinc oxide have also been used to synthesize advanced functional membranes for water treatment. A few nanomaterials have also said to have photocatalytic activity, high antibacterial property, and low toxicity that allow them to be used for new-generation membranes [57]. Adsorptive nanomaterial is one of the potential materials that could be used to remove toxic compounds from wastewater [56].

13.3.1.1 Inorganic Fillers

Inorganic fillers are present as a support via covalent bonding, van der Waals' force, or hydrogen bonding. These inorganic fillers can be produced via different processes namely sol–gel, inert gas volatilization, pulse laser release, fire spark release, ion sparking, spray pyrolysis, photothermal synthesis, plasma thermal synthesis, fire synthesis, reactive low-temperature synthesis, fire extinguisher pyrolysis, mechanical/plant alloying, mechano-chemical synthesis, and electrodeposition. This filler is combined to increase flux and selectivity as well as antibacterial property, and to reduce fouling and bacteria effect toward the membrane during the water purification process. There are various organic fillers that have been used with polymers including silica, zeolite, titanium oxide, CNT, multiwalled CNT (MWCNT), and silver. The inorganic fillers can be combined with polymeric membrane structure via blending with solvent [58]. These inorganic fillers are commonly blended/mixed with membranes to form an adsorptive membrane, which is also known as mixed matrix membrane.

Nanomaterials have been used in water industry for adsorption and removal of pollutants, catalytic degradation, disinfection, and microbe control, as well as desalination. Other than that, combining of polymer with suitable inorganic filler can offer better permeability and selectivity compared to the readily available material [58]. Nanomaterials of silica based are reported to possess a three-dimensional silica structure (SiO_2) that forms pores and gives good mechanical strength as well as high surface area, thermal and water stability that allow metal ion to access the surface at a higher rate. Silica nanoparticle can be modified with other functional groups such as $-NH_2$/phosphonate, polyethylene glycol, $-COOH$, $-SH$, $-NH_2$, octadecyl, and carboxylate/octadecyl, to reduce its aggregation and increase adsorption efficiency. The presence of sylanol (Si–OH) on the surface of the nanosilica could act as chelating agent, to remove organic and inorganic pollutants in wastewater. On the other hand, silica particles functionalized with $-NH_2$ was found to be highly suitable to remove divalent metal cations. This functionalization of nanosilica with $-NH_2$ has proven to increase its chelating ability. Although the nanosilica particles can give a higher adsorption capacity due to their surface property, their commercial application is limited because they tend to aggregate in solution [56].

Next, natural zeolite is one of the best ion-exchange materials that have high affinity toward ammonium ion. Natural zeolite have good physicochemical properties based on its high surface area and pore volume, influence cation selectivity and high cation exchange capacity. Clinoptilolite is one of the natural zeolites that has been widely used for ammonia removal in water. It is also known as an ion exchanger with high affinity for ammonium ion. It owns classical alumino silicate structure and important macro-permeability property. Natural cations present in clinoptilolite are natrium, kalium, and calcium, with ammonium as one of the most favorable cations. Therefore, natural zeolite is believed to possess good ammonia removal [59]. However, it is not that stable in water, as it may dissolve heavily at equilibrium point. In order to use it practically for ammonia removal in wastewater, additional treatment is required to stabilize the particles to avoid from dissolving in solution [3,4]. It was also found that ammonia removal decreased, as the zeolite content increased due to cavity and macro-void, which required optimum amount of the zeolite to be identified prior to adding it in the membrane mixture [4]. Shi et al. [60] have reported on the performance of zeolite adsorptive membranes with loading up to 90 wt% synthesized via facile solvent evaporation for ammonia removal and found a superior ammonia removal trend in comparison to using the common phase inversion method.

Titanium dioxide (TiO_2) nanoparticle is a good agent that can be used for wastewater treatment, as it has advantages of good chemical stability, low cost, photocatalytic behavior, nontoxicity, and easy to synthesize. Most of the TiO_2 nanoparticles are arranged with metal ions or organic polymer to increase their removal efficiency. Since this nanoparticle has pores available and high surface area, it can work with smaller functional group to increase its effectiveness in removing higher ionic valence [56]. Similar to TiO_2, iron (Fe) oxide nanoparticle was found to have bigger prospect to preserve the environment from various pollutants, due to its cheap price, easy to manufacture and modify. Iron oxide has high surface to high volumetric area, superparamagnetism, low toxicity, chemical inertness, biological compatibility, and easy to suppress other adsorbents that are present on the surface. Modification of its surface by binding with other functional group such as –COOH and –NH_2 was found to increase its stability, adsorption efficiency, and surface area to chelate ionic metals from wastewater [56]. Addition of functional group via surface modification is important in removing the toxic pollutants in water. This functional group could also ease the desorption process for reusability [56].

Iron oxide particle can be divided into three categories, namely nanomagnetite, nanomaghemite and nanohematite. Nanomagnetite (Fe_3O_4) is unstable in aqueous environment due to its high surface energy, but the stability can be increased by using various surfactant supports that can cover the particle to avoid aggregation in water-based solution. Maghemite (γ-Fe_2O_3) nanoparticle shows more stable property in water compared to magnetite as well as possesses ferromagnetic property. It is synthesized using various techniques such as chemical vapor deposition (CVD), co-precipitation, spray-fired pyrolysis, sol–gel technique, hydrothermal, and microemulsion. Maghemite nanoparticles are usually used for removal of divalent heavy metal from pentavalent metal or higher. Removal efficiency is directly dependent to its morphology. Morphological handling requires higher energy input that indirectly could increase production cost. In contrary, modification using positively charge functional group could improve its efficiency for heavy metal removal [56]. Hematite (α-Fe_2O_3) nanoparticle is the most stable in water as compared to magnetite and maghemite as well as shows antiferromagnetic behavior. This particle is synthesized via various substrates such as silica and alginate, and it could be used for water treatment. The capacity of this nanoparticle is dependent to its charge, surface area, and pore size. The performance can be increased via moderate surface modification in order to increase its adsorption capacity [56].

Next, CNT has a uniqueness to form long carbon-to-carbon chains and complexes, in the form of double and triple bond, and also atomic accumulation in a different geometrical arrangement. Carbon nanoparticle has been widely used to remove heavy metal due to its high adsorption potential and nontoxicity. CNT structure was found by Iijima [61] by using discharged arc technique. It can also be produced via CVD and laser release techniques [62]. CNT can be divided into two types, namely single-walled CNT and MWCNT. MWCNT is believed to possess higher adsorption capacity than the CNT, which requires surface functionalization to adsorb specific component.

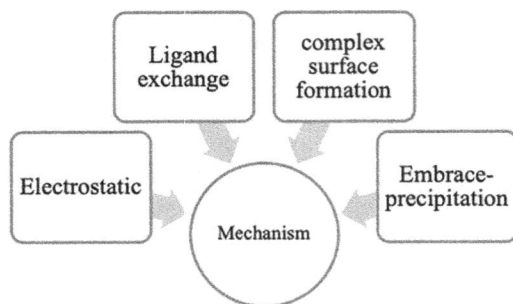

FIGURE 13.6 The mechanism of heavy metal and pollutants removal using CNT.

Surface modification of CNT using functional groups such as –COOH, –NH$_2$, and –OH has been done via chemical pathway, which was found to increase its adsorption capability. Figure 13.6 shows a few mechanisms that can be used to remove heavy metal and pollutants including ammonia using CNT.

Meanwhile, high surface area to volume ratios of nonfunctional CNT plays an important role in producing high-performance adsorbents. Apart from that, several layer and diameter of the CNT also become the main factors in the metal adsorption. Due to this reason, the CNT is capable to remove metal without external energies, even at low dimensions [56]. CNT-based membranes were used for water treatment including seawater or brine desalination, oil–water separation, and removal of heavy metals and organic pollutants [63]. These adsorptive membranes demonstrated an improved mechanical strength, tailored porosity, increased superhydrophobicity, better permeability, and long-term operation stability in comparison to their pristine membranes. However, these membranes have several issues such as high cost, process complexity, membrane fouling, and low permeate flux, which should be further studied. Meanwhile, MWCNT is efficient to remove heavy metals at a suitable reactive pH, and this could be further enhanced through functionalization of its surface by changing its surface charge. Functional CNT could increase its surface area, pore filling, and surface charge, which in turn could influence its adsorption capability, even at acidic pH and in short-term period. There are drawbacks that need to be overcome, such as difficulty in recycling from wastewater for pollutant removal as it will increase secondary pollution. Mixing of photo-assisted multifunctional CNT fillers in NF membranes, i.e., graphitic carbon nitride (g-C$_3$N$_4$), titanium dioxide (TiO$_2$), and graphene oxide (GO) have shown efficient removal of ammonia (50%), antibiotic (80%), and bisphenol A (82%) in water. This attempt was proposed to mitigate the long-standing challenge of GO-based NF membranes in large-scale application by integrating adsorbents, photocatalysis, and NF technologies [64].

13.3.1.2 Organic Fillers

Organic filler-based membranes are modern mixed matrix membranes in which filler materials such as cyclodextrin, polypyrrole (PPY), polyaniline (PANI), chitosan beads, and semi-paired polymeric nanoparticle networks, which are combined in a polymer matrix through mixing and inverse phase methods. Organic fillers have distinct advantages such as having more functional groups, thus making them more adaptable compared to inorganic fillers. Organic fillers can be combined with polymer matrices by chemical reaction or bonding especially with hydrophobic surfaces, making them a better choice to produce membranes specifically for rejection as well as having other properties such as anti-fouling effect, high hydrophilic properties, specific rejection to selective compounds, and higher permeability [58]. The findings of various organic filler materials used for mixed matrix materials are presented in Table 13.9.

TABLE 13.9

Applications of Different Organic Filler Materials Incorporated in Mixed Matrix Membranes

Filler Material	Polymer Matrix	Findings
PEGylated polyethyleneimine (PEI) nanoparticle	PVDF	Anti-fouled membranes are produced, and high-quality water purification achieved.
PANI	PSF	The resulted membrane has high permeability, more hydrophilic in nature, and increased permeability.
PPY nanosphere	PP	The water permeability of the membrane is increased; it is more hydrophilic and obtains a high percentage of bovine serum albumin (BSA) rejection.
Chitosan montmorillonite (MMT)	Polyethersulfone (PES) and polyvinylpyrrolidone	Obtained high pure water flux value, high flux recovery ratio, and high tensile strength.
Quarternized PEI	PES	The high flux value of the dye solution and the high flux recovery ratio (94.5%) prove that the resulting membrane has good anti-fouling properties.

Source: [58].

13.3.1.3 Bio-Based Fillers

Bio-nanomaterials are often used in several applications such as tissue engineering and bio-adsorbents. Among the bio-filler materials available are chitin, aquaporin, lignin, and other amphiphilic compounds. This bio-based nanomaterial is one of the innovative techniques created to increase the effectiveness of membrane technology. Mixed matrix membranes based on bio-filler materials can provide better permeability, anti-fouling ability, and certain functions such as mechanical reinforcement effect on existing membranes [58]. The combination of organic and inorganic materials can provide stability as well as effective adhesive ability to remove ammonia and metal ions. This blending/mixing can be adjusted based on particle morphology, organic phase thickness, and location-specific conditions [56]. For instance, aquaporin is used as a membrane due to its high water permeability and selectivity to ions and large molecules. It has been reported to be impermeable in ammonia than in water, allowing the ammonia to be removed from the ammonia-water mixture [65]. Chitin, lignin, cellulose, and chitosan are other bio-based fillers that can be used to remove ammonia, nitrate, and heavy metals in water. Addition of cellulose in PSF membrane at different loadings could remove 99% and 92% ammonia from 5 and 10 ppm ammonia feed solutions, respectively, with efficient water flux [66].

13.3.1.4 Graphene-Based Fillers

Graphene has good mechanical properties, electrical properties, and thermal conductivity. Similarly, GO has the same versatility as graphene and has a wide range of properties and functions for a variety of applications. GO can be obtained by chemical oxidation of graphite. The resulting GO is a single atomic layered carbon material with a honeycomb lattice arrangement [67]. GO has been reported as one of the most frequently used materials in producing nano-hybrids due to its 2D carbon nanostructure and high surface area [68]. Graphene is available in various forms such as GO and reduced graphene oxide (r-GO) which can be used for the removal of heavy metals from wastewater. GO contains numerous oxygen groups on its surface that can support its interaction with water [69]. GO also contains various functional groups such as carbonyl, epoxy (C–O–C), hydroxyl (OH), and carboxyl (COOH) groups that are capable of binding heavy metal ions and aromatic pollutants. Although r-GO is a product of GO reduction and has many structural disadvantages compared to

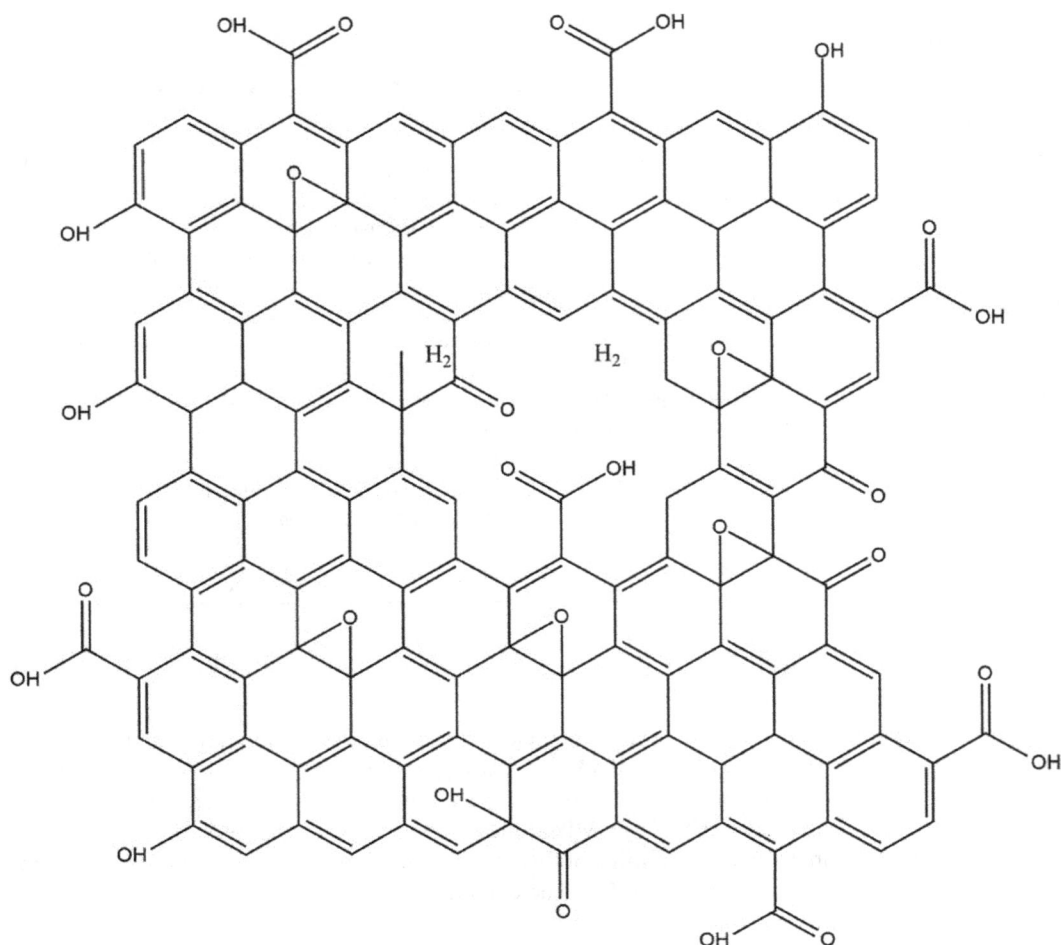

FIGURE 13.7 Chemical structure of graphene oxide. (Source: [67].)

graphene, r-GO can be easily modified by functional groups such as –OH, –NH$_2$, and –COOH. GO can also be easily modified by adding various functional groups for increased adsorption. Moreover, functionalization using metal oxides can largely increase the electronegative charge on GO and in turn improve the removal efficiency of GO [56,70]. Figure 13.7 shows the chemical structure of GO.

High antibacterial activity and hydrophilic functional group presence in the GO nanosheet allows it to be one of the chosen materials to be combined with membrane, in order to increase its hydrophilicity, water flux, and anti-fouling performance [71]. Meanwhile, for ammonia pollutant that is present in wastewater, GO is proved to increase its removal. In a recent report by Zomorodkia et al. [72], the presence of GO in polymeric mixed matrix composite membrane did not lead to good separation of polar molecules (ammonia and water) but rather give good separation to non-polar molecule (i.e. dichloromethane) only. A similar finding was reported by Zaman et al. upon incorporating GO into the polyimide membrane for concentrating succinic acid in the organic mixture, which required a crosslinker to bind the structure for enhancing the separation performance [73].

13.3.1.5 Hybrid Filler Materials

Hybrid fillers are the latest additive fillers in membrane technology. Mixed matrix membranes mixed with hybrid filler materials can be combined through two different ways, namely independently or in

TABLE 13.10

Applications of Different Hybrid Filler Materials Incorporated in Mixed Matrix Membranes

Hybrid Filler Material	Polymer Matrix	Findings
Ag/GO nanoparticle	PSF	GO improved the hydrophilic properties of the membrane and no trace of impurities is detected, making the permeability of the membrane and the flux value of pure water increased.
Functional MWCNT-Ag	PVDF	Increased in tensile strength and membrane permeability as well as salt rejection.
Fe_2O_3/MWCNT	PVDF	Improved anti-fouling ability and fouling agent rejection efficiency.
Halloysite/copper (Cu) nanoparticle	PES	Hydrophilic membranes are produced and bacterial growth on the membrane surface can be inhibited due to increased antibacterial activity.
Hybrid chitosan	PES	Enhanced the adsorption of toxins such as copper (Cu) and anionic dyes, as well as improved the mechanical properties and acid-alkali resistance of the membrane.

Source: [58].

composite form. The purpose of mixing with these hybrid filler materials is to increase the overall effectiveness of the membrane synthesis process as well as to produce membranes with specific characteristics to a particular use [58]. Table 13.10 summarizes different applications of some hybrid filler materials incorporated in mixed matrix membranes. Generally, the membranes incorporated with fillers with functional/multifunctional nanofiltration properties were found to follow Donnan exclusion mechanism, where charged materials played an important role in the solution diffusion through the membranes. The presence of abundant co-ions of similar charges (especially upon adding photochemical functional groups) caused a higher increase in the rejection rate of pollutant through the membranes. The flux and permeation through the membranes were believed to be accelerated in the presence of the multifunctional groups in the adsorptive fillers/polymeric structures [64].

13.3.2 ADSORPTIVE MEMBRANE PERFORMANCE

Adsorptive membrane is created by including functional groups at the surface and pore wall of the polymer membranes, and the target pollutants are selectively adsorbed to the functional group. The adsorptive membrane efficiently integrates the filtration performance of the membrane. Surface-derived reactions on interfaces and adsorption phenomena are primarily necessary processes in environmental chemistry, energy, catalysis, and materials processing, and they serve an important role in membrane performance control. Adsorption processes (either molecular physisorption or dissociative chemisorption) play a crucial role in membrane applications, and interfaces influence membrane performance. Adsorption can be a key determinant of membrane efficiency (i.e., selectivity, permeability, reactivity, stability/durability) or a negative influencing phenomenon (i.e., fouling) that must be managed and potentially prevented to ensure performance stability. The efficiency of different adsorptive membranes from polymer based in removing ammonia have been reported in many literatures. Adam et al. [59] have reviewed on the adsorptive materials that were mixed with inorganic of mainly ceramic, which are mostly different from the current study. Based on the summary presented in Table 13.11, it can be concluded that different blend of polymer with adsorbent/filler to form an adsorptive membrane will create a membrane of different ammonia removal properties. The range of absorbent to polymer ratio as well as the method used to synthesize the membrane will surely influence the membrane properties, especially when it is used for different applications such as removal of ammonia, metals, and nitrogen-based sources.

TABLE 13.11

Removal Efficiency and Other Parameters of Adsorbent Materials for Ammonia/N-Containing Pollutant Removal

Adsorbent Materials	Polymer Membrane	Ratio	Membrane Type	Ammonia Removal Capacity (mg/g adsorbent)	Ammonia Removal (%)	Feed (mg/L)	Flux (L/m²·h)	Reference
Chitosan	PAA (polyacrylic acid/rectorite (REC)	10%–30% in PAA/REC	Composite hydrogel	42	–	100	–	[74]
Zeolite 13X	PES	30% in PES	Mixed matrix	19	–	50	280	[75]
Zeolite	PSF	50% in 15%	Composite Fiber	5.4	–	7	–	[3]
Zeolite	PSF	50% in 15%	Mixed matrix	19.2	75	7	–	[3]
Multifunctional GO	G-C₃N₄/TiO₂-CNT	–	NF	–	12	1.53	16/bar	[64]
Chitosan	PES	0.2% in PES	NF	–	100% for N-containing pollutant	20	22	[76]
Polyethylene glycol (PEG)	PES/ZnO	5% in 17 wt%	Phase inversion	–	54	25.6	14.5	[77]
ZnO	PES	1% in 17%	Phase inversion	–	60%	25.6	7.5	[77]
GO	2-Acrylamido-2-methyl propane sulfonic acid (AMPS) polymer	20% in AMPS	Casting	–	0 (negligible permeability)	25 (wt%)	–	[78]
GO	Polyurethane	20% in PU	Casting	–	0 (negligible permeability)	25 (wt%)	–	[78]
Gypsum	PVDF	5% in 20%	NF	–	27	1	190 at 10 bar	[11]
Zeolite	PVDF	5% in 20%	NF	–	50	1	160 at 10 bar	[11]
Zeolite 13X	PES	100% in 20%	Electrospun	800 ml @>90%	52	6	70	[79]

13.4   CURRENT STATUS AND FUTURE CHALLENGES

Wastewater from different industrial sectors, municipal, agriculture, or even domestic contains pollutants that could be harmful to the environment, especially to aquatic biota. High concentration of ammonia in the wastewater increases eutrophication, which must be treated before being discharged to open sources. This water source may eventually be consumed by human being after undergoing treatment for producing drinking water. Conventional treatment methods such as biological treatment, ion exchange, adsorption, and air stripping have been used over the last decades for removing ammonia in the wastewater. However, certain treatment approaches such as biological approach commonly take a longer time to remove ammonia, and their efficacy is questionable at high ammonia levels. Thus, an alternative technology is explored to replace the existing technologies for removing ammonia in the wastewater. This book chapter focused on the recent trend and development of utilizing various types of membrane especially of nano-range filtration for eliminating ammonia in wastewater. Membrane technology has received numerous attention due to its effectiveness to remove ammonia. For instance, MD is able to produce highly pure water, and it is caused by the volatility of each solute, with the difference in vapor pressure acting as the driving force, whereas MC could eliminate volatile components like ammonia through liquid–liquid or liquid–gas mass transfer without dispersing one phase to another phase. Meanwhile, utilization of pressure filtration membrane of nano-range can also remove ammonia from the wastewater at low treatment cost. Recently, adsorptive nanomembrane has been under the spotlight since this technique integrates the membrane and adsorption process, and it has become an efficient way to remove ammonia from aqueous solutions. Apart from the Donnan exclusion and non-sieving effect in the membrane for separating ammonia in water, reactive functional groups present in the adsorbent enhance binding of ammonia via surface complexation or ion-exchange mechanism. Introduction of different adsorbent and polymeric membrane materials to form the selective membrane structure could be explored in influencing the filtration process. Lastly, the adsorptive membrane is foreseen as a promising approach to treat wastewater containing ammonia based on the combination of two techniques that have been effectively proven for ammonia removal.

13.5   SUMMARY

Blending membranes with various types of adsorbents of organic and inorganic types such as zeolite, gypsum, activated carbon, graphene, and chitosan has emerged as one of the interesting current membrane development. This is mainly attributable to its excellent characteristics for membrane performance enhancement in comparison with other modification methods. The modification of the fabricated membranes with adsorption capability also provides an added advantage to the pressure filtration membrane. The fabricated membranes are potentially good for the removal of ammonia from raw water and wastewater. However, there are limitations of membrane adsorption ability to remove ammonia. Therefore, the synthesis and evaluation of the membrane adsorptive properties with ammonia are very much needed to enhance ammonia removal. To the best of our knowledge, there is very limited literature on various adsorptive membranes of polymeric support reported for the said purpose. Furthermore, the effects of adding adsorbents on the polymeric membrane structure were also important to be investigated by performing necessary characterizations, to monitor the adsorptive mechanism/membrane behavior during filtration. Finally, the membrane performance can be aimed to be significantly improved by blending the membranes with adsorbents in terms of several aspects such as permeability, selectivity, hydrophilicity, rejection tendency, and fouling resistance.

REFERENCES

1. A. Kader, S. Zubaidah, Legal and institutional framework for pollution prevention in Malaysian rivers, *Malay Law J*, 76 (2008) 4.

2. L. Ma, Y. Liu, X. Zhang, Y. Ye, G. Yin, B.A. Johnson, Deep learning in remote sensing applications: A meta-analysis and review, *ISPRS J Photogram Remote Sens*, 152 (2019) 166–177.

3. P. Ahmadiannamini, S. Eswaranandam, R. Wickramasinghe, X. Qian, Mixed-matrix membranes for efficient ammonium removal from wastewaters, *J Membr Sci*, 526 (2017) 147–155.

4. P. Moradihamedani, A.H. Abdullah, Ammonia removal from aquaculture wastewater by high flux and high rejection polysulfone/cellulose acetate blend membrane, *Polym Bull*, 76 (2019) 2481–2497.

5. N. Brautbar, M.P. Wu, E.D. Richter, Chronic ammonia inhalation and interstitial pulmonary fibrosis: A case report and review of the literature, *Arch Environ Health: An Int J*, 58 (2003) 592–596.

6. D.J. Randall, T. Tsui, Ammonia toxicity in fish, *Marine Pollut Bull*, 45 (2002) 17–23.

7. M.M. Hossain, O. Chaalal, Liquid–liquid separation of aqueous ammonia using a hollow-fibre membrane contactor, *Desal Water Treat*, 57 (2016) 21770–21780.

8. A. Fakhru'l-Razi, A. Pendashteh, L.C. Abdullah, D.R.A. Biak, S.S. Madaeni, Z.Z. Abidin, Review of technologies for oil and gas produced water treatment, *J Hazard Mater*, 170 (2009) 530–551.

9. V.K. Gupta, I. Ali, T.A. Saleh, A. Nayak, S. Agarwal, Chemical treatment technologies for waste-water recycling – An overview, *RSC Adv*, 2 (2012) 6380–6388.

10. A. Hasanoğlu, J. Romero, B. Pérez, A. Plaza, Ammonia removal from wastewater streams through membrane contactors: Experimental and theoretical analysis of operation parameters and configuration, *Chem Eng J*, 160 (2010) 530–537.

11. R. Rohani, I.I. Yusoff, N.K. Zaman, A.M. Ali, N.A.B. Rusli, R. Tajau, S.A. Basiron, Ammonia removal from raw water by using adsorptive membrane filtration process, *Sep Purif Technol*, 270 (2021) 118757.

12. P. Parhi, Supported liquid membrane principle and its practices: A short review, *J Chem*, 2013 (2013).

13. C. Wu, H. Yan, Z. Li, X. Lu, Ammonia recovery from high concentration wastewater of soda ash industry with membrane distillation process, *Desal Water Treat*, 57 (2016) 6792–6800.

14. M. Darestani, V. Haigh, S.J. Couperthwaite, G.J. Millar, L.D. Nghiem, Hollow fibre membrane contactors for ammonia recovery: Current status and future developments, *J Environ Chem Eng*, 5 (2017) 1349–1359.

15. A. Taşdemir, İ. Cengiz, E. Yildiz, Y.K. Bayhan, Investigation of ammonia stripping with a hydrodynamic cavitation reactor, *Ultrason Sonochem*, 60 (2020) 104741.

16. L. Cao, J. Wang, S. Xiang, Z. Huang, R. Ruan, Y. Liu, Nutrient removal from digested swine wastewater by combining ammonia stripping with struvite precipitation, *Environ Sci Pollut Res*, 26 (2019) 6725–6734.

17. K. Fang, H. Gong, W. He, F. Peng, C. He, K. Wang, Recovering ammonia from municipal wastewater by flow-electrode capacitive deionization, *Chem Eng J*, 348 (2018) 301–309.

18. M. De Graaff, N. Vieno, K. Kujawa-Roeleveld, G. Zeeman, H. Temmink, C. Buisman, Fate of hormones and pharmaceuticals during combined anaerobic treatment and nitrogen removal by partial nitritation-anammox in vacuum collected black water, *Water Res*, 45 (2011) 375–383.

19. I.B. Hariz, A. Halleb, N. Adhoum, L. Monser, Treatment of petroleum refinery sulfidic spent caustic wastes by electrocoagulation, *Sep Purif Technol*, 107 (2013) 150–157.

20. E. Üresin, H.İ. Saraç, A. Sarıoğlan, Ş. Ay, F. Akgün, An experimental study for H_2S and CO_2 removal via caustic scrubbing system, *Proc Safe Environ Protect*, 94 (2015) 196–202.

21. M. Anjum, N.H. Al-Makishah, M. Barakat, Wastewater sludge stabilization using pre-treatment methods, *Proc Safe Environ Protect*, 102 (2016) 615–632.

22. H.A. Hasan, S.R.S. Abdullah, S.K. Kamarudin, N.T. Kofli, Problems of ammonia and manganese in Malaysian drinking water treatments, *World Appl Sci J*, 12 (2011) 1890–1896.

23. A. Mandowara, P.K. Bhattacharya, Simulation studies of ammonia removal from water in a membrane contactor under liquid–liquid extraction mode, *J Environ Manage*, 92 (2011) 121–130.

24. S. Ashrafizadeh, Z. Khorasani, Ammonia removal from aqueous solutions using hollow-fiber membrane contactors, *Chem Eng J*, 162 (2010) 242–249.

25. E. Ardern, W.T. Lockett, Experiments on the oxidation of sewage without the aid of filters, *J Soc Chem Ind*, 33 (1914) 523–539.

26. H. Cruz, Y.Y. Law, J.S. Guest, K. Rabaey, D. Batstone, B. Laycock, W. Verstraete, I. Pikaar, Mainstream ammonium recovery to advance sustainable urban wastewater management, *Environ Sci Technol*, 53 (2019) 11066–11079.

27. Y. Dong, H. Yuan, R. Zhang, N. Zhu, Removal of ammonia nitrogen from wastewater: A review, *Trans ASABE*, 62 (2019) 1767–1778.

28. M.R. Adam, T. Matsuura, M.H.D. Othman, M.H. Puteh, M.A.B. Pauzan, A. Ismail, A. Mustafa, M.A. Rahman, J. Jaafar, M.S. Abdullah, Feasibility study of the hybrid adsorptive hollow fibre ceramic membrane (HFCM) derived from natural zeolite for the removal of ammonia in wastewater, *Proc Safe Environ Protect*, 122 (2019) 378–385.

29. Q. Yin, R. Wang, Z. Zhao, Application of Mg–Al-modified biochar for simultaneous removal of ammonium, nitrate, and phosphate from eutrophic water, *J Clea Prod*, 176 (2018) 230–240.
30. I.A. Talalaj, P. Biedka, Impact of concentrated leachate recirculation on effectiveness of leachate treatment by reverse osmosis, *Ecol Eng*, 85 (2015) 185–192.
31. B.-S. Xing, Q. Guo, Z.-Z. Zhang, J. Zhang, H.-Z. Wang, R.-C. Jin, Optimization of process performance in a granule-based anaerobic ammonium oxidation (anammox) upflow anaerobic sludge blanket (UASB) reactor, *Bioresour Technol*, 170 (2014) 404–412.
32. R.R. Karri, J.N. Sahu, V. Chimmiri, Critical review of abatement of ammonia from wastewater, *J Mol Liq*, 261 (2018) 21–31.
33. M.H. Hakimi, V. Jegatheesan, D. Navaratna, The potential of adopting struvite precipitation as a strategy for the removal of nutrients from pre-AnMBR treated abattoir wastewater, *J Environ Manag*, 259 (2020) 109783.
34. C. Peng, L. Chai, C. Tang, X. Min, Y. Song, C. Duan, C. Yu, Study on the mechanism of copper–ammonia complex decomposition in struvite formation process and enhanced ammonia and copper removal, *J Environ Sci*, 51 (2017) 222–233.
35. Y. Miao, N.W. Johnson, T. Phan, K. Heck, P.B. Gedalanga, X. Zheng, D. Adamson, C. Newell, M.S. Wong, S. Mahendra, Monitoring, assessment, and prediction of microbial shifts in coupled catalysis and biodegradation of 1, 4-dioxane and co-contaminants, *Water Res*, 173 (2020) 115540.
36. X. Yao, X. Hu, Y. Liu, X. Wang, X. Hong, X. Chen, S.C. Pillai, D.D. Dionysiou, D. Wang, Simultaneous photocatalytic degradation of ibuprofen and H_2 evolution over Au/sheaf-like TiO_2 mesocrystals, *Chemosphere*, 261 (2020) 127759.
37. K. Vikrant, K.-H. Kim, F. Dong, D.A. Giannakoudakis, Photocatalytic platforms for removal of ammonia from gaseous and aqueous matrixes: Status and challenges, *ACS Catal*, 10 (2020) 8683–8716.
38. I.I. Yusoff, R. Rohani, L.Y. Ng, A.W. Mohammad, Conductive polyelectrolyte multilayers PANI membranes synthesis for tunable filtration ranges, *J Mater Sci*, 54 (2019) 12988–13005.
39. Y.H. Teow, L.M. Kam, A.W. Mohammad, Synthesis of cellulose hydrogel for copper (II) ions adsorption, *J Environ Chem Eng*, 6 (2018) 4588–4597.
40. R. Rohani, M. Hyland, D. Patterson, A refined one-filtration method for aqueous based nanofiltration and ultrafiltration membrane molecular weight cut-off determination using polyethylene glycols, *J Membr Sci*, 382 (2011) 278–290.
41. N.K. Zaman, R. Rohani, A.W. Mohammad, A.M. Isloor, Polyimide-graphene oxide nanofiltration membrane: Characterizations and application in enhanced high concentration salt removal, *Chem Eng Sci*, 177 (2018) 218–233.
42. A. Mautner, Nanocellulose water treatment membranes and filters: A review, *Polym Int*, 69 (2020) 741–751.
43. E. Nagy, Membrane materials, structures, and modules, basic equations of mass transport through a membrane layer (2019) 11–19.
44. M. Barakat, New trends in removing heavy metals from industrial wastewater, *Arab J Chem*, 4 (2011) 361–377.
45. P. Goh, T. Matsuura, A. Ismail, N. Hilal, Recent trends in membranes and membrane processes for desalination, *Desalination*, 391 (2016) 43–60.
46. A.G. Fane, R. Wang, M.X. Hu, Synthetic membranes for water purification: Status and future, *Angew Chem Int Ed*, 54 (2015) 3368–3386.
47. Z. Yang, Y. Zhou, Z. Feng, X. Rui, T. Zhang, Z. Zhang, A review on reverse osmosis and nanofiltration membranes for water purification, *Polymers*, 11 (2019) 1252.
48. A. Ashadullah, N. Kishimoto, Influence of operational parameters on rapid nitrate removal using an electrochemical flow cell, *Int J Environ Sci Develop*, 7 (2016) 499.
49. P. Mikulášek, J. Cuhorka, Removal of heavy metal ions from aqueous solutions by nanofiltration, *International Conference on Nanotechnology Based Innovative Applications for the Environment*, AIDIC Servizi srl, 2016.
50. B. Cancino-Madariaga, J. Aguirre, Combination treatment of corn starch wastewater by sedimentation, microfiltration and reverse osmosis, *Desalination*, 279 (2011) 285–290.
51. P. Dżygiel, P.P. Wieczorek, *Supported Liquid Membranes and Their Modifications: Definition, Classification, Theory, Stability, Application and Perspectives, Liquid Membranes*, Elsevier, 2010, pp. 73–140.
52. L. Lozano, C. Godínez, A. De Los Rios, F. Hernández-Fernández, S. Sánchez-Segado, F.J. Alguacil, Recent advances in supported ionic liquid membrane technology, *J Membr Sci*, 376 (2011) 1–14.
53. S. Srisurichan, R. Jiraratananon, A. Fane, Mass transfer mechanisms and transport resistances in direct contact membrane distillation process, *J Membr Sci*, 277 (2006) 186–194.

54. Z. Xie, T. Duong, M. Hoang, C. Nguyen, B. Bolto, Ammonia removal by sweep gas membrane distillation, *Water Res*, 43 (2009) 1693–1699.

55. M. El-Bourawi, M. Khayet, R. Ma, Z. Ding, Z. Li, X. Zhang, Application of vacuum membrane distillation for ammonia removal, *J Membr Sci*, 301 (2007) 200–209.

56. D. Qu, D. Sun, H. Wang, Y. Yun, Experimental study of ammonia removal from water by modified direct contact membrane distillation, *Desalination*, 326 (2013) 135–140.

57. Z. Ding, L. Liu, Z. Li, R. Ma, Z. Yang, Experimental study of ammonia removal from water by membrane distillation (MD): The comparison of three configurations, *J Membr Sci*, 286 (2006) 93–103.

58. C. Yang, X. Peng, Y. Zhao, X. Wang, L. Cheng, F. Wang, Y. Li, P. Li, Experimental study on VMD and its performance comparison with AGMD for treating copper-containing solution, *Chem Eng Sci*, 207 (2019) 876–891.

59. H.H. Park, B.R. Deshwal, H.D. Jo, W.K. Choi, I.W. Kim, H.K. Lee, Absorption of nitrogen dioxide by PVDF hollow fiber membranes in a G–L contactor, *Desalination*, 243 (2009) 52–64.

60. E. Licon, M. Reig, P. Villanova, C. Valderrama, O. Gibert, J.L. Cortina, Ammonium removal by liquid–liquid membrane contactors in water purification process for hydrogen production, *Desal Water Treat*, 56 (2015) 3607–3616.

61. G.K. Agrahari, N. Verma, P.K. Bhattacharya, Application of hollow fiber membrane contactor for the removal of carbon dioxide from water under liquid–liquid extraction mode, *J Membr Sci*, 375 (2011) 323–333.

62. C.-L. Lai, S.-h. Chen, R.-M. Liou, Removing aqueous ammonia by membrane contactor process, *Desal Water Treat*, 51 (2013) 5307–5310.

63. Y. Ahn, Y.-H. Hwang, H.-S. Shin, Application of PTFE membrane for ammonia removal in a membrane contactor, *Water Sci Technol*, 63 (2011) 2944–2948.

64. G.K. Agrahari, S.K. Shukla, N. Verma, P.K. Bhattacharya, Model prediction and experimental studies on the removal of dissolved NH_3 from water applying hollow fiber membrane contactor, *J Membr Sci*, 390 (2012) 164–174.

65. E. Salehi, P. Daraei, A.A. Shamsabadi, A review on chitosan-based adsorptive membranes, *Carbohydr Polym*, 152 (2016) 419–432.

66. T.S. Vo, M.M. Hossain, H.M. Jeong, K. Kim, Heavy metal removal applications using adsorptive membranes, *Nano Converg*, 7 (2020) 1–26.

67. L. Sellaoui, T. Depci, A.R. Kul, S. Knani, A.B. Lamine, A new statistical physics model to interpret the binary adsorption isotherms of lead and zinc on activated carbon, *J Mol Liq*, 214 (2016) 220–230.

68. P. Moradihamedani, Recent developments in membrane technology for the elimination of ammonia from wastewater: A review, *Polym Bull*, 78 (2021) 5399–5425.

69. M. Buonomenna, G. Golemme, S.-H. Choi, J. Jansen, M. De Santo, E. Drioli, Surface skin layer formation and molecular separation properties of asymmetric PEEKWC membranes, *Sep Purif Technol*, 77 (2011) 104–111.

70. M. Wang, Y. Zhang, L. Lin, X. Ding, H. Li, Preparation and characterization of 13X zeolites/PES membrane adsorbent for ammonia nitrogen removal, *Chinese J Environ Eng*, 7 (2013) 3749–3754.

71. J. Yin, G. Zhu, B. Deng, Multi-walled carbon nanotubes (MWNTs)/polysulfone (PSU) mixed matrix hollow fiber membranes for enhanced water treatment, *J Membr Sci*, 437 (2013) 237–248.

72. A. Jawed, V. Saxena, L.M. Pandey, Engineered nanomaterials and their surface functionalization for the removal of heavy metals: A review, *J Water Proc Eng*, 33 (2020) 101009.

73. Q. Li, S. Mahendra, D.Y. Lyon, L. Brunet, M.V. Liga, D. Li, P.J. Alvarez, Antimicrobial nanomaterials for water disinfection and microbial control: Potential applications and implications, *Water Res*, 42 (2008) 4591–4602.

74. D. Qadir, H. Mukhtar, L.K. Keong, Mixed matrix membranes for water purification applications, *Sep Purif Rev*, 46 (2017) 62–80.

75. W. Shi, F. Gao, X. Li, Z. Wang, High zeolite loading mixed matrix membrane for effective removal of ammonia from surface water, *Water Res* (2022) 118849.

76. S. Iijima, Helical microtubules of graphitic carbon, *Nature*, 354 (1991) 56–58.

77. S. Mukhtar, W. Asghar, Z. Butt, Z. Abbas, M. Ullah, R. Atta-Ur-Rehman, RETRACTED ARTICLE: Development and characterization of hot dip aluminide coated stainless steel 316L, *J Centr South Univ*, 25 (2018) 2578–2588.

78. M. Barrejón, M. Prato, Carbon nanotube membranes in water treatment applications, *Adv Mater Interf*, 9 (2022) 2101260.

79. Q. Zhang, S. Chen, X. Fan, H. Zhang, H. Yu, X. Quan, A multifunctional graphene-based nanofiltration membrane under photo-assistance for enhanced water treatment based on layer-by-layer sieving, *Appl Catal B: Environ*, 224 (2018) 204–213.

14 Application of Nanofiltration for Reclamation and Reuse of Wastewater and Spent Geothermal Fluid

Yakubu A. Jarma
Ege University and University of California Los Angeles

Aydın Cihanoğlu
Ege University

Enver Güler
Atılım University

Barbara Tomaszewska
Mineral and Energy Economy Research
Institute, Polish Academy of Science
AGH-University of Science and Technology

Aleksandra Kasztelewicz
Mineral and Energy Economy Research
Institute, Polish Academy of Science

Alper Baba
Izmir Institute of Technology

Nalan Kabay
Ege University

CONTENTS

DOI: 10.1201/9781003261827-14

14.1   INTRODUCTION

Dwindling water resources, increasing consumption patterns and farming practices probably will cause in the nearest future that water availability will be insufficient to fulfill the world's demand [1]. Over the past years, the consecutive occurrence of the drought periods and globally increasingly limited sources of drinking water and water intended for irrigation challenge for many regions, including Europe, to find new effective solutions. Water stress in Europe results in crucial problems such as pollution, water shortages, and ecosystem damage. One-third of European countries have less than 5,000 cubic meters of water per person per year. This amount is relatively low and southern countries are particularly affected. On the other hand, the severely populated countries of northern Europe with moderate rainfall also come within the low availability group. Two-thirds of people in Europe including Poland rely on groundwater for their supplies of drinking and other water needs [2]. In Figure 14.1, water stress as the ratio of total water withdrawals to available renewable surface and groundwater supplies which include domestic, industrial, irrigation, and livestock consumptive and no consumptive uses is depicted [3]. The map shows that water stress is prevalent in the whole world, and areas of the higher water stress are common on every continent, particularly in Africa, and South and Central Asia. Additionally, water stress does not only apply to dry regions, e.g., in many parts of Europe show medium to high levels of water risk. Although these regions are not dry, they experience significant water stress because demand is increasing, and supply depends on climate change and other factors. The water stress will be more severe shortly due to climate change and increased demand for food and water [4].

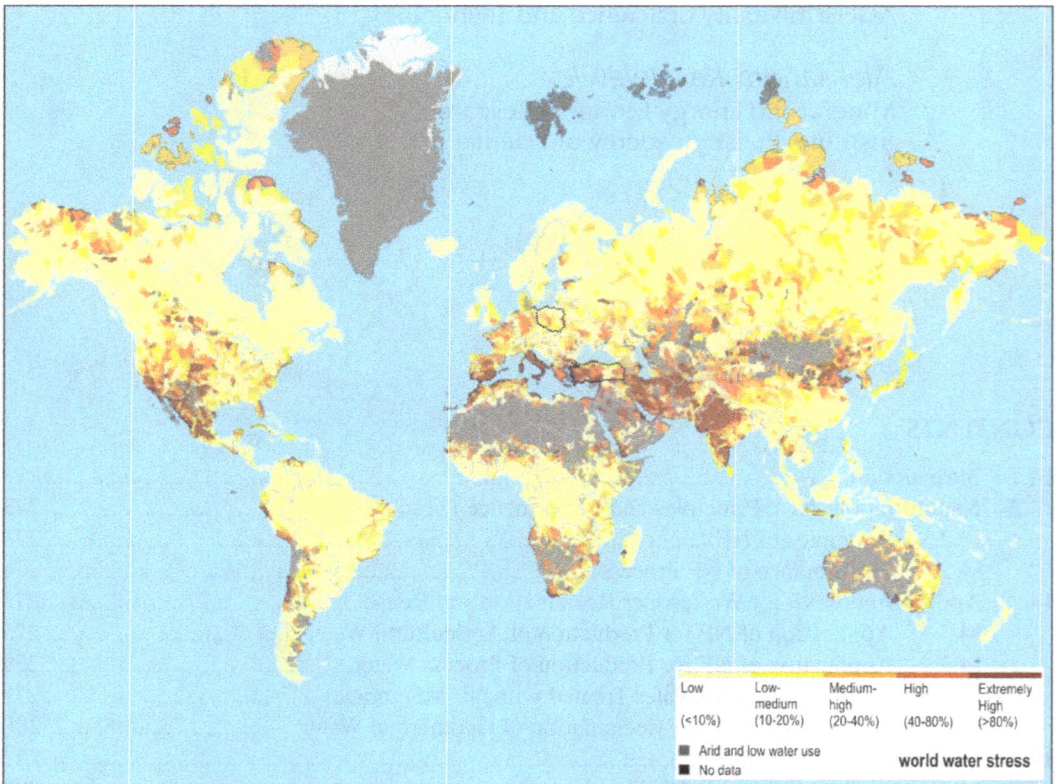

FIGURE 14.1   Water stress worldwide. (On the basis of Ref. [3].)

The productivity of agriculture depends heavily on the land, energy, and water resources. The requirement of water for production depends on per area and time, and the quality of used water is also essential because the sensitivity of crops to ions and compounds in water vary. In recent years, the water deficit and the related drought phenomenon have increased. This is due to both increasing anthropogenic and climate change. The common urbanization process and the accompanying sealing of biologically active surfaces in the catchment area will contribute to an adverse change in the balance of water resources.

Geothermal energy is a cost-effective, technically proven, reliable, and safe one, and it has been operated in various fields and applications for many decades. These energy sources can be utilized directly or by converting them to other types of energy. Using geothermal water for different purposes such as power, heating, cooling, greenhouses, dry food thermal tourism, fishing, and mineral recovery has become widespread in many countries. Practices of using geothermal waters for irrigation are rather rare due to chemical composition and high salinity. A successful application of geothermal water as irrigation is in the Kebili region of Tunisia, where 71% of low-enthalpy geothermal water is used for the irrigation of oases after cooling [5]. Both in Turkey and in Poland, there are cases where geothermal resources are in the close proximity of agricultural areas. Hence, geothermal sources can be considered as a viable source of energy and water for crops [6,7]. The sustainability of geothermal resources is evaluated in terms of optimising energy production and managing water resources, as it is both an energy source and a valuable raw material [8].

For several years, desalination technologies such as membrane filtration, ion exchange, and adsorption have been studied by Kabay et al. [9–13] and Kaya et al. [14]. Water treatment processes, used for seawater desalination and, in recent years, also for geothermal water treatment, due to their relatively high efficiency, have the potential to be used as a treatment technology for cooled geothermal waters for irrigation purposes. Studies on use and desalination of geothermal water have been published by Kabay et al. [15,16], Tomaszewska and Bodzek [17], Tomaszewska and Dendys [8], Tomaszewska et al. [18,19], and Tomaszewska [20].

14.2 NANOFILTRATION (NF): PRINCIPLES AND PERFORMANCE PARAMETERS

14.2.1 PRINCIPLES OF NF

NF is a transition process between ultrafiltration (UF) and reverse osmosis (RO). Thus, it is sometimes difficult to set a clear definition for NF. In NF, like all other pressure-driven membrane processes due to size exclusion, the solvent and some low-molecular-weight components can permeate through the membrane, and all others can be retained. Nevertheless, the main difference between NF and UF is the pore size of the membranes and the molecular weight of the components that can be rejected by the membrane. UF membranes have pore sizes ranging between 2 and 10 nm, whereas NF membranes have pore sizes between 0.2 and 2 nm. Smaller pore sizes of NF membranes result in larger applied pressure in practice. However, this required pressure to produce permeate (i.e., product water) is lower than that of RO. The transport mechanism of NF is mainly determined by transport through the pores, whereas, in RO, it is based on both diffusion and solution of components. One other distinguishing feature of NF is that membrane can possess positive or negative fixed charges. That means NF membranes utilize electrical charges for separation, especially the monovalent ions from multivalent ones or from a neutral medium.

In NF process, mass transport (component flux, J_i) is controlled by two main parameters, namely one due to chemical potential gradient $\left(\dfrac{d\mu_i}{dz}\right)$ and the other due to applied pressure $\left(\dfrac{dp}{dz}\right)$ as described in Equation (14.1) [21]:

$$J_i = L_i \frac{d\mu_i}{dz} + L_p C_{i,\,m} \frac{dp}{dz},$$

(14.1)

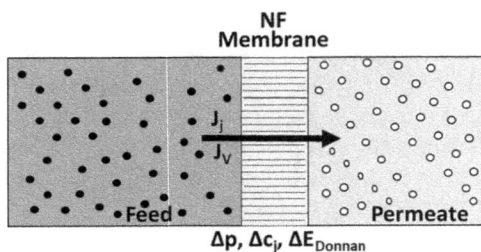

FIGURE 14.2 Scheme of the mass transport through an NF membrane.

where L_i and L_p are the coefficients due to diffusion and permeability, respectively. In addition, the volume flux (J_v) through the membrane can be determined as the sum of fluxes of individual components multiplied by their molar volumes (V_i):

$$J_v = \sum_i J_i V_i, \tag{14.2}$$

$C_{i,\,m}$ in Equation (14.1) represents the component concentration in the membrane, which is related to the one in the solution ($C_{i,\,s}$) such that:

$$C_{i,m} = k_i C_{i,s}, \tag{14.3}$$

where k_i is the partition coefficient. The partition coefficient can be defined as the product of two partition coefficients that are due to exclusions of size and charge (in the case of an NF membrane containing electrical charges). The latter is defined as a very well-known Donnan exclusion depending on the Donnan potential (E_{Donnan}) defined between the charged membrane and the electrolyte solution in contact [21]:

$$E_{Donnan} = \sum_i \frac{1}{z_i F}\left[RT \ln \frac{C_{i,s}}{C_{i,m}} \right], \tag{14.4}$$

where z is the valence of the individual component and F is the Faraday constant. Thus, the effect of this potential should be taken into account in mass transport through NF membranes as well. An illustration of the mass transport principles in NF is shown in Figure 14.2.

14.2.2 PERFORMANCE OF NF PROCESS

Same as all other pressure-driven filtration processes, three parameters can define the performance of an NF process: flux, rejection, and recovery. Considering only the impact of viscosity and membrane porosity properties for a porous NF membrane, flux can be simply determined by the Hagen–Poiseuille equation [22,23]:

$$J = \frac{\varepsilon r^2}{8\eta\tau}\left(\frac{\Delta P}{\Delta x} \right), \tag{14.5}$$

where ε is the surface porosity, r is the pore radii, η is the viscosity, and τ is the tortuosity. The equation shows that relatively lower membrane thickness (Δx), lower viscosity, and lower tortuosity favor high levels of flux. It should also be noted that pressure difference terms should be corrected for solutions with significant salinity (i.e., the osmotic pressure should be excluded from applied pressure).

Equation (14.6) represents the rejection of a solute (R_i) based on the concentrations of the feed $(C_{i,f})$ and permeate $(C_{i,p})$, respectively:

$$R_i = 1 - \frac{C_{i,p}}{C_{i,f}}, \tag{14.6}$$

The rejection is mainly determined by solute size, charge, and hydrophobicity. It basically shows the separation ability of the membrane. It is a general trend that rejection increases with pressure but decreases with increasing concentration. Nevertheless, molecular weight cut-off (MWCO) is another useful term to indicate this separation performance. MWCO is defined as the molecular weight for which 90% of the solute is rejected. For NF, MWCO ranges from 200 to 1,000 Da corresponding to the pore size of between 0.2 and 2.0 nm [23]. For high levels of rejection, MWCO should be carefully determined along with solute physicochemical properties. For instance, hydrophobic species having large MWCO have lower tendency for retention by NF membranes. Similarly, recovery is influenced by membrane and solute properties but when design parameters of the process are considered, these effects become less significant.

14.3 APPLICATION OF NF FOR WASTEWATER RECLAMATION AND REUSE

14.3.1 APPLICATION OF NF FOR PRODUCTION OF AGRICULTURAL WATER

Some concerns have been raised across the globe about issues like global warming, which is related to greenhouse gas emission, food security, as well as water scarcity [24]. Indeed, food and water scarcity problem are directly interwoven. Therefore, these two issues must be addressed via a sustainable way; otherwise, present and the next generation will have to battle with them for their survival. Over 70% of the world's freshwater resources, such as rivers, lakes, and aquifers, are depleted as a direct result of irrigation practices used in agricultural production [25]. In addition to the consequences of climate change, demographic change, and rapid urbanization, there is also an exponential increase in the amount of competition between agriculture and other water-consuming sectors regarding water allocation. In light of all these factors, agriculture needs to become more productive, more efficient with resources, and much more sustainable [25].

To meet high food production demands while maximizing agricultural output alongside minimizing the impacts on the available natural water resources, many countries began to use water produced from various saline or brackish water sources for agricultural irrigation [26]. Municipal wastewater (MWW) reclamation and reuse can also be a reliable source of the water for different sectors. To increase food growth, on the other hand, irrigation is becoming a very important option. Having said that, a promising, sustainable, as well as economic technology must be selected in order to tailor the reclaimed wastewater for agricultural irrigation purpose. However, implementation of membrane technology instead of classical wastewater reclamation and reuse comes down to money.

Membrane-based technologies are considered as a low energy-consumption and high-efficiency potential alternative in an industrial application to resolve many of the evolving, social, and environmental problems [27,28]. Membrane-based treatment processes have presented venues to resolve water resource scarcity [29,30]. NF membranes of molecular weight cut off (MWCO) typically range from 0.20 to 2.00 kDa are getting more popular among many membrane technologies due to the relatively low energy usage as well as a distinct and unique ability to separate a variety of salts and small organic molecules, including organic dyes, mostly through size exclusion and Donnan effects [31–33]. When compared to RO membranes, NF membranes can be operated under lower hydraulic pressure, which results in reduced energy consumption and hence operational cost [34].

Some potential solutions include developing low-cost, climate-independent fertigation water resources, which are associated with desalination technologies. Fertigation is a fertilizer application method in which fertilizer is incorporated into irrigation water via the drip system. During agricultural irrigation, this system distributes fertilizer solution evenly. Because nutrients are readily available, efficiency is increased. Water desalination and wastewater reclamation, therefore, play a vital role for efficient desalination technologies as well as for irrigated agriculture [35]. To meet water demands, many countries have begun to use the desalinated water especially for agricultural purposes. For example, Spain used 22% of desalination capacity with 1.4 million m^3/day for fertigation, whereas Kuwait has a desalination capacity greater than 1 million m^3/day and 13% of this water is allocated for fertigation [26]. Nonetheless, only 0.5% of desalinated water is presently used for fertigation. Italy and Bahrain put in place desalination capacities of 64,700 and 620,000 m^3/day, respectively, while using only a small percentage of desalinated water for agriculture (1.5% and 0.4%). Only 1.3% and 0.1% of desalinated water were used for agricultural purposes in the United States and Qatar, respectively.

Compared to RO membranes, NF membranes have higher water permeability and can be operated at lower pressures, thus reducing the specific energy consumption and treatment-related costs. NF membranes were tested by Jamil et al. [36], to investigate whether the currently used RO process in many countries can be replaced by the NF process (or used as the first process prior to RO) and achieve an equivalent quality of permeate water [36].

Depending on the water source NF membranes are preferred compared to RO membranes due to their: (i) high divalent ions rejection ; (ii) ability of knocking down total dissolved solids as well as their potential to cost-effective alternative to irrigation with brackish water; (iii) relatively high tolerance (in broad sense) for foulants, in contrast with RO membranes; (iv) ability of removal of certain impurities in contaminated water without severe fouling; and (v) lower operational energy requirement. Hence, NF membranes are potential alternatives for agricultural desalination from various water sources [37]. In spite of the fact that NF is rarely used to desalinate water with high salinity, it is effective in agriculture since it eliminates divalent and multivalent ions at higher percentage (necessary in agriculture at certain concentrations) and partially monovalent ions (undesirable for agriculture). NF membrane treatment of surface brackish water resulted in a 40% reduction in energy usage, a 34% reduction in groundwater desalination, as well as 18% improvement in biomass production in comparison to RO systems according to some literature [38].

Soil salinity alleviation through wastewater treatment with NF is a viable technology since it improves irrigation water quality and reduces soil salinity, lowers root zone salinity and lessens production failure risk, improves biological pest/disease control, and increases beneficial species (soil biodiversity) [39].

A reduction in the amount of potable water used for irrigation, as well as the over-abstraction of surface and groundwater and the consequent scarcity and stress on water resources, could be achieved through the use of NF membranes in the reclamation of MWW, industrial wastewater (IWW), or spent geothermal fluid for agricultural irrigation. However, before reclaimed water may be used properly for irrigation, various considerations must be made. Table 14.1 shows some of the water quality variables to consider when evaluating irrigation water quality, such as salinity, electrical conductivity (EC) value, pH, ion toxicity (such as excessive levels of boron, sodium, arsenic, etc.), and sodium dangers defined by a dimensionless sodium adsorption ratio (SAR) on soil infiltration.

The possibilities of utilizing commercially available membranes employed for wastewater reuse are discussed in this section. Table 14.2 shows the average water quality obtained during filtration of different wastewaters by different commercially available NF membranes compared with irrigation water standards. The water produced with NF membranes is appropriate for use in irrigation regarding EC, total dissolved solids (TDS), pH, Ca^{2+}, and Mg^{2+}. It can clearly be seen that, some of the NF product water failed to meet irrigation water standard with respect to Cl^-, Na^+, and boron. The major factor here is source of the water employed as feed to NF membranes and properties of a particular membrane. Hence, the problem of wastewater sources can only be fixed by carefully examining the rejection properties of a membrane and comparing with the available water properties to be treated. For example, NF270 membrane shows a poor rejection for monovalent ions like Na^+ and Cl^- compared to tight NF membranes like NF90 which sometimes could be an alternative to some of the

TABLE 14.1

WHO Irrigation Water Guidelines [40]

Parameter		Unit	Degree of Restriction on Use		
			None (Quality I)	Slight to Moderate (Quality II)	Severe (Quality III)
Salinity	EC	mS/cm	<0.7	0.7–3.0	>3.0
	TDS	mg/L	<450	450–2,000	>2,000
Infiltration (affects infiltration rate of water into soil. Evaluate using EC_w and SAR together)					
SAR	0–3	EC	>0.7	0.2–0.7	<0.2
	3–6	(mS/	>1.2	0.3–1.2	<0.3
	6–12	cm)	>1.9	0.5–1.9	<0.5
	12–20		>2.9	2.9–1.3	<1.3
	20–40		>5.0	2.9–5.0	<2.9
Na^+	Surface irrigation	mg/L	<69	69–207	>207
	Sprinkle irrigation		<69	>69	
Cl^-	Surface irrigation		<142	142–354	>354
	Sprinkle irrigation		<106.5	>106.5	
Boron			<0.7[a]	0.7–3.0[a]	>3.0[a]
pH		–		6.5–8.5[a]	
Other parameters					
SO_4^{2-}		mg/L		960	
Mg^{2+}				61	
Ca^{2+}				400	
Arsenic		µg/L		<10[a]	

[a] TR-MEU, 2010.

currently used RO membranes [41]. Hence, membrane selection will play a significant role depending on the salinity of the available wastewater [42]. Boron concentration in NF product water as reported by Falizi et al. [43] falls under Quality II of irrigation water qualities. Therefore, this water can also be applied as irrigation water in terms of boron. The NF product water in the study conducted by Jarma et al. [34,44] is not suitable for irrigation in terms of boron as can be seen in Table 14.2. Apart from the study conducted by Elazhar et al. [45], the product water obtained from all NF membranes cannot be used directly for irrigation in terms of SAR value. The sodium hazard on soil infiltration is assessed by the SAR value, a dimensionless variable calculated by comparing the Na^+ ion concentration in irrigation water to the sum of Ca^{2+} and Mg^{2+} ion concentrations in irrigation water. Irrigation waters are classified as Quality I, II, or III (Table 14.1). Quality I indicates that this water will have no negative impact on crops and on soil when employed for agricultural irrigation. Water in Quality II may have some negative impacts, but it can be used for irrigation with caution, whereas water in Quality III will cause major infiltration difficulties on the soil as well as negative plants growth. It is, therefore, very important to take SAR values into consideration while assessing irrigation water quality. Although WHO or Turkish Ministry of Environment and Urbanization (TR-MEU) does not specify F^- or SiO_2 restrictions in irrigation water standards, they do have, however, an impact on the irrigation equipment. When the concentration of silica in irrigation water is high, it can cause major scaling problems on piping materials as well as on the pumps as mentioned by Jarma et al. [34]. As a result, when considering irrigation water potentials and quality, one must also take irrigation water equipment into account.

TABLE 14.2

Characteristics of Different NF Membrane Product Water Compared with Irrigation Water Standards

Parameter	Unit	Jarma et al. [44]	Gündoğdu et al. [41]	Gündoğdu et al. [41]	Elazhar et al. [45]	Jamil et al. [36]	Czuba et al. [46]	Racar et al. [47]	Racar et al. [47]	Lew et al. [38]	Lebron et al. [48]	Jarma et al. [34]	Hacıfazlıoğlu et al. [49]	Falizi et al. [43]	Jarma et al. [42]	Azaïs et al. [50]
Salinity																
EC	mS/cm	0.86[a]	2.05[b]	2.65[b]	1.12[b]	0.24[a]	0.52[a]	0.07[a]	0.44[a]	2.58[b]	0.28[a]	0.06[a]	0.12[a]	2.27[b]	0.15[a]	0.93[b]
TDS	mg/L	346.0[a]	1,037[b]	1,373[b]	725[b]	NG	333.4[a]	NG	NG	NG	NG	26.2[a]	57.1[a]	1,300[b]	70.8[a]	NG
Permeability																
SAR	–	254.6[c]	76.53[c]	66.18[c]	9.65[b]	4.43[c]	NG	12.10[c]	9.92[c]	42.34[c]	NG	27.5[c]	5.16[c]	25.7[c]	24.3[c]	19.01[c]
Specific Ion Toxicity																
Na+	mg/L	180.0[a]	363[c]	434[c]	45[a]	34.3[a]	45.3[a]	9.41[a]	40.87[a]	276[a]	NG	20.9[a]	26.5[a]	533[c]	21.4[a]	84.8[a]
Ca2+	mg/L	0.5	41[a]	72[a]	35.2[a]	2.34[a]	NG	1.06[a]	30.51[a]	52	3.1	5.2[a]	47.9[a]	527[a]	10.3[a]	38.9[a]
Mg2+	mg/L	0.5	4[a]	14[a]	8.26[a]	1.32[a]	1.4[a]	0.15[a]	3.44[a]	33[a]	2.6[a]	<0.5[a]	4.73[a]	28.3[a]		0.88[a]
Cl-	mg/L	129.0[a]	677[c]	824[c]	404.6[c]	158.6[b]	NG	3.55[a]	NG	641[c]	39[a]	<0.5[a]	0.03[a]	2.42[a]		297.2[b]
Boron	mg/L	8.7[c]	NG	NG	NG	NG	NG	NG	NG	NG	NG	8.6[c]	NG	2.56[b]	8.2[c]	NG
pH	–	6.9[a]	7.23[a]	7.67[a]	7.3[a]	NG	7.3[a]	NG	7.82[a]	7.2[a]	NG	7.1[a]	6.61[a]	7.36[a]	6.5[a]	7.50[a]
Arsenic	µg/L	<23.0[c]	NG	NG	NG	<10[a]	NG	NG	NG	NG	NG	<10[a]	NG	NG	<10[a]	NG
SO4 2–	mg/L	2.6[a]	<40[a]	112.3[a]	20.81[a]	12.0[a]	NG	0.22[a]	0.46[a]	133[a]	3.4[a]	1.49[a]	1.58[a]	NG	2.2[a]	1.25[a]
F-	mg/L	2.4[ns]	NG	NG	NG	NG	NG	0.05	0.28[ns]	NG	NG	0.36[ns]	NG	<0.5[ns]	0.55[ns]	NG
SiO2	mg/L	93.8[ns]	5.93[ns]	6.03[ns]	NG	NG	NG	NG	NG	NG	NG	30[ns]	NG	NG	NG	NG

[a]Quality I, [b]Quality II, [c]Quality III, [ns]No standard. NG, Not given.

14.3.2 Application of NF for Production of Process Water

Indeed, water produced by NF technology properties (physical and chemical) can be suitably used in the various industrial operations. Many sectors have embraced the NF membrane technology due to its effectiveness, small footprint, greater productivity, scalable mechanical design, and environmentally friendly technology [51]. Yang et al. [52] authored a comprehensive study on the utilization of membrane technologies for water recycling and reuse, with an emphasis on municipal sewage plant effluent. Other papers have been published on the use of membrane technology in conjunction with some other processes for wastewater recycling and reuse in various industries, including oil and gas [53], pulp and paper [54], dairy and soy processing [55], and acid mine drainage [56]. However, throughout the bleaching process, a concentration of transition metals, most notably iron, copper, and manganese, which react with hydrogen peroxide rather than with the pulp, can be the possible drawback on the use of NF produced as a process water, especially bleaching process [57]. Textile, mining, food, tannery, oil and gas, pharmaceutical, and pulp and paper industries are among few examples of industries which use a lot of water and hence produce large volume of wastewater. The wastewater produced from aforementioned (Textile, mining, food, tannery, oil and gas, pharmaceutical, and pulp and paper)industries has indeed a negative impact on the ecosystem, receiving water bodies, soil, as well as human health [51]. NF technology will therefore play a crucial role in addressing the some of the environmental concerns. NF technology can also be used to recover some of the valuable elements contained in the produced wastewater.

Ahmad et al. [51], integrated NF technology with numerous processes such as microfiltration (MF)+membrane bioreactor (MBR)+NF, MBR+NF+UV, MBR+NF, MBR+NF+ advanced oxidation process (AOP), MBR+AOP+NF, and UF+NF to produce the process water for textile industry. Other processes were integrated with NF technology in order to reduce the fouling propensity in NF processes, e.g., NF+RO or NF+ forward osmosis+RO [58,59]. It was also said that water produced from such process can be utilized as process water in different steps of textile industry. Gündoğdu et al. [41] tried to reclaim MWW and IWW from the industrial organized industrial zone by employing different commercially available NF membranes. The MBR effluent discharged from the wastewater treatment plant was used as feed to NF membranes. Based on the results, they obtained that the quality of the NF90 product water suits well to cooling water and boiler water quality standards. It was also mentioned that the NF90 product water satisfies brown and white paper-grade water qualities as well as textile industry water standards.

The NF membrane can be also used as pre-treatment prior to RO processes when salinity of the feed water is high. In the study conducted by Parlar et al. [60], it was found that the NF product water alone is not suitable for industrial cooling processes, boiler water, or steam generation due to the high salinity. However, the product water became suitable for process water when NF was used as pre-treatment for RO processes.

14.3.3 Polishing of Wastewater Treated with NF Membranes

As demonstrated in Table 14.2, the water reclaimed using NF membrane is generally not suitable for agricultural irrigation. Indeed, Ca^{2+} and Mg^{2+} ions are highly removed with NF compared to Na^+ ions when NF membrane is employed. Nevertheless, ratio of Na^+ to Ca^{2+} and Mg^{2+} plays a vital role for both soil permeability as well as plant growth. As a result, it is critical that the water produced by NF membranes be adjusted to meet the quality requirements for irrigation water quality. Irrigation water produced from a pressure-driven membrane operation can be tailored according to the plant(s) intended to be irrigated to maximize agricultural productivity, which includes: (i) direct addition of CO_2 or lime for pH adjustment [61], (ii) direct addition of missing elements into the treated wastewater through the use of Mg^{2+} and Ca^{2+} ion-rich substance such as dolomite or calcite bed, (iii) employing the NF concentrate stream as a source of divalent ions after suitable treatment

[62], and (iv) blending the NF concentrate stream with another water source that is free of unwanted ions like well water as demonstrated by Jarma et al. [34] and Meric et al. [63].

One must carefully analyze the chemical properties of a water source that is intended to be used for remineralization. Also, it is good to keep in mind that the NF concentrate in some cases might contain unwanted elements such as arsenic, boron, SiO_2, and heavy metals (Ni, Rb, Cr, Hg, Mn, Ag, Fe, Cu, Pb, Cs, etc.) depending on the water sources. In the study conducted by Jarma et al. [34,44], remineralization was thought to be possible by blending the produced water from NF membranes with well water. According to the chemical analysis of the well water, boron levels in well water were extremely low, while arsenic levels were similarly substantially below the maximum permitted limit for drinking or irrigation water (10 µg/L). As a result, tailoring irrigation water from the NF membrane product water will require blending the permeate water of NF membranes with well water at a specific ratio.

In other studies, conducted by Gündoğdu et al. [41] and Falizi et al. [43], the MBR effluent was proposed to be source of the missing nutrient as well as infiltration problem of the product water of the NF membrane. However, improper blending of either NF concentrate or MBR effluent water with NF product water can lead to another problem like specific ion toxicity and high or low salinity. Hence, it of paramount importance to determine an optimum blending ratio that will solve both missing ion problem as well as infiltration problem. Furthermore, pH, arsenic and boron concentration of the product water must also be taken into consideration while determining an optimum blending ratio. Jarma et al. [44] found an optimum blending ratio of 50% NF product water with 50% well water while tailoring irrigation water produced from the spent geothermal fluid. In another study, 30% of well water was thought to be optimum blending ratio with 70% of NF product water to utilize the spent geothermal water for agricultural irrigation [34,42].

Gündoğdu et al. [41] also determined the optimum blending ratio of different commercially available NF membrane product water with MBR effluent. The optimum theoretical blending ratios of MBR effluent and NF membrane product water were found as 0.2:0.8, 0.6:0.4, and 0.3:0.7 for TR60, NF270 and NF90 membranes, respectively. By blending the product water of the NF membrane with appropriate mineral-rich water source (well water and MBR effluent in this case), infiltration problem can be solved as seen in Table 14.3. However, boron concentration was still high as reported in the study conducted by Jarma et al. [44] even after remineralization. Hence, tolerance limit of crop(s) that is intended to be irrigated is also very important. For example, tomato has boron tolerance up to 6 mg/L as reported in Regis et al. [64]. Other plants that tolerate low to moderate boron concentration include alfalfa, cauliflower, sugar beet, cabbage, oats, bluegrass, turnip, sorghum, bluegrass, cowpea, and barley.

14.3.4 Utilization of NF for Reclamation of Geothermal Water

Due to continued thermal gradient as well as yields, geothermal energy is often stable and reliable when compared to other renewable sources such as wind, hydropower, and solar power [65]. Geothermal resources vary from one location to another, depending on different factors such as geothermal well efficiency, rock chemistry, geothermal fluid temperature, and geofluid accessibility [66]. However, geothermal fluid contains some undesirable elements such as B, As, NH_3, and SiO_2 in their neutral state and some noble gases along with Ca^{2+}, Mg^{2+}, Li^+, HCO_3^-, SO_4^{2-}, CO_3^{2-}, F^-, I^-, K^+, Na^+, and Cl^- ions at high concentrations. There are also some heavy metals such as Rb, Cr, Hg, Ag, Mn, Fe, Cu, Pb, Ni, and Cs making spent geothermal fluid be inappropriate for domestic, industrial, or agricultural irrigation use [67]. It is, therefore, very important to find low energy and overall cost and efficient technology to get rid of the unwanted elements present in spent geothermal fluid before employing it as either process or irrigation water source.

The spent geothermal fluid of a geothermal heating facility in Izmir, Turkey, contains the undesirable substances such as arsenic and boron at high concentration above irrigation water standard [44]. The concentrations of arsenic and boron in the spent geothermal fluid were 150–180 µg/L

TABLE 14.3

Evaluation of the Polished Product Water of NF Membrane with Respect to Irrigation Water Standards

Parameter	Unit	Jarma et al. [44]	Gündoğdu et al. [41]	Gündoğdu et al. [41]	Gündoğdu et al. [41]	Jarma et al. [34]	Jarma et al. [34]
Salinity							
EC	mS/cm	1.31[b]	2.89[b]	2.82[b]	1.16[b]	0.570[a]	0.605[a]
TDS	mg/L	704[b]	1,477[b]	1,439[b]	594[b]	436[a]	480[a]
Permeability							
SAR	–	16.3[b]	11.8[b]	11.7[b]	5.10[b]	10.6[b]	9.9[b]
Specific Ion Toxicity							
Na^+	mg/L	148.3[b]	456[c]	439[c]	173[b]	45.1[a]	43.5[a]
Cl^-	mg/L	169.0[b]	778[c]	825[c]	273[c]	86.4[a]	90.0[a]
Boron	mg/L	5.6[c]	NG	NG	NG	6.0[c]	5.7[c]
pH	–	7.0[a]	NG	NG	NG	7.1[a]	6.5[a]
Arsenic	μg/L	<10[a]	NG	NG	NG	<10[a]	<10[a]
SO_4^{2-}	mg/L	91.3[a]	NG	NG	NG	56.5[a]	56.2[a]
F^-	mg/L	1.2[ns]	NG	NG	NG	0.3[ns]	0.4[ns]
SiO_2	mg/L	46.9[ns]	NG	NG	NG	21.0[ns]	22.4[ns]

[a]Quality I, [b]Quality II, [c]Quality III, [ns]No standard, NG, Not given.

and 10–12 mg/L, respectively. While the recommended maximum value for arsenic concentration in irrigation water is 10 μg/L, boron concentration should be 0.7–3 mg/L in irrigation water based on TR-MEU criteria (Table 14.1). Because of their flexibility and ability to treat both surface water and groundwater, NF technologies were also used for desalting geothermal fluid [34,42,44,68]. Geothermal fluid treatment using NF technology is proven to be lower in term of process cost compared to RO technology. The rejection mechanisms of pollutants from geothermal fluids involved hysterical hindrance (sieving) as well as electrostatic repulsions (charge effects). The removal efficiency of NF membrane is also favored by its negatively charged active surface area [69]. Therefore, negatively charge monovalent ions like Cl^- are also eliminated from the feed water due to charge effect. Jarma et al. [42] investigated the effect of different commercially available NF and RO membranes for the simultaneous removal of arsenic and boron from spent geothermal fluid. The arsenic concentration in the spent geothermal fluid was reduced to the WHO and TR-MEU irrigation water standard (<10 μg/L). The NF membrane therefore plays a vital role in the treatment of spent geothermal water for agricultural irrigation purpose. A preliminary study was done in a mini pilot-scale NF and integrated NF+RO system installed to test different commercially available membranes (Figure 14.3). The results obtained from the mini pilot system were promising as shown in Table 14.3. A big pilot system (with capacity up to 3 m³/h) was considered by implementing the optimized conditions obtained from mini pilot NF membrane system (with capacity up to 0.12 m3/h) studies to produce agricultural irrigation water [44]. Successful application of the NF product water (after blending with well water) for agricultural irrigation was carried out by Jarma et al. [44]. Nevertheless, further analysis is needed to have a clear picture on the effect of NF product water on plant growth and yield as well as possible accumulation of arsenic or boron residual in the plants. Gonzalez et al. [70] also investigated the effect of temperature on arsenic removal from the geothermal fluid using the NF membrane. Their results showed that NF270 membrane gave mono and divalent As^{5+} removals of 97% and 93%, respectively (at 50°C of operational temperature). Another

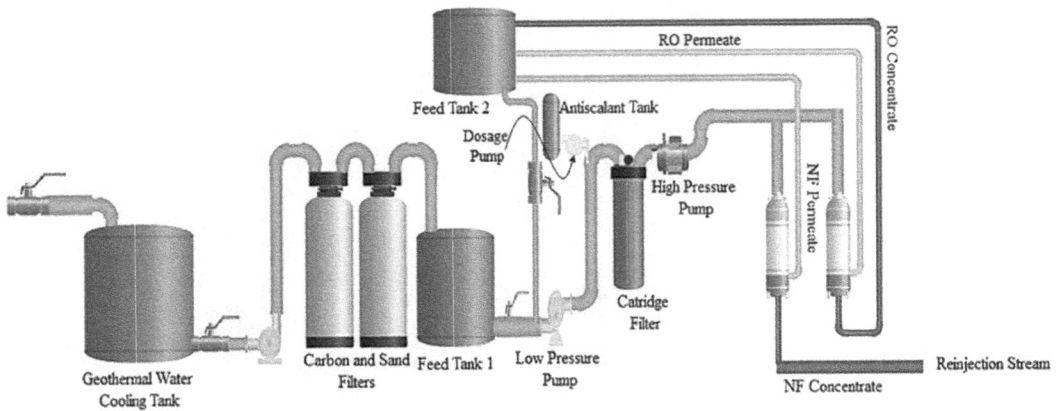

FIGURE 14.3 Flow diagram of the mini pilot-scale-integrated NF+RO test system employed for the treatment of spent geothermal fluid at geothermal heating center. (Reproduced with permission from [44]. Copyright of Elsevier)

NF membrane showed 86% and 94% of mono and divalent As^{5+} removals, respectively (at 50°C of operational temperature).

In Poland, the first tests focused on the treatment of geothermal water with mineralization of 2.4 g/L, temperature of 30°C, boron content of 9.56 mg/L, and arsenic with a concentration of 0.006 mg/L. The irrigation of crops was carried out using two different configurations: (i) pre-filtration, iron removal using a catalyst bed, UF, NF, and RO (Figure 14.4a) and (ii) pre-filtration, iron removal using a catalyst bed, UF, two steps of RO (RO-1 and RO-2 connected in series) (Figure 14.4b). In the second case, pH of water before RO-1 was lowered to about 5.5±0.5 by dosing small amounts of hydrochloric acid, which effectively prevented membrane scaling and pH of permeate obtained after RO-1 was corrected to about 10±0.5 and put to further filtration in RO-2 step, for effective boron removal [17]. The result of research showed a high quality of water can be obtained. However, in comparison with the recommendations for the assessment of irrigation water quality after using UF-NF-RO, very low EC of water was achieved, but also low concentrations of Ca, Mg, Fe, Mn, Zn, and HCO_3 and very high boron concentration (4.6 mg/L) were obtained. In case UF-RO-RO configuration, EC of water and concentration of Ca, Mg, Fe, Mn, Zn, and SO_4 ions were too low but reduction of boron to 0.2 mg/L was achieved. It is therefore necessary to conduct further research, because first tests showed that an increased content of boron in feed water, at a level exceeding 4.5 mg/L, requires the use of additional technical solutions, apart from membranes, that would reduce the content of this element in water.

An assessment of the quality of the irrigation water should be carried out regarding the type of irrigation system. Generally, the boron content in geothermal waters is higher than that in seawater and brackish water. If it is directly discharged, boron accumulates in the soil and changes the chemical, physical, and biological properties of soils. Furthermore, boron has a significant impact on groundwaters, surface waters, aquatic life, and vegetation [71]. That is why the selection of water desalination technology should to a large extent be considered the possibility of achieving the required boron concentration in the permeate. Proper configuration of the water treatment system allows for obtaining the appropriate permeate that meets the presented requirements. However, desalination processes produce large amounts of a concentrate, which may be increased in temperature, suspended solids, heavy metals, salts, and cleaning chemicals. That is why one of the most important aspects is also the proper utilization of the concentrate obtained, because it may have concentration of salts from 20% even up to more than 70% of feed water. The concentrate quality depends on recovery rate and specific process parameters [72–75] and may also contain chemicals like antiscalants, biocides, coagulants, antifoaming agents, or cleaning chemicals.

(a)

(b)

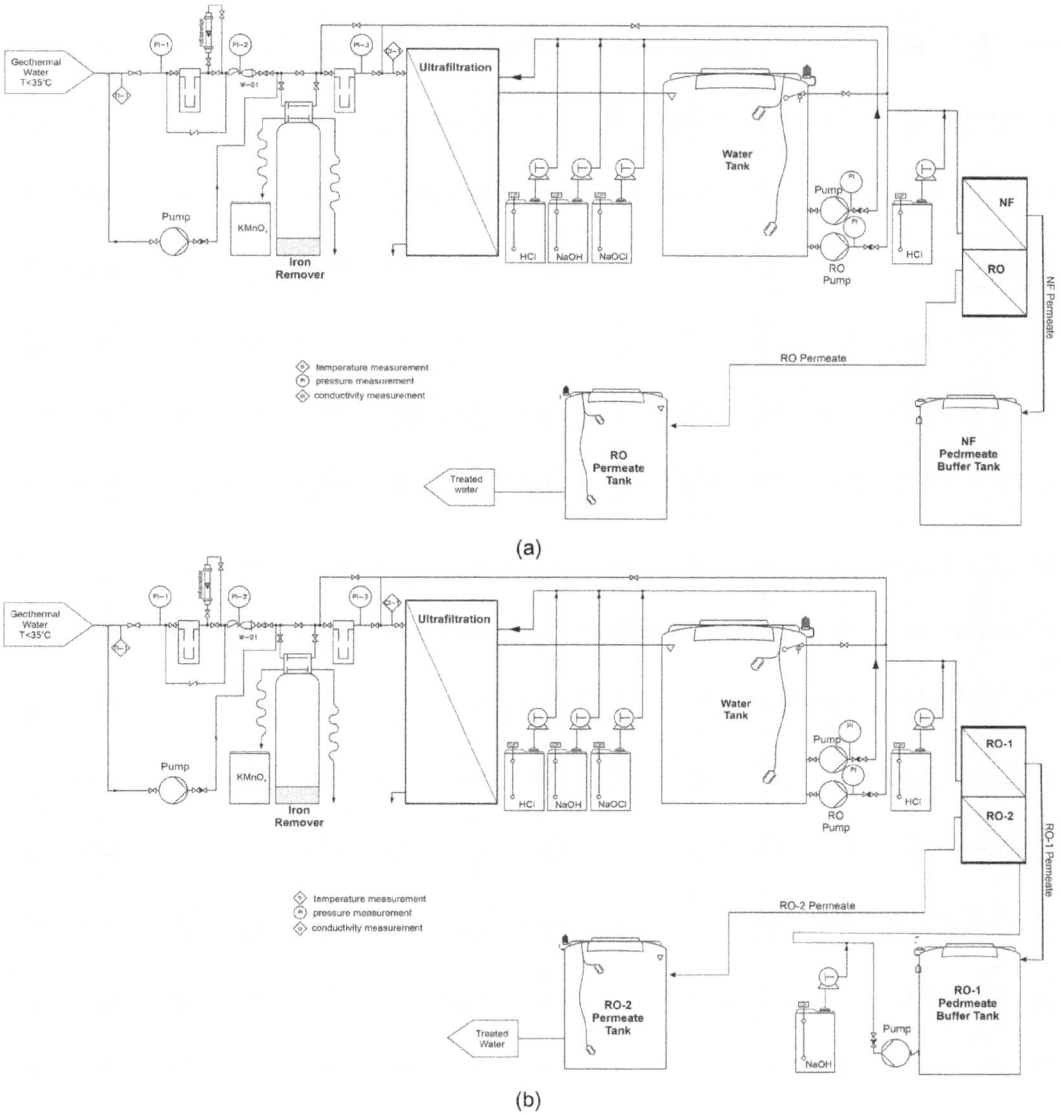

FIGURE 14.4   Scheme of the water desalination configurations: (a) pre-filtration, iron removal using a catalyst bed, UF, NF, and RO; (b) pre-filtration, iron removal using a catalyst bed, UF, two steps of RO (RO-1 and RO-2 connected in series).

14.4   FOULING IN NF MEMBRANES

The membrane is the main component in membrane-based separation systems, and its cost accounts for 20%–25% of the total cost [76]. Therefore, operating a membrane usefully for a long time is one of the most effective strategies to reduce operating costs. However, membrane fouling is an inevitable phenomenon, and it causes a significant flux reduction and an increased operating pressure in all pressure-driven membrane processes due to the accumulation of foulants on the membrane surface and inside pores. In detail, flux reduction is based on four main reasons: concentration polarization resistance, adsorption resistance, cake layer resistance, and membrane hydraulic resistance. Considering all these factors, the permeate flux can be expressed by Equation (14.7):

$$J = \frac{\Delta P}{\mu_p \left(R_m + R_p + R_a + R_c \right)},$$

(14.7)

where ΔP is the applied pressure (Pa), μ_p is the viscosity of the permeate (Pa s), and R_m, R_p, R_a, and R_c are the resistance of the membrane itself (m^{-1}), the concentration polarization (m^{-1}), the adsorption (m^{-1}), and the cake layer (m^{-1}), respectively [77]. The fouling that occurs in membrane structure categorizes as reversible and irreversible foulings. The concentration polarization and cake layer cause the formation of reversible fouling, while the adsorption is responsible for the formation of irreversible fouling. Reversible fouling can be easily eliminated after the rinsing steps. However, irreversible fouling cannot be removed by rinsing steps due to the strong interaction between the foulants and membrane materials. Therefore, chemical cleaning steps are needed to eliminate irreversible fouling.

Further, fouling reduces rejection efficiency and membrane lifespan and increases chemical cleaning frequency, and accordingly, the operating cost increases [78–80]. Furthermore, fouling blocks the surface properties such as pore size distribution, hydrophilicity, and charge that help to enhance the separation efficiencies [81]. In the literature, several studies have been carried out to understand fouling mechanisms in pressure-driven membranes and identified with mathematical models based on complete pore blocking, internal pore blocking, partial pore blocking, and cake filtration [82,83].

There are three main components creating the fouling layers on NF membrane surfaces and inside pores. These are organic, inorganic, and biological foulants (Figure 14.5).

The affecting parameters for the formation of fouling can be classified into three parts, membrane properties, characteristics of the feed solution, and operating conditions [84]. The membrane properties related to fouling propensity are charge, roughness, and hydrophobicity of the surface [85]. On the other hand, the physical and chemical properties of feed solution are the foulant stoke radius, charge, pH, diffusion coefficient, hydrophobicity, and molecular conformation [86–88]. These properties also directly influence the fouling behavior. Finally, the operating conditions, such as applied pressure, feed flow rate, and operating temperature, are other parameters that affect foulant accumulation on membrane surfaces and inside pores [89,90]. Among all these properties, charge, hydrophobicity of the membrane surface and foulant, and surface roughness of the membrane are the main driving forces for the membrane–foulant and foulant–foulant interactions. Boussu et al. [86] reported that negatively charged membrane surfaces showed resistance against the negatively charged molecules due to the electrostatic repulsion forces between them. On the other hand, the same negatively charged surface was prone to foul by positively charged molecules since the electrostatic interaction occurred between the oppositely charged molecules. Hobbs et al. [91] investigated the effect of NF membrane roughness on their flux behavior using high organic surficial groundwater. They concluded that fouling increased with increasing surface roughness. Vrijenhoek et al. [92] studied the effect of surface roughness on the colloidal fouling behavior of NF membranes. The atomic force microscopy (AFM) analysis results indicated that a rough surface

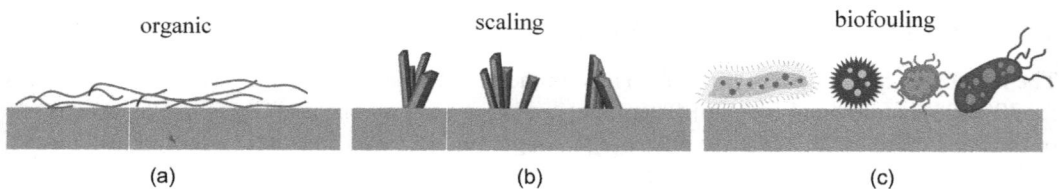

organic   scaling   biofouling

(a)   (b)   (c)

FIGURE 14.5 Classification of fouling on NF membranes.

was suitable for the deposition of colloidal particles due to its valley structure. They identified this phenomenon as valley clogging, causing a more severe flux decline than smooth membranes.

Besides membrane and foulant's physical and chemical properties, the operating conditions such as pressure, flow rate, and temperature affect the fouling rate. In this chapter, we investigated the effect of operating conditions on the inorganic fouling occurring in the membrane structure since inorganic foulants have the potential for physical damage to the membrane structure, reasoning selectivity losses. Common inorganic foulants frequently encountered in applications are silica, $CaSO_4$, $BaSO_4$, $CaCO_3$, $Ca_3(PO_4)_2$, $Fe(OH)_3$, and $Al(OH)_3$. It is known that inorganic foulants create a cake layer on the surface or block inside pores due to nucleation and crystal growth of inorganic particles [93]. For instance, Lin et al. [94] investigated the effect of operating pressure and flow rate on the $CaSO_4$ fouling rate in the NF membrane structure. The results showed that increasing operating pressure increased the feed water recovery. However, this caused the increasing the concentration polarization on the membrane surface, resulting in sharp flux reduction. In summary, increasing operating pressure and reducing flow velocity enhanced the formation of $CaSO_4$ nucleation and crystallization in the membrane. Similarly, Lee et al. [95] studied the effect of transmembrane pressure and flow velocity on the $CaSO_4$ scale formation mechanisms in NF membranes for water softening. The scanned electron microscope (SEM) images of the fouled membranes showed that two fouling mechanisms, surface and bulk crystallization, were carried out in the NF membrane. The surface crystallization, known as concentration polarization, increased with increasing operating pressure but reduced with the increasing flow velocity. On the other hand, bulk crystallization seemed to be the most significant at moderate crossflow velocity and higher operating pressure. Besides, feed temperature is one of the crucial parameters that affect the scaling formation on the NF membranes since the thermodynamic solubility of inorganic particles depends on the solution temperature [96]. Her et al. [97] investigated the effect of temperature effect on inorganic fouling in NF membranes. They observed that temperature strongly affected inorganic salt precipitation. At high temperatures, the solubility of $CaCO_3$ and $CaSO_4$ decreased, and homogeneous crystallization occurred due to high supersaturation on the surface, leading to moderate flux decline. At low temperatures, the solubility of salt ions increased, but the complex occurred between salt ions and natural organic matter to form heterogeneous crystallization. The formed heterogeneous crystallization caused a more critical flux decline than homogeneous crystallization. Andritsos et al. [98] investigated the affecting parameters on the scaling formation of $CaCO_3$. They reported that the crystal structure of $CaCO_3$ was strongly affected by temperature. At less than 25°C, the only form of $CaCO_3$ was observed, whereas at higher temperatures, the aragonite form, kinetically stabilized, became dominant. Besides the operating parameters, Lee et al. [99] investigated the effect of NF membrane modules with different configurations on scale formation mechanisms of $CaSO_4$. The total resistance (reversible and irreversible resistances) analysis results indicated that the spiral wound module has the lowest fouling tendency of other modules.

14.4.1 Fouling Control in NF Membranes

It is well known that membrane fouling decreases separation efficiency and increases the operating cost. Therefore, the prevention of fouling or cleaning it is an inevitable fact in membrane-based separation processes. Three different strategies are applied to reduce the fouling tendency of NF membranes. These are pre-treatment of the feed solution, optimizing the operating conditions, and modification of membrane surface. Even after using the mentioned strategies above, membrane fouling still occurs, and chemical or physical cleaning applies.

Feed water quality is one of the most significant factors, and the application of pre-treatment before feeding it to the NF membrane unit helps mitigate the fouling layer. Among the pre-treatment methods, coagulation, adjustment of feed pH, and membrane filtration (MF or UF) are the most widely used techniques for preventing fouling [100–106]. In the coagulation technique, the concentration and dosage of the coagulant should be well adjusted to precipitate the foulants. Otherwise, the coagulant itself can act as a foulant [100,102]. The membrane-based filtration, especially UF membranes, is widely used in pre-treatment since it does not require chemicals and has excellent efficiency [101]. Finally, adjusting the pH of the feed solution is the most critical strategy to prevent the formation of a complex fouling layer. For instance, the adjustment of feed solution to alkaline pH prevents the formation of organic fouling due to the electrostatic repulsion between organic foulants such as humic acid, sodium alginate, and bovine serum albumin and the membrane surface. The reason is the surface of the membrane, and organic foulants become negatively charged at high pH values. However, it should be noted that inorganic foulants, such as $CaCO_3$ and $Mg(OH)_2$, tend to become together to form aggregates that precipitate on the membrane surface at high pH values. Therefore, the pH of the feed solution should be carefully adjusted considering the quality of the raw water [103–106].

The optimization of operating conditions is another critical parameter that controls fouling. It is emphasized that high pressure and low flow velocity trigger the fouling due to the formation of concentration polarization on the membrane surface. Therefore, determination of the critical flux and the operation of the system under this critical flux are essential for fouling control. Lastly, the reflux ratio of the system should be optimized. A high reflux ratio improves the permeation flux but helps to increase the formation of concentration polarization and to scale on the membrane surface [107–109].

The coating of membrane surfaces with hydrophilic polymers or modification of them with nanoparticles is widely used for fouling mitigation. The main purpose is to increase the surface hydrophilicity and decrease the surface roughness of membranes. Improvement of surface properties prevents the interaction between foulant and surface and the accumulation of foulant on the membrane surface [110–115].

Although all these strategies for fouling mitigation can work, membranes need chemical or physical cleaning after a certain time of operation, depending on the applications. The physical cleaning methods often include hydraulic cleaning, gas–liquid pulse, scrubbing, backwashing, air sparging, and ultrasonic vibration [116,117]. This method is generally used for reversible fouling cleaning and is environmentally friendly. However, when starting irreversible fouling after a long-term operation, physical cleaning does not work, and the membrane needs a chemical cleaning. On the other hand, chemical cleaning uses acid, base, oxidants, and surfactant solutions. Each cleaning solutions work on a different foulant. For instance, an acidic solution is suitable for preventing precipitated salts, scales, and inorganic matter, and hydrochloric acid, citric acid, nitric acid, and phosphoric acid are widely used for cleaning [118]. An alkaline solution is appropriate for organic and inorganic foulants [119]. Generally, sodium hydroxide and sodium hypochlorite are used to prepare alkaline solutions. The most critical factor in the chemical cleaning application is to consider the tolerance of the membrane materials against these chemical cleaning solutions [120].

14.5 CONCLUSIONS

Recently, membrane technologies provide new challenges and opportunities in various processes from water treatment to resource recovery. Among the membrane technologies, NF is widely used for water and wastewater treatment in addition to other applications such as desalination and cheese whey demineralization. Its application plays an important role to partially replace RO by reducing the process costs. The NF process has the similar principle to the RO process. But the pressure applied in the NF systems is less than that applied in the RO systems. The univalent ions are passed through the NF membranes, while the polyvalent ions are retained. Therefore, NF membranes are used to remove divalent ions giving hardness and dissolved organic substances. NF can be also integrated with other membrane processes to be efficient in water and wastewater treatment. In the recent literature, there is a growing trend in NF research and this explains that NF technology is important and this interest should be directed to further applications for a robust and sustainable future.

In this chapter, a comprehensive overview of the use of NF technology for the recovery and reuse of wastewater and spent geothermal fluid is presented.

ACKNOWLEDGMENTS

The authors acknowledge editors of this book for the kind invitation to write this chapter. Some of the research work on geothermal water reclamation for irrigation using membrane technologies was supported by TUBITAK-NCBR international research project (Project No. 118Y490-POLTUR3/Geo4Food/4/2019). Izmir Geothermal Inc., located in Izmir, is highly appreciated for giving us permission to install and conduct our pilot-scale membrane treatment studies.

A PhD fellowship was provided to Dr. Y. A. Jarma and he would like to express his gratitude to the Presidency of the Turkish Abroad and Associated Communities (YTB). Dr. A. Cihanoğlu would also like to express his gratitude to TUBITAK for their financial support, which was provided through the National Postdoc initiative (Project No. 118C549).

REFERENCES

1. Caldera, U., Breyer, C., 2020. Strengthening the global water supply through a decarbonised global desalination sector and improved irrigation systems. *Energy* 200, 117507.
2. European Environmental Agency. https://www.eea.europa.eu/.
3. WRI Aqueduct, 2019. https://www.wri.org/aqueduct.
4. Bond, N.R., Burrows, R.M., Kennard, M.J., Bunn, S.E., 2019. Water scarcity as a driver of multiple stressor effects. *Multiple Stressors in River Ecosystems*, 111–129.
5. Mohamed, M.B., 2015. *Geothermal Energy Development: The Tunisian Experience, Proceedings World Geothermal Congress*. Melbourne, Australia, 19–25 April 2015. https://pangea.stanford.edu/ERE/db/WGC/papers/WGC/2015/28008.pdf.
6. Bundschuh, J., Chen, G., Tomaszewska, B., Ghaffour, N., Mushtaq, S., Hamawand, I., Reardon-Smith, K., Maraseni, T., Banhazi, T., Mahmoudi, H., Goosen, M., Antille, L.D., 2017. Solar, wind and geothermal energy applications in agriculture: Back to the future? In: *Geothermal, Wind and Solar Energy Applications in Agriculture and Aquaculture* (eds. J. Bundschuh, G. Chen, D. Chandrasekharam, J. Piechocki). London: CRC Press; Taylor & Francis Group, (Sustainable Energy Developments; ISSN 2164-0645). ISBN: 978-1-138-02970-5; e-ISBN: 978-1-315-15896-9, pp. 1–32.
7. Bundschuh, J., Tomaszewska, B., Ghaffour, G., Hamawand, I., Mahmoudi, H., Goosen, M., 2018. Coupling geothermal direct heat with agriculture. In: *Geothermal Water Management* (eds. J. Bundschuh, B. Tomaszewska). CRC Press. Taylor & Francis Group, (Sustainable Water Developments: Resources, Management, Treatment, Efficiency and Reuse); ISSN 2373-7506; vol. 6). ISBN: 978-1-138-02721-3; e-ISBN: 978-1-315-73497-2, pp. 277–300.
8. Tomaszewska, B., Dendys, M., 2018. Zero-waste initiatives – Waste geothermal water as a source of medicinal raw material and drinking water. *Desalination and Water Treatment* 112, 12–18.
9. Kabay, N., Yilmaz-Ipek, I., Soroko, I., Makowski, M., Kirmizisakal, O., Yag, S., Bryjak, M., Yuksel, M., 2009. Removal of boron from Balcova geothermal water by ion Exchange-microfiltration hybrid process. *Desalination* 241, 167–173.
10. Kabay, N., Köseoğlu, P., Yapıcı, D., Yüksel, Ü., Yüksel, M., 2013. Coupling ion exchange with ultrafiltration for boron removal from geothermal water-investigation of process parameters and recycle tests. *Desalination* 316, 17–22.
11. Kabay, N., Bryjak, M., Hilal, N., 2015. *Boron Separation Processes*. Amsterdam: Elsevier.
12. Kabay, N., Sözal, P.Y., Yavuz, E., Yuksel, M., Yuksel, Ü., 2018a. Treatment of geothermal waters for industrial and agriculture purposes. In: *Geothermal Water Management* (eds. J. Bundschuh, B. Tomaszewska). CRC Press. Taylor & Francis Group, (Sustainable Water Developments: Resources, Management, Treatment, Efficiency and Reuse; ISSN 2373-7506, vol. 6). ISBN: 978-1-138-02721-3; e-ISBN: 978-1-315-73497-2, pp. 113–134.
13. Kabay, N., Ipek, I.Y., Yilmaz, P.K., Samatya, S., Bryjak, M., Yosizuka, K., Tuncel, S.A., Yuksel, U, Yuksel, M., 2018b. Removal of boron and arsenic from geothermal water by ion-exchange. In: *Geothermal Water Management* (eds. J. Bundschuh, B. Tomaszewska). CRC Press. Taylor & Francis Group, (Sustainable Water Developments: Resources, Management, Treatment, Efficiency and Reuse; ISSN 2373–7506; vol. 6). ISBN: 978-1-138-02721-3; e-ISBN: 978-1-315-73497-2, pp. 135–156.
14. Kaya, C., Sert, G., Kabay, N., Arda, M., Yuksel, M., Egeman, O., 2015. Pre-treatment with nanofiltration (NF) in seawater desalination-Preliminary integrated membrane test in Urla, Turkey. *Desalination* 369, 10–17.

15. Kabay, N., Yilmaz, I., Yamac, S., Samatya, S., Yuksel, M., Yuksel, U., Arda, M., Saglam, M., Iwanaga, T., Hirowatari, K., 2004a. Removal and recovery of boron from geothermal wastewater by selective ion exchange resins. I. Laboratory tests, *Reactive and Functional Polymers* 60, 163–170.

16. Kabay, N., Yilmaz, I., Yamac, S., Yuksel, M., Yuksel, U., Yildirim, N., Aydogdu, O., Iwanaga, T., Hirowatari, K., 2004b. Removal and recovery of boron from geothermal wastewater by selective ion-exchange resins. II. Field tests. *Desalination* 167, 427–438.

17. Tomaszewska, B., Bodzek, M., 2013. The removal of radionuclides during desalination of geothermal waters containing boron using the BWRO system. *Desalination* 309, 284–290.

18. Tomaszewska, B., Rajca, M., Kmiecik, E., Bodzek, M., Bujakowski, W., Tyszer, M., Wątor, K., 2017. Process of geothermal water treatment by reverse osmosis. The research with antiscalants. *Desalination and Water Treatment* 73, 1–10.

19. Tomaszewska, B., Tyszer, M., Bodzek, M., Rajca, M., 2018. The concept of multivariant use of geothermal water concentrates. *Desalination and Water Treatment* 128, 179–186.

20. Tomaszewska, B., 2018. New approach to the utilisation of concentrates obtained during geothermal water desalination. *Desalination and Water Treatment* 128, 407–413.

21. Strathmann, H., Giorno, L., Drioli, E., 2011. *An Introduction to Membrane Science and Technology.* Weinheim, Germany: Wiley-WCH, p. 544.

22. Uragami, T., 2017. *Science and Technology of Separation Membranes-Nanofiltration*, First edition. John Wiley & Sons Ltd, pp. 297–323.

23. Warsinger, D.M., Chakraborty, S., Tow, E.W., Plumlee, M.H., Bellona, C., Loutatidou, S., Karimi, L., Mikelonis, A.M., Achilli, A., Ghassemi, A., Padhye, L.P., Snyder, S.A., Curcio, S., Vecitis, C.D., Arafat, H.A., Lienhard, J.H., 2018. A review of polymeric membranes and processes for potable water reuse. *Progress in Polymer Science* 81, 209–237.

24. Aghamohammadi N., Reginald S.S, Shamiri A., Zinatizadeh A.A., Wong L.P., Meriam N., Sulaiman B.N., 2016, An investigation of sustainable power generation from oil palm biomass: A case study in Sarawak. *Sustainability* 8, 416.

25. World Bank Group, 2021. Water in agriculture. https://www.worldbank.org/en/topic/water-in-agriculture.

26. Barron, R.A.O., Hodgson, G., Smith, D., Qureshi, E., McFarlane, D., Campos, E., Zarzo, D., 2015. Feasibility assessment of desalination application in Australian traditional agriculture. *Desalination* 364, 33–45.

27. Ridgway, H.F., Orbell, J., Gray, S., 2017. Molecular simulations of polyamide membrane materials used in desalination and water reuse applications: Recent developments and future prospects. *Journal of Membrane Science* 524, 436–448.

28. Lau, W.J., Gray, S., Matsuura, T., Emadzadeh, D., Chen, J.P., Ismail, A.F., 2015. A review on polyamide thin film nanocomposite (TFN) membranes: History, applications, challenges and approaches. *Water Research* 805, 37.

29. Wang, Z., Wu, A., Colombi Ciacchi, L., Wei, G., 2018. Recent advances in nanoporous membranes for water purification. *Nanomaterials* 8, 65.

30. Tang, C., Yang, Z., Guo, H., Wen, J., Nghiem, L.D., Cornelissen, E., 2018. Potable water reuse through advanced membrane technology. *Environmental Science Technology* 52, 10215–10223.

31. Yang, Z., Ma, X., Tang, C., 2018. Recent development of novel membranes for desalination. *Desalination* 434, 37–59.

32. Zhou, D., Zhu, L., Fu, Y., Zhu, M., Xue, L., 2015. Development of lower cost seawater desalination processes using nanofiltration technologies – A review. *Desalination* 376, 109–116.

33. Zhang, H., Sun, J.Y., Zhang, Z.L., Xu, Z.L., 2021a. Hybridly charged NF membranes with MOF incorporated for removing low-concentration surfactants. *Separation and Purification Technology* 258, 118069.

34. Jarma, Y.A., Karaoğlu, A., Tekin, O., Senan, I.A., Baba, A., Kabay, N., 2022a. The use of integrated pressure-driven membrane separation processes for the production of agricultural irrigation water from spent geothermal Water-Mini-pilot tests at geothermal heating center. *Desalination* 523, 115428.

35. Quist-Jensen, F.M.C.A., Drioli, E., 2015. Membrane technology for water production in agriculture: Desalination and wastewater reuse. *Desalination* 364, 17–32.

36. Jamil, S., Loganathan, P., Khan, S.J., Mc Donald, J.A., Kandasamy, J., Vigneswaran, S., 2021. Enhanced nanofiltration rejection of inorganic and organic compounds from a wastewater-reclamation plant's micro-filtered water using adsorption pre-treatment. *Separation and Purification Technology* 260, 118207.

37. Gonzalez, E., Ruigomez, I., Gomez, J., Marrero, M.C., Diaz, F., Delgado, S., Vera, L., 2018. Critical assessment of the nanofiltration for reusing brackish effluent from an anaerobic membrane bioreactor. *Environmental Progress Sustainable Energy* 37(1), 383–390.

38. Lew, B., Tarnapolski, O., Afgin, Y., Portal, Y., Ignat, T., Yudachev, V., Bick, A., 2020. Exploratory ranking analysis of brackish groundwater desalination for sustainable agricultural production: A case study of the Arava Valley in Israel. *Journal of Arid Environments* 174, 104078.
39. Panagea, I.S., Daliakopoulos, I.N., Tsanis, I.K., Schwilch, G., 2016. Evaluation of promising technologies for soil salinity amelioration in Timpaki (Crete): A participatory approach. *Solid Earth* 7, 177–190.
40. WHO, 2006. *Wastewater Use in Agriculture: Guidelines for the Safe Use of Wastewater, Excreta and Greywater, II.* Geneva: World Health Organization.
41. Gündoğdu, M., Jarma, Y.A., Kabay, N., Pek, T.O., Yüksel, M., 2019. Integration of MBR with NF/RO processes for industrial wastewater reclamation and water reuse-effect of membrane type on product water quality. *Journal of Water Process Engineering* 29, 100574.
42. Jarma, Y.A., Karaoglu, A., Tekin, O., Baba, A., Okten, H.E., Tomaszewska, B., Bostancı, K., Arda, M., Kabay, N., 2021. Assessment of different nanofiltration and reverse osmosis membranes for simultaneous removal of arsenic and boron from waste geothermal water. *Journal of Hazardous Materials* 405, 124129.
43. Falizi, N.J., Hacıfazlıoğlu, M.C., Parlar, I., Kabay, N., Pek, T.O., Yüksel, M., 2018. Evaluation of MBR treated industrial wastewater quality before and after desalination by NF and RO processes for agricultural reuse. *Journal of Water Process Engineering* 22, 103–108.
44. Jarma, Y.A., Karaoglu, A., Senan, I.R.A., Meriç, M.K., Kukul, Y.S., Özçakal, E., Barlas, N.T., Çakıcı, H., Baba, A., Kabay, N., 2022b. Utilization of membrane separation processes for reclamation and reuse of geothermal water in agricultural irrigation of tomato plants-pilot membrane tests and economic analysis. *Desalination* 528, 115608.
45. Elazhar, F., Elazhar, M., El-Ghzizel, S., Tahaikt, M., Zait, M., Dhiba, D., Elmidaoui, A., Taky, M., 2021a. Nanofiltration-reverse osmosis hybrid process for hardness removal in brackish water with higher recovery rate and minimization of brine discharges. *Process Safety and Environmental Protection* 153, 376–383.
46. Czuba, K., Bastrzyk, A., Rogowska, A., Janiak, K., Pacyna, K., Kossińska, N., Kita, M., Chrobot, P., Podstawczyk, D., 2021. Towards the circular economy-A pilot-scale membrane technology for the recovery of water and nutrients from secondary effluent. *Science of the Total Environment* 791, 148266.
47. Racar, M., Dolar, D., Karadakić, K., Čavarović, N., Glumac, N., Ašperger, D., Košutić, K., 2020. Challenges of municipal wastewater reclamation for irrigation by MBR and NF/RO: Physico-chemical and microbiological parameters, and emerging contaminants. *Science of the Total Environment* 722, 137959.
48. Lebron, Y.A.R., Moreira, V.R., Furtado, T.P.B., da Silva, S.C., Lange, L.C., Amaral, M.C.S., 2020. Vinasse treatment using hybrid tannin-based coagulation-microfiltration-nanofiltration processes: Potential energy recovery, technical and economic feasibility assessment. *Separation and Purification Technology* 248, 117152.
49. Hacıfazlıoğlu, M.C., Tomasini, H.R., Bertin, L., Pek, T.Ö., Kabay, N., 2019. Concentrate reduction in NF and RO desalination systems by membrane-in-series configurations-evaluation of product water for reuse in irrigation. *Desalination* 466, 89–96.
50. Azaïs, A., Mendret, J., Gassara, S., Petit, E., Deratani, A., Brosillon, S., 2014. Nanofiltration for wastewater reuse: Counteractive effects of fouling and matrice on the rejection of pharmaceutical active compounds. *Separation and Purification Technology* 133, 313–327.
51. Ahmad, N.N.R., Anga, W.L., Teowa, Y.H., Mohammad, A.W., Hilal, N., 2022. Nanofiltration membrane processes for water recycling, reuse and product recovery within various industries: A review. *Journal of Water Process Engineering* 45, 102478.
52. Yang, J., Monnot, M., Ercolei, L., Moulin, P., 2020. Membrane-based processes used in municipal wastewater treatment for water reuse: State-of-the-art and performance analysis. *Membranes* 10, 131.
53. Zolghadr, E., Firouzjaei, M.D., Amouzandeh, G., LeClair, P., Elliott, M., 2021. The role of membrane-based technologies in environmental treatment and reuse of produced water. *Frontiers in Environmental Science* 9, 71.
54. Mänttäri, M., Kallioinen, M., Nyström, M., 2015. *Membrane Technologies for Water Treatment and Reuse in the Pulp and Paper Industries. Advances in Membrane Technologies for Water Treatment.* Elsevier, pp. 581–603.
55. Wang, Y., Serventi, L., 2019. Sustainability of dairy and soy processing: A review on wastewater recycling. *Journal of Cleaner Production* 237, 117821.
56. Naidu, G., Ryu, S., Thiruvenkatachari, R., Choi, Y., Jeong, S., Vigneswaran, S., 2019. A critical review on remediation, reuse, and resource recovery from acid mine drainage. *Environmental Pollution* 247, 1110–1124.

57. Cristina, D. Caldeira, D., Silva, C.M., Colodette, J.L., Rodrigues, F.A., Da Mata, R.A., Menezes, K.S., Vieira, J.C., Zanuncio, A.J.V., 2021. A case study on the treatment and recycling of the effluent generated from a thermo-mechanical pulp mill in Brazil after the installation of a new bleaching process. *Science of the Total Environment* 763, 142996.

58. Altaee, A., Hilal, N., 2015. High recovery rate NF-FO-RO hybrid system for inland brackish water treatment. *Desalination* 363, 19–25.

59. Gutman, J., Fox, S., Gilron, J., 2012. Interactions between biofilms and NF/RO flux and their implications for control-A review of recent developments, *Journal of Membrane Science* 421, 1–7.

60. Parlar, I., Hacıfazlıoğlu, M., Kabay, N., Pek, T.Ö., Yüksel, M., 2019. Performance comparison of reverse osmosis (RO) with integrated nanofiltration (NF) and reverse osmosis process for desalination of MBR effluent. *Journal of Water Process Engineering* 29, 100640.

61. Karabelas, A.J., Koutsou, C.P., Kostoglou, M., Sioutopoulos, D.C., 2018. Analysis of specific energy consumption in reverse osmosis desalination processes. *Desalination* 431, 15–21.

62. Hasson, D., Fine, L., Sagiv, A., Semiat, R., Shemer, H., 2017. Modeling remineralization of desalinated water by micronized calcite dissolution. *Environmental Science and Technology* 51, 12481–12488.

63. Meric, M.K., Kukul, Y.S., Ozcakal, E., Barlas, N.T., Cakıcı, H., Jarma, Y.A., Kabay, N., Baba, A., 2021. Use of geothermal fluid for agricultural irrigation: Preliminary field tests prior to irrigation studies at Balçova – Narlıdere Geothermal Field (Turkey). *Turkish Journal of Earth Science* 30, 1186–1199.

64. Regis, A.O., Vanneste, J., Acker, S., Martínez, G., Ticona, T., García, V., Alejo, F.D., Zea, J., Krahenbuhl, R., Vanzin, G., Sharp, J.O., 2022. Pressure-driven membrane processes for boron and arsenic removal: pH and synergistic effects. *Desalination* 522, 115441.

65. Loutatidou, S., Arafat, H.A., 2015. Techno-economic analysis of MED and RO desalination powered by low-enthalpy geothermal energy. *Desalination* 365, 277–292.

66. Prajapati, M., Shah, M., Soni, B., 2021. A review of geothermal integrated desalination: A sustainable solution to overcome potential freshwater shortages. *Journal of Cleaner Production* 326, 129412.

67. Shah, M., Vaidya, D., Sircar, A., 2018. Using Monte Carlo simulation to estimate geothermal resource in Dholera geothermal field, Gujarat, India, multiscale and multidisciplinary modeling. *Experiments and Design* 1(2), 83–95.

68. Jarma, Y.A., Karaoglu, A., Senan, I.R.A., Baba, A., Kabay, N., 2022c. Brine minimization in desalination of the spent geothermal reinjection fluid by pressure-driven membrane separation processes. *Desalination* 535, 115840.

69. Moreira, V.R., Lebron, Y.A.R., Santos, L.V.S., Coutinho de Paula, E., Amaral, M.C.S., 2021. Arsenic contamination, effects and remediation techniques: A special look onto membrane separation processes. *Process Safety and Environmental Protection* 148, 604–623.

70. Gonzalez, B., Heijman, S.G.J., Rietveld, L.C., van Halem, D., 2019. As(V) rejection by NF membranes using high temperature sources for drinking water production. *Groundwater for Sustainable Development* 8, 198–204.

71. Gude. V.G., 2016. Geothermal source potential for water desalination-Current status and future perspective. *Renewable and Sustainable Energy Reviews* 57, 1038–1065.

72. Mezher, T., Fath, H., Abbas, Z., Khaled, A., 2011. Techno-economic assessment and environmental impacts of desalination technologies. *Desalination* 266, 263–273.

73. Tomaszewska, B., Pająk, L., Bodzek, M., 2014. Application of hybrid UF-RO process to geothermal water desalination, concentrate disposal and costs analysis. *Archives of Environmental Protection* 40(3), 137–151.

74. Tomaszewska, B., Tyszer, M., 2017. Assessment of the influence of temperature and pressure on the prediction of the precipitation of minerals during the desalination process. *Desalination* 424, 102–109.

75. Tyszer, M., Tomaszewska, B., Kabay, N., 2021. Desalination of geothermal wastewaters by membrane processes: Strategies for environmentally friendly use of retentate streams. *Desalination* 520, 115330.

76. Al-Amoudi, A., Lovitt, R.W., 2007. Fouling strategies and the cleaning system of NF membranes and factors affecting cleaning efficiency. *Journal of Membrane Science* 303, 4–28.

77. Zhang, R., Liu, Y., He, M., Su, Y., Zhao, X., Elimelech, M., Jiang, Z., 2016. Antifouling membranes for sustainable water purification: Strategies and mechanisms. *Chemical Society Reviews* 45, 5888.

78. Aguiar, A., Andrade, L., Grossi, L., Pires, W., Amaral, M., 2018. Acid mine drainage treatment by nanofiltration: A study of membrane fouling, chemical cleaning, and membrane ageing. *Separation and Purification Technology* 192, 185–195.

79. Tin, M.M.M., Anioke, G., Nakagoe, O., Tanabe, S., Kodamatani, H., Nghiem, L.D., Fujioka, T., 2017. Membrane fouling, chemical cleaning and separation performance assessment of a chlorine-resistant nanofiltration membrane for water recycling applications. *Separation and Purification Technology* 189, 170–175.

80. Ilyas, S., de Grooth, J., Nijmeijer, K., De Vos, W.M., 2015. Multifunctional polyelectrolyte multilayers as nanofiltration membranes and as sacrificial layers for easy membrane cleaning. *Journal of Colloid Interface Science* 446, 365–372.

81. Zhao, Y.Y., Wang, X.M., Yang, H.W., Xie, Y.F.F., 2018. Effects of organic fouling and cleaning on the retention of pharmaceutically active compounds by ceramic nanofiltration membranes. *Journal of Membrane Science* 563, 734–742.

82. Field, R., 2010. *Fundamentals of Fouling, Membranes for Water Treatment*, Volume 4. Wiley-VCH Verlag GmbH & Co. KGaA Weinheim, pp. 1–23.

83. Goosen, M.F.A., Sablani, S.S., Roque-Malherbe, R., 2009. Chapter 11: Membrane fouling: Recent strategies and methodologies for its minimization. In: *Handbook of Membrane Separations: Chemical, Pharmaceutical, and Biotechnological Applications*.

84. Li, Q., Elimelech, M., 2004. Organic fouling and chemical cleaning of polymeric Membranes: Measurements and mechanisms. *Environmental Science & Technology* 38, 4683–4693.

85. Zhu, X., Elimelech, M., 1997. Colloidal fouling of reverse osmosis membranes: Measurements and fouling mechanisms. *Environmental Science & Technology* 31, 3654–3662.

86. Boussu, K., Belpaire, A., Volodin, A., Van Haesendonck, C., Van der Meeren, P., Vandecasteele, C., Van der Bruggen, B., 2007. Influence of membrane and colloid characteristics on fouling of nanofiltration membranes. *Journal of Membrane Science* 289, 220–230.

87. Listiarini, K., Sun, D.D., Leckie, J.O., 2009. Organic fouling of nanofiltration membranes: Evaluating the effects of humic acid, calcium, alum coagulant and their combinations on the specific cake resistance. *Journal of Membrane Science* 332, 56–62.

88. Mahlangu, T.O., Thwala, J.M., Mamba, B.B., D'Haese, A., Verliefde, A.R.D., 2015. Factors governing combined fouling by organic and colloidal foulants in cross-flow nanofiltration. *Journal of Membrane Science* 491, 53–62.

89. Abdelkader, B.A., Antar, M.A., Khan, Z., 2018. Nanofiltration as a pretreatment step in seawater desalination: A review. *Arabian Journal for Science and Engineering* 43, 4413–4432.

90. Du, Y., Pramanik, B.K., Zhang, Y., Dumée, L., Jegatheesan, V., 2022. Recent advances in the theory and application of nanofiltration: A review. *Current Pollution Reports* 8, 51–80.

91. Hobbs, C., Hong, S., Taylor, J., 2006. Effect of surface roughness on fouling of RO and NF membranes during filtration of a high organic surficial groundwater. *Journal of Water Supply: Research and Technology* 55, 559–570.

92. Vrijenhoek, E.M., Hong, S., Elimelech, M., 2001. Influence of membrane surface properties on initial rate of colloidal fouling of reverse osmosis and nanofiltration membranes. *Journal of Membrane Science* 188, 115–128.

93. Schäfer, A., Andritsos, N., Karabelas, A.J., Hoek, E.M.V., Schneider, R., Nyström, M., 2004. *Fouling in Nanofiltration*. Edinburgh: Elsevier.

94. Lin, C.J., Shirazi, S., Rao, P., Agarwal, S., 2006. Effects of operational parameters on cake formation of CaSO₄ in nanofiltration. *Water Research* 40, 806–816.

95. Lee, S., Lee, C.H., 2000. Effect of operating conditions on CaSO₄ scale formation mechanism in nanofiltration for water softening. *Water Research* 34, 3854–3866.

96. Sheikholeslami, R., 2003. Mixed salts-scaling limits and propensity. *Desalination* 154, 117–127.

97. Her, N., Amy, G., Jarusutthirak, C., 2000. Seasonal variations of nanofiltration (NF) foulants: Identification and control. *Desalination* 132, 143–160.

98. Andritsos, N., Kontopoulou, M., Karabelas, A.J., Koutsoukos, P.G., 1996. Calcium carbonate deposit formation under isothermal conditions. *The Canadian Journal of Chemical Engineering* 74, 911–919.

99. Lee, S., Kim, J., Lee, C.H., 1999. Analysis of CaSO₄ scale formation mechanism in various nanofiltration modules. *Journal of Membrane Science* 163, 63–74.

100. Ang, W.L., Mohammad, A.W., Benamor, A., Hilal, N., Leo, C.P., 2016. Hybrid coagulation-NF membrane process for brackish water treatment: Effect of antiscalant on water characteristics and membrane fouling. *Desalination* 393, 144–150.

101. Fan, G., Li, Z., Yan, Z., Wei, Z., Xiao, Y., Chen, S., Shang, H. Lin, H. Chang, H., 2020. Operating parameters optimization of combined UF/NF dual-membrane process for brackish water treatment and its application performance in municipal drinking water treatment plant. *Journal of Water Process Engineering* 38, 101547.

102. Elazhar, F., Elazhar, M., El Filali, N., Belhamidi, S., Elmidaoui, A., Taky, M., 2021a. Potential of hybrid NF-RO system to enhance chloride removal and reduce membrane fouling during surface water desalination. *Separation and Purification Technology* 261, 118299.

103. Zhang, H., Tian, J., Hao, X., Liu, D., Cui, F., 2021b. Investigations on the fouling characteristic of humic acid and alginate sodium in capacitive deionization. *Journal of Water Reuse and Desalination* 11, 160–176.

104. Vicini, S., Mauri, M., Wichert, J., Castellano, M., 2017. Alginate gelling process: Use of bivalent ions rich microspheres. *Polymer Engineering Science* 57, 531–536.

105. Eichinger, S., Boch, R., Leis, A., Koraimann, G., Grengg, C., Domberger, G., Nachtnebel, M., Schwab, C., Dietzel, M., 2020. Scale deposits in tunnel drainage systems-A study on fabrics and formation mechanisms. *Science of the Total Environment* 718, 137140.

106. Andreeva, M.A., Gil, V.V., Pismenskaya, N.D., Dammak, L., Kononenko, N.A., Larchet, C., Grande, D., Nikonenko, V.V., 2018. Mitigation of membrane scaling in electrodialysis by electroconvection enhancement, pH adjustment and pulsed electric field application. *Journal of Membrane Science* 549, 129–140.

107. Mitko, K., Laskowska, E., Turek, M., Dydo, P., Piotrowski, K., 2020. Scaling risk assessment in nanofiltration of mine waters. *Membranes* 10, 288.

108. Lewis, W.J.T., Mattsson, T., Chew, Y.M.J., Bird, M.R., 2017. Investigation of cake fouling and pore blocking phenomena using fluid dynamic gauging and critical flux models. *Journal of Membrane Science* 533, 38–47.

109. Braghetta, B.A., DiGiano, F.A., Ball, W.P., 1998. NOM accumulation at NF membrane surface: Impact of chemistry and shear. *Journal of Environmental Engineering* 124, 1087–1098.

110. Bai, L., Wu, H., Ding, J., Ding, A., Zhang, X., Ren, N., Li, G., Liang, H., 2020. Cellulose nanocrystal-blended polyethersulfone membranes for enhanced removal of natural organic matter and alleviation of membrane fouling. *Chemical Engineering Journal* 382, 122919.

111. Chae, H., Lee, J., Lee, C., Kim, I., Park, P., 2015. Graphene oxide-embedded thin-film composite reverse osmosis membrane with high flux, anti-biofouling, and chlorine resistance. *Journal of Membrane Science* 483, 128–135.

112. Masheane, M.L., Nthunya, L.N., Malinga, S.P., Nxumalo, E.N., Mamba, B.B., Mhlanga, S.D., 2017. Synthesis of Fe-Ag/f-MWCNT/PES nanostructured-hybrid membranes for removal of Cr(VI) from water. *Separation and Purification Technology* 184, 79–87.

113. Low, Z.X., Ji, J., Blumenstock, D., Chew, Y.M., Wolverson, D., Mattia, D., 2018. Fouling resistant 2D boron nitride nanosheet-PES nanofiltration membranes. *Journal of Membrane Science* 563, 949–956.

114. Weinman, S.T., Husson, S.M., 2016. Influence of chemical coating combined with nanopatterning on alginate fouling during nanofiltration. *Journal of Membrane Science* 513, 146–154.

115. Wu, H., Tang, B., Wu, P., 2014. Development of novel SiO_2-GO nanohybrid/polysulfone membrane with enhanced performance. *Journal of Membrane Science* 451, 94–102.

116. Dana, A., Hadas, S., Ramon, G.Z., 2019. Potential application of osmotic backwashing to brackish water desalination membranes. *Desalination* 468, 114029.

117. Regula, C., Carretier, E., Wyart, Y., G'esan-Guiziou, G., Vincent, A., Boudot, D., Moulin, P., 2014. Chemical cleaning/disinfection and ageing of organic UF membranes: A review. *Water Research* 56, 325–365.

118. Al-Amoudi, A., 2016. Nanofiltration membrane cleaning characterization. *Desalination and Water Treatment* 57, 323–334.

119. Mo, Y., Chen, J., Xue, W., Huang, X., 2010. Chemical cleaning of nanofiltration membrane filtrating the effluent from a membrane bioreactor. *Separation and Purification Technology* 75, 407–414.

120. Van der Bruggen, B., Manttari, M., Nystrom, M., 2008. Drawbacks of applying nanofiltration and how to avoid them: A review. *Separation and Purification Technology* 63, 251–263.

15 Water Reclamation and Heavy Metals Recovery from Industrial Wastewater

Jing Yao Sum
UCSI University

Leow Hui Ting Lyly
Universiti Kebangsaan Malaysia

Li Sze Lai
UCSI University
UCSI-Cheras Low Carbon Innovation Hub Research Consortium

Pui Vun Chai
UCSI University

CONTENTS

DOI: 10.1201/9781003261827-15

15.1 INTRODUCTION

Rapid growth of industrial activities produces a tremendous amount of effluent-containing heavy metals. The effluent must be properly treated to reduce the amount of toxic compounds to below the permissible limit before discharge into water body. Conventional heavy removal techniques such as chemical precipitation, adsorption, coagulation, flotation and ion exchange might be insufficient to produce discharge effluent that fulfils the regulation need. Nanofiltration (NF) edges over other conventional techniques in terms of permeate quality and its versatility in handling the feed of varying conditions. It could handle a feed with low concentrations that are particularly difficult to be handled by any other conventional techniques [1]. Previously, membrane filtration served the purpose to reduce the volume of effluent and produce a clean permeate just good enough for direct discharge to the water body. An additional process is needed to reduce the secondary effluent produced during the filtration process (the concentrated retentate stream). In recent decades, the intention has been shifted to reclaim water and simultaneous recovery valuable minerals such as precious metals, acids and other useful organic compounds like phosphorous compounds for further reuse in the process.

NF is a pressure-driven membrane with a charged surface, and pore size ranging from 0.2 to 2 nm (equivalent to a molecular weight cut-off of 200–1,000 Da). It has been widely used to treat wastewater containing heavy metals. The membrane could retain the multivalent heavy metal ions at the retentate and allow water and monovalent ions and small neutral solutes to permeate across. Understanding the NF properties and feed characteristics under various operating variables is crucial to decide the suitable NF membrane and the subsequent design of NF system. Therefore, this chapter offers a guide to select a suitable NF membrane for treating a feed containing heavy metals, covering the membrane material, modules and configuration, followed by the design of the NF system (a standalone NF and integrated NF process). The challenges of the NF in treating wastewater containing heavy metals will be addressed too.

15.2 SOURCES OF WASTEWATER CONTAINING HEAVY METALS

Heavy metals have been described as metallic chemical elements and metalloids which are hazardous. Most heavy metals have a density greater than 4 ± 1 g/cm³ [2], while some toxic metalloids are lighter metals, such as arsenic (yellow form) and aluminium, around 2.0–2.7 g/cm³.

Heavy metals can be released into the environment in natural and anthropogenic ways. They can present in several forms, such as hydroxides, oxides, sulphides, sulphates, phosphates, silicates or organic-metallic compounds. The sources of the natural emission include volcanic eruptions, sea-salt sprays, forest fires, rock weathering, biogenic sources and wind-borne soil particles [3]. Meanwhile, the anthropogenic way refers to the emission route under the influence of human activities in nature. Table 15.1 tabulates the examples of the anthropogenic sources. Some activities might generate wastewater that contains heavy metals, such as metalliferous mining and smelting, industrial processes, atmospheric deposition, agriculture activities and waste disposal.

Table 15.1 summarizes the heavy metals commonly found in industrial effluent, including cadmium, zinc, lead, chromium, nickel, copper, vanadium, platinum, silver, titanium and arsenic [4,5]. Industrial activities such as electroplating, landfills, centralized waste treatment and manufacturing inorganic chemicals, iron, steel, glass, battery, rubber, etc., discharge a tremendous amount of

TABLE 15.1

List of Heavy Metals with Their Source of Origin and Allowable Level Based on Standard A [4] and ELG Database [5]

Elements	Source of Origin	Subcategories	Allowable Level (Standard A)	Daily Allowable Level (EPA)
Cadmium	Inorganic chemical manufacturing process	Cadmium pigments and salts production	0.01 mg/L	0.84 mg/L
	Electroplating	Common metals, precious metals, anodizing, coatings, chemical etching and milling, electroless plating, PCB		1.2 mg/L
Zinc	Electroplating	Common metals, precious metals, anodizing, coatings, chemical etching and milling, electroless plating, PCB	2.0 mg/L	4.2 mg/L
	Organic chemicals, plastics and synthetic fibres (OCPSF)	Rayon fibres, other fibres, thermoplastic resins, thermosetting resins, commodity organic chemicals, bulk organic chemicals, specialty organic chemicals		6,796 µg/L
	Inorganic chemicals manufacturing	Production of chrome pigments, ferric chloride, aluminium sulphate, sodium bisulphite, cadmium pigments and salts, zinc chloride, hydrofluoric acid, sodium sulphite		11.4 mg/L
	Iron and steel manufacturing	Sintering, ironmaking, steelmaking, vacuum degassing, continuous casting, acid pickling, cold forming, hot coating		0.327 kg/day
	Nonferrous metals manufacturing	Secondary aluminium smelting, primary electrolytic copper refining, secondary copper, primary lead, primary zinc, metallurgical acid plants, primary tungsten, primary columbium-tantalum, secondary silver, secondary lead, primary and secondary germanium and gallium, secondary indium, primary precious metals and mercury, secondary precious metals, secondary tantalum		10 mg/L
	Steam electric power generating	–		1.0 mg/L
	Rubber manufacturing	Latex foam		0.058 kg/kg of product
	Pulp, paper, and paperboard	Mechanical pulp		0.30 kg/kg of product
	Landfills	Hazardous waste landfill, non-hazardous waste landfill		0.535 mg/L
	Battery manufacturing	Cadmium, leclanche, zinc		1.47 mg/L
	Centralized waste treatment	Treatment and recovery for metals, oils organics, multiple waste streams		8.26 mg/L

(Continued)

TABLE 15.1 (*Continued*)
List of Heavy Metals with Their Source of Origin and Allowable Level Based on Standard A [4] and ELG Database [5]

Elements	Source of Origin	Subcategories	Allowable Level (Standard A)	Daily Allowable Level (EPA)
Lead	Electroplating	Common metals, precious metals, anodizing, coatings, chemical etching and milling, electroless plating, PCB	0.10 mg/L	0.6 mg/L
	OCPSF	Rayon fibres, other fibres, thermoplastic resins, thermosetting resins, commodity organic chemicals, bulk organic chemicals, specialty organic chemicals		690 µg/L
	Inorganic chemicals manufacturing	Production of chrome pigments, ferric chloride, aluminium sulphate, sodium bisulphite, cadmium pigments and salts, zinc chloride, hydrofluoric acid, sodium sulphite		3.4 mg/L
	Iron and steel manufacturing	Sintering, ironmaking, steelmaking, vacuum degassing, continuous casting, acid pickling, cold forming, hot coating		0.245 kg/day
	Nonferrous metals manufacturing	Secondary aluminium smelting, primary electrolytic copper refining, secondary copper, primary lead, primary zinc, metallurgical acid plants, primary tungsten, primary columbium-tantalum, secondary silver, secondary lead, primary and secondary germanium and gallium, secondary indium, primary precious metals and mercury, secondary precious metals, secondary tantalum		0.0006 kg/kg of product
	Glass manufacturing	Television picture tube envelope manufacturing, hand pressed and blown glass manufacturing		0.2 mg/L
	Rubber manufacturing	General moulded, extruded, and fabricated rubber plants		5.8 kg/kg
	Metal finishing	–		0.69 mg/L
	Pesticide chemicals	Organic pesticide chemicals manufacturing		690 µg/L
	Battery manufacturing	Lead, lithium, magnesium		22.1 mg/kg of cells produced
	Centralized waste treatment	Treatment and recovery for metals, oils organics, multiple waste streams		1.32 mg/L
Chromium	Electroplating	Common metals, precious metals, anodizing, coatings, chemical etching and milling, electroless plating, PCB	0.05 mg/L (hexavalent); 0.20 mg/L (trivalent)	7.0 mg/L
	Textile mills	Wool scouring, wool finishing, woven fabric finishing, knit fabric finishing, carpet finishing, stock and yarn finishing, nonwoven manufacturing, felted fabric processing		0.22 kg/kg

(Continued)

TABLE 15.1 (*Continued*)

List of Heavy Metals with Their Source of Origin and Allowable Level Based on Standard A [4] and ELG Database [5]

Elements	Source of Origin	Subcategories	Allowable Level (Standard A)	Daily Allowable Level (EPA)
	Inorganic chemicals manufacturing	Production of chrome pigments, ferric chloride, sodium bisulphite, sodium chlorate, potassium dichromate, sodium dichromate and sodium sulphate, sodium sulphite, titanium dioxide, aluminium fluoride		3 mg/L
	Petroleum refining	Topping, cracking, petrochemical, lube, integrated		1 mg/L
	Rubber manufacturing	Latex-dipped, latex-extruded, and latex-moulded rubber		0.0086 kg/kg
	Landfills	Hazardous waste landfill		1.1 mg/L
	Battery manufacturing	Lithium, zinc subcategory		3.85 mg/kg of cell
Nickel	Electroplating	Electroplating of common metals and precious metals, anodizing coatings, chemical etching and milling, electroless plating, PCB	0.20 mg/L	4.1 mg/L
	OCPSF	Rayon fibres, other fibres, thermoplastic resins, thermosetting resins, commodity organic chemicals, bulk organic chemicals, specialty organic chemicals		3,980 µg/L
	Inorganic chemicals manufacturing	Production of copper salts, ferric chloride, nickel salts, cobalt salts, chlorine and sodium or potassium hydroxide, hydrofluoric acid, sodium dichromate and sodium sulphate, titanium dioxide, aluminium fluoride		6.4 mg/L
	Metal finishing	Metal finishing		3.98 mg/L
	Battery manufacturing	Cadmium, zinc		1,916.2 mg/kg
Copper	Electroplating	Electroplating of common metals, electroplating of precious metals, anodizing coatings, chemical etching and milling, electroless plating, PCB	0.20 mg/L	4.5 mg/L
	Inorganic chemicals manufacturing	Production of copper salts, ferric chloride, nickel salts, cobalt salts, chlorine and sodium or potassium hydroxide		3.3 mg/L
	Steam electric power generating	Steam electric power generating		1.0 mg/L
	Metal finishing	Metal finishing		3.38 mg/L
	Centralized waste treatment	Metals treatment and recovery, oils treatment and recovery, organics treatment and recovery, multiple wastestreams		4.14 mg/L
	Battery manufacturing	Lead		0.026 mg/kg
Vanadium	Centralized waste treatment	Metals treatment and recovery, multiple wastestreams	N/A	0.218 mg/L

(*Continued*)

TABLE 15.1 (*Continued*)
List of Heavy Metals with Their Source of Origin and Allowable Level Based on Standard A [4] and ELG Database [5]

Elements	Source of Origin	Subcategories	Allowable Level (Standard A)	Daily Allowable Level (EPA)
Platinum	Nonferrous metals manufacturing	Secondary precious metals	N/A	0.064 mg/troy ounce
Silver	Electroplating	Electroplating of precious metals	0.1 mg/L	1.2 mg/L
	inorganic chemicals manufacturing	Silver nitrate production		1 mg/L
	Metal finishing	Metal finishing		0.43 mg/L
	Centralized waste treatment	Metals treatment and recovery, multiple wastestreams		0.12 mg/L
	Photographic	Photographic processing		0.14 kg/1,000 m^2
	Battery manufacturing	Cadmium, magnesium, zinc		21.6 mg/kg of cells
Titanium	Nonferrous metals manufacturing	Primary and secondary titanium	N/A	2.125 mg/kg of scrap milled
	Centralized waste treatment	Metals treatment and recovery, multiple wastestreams		0.0947 mg/L
	Waste combustors	Commercial hazardous waste combustor		60 µg/L
Arsenic	Inorganic chemicals manufacturing	Production of boric acid, zinc chloride	0.05 mg/L	3 mg/L
	Steam electric power generating	Steam electric power generating		18 µg/L
	Centralized waste treatment	Metals treatment and recovery, oils treatment and recovery, multiple waste streams		2.95 mg/L
	Waste combustors	Commercial hazardous waste combustor		84 µg/L
	Landfills	Hazardous waste landfill		1.1 mg/L
	Electrical and electronic components	Electronic crystals		2.09 mg/L

effluent-containing heavy metals into the environment. It is worth highlighting that the electroplating process involved common metals, precious metals, anodising, coatings, chemical etching and milling, electroless plating and printed circuit board (PCB) that could generate effluent-containing cadmium, zinc, lead, chromium, nickel, copper, and silver [6]. Besides that, the wastewater effluent generated in the electroplating industry contains toxic cyanide (CN), oils and greases, and organic solvents, which complicates the subsequent treatment process. Manufacturing processes of the chemicals and metals have also generated wastewater containing heavy metals, such as arsenic, titanium, silver, vanadium, copper, nickel, lead, zinc and cadmium. Compounds containing arsenic have been used to manufacture agrochemicals such as herbicides, insecticides, wood preservatives and dye stuffs [7]. For cadmium, the major industrial applications include the production of alloys, pigments and batteries [8]. Besides, chromium is widely used in industrial processes, including metal processing, tannery facilities, chromate production, stainless steel welding, ferrochrome and chrome pigment production [7]. Chromium could be presented in two different oxidation states, with the hexavalent form [Cr(VI)] known to be carcinogenic and more toxic than its trivalent form [Cr(III)]. Meanwhile, lead has been majorly used in producing lead-acid batteries, followed by ammunition, metal products and device to shield X-rays [9].

Heavy metal causes impacts on the environment and human health. Soil and water contamination by heavy metals might destroy the entire ecosystem. The non-degradable metal ion will reside in the food chains. The presence of heavy metals will affect the properties of soil, such as the soil pH, porosity, colour and natural chemistry. Heavy metal cannot be degraded or destroyed. It can only be accumulated in soils, water and sediments. As a result, it can be frequently absorbed by the plants and affect other organisms by entering food chains [10]. Besides, exposure to heavy metals affects human health. For instance, chromium (Cr) is carcinogenic when it is presented as CrO_4^{2-}. It can irritate skin and mucous membrane. Exposure to high concentrations of Cr and arsenic (As) can lead to systemic health impacts, involving kidney, liver, gastrointestinal tract and circulatory system and promote carcinogenesis. Arsenic can even cause the impairment of cellular respiration. On the other hand, ingestion of high levels of zinc (Zn) may cause anaemia and damage to the pancreas. Cadmium (Cd) poisoning leads to kidney damage and obstructive lung disease. Besides, palladium (Pb) impairs nerves and leads to blood and brain disorders [7].

Heavy metals are extremely toxic and harmful to living life, even in small amounts. Therefore, the industrial effluent-containing heavy metals should be properly treated before being discharged into the water bodies. Table 15.1 summarizes the maximum allowable levels for each heavy metal in the industrial effluent based on Malaysia Environmental Quality Act 1974 and effluent limitations guidelines and standards (ELG) database based on the United States Environmental Protection Agency. A lower limit of maximum allowable levels can be seen for Standard A as it applies to the discharges into any inland waters within catchment areas that require higher quality of water.

15.3 MECHANISM OF HEAVY METALS REMOVAL USING NF

Different from other classes of the pressure-driven membrane process, the separation in NF involves electrostatic interaction between the metal cations and membrane surface. Referring to Table 15.2, most heavy metals have an ionic radius and hydrated radius falls under the sub-nanometer. Despite having a smaller radius than the membrane's pore opening, the NF membrane would still retain most of the heavy metals. This is mainly attributed to the transport of solute along with membrane pores that involve diffusion, advection and electro-migration [11]. Hence, the separation of heavy metals involves a combination of size or Donnan exclusion effect, and occasionally adsorption mechanism. Each separation mechanism will be described in the following sections.

15.3.1 SIZE EXCLUSION/STERIC HINDRANCE

Size exclusion or steric hindrance is a mechanism based on the relative size of the solutes in feed to the membrane pores opening. The separation mechanism of heavy metals through size exclusion effect is visualized in Figure 15.1. In size exclusion, solutes larger than membrane pores will be

TABLE 15.2
Properties of Selected Heavy Metal Ions [12]

Elements	Metal Ions	Ionic Radius (Å)	Hydrated Ionic Radius (Å)
Cadmium	Cd^{2+}	0.95	4.26
Zinc	Zn^{2+}	0.74	4.30
Lead	Pb^{2+}	1.19	4.01
Chromium	Cr^{3+}	0.52	4.61
Nickel	Ni^{2+}	0.71	4.04
Copper	Cu^{2+}	0.73	4.19
Silver	Ag^+	1.75	3.41

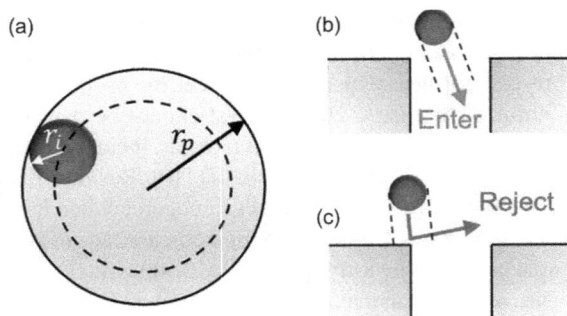

FIGURE 15.1 (a) Comparison between ion size (r_i) and pore size (r_p). (b) Transport of solute across the membrane when the direction of its trajectory falls within the pore region. (c) Hindrance of ion when the direction of its trajectory is not completely within the pore region [13].

hindered from passing across the membrane, while smaller solutes will be diffused across membrane pores. In treating an industrial effluent-containing heavy metal using NF, the metal cations are retained while the water permeates. Therefore, the precious metal could be retained while a clean permeate could be discharged or reused back in the manufacturing process.

Selection of the membrane according to its molecular weight cut-off (MWCO) and the sizes of the solutes in the feed determine the effectiveness of the size exclusion and hence the separation efficiency of the membrane. A comparison was made between two commercial NF membranes of different MWCO, NF90 and NF270 manufactured by Dow FilmTec in treating a feed containing heavy metals [14]. The looser NF270 demonstrated slightly lower rejection on Ni^{2+} (98.7%) and Cr^{6+} (95.7%), while the tighter NF90 has slightly higher rejection rates at 99.2% and 96.5% for Ni^{2+} and Cr^{6+}, respectively.

In separating a feed containing a mixture of heavy metals of similar valency, the selectivity will be determined by the steric hindrance [15,16]. A heavy metal cation forms a hydration shell in a solvation state. Metal cations with smaller ionic radius tend to hold the water molecules tightly to their hydration shell, forming larger hydrated radii and thus easier to be size excluded [17]. Moradi et al. [15] reported that the tetrathioterephthalate filler incorporated polyethersulfone (PES) NF membrane showed a high rejection rate of 99.2% for Zn^{2+}, 98.5% for Cu^{2+} and 97.4% for Pb^{2+}. The trend of rejection ($Zn^{2+} > Cu^{2+} > Pb^{2+}$) was in accordance with the hydrated radius of metal cations [Zn^{2+} (4.30 Å) > Cu^{2+} (4.19 Å) > Pb^{2+} (4.01 Å)]. Besides, Zhang et al. [16] reported the removal of heavy metals using phytic-acid–incorporated polyamide thin-film composite (TFC) NF membrane. The authors found that rejection rate ($Cu^{2+} \approx Cd^{2+} > Pb^{2+} > Ni^{2+} > Mg^{2+} > Ca^{2+}$) was consistent with the bivalent metal ions hydrate radii which is in descending order of Mg^{2+} (4.28 Å) > Cd^{2+} (4.26 Å) > Cu^{2+} (4.19 Å) > Ca^{2+} (4.12 Å) > Ni^{2+} (4.04 Å) > Pb^{2+} (4.01 Å).

It has been reported that the addition of a chelating agent enhances the rejection of metal ions attributed to the larger size of the chelate as compared to the free metal ions [18]. Few commonly used chelating agents include ethylenediaminetetraacetic acid (EDTA), diethylenetriaminepenta-acetic acid and biodegradable ligands such as [S, S]-ethylenediaminedisuccinic acid (EDDS) and citrate. The addition of Na_2EDTA as chelating agent increases the size of the metal cations, resulting in high recovery capacities of 99% of Cu^{2+} and Pd^{2+} [19]. This was mainly attributed to the binding or inactivation of the mobile cation that enhances the interaction between metal salts and the polyamide functional groups of the membrane. When the pH increased, metal ion rejection will also be increased due to the formation of a larger complex.

On the other hand, the membrane's pores can be affected by several factors, such as the pH of the solution, temperature and the presence of inorganic salts [20]. High pH and concentration of

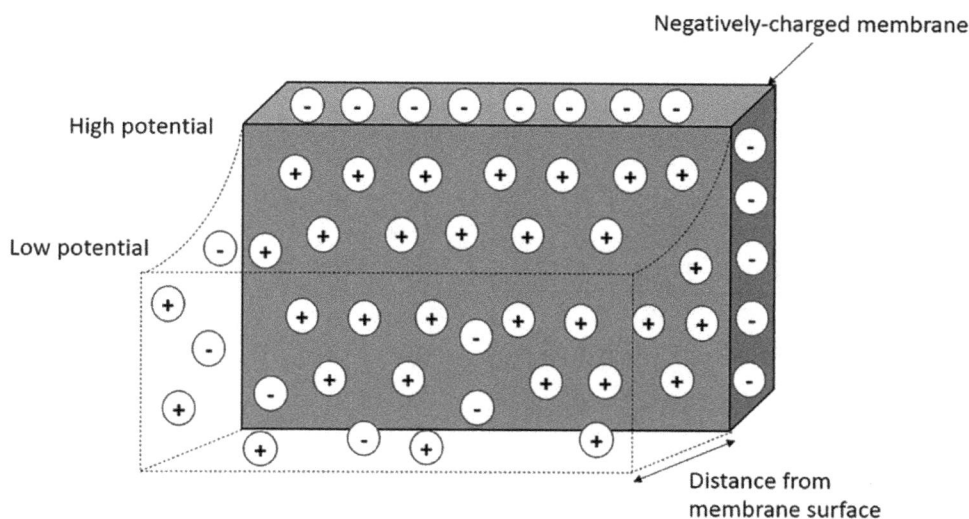

FIGURE 15.2 Schematic diagram of potential distribution as a function of distance on the surface of membrane containing fixed negative charge groups.

salt ions cause higher surface charges in the membrane pores; thus, the number of counter ions also increases. The higher charge density induces strong electrostatic repulsion within the membrane pores. As a result, the membrane will be swelled, resulting in increasing pore size and the effective thickness of the membrane [21]. Apart from that, when the temperature increases from 5°C to 41°C, expansion of membrane pores up to 21% can be observed [22]. High temperature also contributes to the reduction of the kinematic viscosity and increment of solute diffusivity. Eventually, the efficiency of heavy metal removal will be declined.

15.3.2 DONNAN EXCLUSION

Apart from the size exclusion effect, Donnan exclusion also plays an important role in heavy metal ions removal. NF membrane made from polyamide is usually attributed with functional groups such as carboxylic acid and amine. Ionization of these functional groups enriches the charges on the membrane surface. Hence, the ions in the feed could be repelled by the charges on the membrane surface although their hydrated radius is smaller than the membrane pores. Figure 15.2 illustrates the concept of Donnan exclusion based on the ionic distribution. The surface of the membrane usually contains a high concentration of counter ions due to the attraction between counter ions and fixed charges on the membrane surface. As a result, the concentration of the counter ions near the membrane surface is decreased. A potential difference is thus created at the membrane-solution interface and this is called Donnan potential. Donnan potential developed on the membrane surface tends to repel the heavy metal ions in the feed. Figure 15.3 shows the concentration profile along the pore with a counter ion transported across the NF membrane active layer. Different spatial distributions of the concentration along the pore can be seen.

The changes in pH of the feed may alter the membrane charge properties and ionic speciation of the solutes [20]. The surface of the membrane will be positively charged when the pH values are below the isoelectric point (IEP). The surface charge turns negative at pH values higher than IEP. Al-Rashdi [23] reported that pH can affect the metal ion rejection based on Donnan exclusion and adsorption mechanism that led to electrostatic repulsion or deposition of metal ions onto the membrane. It was reported that the commercial NF270 membrane has an IEP in the pH range of 3.3–4.

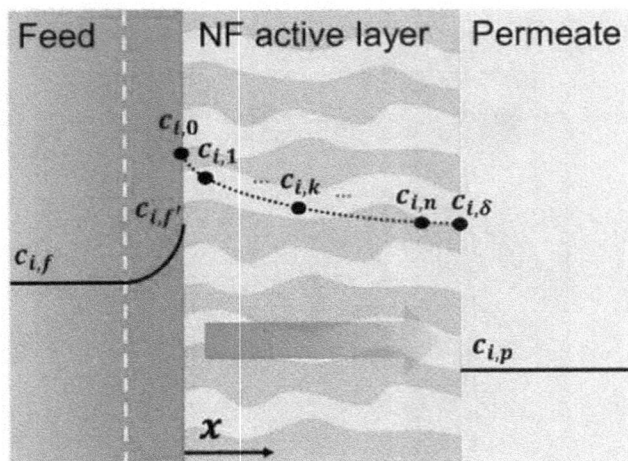

FIGURE 15.3 Concept illustration of ion transported across NF membrane active layer. $c_{i,f}$ is the concentration of ion I in the bulk feed solution, while $c_{i,f'}$ is the concentration at the membrane surface considering CP [13].

Hence, the surface of the membrane is positively charged when the solution pH is less than 4 and negatively charged when the solution pH is higher than 4. At pH around 1.50, the positive charge membrane surface enhances the repulsion towards cations. As a result, when the feed pH was lower than IEP, rejection was higher.

It was also reported that an increase in pH from 3 to 8 might cause a higher negative charge of the membrane, which was attributed to the increase in the –OH group that attracts heavy metal ions on the surface of the TFC membrane [24]. A negatively charged membrane surface will repel the multivalent anion and thus, results in a high rejection of the salt with a high valent anion. Poly(m-phenylene isophthalamide) membrane demonstrated electrolytes rejection in a sequence of $Na_2SO_4 > MgSO_4 > NaCl > MgCl_2$ based on their electrostatic action. Besides, in an acidic environment with pH 3.0, the membrane showed lower rejection towards monovalent anion $HCrO_4^-$. At higher pH, rejection up to 98% was achieved with the conversion of chromium species to CrO_4^- [25].

Donnan exclusion works well under a few conditions. As mentioned previously, Donnan exclusion is negligible at IEP since the membrane is uncharged. Secondly, the composition of the solution and operating condition will determine the effectiveness of Donnan exclusion. It was also reported that the Donnan exclusion effect plays a dominant role at lower heavy metal ionic concentrations (<100 ppm), whereas, at higher ionic concentrations, the solvation energy barrier effect plays a major role [26]. To have an effective Donnan exclusion effect, the concentration in the feed will need to be low, while the concentration of the fixed charge is high [27].

Most commercial membranes are negatively charged. The IEP of NF270 is around pH 3, indicating the membranes are negatively charged in a neutral feed. The negative charge is due to the surplus of anionic carboxyl (–COO⁻) over the cationic (–NH₃⁺). Therefore, membrane surface modification using amine-rich additives is needed to convert it into a positively charged surface that targets to repel heavy metal cations. Polyethylenimine (PEI) has been grafted on the membrane surface to incorporate a free carboxylic acid group to a positively charged NF membrane in the presence of 2-chloro-1-methyliodopyridine [28]. A greater amount of meshed or aggregated pores were produced on the membrane surface. It was reported that as the concentration of PEI increases, the membrane repulsion of inorganic salts also increases significantly. However, upon reaching 1.0 g/L, the water permeability gradually decreases due to the Donnan effect.

TABLE 15.3
Element with Different Properties [31]

Acid/Base	Properties	Element
Acid	Hard	Na, Ca, Mg, Ti, V, Cr, Al, Si, As, Sn, Sb, Ge, Ga
	Intermediate	Mn, Fe, Co, Ni, Cu, Zn
	Soft	Ru, Pd, Rh, Pt, Ag, Au, Cd, Hg, TI
Base	Hard	F, O, N
	Intermediate	Cl
	Soft	C, P, As, S, Sb, Se, Te, Br, I

15.3.3 ADSORPTION

The heavy metal ion is selectively adsorbed on the membrane surface, which mainly depended on the surface complexation, electrostatic interaction or ion-exchange mechanisms. Adsorption of heavy metal ions on the membrane surface promotes the diffusion of ions across the membrane due to a higher concentration gradient. By adjusting the initial metal concentration, initial temperature, pH of the solution, flow rate, sorbent mass and contact time, the adsorption performance of the membrane can be improved [29]. For instance, alteration of pH might change the surface charge or zeta potential of the sorbent, thus affecting the adsorption capacity. A lower temperature reduces the rate of molecular interaction and solubility, which might increase the chemical potential across the membrane with lower dissolution of adsorbate. As a result, the adsorption might be enhanced. However, the adsorption of the ions will also create a concentration polarization (CP) layer near the membrane surface, which hinders the membrane's separation performance [20]. Active functional groups such as carboxyl ($-COOH$), amine ($-NH_2$), or sulphonate ($-SO_3H$) usually act as the active sites for adsorption. For example, the adsorption of Cr(VI) can be enhanced with the protonation of $-COOH$, $-NH_2$ or $-SO_3H$ on the surface of adsorbents and causing the adsorption of $HCrO_4^-$ by the electrostatic attraction. At lower pH, more adsorption takes place with a larger amount of H^+ ions on the surface of the adsorbent. Thus, electrostatic attraction between positively charged adsorbent surface and chromate ion will be stronger.

Based on Hard-Soft Acid-Base Theory, hard acid tends to react with hard bases [30]. Soft acids and bases were those carried with high polarizability due to a small energy gap. Hence, soft acids and bases were largely covalently bound and hard acids and bases were ionically bound. Table 15.3 shows examples of acids and bases with different properties. Based on the theory, it is expected that hard metal ions, like Ca^{2+} and Mg^{2+}, will be preferentially bonded on the surface of the membrane, such as phytic-acid–incorporated polyamide-TFC membrane, with a higher rejection rate as compared to intermediate ions like Ni^{2+} or Cu^{2+}. However, the removal efficiency will also be affected by the hydrated radii of the metal ion [16].

Few works demonstrated the usefulness of the adsorptive removal of heavy metals. Abdi [32] reported that the functional groups presented on graphene oxide (GO) nanosheets enhanced the electrostatic interaction between copper ions and lone pairs electron of oxygen and amine groups. The characteristics of magnetic-graphene–based composite hybrid enhanced the hydrophilicity of membrane surface and, thus, increased the active sites that are available for copper ions adsorption via a better dispersion. The separation efficiency of the membrane can be enhanced via adsorption by incorporating metal complexes or adsorptive polymers. However, several limitations will be encountered, such as poor selectivity, increased membrane fouling and unrecyclable capability. Halloysite nanotube (HNT)-blended polyvinylidene fluoride (PVDF) membrane has been used to remove heavy metal ions via adsorption [33]. After 3-aminopropyltriethoxysilane (APTES) modification, some amino groups were attached on the HNT and created chelation reactions with cationic metal with more effective adsorption sites.

15.4 SELECTION OF NF MEMBRANE

NF has been employed in various industrial applications and water and wastewater treatment for the removal of organic substances and ions. NF is a promising membrane technology for recovery of valuable products from industrial waste and water reclamation due to its high selectivity on multivalent ions and low selectivity on monovalent ions. In this section, NF membrane configurations are reviewed followed by the discussion of membrane materials that are employed in NF.

15.4.1 MEMBRANE CONFIGURATION

Membrane configuration indicates the membrane's shape and membrane's arrangement in the module with the fluid feed flow and permeate flow [34]. There are four types of membrane modules for the pressure-driven separation process which are flat sheet module, spiral wound module, tubular module and hollow fibre module [35]. The most key properties of a module are uniformity in velocity distribution, high packing density, low resistance during tangential flow, high degree of turbulence on retentate flow to promote mass transfer and mitigate fouling, low operating cost per unit membrane area and convenience for cleaning and maintenance [36]. The working principles and properties of the membrane configurations are tabulated in Table 15.4.

Plate-and-frame membrane modules were the earliest version of modules which consists of a flat membrane, membrane support or spacer and feed distribution plate [37]. The membrane can be in circular or square shape and arranged horizontally or vertically. Two flat membranes are in a sandwich-like arrangement and in between the two membranes, a spacer is placed for permeate removal [38]. The repeating membrane sets units in a module resulting in plate-and-frame stacks. The quantity of repeating membrane set in a module is based on the membrane area needed for the separation process. Regardless of the quantity of repeating units to form plate-and-frame stacks, the whole module is equipped with two endplates and sealed with rings [39]. To establish a uniform distribution within the module, baffles have been introduced in the system to avoid the flow by-passing some of the membrane area [40]. Plate-and-frame module has its disadvantage in UF process study due to the design of the module having a low surface area to volume ratio (low packing density) with a range of $100-400 \, m^2/m^3$ and it cannot withstand high pressure.

Spiral wound membrane module was developed for RO application initially and it was now applied in gas permeation, NF and UF application [41,42]. A spiral wound membrane is almost similar to plate-and-frame membrane modules where it is a plate-and-frame module wrapped around a collection pipe located at the centre with active sides of the membrane facing outwards [43]. Two membranes are glued along three edges with the fourth edge open towards the collection pipe to form a membrane envelope [44]. Briefly, feed flows from the feed spacer in an axial direction while permeate is first collected in the membrane envelope and spirals into the collection pipe [45]. Compared with the plate-and-frame module, this spiral wound has a higher packing density with a range of $300-1,000 \, m^2/m^3$ [46]. The packing density in spiral wound design is affected by the channel height which is determined by feed spacer material and thickness of permeate stream [47]. Generally, net-type spacers are used in this module. The purpose of the net-type spacer is to enhance mass transfer of spiral membrane module with minimum pressure drop [48]. Additionally, the presence of feed spacer can partially block the suspended particles during operation. Therefore, factors such as hydrodynamic angle, void, filament diameter, mesh size and flow position need to be considered to achieve good module performance [49]. The disadvantage for the spiral module is its difficulty in cleaning when membrane is fouled or scaled which could affect the membrane performance [50]. Chemical cleaning are necessary for the fouled NF membrane; however, chemical cleaning would definitely affect the membrane lifespan [51].

Tubular membrane module is similar to a shell and tube heat exchanger. It consists of a number of porous membrane tubes where the inside of each membrane tube was cast with membrane. All the casted membrane tubes are joined to the end plated and arranged in parallel bundles in a shell [52].

TABLE 15.4

Summary of Membrane Configurations, Their Working Principle and Properties [40,44,54,60,61]

Membrane Configuration	Working Principle	Properties	
Plate-and-frame module	Two flat membranes are in a sandwich-like arrangement and in between the two membranes, a spacer is placed for permeate removal	Feed Flow	$0.25–0.50\,m^3/m^2\text{-}s$
		Feed pressure	43–85 psi
		Packing density	Low ($100–400\,m^2/m^3$)
		Channel gaps	0.5–10 mm
		Lengths	10–60 cm
		Pretreatment particle size in feed	150 μm
Spiral wound module	Feed flows from the feed spacer in an axial direction while permeate is first collected in the membrane envelope and spirals into the collection pipe	Feed flow	$0.25–0.50\,m^3/m^2\text{-}s$
		Feed pressure	43–85 psi
		Packing density	High $300–1,000\,m^2/m^3$
		Spacers thickness	0.56–3 mm
		Pressure drop	High
Tubular module	Feed flows inside the tube from the beginning of the membrane tube, permeate is collected at the shell during a filtration process and retentate flows till the end of the membrane tube	Feed flow	$1.0–5.0\,m^3/m^2\text{-}s$
		Feed pressure	28–43 psi
		Packing density	Low (less than $300\,m^2/m^3$)
		Eternal diameters	5–25 mm
		Lengths	0.6–6 m
		Tube spacing	0.5–2 cm
		Holdup volume	High
Hollow fibre	Hollow fibre membrane module has two different types of arrangement operation mode which are inside out mode where the feed enters the hollow fibre lumen and permeate is collected outside of the hollow fibre and outside in mode where the feed enters from the shell and the permeate is collected at the hollow bibber lumen	Feed Flow	$0.005\,m^3/m^2\text{-}s$
		Feed pressure	1.4–4.3 psi
		Packing density	High (up to $30,000\,m^2/m^3$)
		Fibre diameter	0.2–3 mm
		Fibre length	18–120 cm
		Allowed pressure	Max. 2.5 bar
		Pretreatment particle size in feed	100 μm

Briefly, feed flows inside the tube from the beginning of the membrane tube, permeate is accumulated at the shell during a filtration process and retentate flows until the end of the membrane tube. The material of membrane tubes could be porous ceramic, stainless steel, plastic or fabric to withstand the applied pressure [53]. Due to the large internal diameter (5–25 mm), a tubular membrane module is often applied to deal with feed that consists of large particles [54]. Additionally, tubular module is easy to regenerate chemically and mechanically by flowing the cleaning agent from the shell to unclog the membrane inside the tube [55]. Among the membrane configurations, the tubular module has the lowest surface area to volume ratio (less than $300\,m^2/m^3$) [56]. Tubular modules are normally operated under tubular flow which required a large pumping capacity during operation. Nevertheless, the tubular membrane module requires a large area to operate due to high holdup volume [54]. Generally, tubular membrane modules are limited to NF application due to their high resistance towards membrane fouling [57].

Hollow fibre membrane module consists of 50–3,000 single hollow fibres with each end sealed with epoxy resin and assembled in a shell and tube exchanger [54]. Hollow fibre membrane module has two different types of arrangement operation mode which are inside out mode where the feed enters the hollow fibre lumen and permeate is collected outside of the hollow fibre and outside in mode where the feed enters from the shell and the permeate is collected at the hollow bibber lumen. The operation modes are based on the application. Hollow fibre membrane module has the highest packing density which can achieve up to 30,000 m²/m³ [58]. Hollow fibre membrane module has lower manufacturing cost compared to other module configurations due to hollow fibres being prepared with automated high-speed equipment; therefore, the manufacturing cost of hollow fibre modules per square metre is low. In addition, hollow fibre module has low energy consumption due to low-pressure drop and low crossflow rate [59]. It is easy to clean the hollow fibre module by backwashing because hollow fibres are self-supported. Nevertheless, the module has a few limitations such as low mass transfer due to laminar flow, low-pressure resistance and being prone to block by feed with large particles during inside out mode [54].

15.4.2 Membrane Materials

One of the important keys to the development of NF membrane is the fabrication of good performance membrane through methods such as incorporation of nanoparticles, interfacial polymerization, UV treatment, etc. These methods are utilized to develop better membranes in terms of antifouling, high rejection and high selectivity. There are three types of membrane for NF application which include organic/polymer membrane, inorganic/ceramic membrane and hybrid/mixed matrix membrane. Selection of membrane type for NF membrane fabrication often depends on the characteristics of the feed [62].

15.4.2.1 Polymeric Membrane

Organic NF membranes are popular with their low cost and excellent performance and thus dominate the market. Commercially, there are two types of NF membranes for the NF process, which are TFC membranes manufactured via interfacial polymerization technique and asymmetric membranes synthesized via phase inversion method [63]. Two important polymers for NF membranes fabrication are polyamide which consists of amide (–CO-NH–) group and cellulose acetate (CA). TFC membrane consists of porous, thick and supportive substrate that is covered by a thin active layer. TFC membrane has a good solute selection, high solvent permeability and stable composite structure compared to the asymmetric membrane [64,65]. However, TFC membranes are not stable in acidic and basic condition and still face challenges such as membrane fouling and low separation efficiency [66]. An asymmetric membrane is generally formed by phase inversion using polymers or CA and it is easy to prepare [67,68]. Nevertheless, the prepared asymmetric membrane normally faces limited rejection and flux; therefore, research was directed to improve NF performance such as flux, rejection, fouling mitigation and selectivity and novel modification of NF membrane to adapt to challenging feeds [69]. Various methods were reported for NF membrane preparation such as electron-beam irradiation, interfacial polymerization, layer by layer, nanoparticle incorporation, plasma treatment and UV/photo-grafting [70].

Table 15.5 shows the commercial NF membrane with different manufacturers and their specifications. Dow FilmTec™ provides polyamide-TFC membrane with different characteristics where NF has high rejection, NF90 required low energy and low pressure and NF270 is suitable in organics removal and softening. All the membrane has an MWCO in the range of 200–400 Da and works well at pH 2–11. In terms of flux performance, NF270 has the highest flux followed by NF90 and NF. In the recent study by Hacifazlioglu et al. [71], NF90 and NF270 membranes were used in wastewater reclamation. Suez™ provides a few types of polyamide-TFC and CA membrane which has an MWCO at a range of 150–300 and ~2,000 Da, respectively. Suez™ NF polyamide-TFC membrane has been applied in the study of lithium extraction and aluminium separation from lepidolite leaching solution [72]. Besides, in the performance study by Religa et al. [73], four types of Suez™ NF membrane, namely DK, DL, HL and CK, were employed to treat tannery wastewater for

TABLE 15.5

Commercial Polymeric NF Membrane

Manufacturer	Membrane	Polymer	Characteristics	pH Range	MWCO (Da)	Flux (GFD)/psi
Dow FilmTec™	NF	Polyamide-TFC	High rejection	2–11	~200–400	26.5–39.5/130
	NF90	Polyamide-TFC	Low energy, low pressure	2–11	~200–400	46.0–60.0/130
	NF270	Polyamide-TFC	Organics removal, softening	2–11	~200–400	72.0–98.0/130
Suez™	Duracid NF	Polyamide-TFC	Acid purification, mineral concentration	0–9	~150–200	10–19/225
	DK	Polyamide-TFC	High rejection	2–10	~150–300	22/100
	DL	Polyamide-TFC	Low energy, low pressure	2–10	~150–300	28/220
	HL	Polyamide-TFC	Softening	3–9	~150–300	39/100
	CK	CA	Chlorine resistant, softening	2–8	~2,000	28/200
Microdyn Nadir™	NP010	PES	Acid/caustic preparation, metal, chemical	0–14	~1,000	>117.60/580
	NP030	PES	Acid/caustic preparation, metal, chemical	0–14	~500	>23.53/580
Synder Filtration™	NDX	Polyamide-TFC	n/a	4–9	~500–700	35–45/110
	NFG	Polyamide-TFC	High flux, softening	4–10	~600–800	65–70/110
	NFW	Polyamide-TFC	Softening	4–10	~300–500	50–55/110
	NFX	Polyamide-TFC	High rejection, softening	3–10.5	~150–300	30–35/110
TriSep™	TS40	Polypiperazineamide	Process	2–11	~200	20/110
	TS80	Polyamide-TFC	Softening	2–11	~150	20/110
	SB90	CA	Chlorine resistant	n/a	~150	30/225
	XN45	Polypiperazineamide	Process	2–11	~500	35/110

the recirculation of chromium(III). Among the four membranes, DL membrane has the highest flux performance and chromium(III) retention (94%–97%). Next, Microdyn Nadir™ provides PES NF membrane, namely NP010 and NP030, with the MWCO of ~1,000 and ~500 Da, respectively. The membrane can be applied to any solution due to its broad pH working range (0–14). Both NP010 and NP030 were employed in the removal of surfactant from solution [74]. The authors have concluded that a smaller value of MWCO of NP030 has a higher retention coefficient for surfactants compared to NP010. Synder Filtration™ also provides four different types of polyamide-TFC membranes, namely NDX, NFG, NFW and NFX. These NF membranes have been applied in air conditioning applications for liquid desiccant regeneration [75]. Trisep™ provides a variety of NF membrane products with different MWCO at a range of 150–500 Da, which includes TS40, TS80, SB90 and XN45. TS40 is a polypiperazineamide NF membrane with an MWCO of about 200 Da, mainly used in food, dairy products and other industrial applications [76]. TS80 is a polyamide-TFC NF membrane with an MWCO of ~150 Da and its use for water softening [77]. SB90 is a high-flux CA membrane and is mainly used for chlorine disinfection [78]. XN45 membrane with an MWCO of ~500 Da and often used in industrial processes and low-pressure water purification [79].

Surface modification is applied to improve the antifouling performance and separation efficiency of TFC membrane [80]. Surface modification of NF membrane could be achieved through surface coating and surface grafting [81]. The purpose of coating on NF membrane is to improve antifouling properties by providing the membrane surface with a hydrophilic and smooth surface. Surface coating is a simple technique; therefore it is easy to adopt by industry. In a recent surface coating using inkjet printing of polydopamine (PDA) and GO on TFC NF membrane studied by Wang et al. [82], the PDA-GO-TFC NF membrane exhibited better antifouling properties and chlorine resistance. In another surface coating of NF experiment performed by Li et al. [83], they modified NF membrane with sulfobetaine-based zwitterion with polyhydroxy group which improved NF water flux, salt rejection and fouling resistance. Besides, Oulad et al. [84] modified NF membrane with tannic-acid–coated boehmite (TA-BM). The blending of TA-BM in PES membrane showed better dye removal capability and exhibited antifouling towards protein. On the other hand, the surface grafting method was adopted to solve the durability problem of surface coating. Chemical vapour deposition [85,86], redox-initiated polymerization grafting [87] and photochemical polymerization grafting [88] are some examples of surface grafting methods on TFC membrane by creating a more hydrophilic surface to improve water permeation and antifouling properties. Additionally, grafting on membrane surfaces has shown better durability compared with surface coating on the membrane.

Membranes are synthesized from synthetic polymers or carbohydrate-based polymers. Synthetic polymers include polyethylene, PES, polypropylene (PP), polysulfone (PSF), polyvinyl alcohol and polyvinylidene fluoride (PVDF) while carbohydrate-based polymers include CA, alginates and chitosan [89–91]. Either synthetic polymers or carbohydrate-based polymers are widely used for membrane fabrication. To prepare an asymmetric NF membrane, non-solvent-induced phase inversion (NIPS) was employed. In NIPS process, polymer casting solution was first cast on a fixed plate and followed by an instant immersion in a non-solvent coagulation bath. The interchange between solvent and non-solvent on the immersed polymer film resulted in the formation of asymmetric membrane [92]. Parameters that need to be considered during NIPS process include polymer concentration, relative humidity, drying time, type of solvent and non-solvent used, immersion time and so on [93]. These parameters would affect the separation performance of NF membrane. Among the parameters, the most feasible method is altering the non-solvent coagulation bath to achieve a high-performance membrane; therefore, novel phase inversion process has been developed by researchers such as alkaline-induced phase inversion [94], chemical modification phase inversion [95], ionic-induced phase inversion [96] and salt-induced phase inversion [97]. Changes in the coagulation bath have an impact on the characteristics of the fabricated membrane such as surface morphology, surface roughness, porosity, permeability and pore size, which then subsequently affect the membrane performance during the filtration process [93].

CA polymer has gained interest among researchers because of it is readily available, biodegradable, hydrophilic and chlorine resistance properties, which could be easy to be functionalized with a different group to improve the membrane performance [98]. CA membrane possessed good hydrophilicity, good chemical stability and mechanical strength, high transport properties, low adsorption of protein and excellent film-forming during fabrication [99]. Hence, it is widely used in the pressure-driven membrane separation process. However, during the phase inversion process of CA membrane fabrication, the exchange between solvent and non-solvent is slower due to its hydrophilic property, which then results in a denser skin layer and low porosity [89]. Consequently, it would affect the membrane flux, rejection ratio and resistless to chemicals and bacteria. To overcome the problems, modifications on CA membranes are required where one of the ways is through the mixing of polymer solution with proper additives. Blending of polymer is a simple and effective method where it shows a wider range of properties [100].

15.4.2.2 Inorganic Membrane

Inorganic membranes, which typically consist of a microporous active layer and a macroporous support layer, have been commercialized for application in wastewater treatment [101]. The inorganic membrane is divided into two classes, which are metal oxide membrane and carbon-based membrane [102]. The inorganic membrane has stable flux performance, is resistant to organic solvent, has good mechanical strength and has high tolerance towards severe conditions compared to the polymeric membrane. Nonetheless, ceramic membranes are difficult to scale up due to their brittle property. Wastewater treatment and desalination using inorganic membrane is an alternative solution for challenging feed where the polymeric membrane is not efficient due to membrane stability and membrane fouling issues [103].

Metal oxides such as alumina (Al_2O_3), silica (SiO_2), titania (TiO_2), and zirconia (ZrO_2) are commonly used in the ceramic membrane [104]. Alpha-alumina (α-Al_2O_3) and gamma-alumina (γ-Al_2O_3) are the most applied inorganic material due to their hydrophilic nature, covalent bonding property and resistance to high transmembrane pressure [36]. A metal oxide inorganic membrane has an asymmetric structure that comprises a thin top layer with a pore size of less than 1 nm, a middle layer with a pore size around 2–5 nm and a thick support layer with a larger pore size around 0.5 μm [105]. During fabrication, elevated temperature sintering is necessary to combine the layer. The multi-layer membrane could provide stronger mechanical strength to the membrane system. To prepare ceramic membrane with metal oxides, the sol-gel method was adopted which changed the solution to a solid membrane. Four steps are involved in the sol-gel method and it is started with precipitation of hydrolysed precursors, followed by peptization where the precipitate change to sol, coating of sol on porous support where gelation occurs during drying and lastly sintering at high temperature [106]. In NF system, alumina membranes are the most studied inorganic membrane with a pore size between 3 and 5 nm and correspond to MWCO of 3,000–10,000 Da. During NF separation performance with metal oxide membrane, the rejection is governed by electrostatic interaction between positively charged inorganic membrane and multivalent cations [107]. Except for metal oxide membrane, current research focuses on the composite membrane which combines two or more metal oxide in ceramic membrane to further improve NF performance.

α-Al_2O_3/γ-Al_2O_3 hollow fibre membranes were fabricated by Wang et al. [108] through the dip-coating process followed by a sintering process. The α-Al_2O_3/γ-Al_2O_3 hollow fibre membranes have a positive charge surface and demonstrated water flux up to 17.4 LMH and high retention to multivalent ions. Guo et al. [109] mixed between TiO_2 and ZrO_2 nanoparticles to fabricate inorganic NF membrane via the sol-gel method and dip-coating method to enhance the membrane thermal stability. Nevertheless, the membrane has low water flux due to low porosity. Inorganic membranes were also modified with TiO_2 and GO to create more a hydrophilic and negative charge membrane surface to mitigate membrane organic fouling [110].

Carbon-based nanostructures such as carbon nanotubes (CNTs) and graphene are selected for application in wastewater treatment due to their high water flux, uniform structure, strong atomic bond, high surface, antimicrobial activity, and tuneable pore size [111,112]. CNTs have diameters that are measured in nanometers and composed of rolled-up cylindrical graphite sheets. Based on the graphene shell layers, CNTs are divided into single-walled CNTs, double-walled CNTs and multi-walled CNTs (MWCNTs) [113]. Nanotubes' length and diameter, morphology and atomic arrangement would affect the properties of nanotubes. In addition, the tunable property in terms of pore size, conductivity and surface chemistry possessed by CNTs offers possibilities in the field of water recovery [114]. CNTs have low selectivity towards specific ions and high cost which then limit the commercialization. Graphene has been discovered as a two-dimension carbon allotrope which comprises a monolayer carbon that is highly permeable for the water recovery process [115].

15.4.2.3 Hybrid/Mixed Matrix Membrane

Hybrid membranes are developed by integrating organic matrix with inorganic fillers [116]. The purpose of a hybrid membrane is to combine the advantages of both polymer and ceramic membranes [117]. Polymer membranes that are outstanding in selectivity, high packing density and low cost are incorporated with a ceramic membrane that has strong mechanical strength, long-term stability and could regenerate. Various methods are employed to produce mixed matrix membranes by blending inorganic metal oxide into polymeric membranes such as dip coating, dispersion cross-linking and interfacial polymerization [105,118]. The incorporation of metal oxide nanoparticles into polymer membrane would change the polymer membrane structure and consequently affect the transportation of solution via membrane pores [119].

Sherugar et al. [120] prepared a hybrid membrane by incorporating hydrophilic-zinc–doped aluminium oxide (Zn:Al$_2$O$_3$) into PSF via diffusion induced phase separation method and applied in heavy metal removal from water and antifouling study. The fabricated membrane has a porous finger-like structure under FESEM has a water flux of 60 LMH. Hybrid membrane showed antifouling properties with BSA feed and obtained a flux recovery of 98.4%. Additionally, the synthesized hybrid membrane was able to remove 87% of arsenic and 98% of lead. In another study by Moradi et al. [121], boehmite curcumin nanoparticles were blended into PES membrane via phase inversion method. The hybrid membrane showed high water flux with the flux within 120–140 LMH and antifouling behaviour. Heavy metals such as Cu^{2+}, Fe^{2+}, Mn^{2+}, Ni^{2+}, Pb^{2+} and Zn^{2+} have rejection up to 99%. Besides, Li et al. [122] synthesized a TiO$_2$/polyamide NF membrane via interfacial polymerization. High water flux of 105.5 LMH was obtained and the composite membrane exhibited antifouling properties. From the study conducted by different researchers, it was found that the combination of metal oxide and polyamide membranes showed antifouling properties, high permeate flux and the ability to remove heavy metals.

Another type of hybrid membrane is the combination of carbon-based inorganic material with a polymeric material. Roy et al. [123] studied the effect of CNT incorporation on membrane surface via interfacial polymerization. It was found that the CNT-polyamide layer has good selectivity. In another study by Grosso et al. [124], CNT that functionalized with –NH$_2$ or –OH group provided the membrane with good dispersion. Farahani et al. [125] prepared a hybrid membrane by functionalizing MWCNT with COOH in a polyamide membrane. It was found that the membrane has an increase in porosity and pore size. Additionally, the membrane showed improvement in liquid transport and sorption. In the later study by Farahani and Chung [126], they have developed an NH$_2$-MWCNT polyamide hollow fibre hybrid membrane for organic solvent NF. The novel membrane has the potential to be applied in the petrochemical, pharmaceutical and food industries due to its high rejection. Even hybrid membrane possesses advantages from both polymeric and ceramic membrane, yet studies of the hybrid membrane with pilot-scale are vital before manufacturing on large scale.

15.5 CHALLENGES OF NF IN HEAVY METAL REMOVAL

It is undeniable that NF membranes have been used in various interesting applications over the past decades due to the attractiveness of their selective separation behaviour. Though the performances of the NF membrane were found to be competitive; however, NF membrane still faces several challenges that limit its operational effectiveness. Hence, it is important to understand the reason behind these challenges for the membrane operation to reach at least desired or better performance [127]. The following section discusses the challenges of NF membrane in treating wastewater containing heavy metals.

15.5.1 Concentration Polarisation

CP remains an inherent challenge that restricts NF membrane to operate at its desired capacity. Generally, when NF membrane selectively removes the solutes in the feed solution during the convective transport towards the membrane, the unremoved solutes or retained solutes will accumulate as a thin layer adjacent to the surface of the membrane. Associated with that, a concentration gradient exists between the thin layer near the membrane surface and the bulk solution thus causing a back diffusion of the solutes near the region of membrane to the bulk solution. As the membrane filtration continues, the polarized layer increases significantly on the vicinity of the membrane surface and eventually it forms a resistive layer that obstructs the mass flow across the membrane.

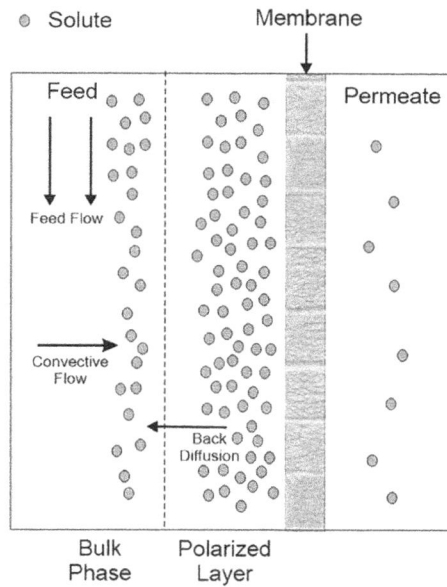

FIGURE 15.4 Graphical illustration of CP [129].

The mechanism of this phenomenon is illustrated in Figure 15.4. This hydrodynamic phenomenon is known as CP and it is highly undesirable as it will critically affect membrane performance parameters such as permeate flux, solute rejection, and increase the risk of fouling that often lead to shorter membrane lifespan. Ultimately, it will add unwanted costs to the membrane filtration operation [128]. Although this is an unavoidable issue for membrane filtration process, the positive side of this phenomenon is the reversibility of the process which can be alleviated by promoting agitation on the bulk feed solution or operating the membrane filtration at higher flow rate to maintain the feed solution's concentration profile at a satisfactory and uniform level [20].

15.5.2 FOULING AND SCALING

Membrane fouling remains as one of the greatest challenges for membrane operation attributed to its negative impact towards the effectiveness of membrane performance. It is an unavoidable phenomenon that often faced by the membrane filtration operation ever since the introduction of membrane technology. This undesirable phenomenon is caused by the deposition, accumulation or adsorption of retained solutes on membrane surface or inside membrane pores wall [130]. The adverse effects of fouling phenomenon often lead to the deterioration of membrane performance in terms of permeate flux and solutes removal efficiency, higher transmembrane pressure, shorter membrane lifespan and higher operating cost [131]. Depending on the types of foulants in the raw water feed, membrane fouling can be classified into four major categories; colloidal fouling, organic fouling, biological fouling, and inorganic fouling/scaling as shown in Figure 15.5. In most of the low pressure–driven membrane such as microfiltration and ultrafiltration (UF), they suffer from internal fouling associated with the adsorption and clogging of pores while for high pressure–driven membrane which is NF and RO membrane, surface fouling is the major contributor [132]. Nevertheless, the treatment of heavy-metal–containing wastewater using NF membrane often resulted into scaling. Generally, scaling refers to the fouling caused by the precipitation and crystallization of inorganic compounds such as calcium and magnesium salts. When the concentration of the dissolved salts exceeds the equilibrium solubility during the membrane filtration process, it will precipitates

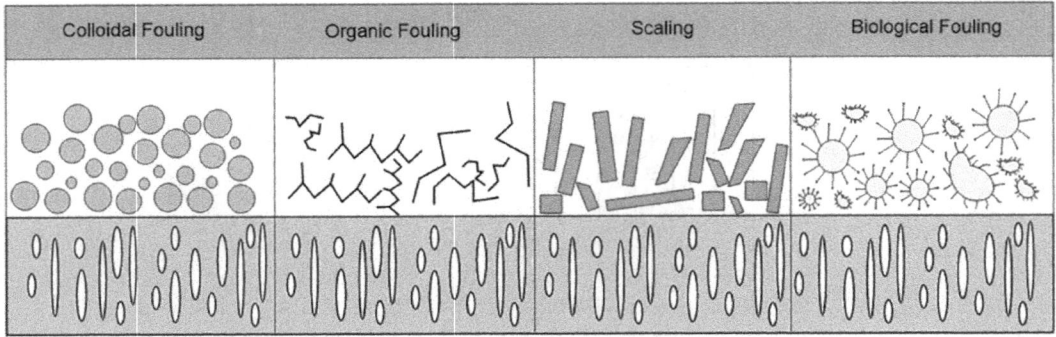

FIGURE 15.5 Types of fouling [136].

and forms a scaling layer on the membrane surface [133]. This is an unavoidable challenge for membrane filtration process, and it must be properly controlled in order to allow membrane operation to operate at is highest potential. Associated with this, significant effort and research have been devoted to alleviate the membrane fouling occurrence which includes feed water pretreatment, physical and/or chemical cleaning and membrane modification [134,135].

15.5.3 SELECTIVITY OF MULTICOMPONENT FEED

Selective removal of solutes using NF membrane has been an attractive interest for various industrial applications including heavy metal–based industrial effluent. However, the selectivity of NF membrane in a multicomponent feed (complex mixture) varies with the condition of the feeds. For instance, Aamer and his co-workers [137] revealed that NF membrane selectivity varies while treating wastewater from different sources namely colour rinse wastewater, alkaline pickling rinse, acidic pickling rinse, and anodizing rinse and mixture of all the wastewater streams. The rejection of sodium in acidic pickling rinse is only 31% while the rejection of sodium in other wastewater is in the range of 57%–70%. Besides, another study by Mullet et al. [138] also revealed the impact of the feed pH and membrane charge on metal recovery from mine wastewater. At different wastewater feed pH, the rejection of multivalent ions (Ca^{2+}, Cu^{2+}, Mg^{2+}, Mn^{3+}, S^{2-}) was different which was attributed to the relationship between membrane charge, feed pH and iso-electric point. Such uncertain performance often resulted in the limitation to adopt NF membrane in treating heavy metal–based industrial effluents. Therefore, it is important to understand the reason behind this behaviour in order for the NF membrane to operate effectively.

In practice, all industrial effluent contains a complex mixture, not just limited to the heavy metals, but other coexisting ions such as monovalent Na^+ and proton (H^+), hardness such as Mg^{2+}, and Ca^{2+} and neutral solutes such as total dissolved solid (TDS), where the composition depends on the origins of the source. In a typical filtration process, the membrane would have a higher rejection of multivalent ions than the monovalent ions due to Donnan effect and dielectric exclusion. Therefore, monovalent ions and small neutral solutes would diffuse across the membrane, leaving the multivalent metal ions at the retentate site [139]. However, this is not the case when a NF membrane is used on multicomponent feed consisting of heavy metals and competing ions. The presence of the competing ions could significantly override the membrane surface charge. For instance, when competing ions such as Na^+, Mg^{2+} and Ca^{2+} were present in the feed, it caused 18%–37% drop in the rejection of Cd^{2+}, Pb^{2+} and Zn^{2+} using a lab-fabricated nano-fibrous metal-organic framework membrane [140]. A similar observation was found whereby the presence of sodium nitrate reduced the rejection of multivalent heavy metal cations [141]. An increase of ionic strength (sodium nitrate) from 1 to 100 mM caused the rejection on Ni^{2+} reduced by 5% for NF200 and 11% for Desal-HL. The decline in rejection was due to the adsorption of Na^+ on the membrane surface partially shield the negatively charged membrane, leading to poorer electrostatic interaction with the multivalent ions in the feed. In a long run, NF membrane has to be assessed in terms of its pH sensitivity and its resistance towards the ionic strength of the wastewater to select the most suitable NF membrane to be used. Worth highlighting also metalloids such as arsenic, which usually coexist with other heavy metals in the metallurgical process will present as uncharged H_3AsO_4 [142]. This uncharged solute could easily permeate across the membrane; therefore, pretreatment using a reducing agent and precipitation is needed to remove the arsenic.

In a feed containing a mixture of heavy metals, most commercial NF membranes could achieve an overall rejection above 90% for all metals. For example, a rejection over 90% on both chromium(VI) and nickel(II) in metal plating effluent could be achieved using NF90 and NF270 under optimal operating conditions (pH 10 and pressure 30 bar) [14]. However, the limitation of NF membrane arises in treating mixture consisting of metal cations of similar valency. In this case, the diffusivity of the metal cations could determine the removal selectivity of the metal. For instance, nickel in a nickel/lead binary mixture has higher rejection than lead due to the lower diffusivity of nickel compared to lead [143]. Sum et al. [144] reported filtration on a ternary metal mixture containing copper, cadmium and chromium using a lab-fabricated polyamide-TFC NF membrane. The authors found that copper and chromium could adsorb onto the membrane surface, forming a more positively charged surface that repels cadmium ions. The rejection of cadmium in the ternary mixture is even higher than filtration on a single metal.

Besides, NF membrane also has limitations dealing with mixture containing different anions associated with the same metal cations. A comparison was made between sulphate (SO_4^{2-}) and chloride (Cl^-) of the same types of metal cations using negatively charged AFC30 and AFC80 NF membranes. The study revealed that the metal salts attached with the SO_4^{2-} ions had a better rejection as compared to their counterparts (Cl^-) which was attributed to the valence of the anions and size of the anions [145]. Similarly, the monovalent nitrate (NO_3^-) could have a higher passage than divalent SO_4^{2-}. Despite the cationic metal ions could be adsorbed onto the membrane surface, it was found that the rejection of sulphate ions was edged over the chloride ion due to stronger electrostatic repulsion between divalent anionic sulphate with the membrane surface [143]. Therefore, the metal ion was simultaneously rejected to maintain the solution's electroneutrality.

15.6 DESIGN AND OPERATION OF NF SYSTEM

An entire NF water treatment system consists of pretreatment, the filtration process and post-treatment which involve membrane cleaning and restoration. In this section, the design of a membrane system for treating effluent-containing heavy metals is addressed. In the meantime, operating variables such as feed pH, circulation rate, temperature, pressure and ionic strength that govern the quality of the permeate are reviewed. This section could provide insights on optimizing the operating conditions that yield desired separation efficiency.

15.6.1 Setting up NF System for Treating Feeds Containing Heavy Metals

Figure 15.6 shows a standalone NF system consisting high-pressure feed pump to circulate the feed, a filtration cell that holds the membrane, a feed tank and instrumentation such as flowmeter and pressure gauge which can be used to monitor the flow rate and regulate the pressure across the membrane filtration cell. A feed is continuously pumped at elevated pressure to the membrane system. Within the membrane system, the solutes in the feed would be retained over the retentate side, while the water is allowed to pass through the membrane forming a clean permeate stream.

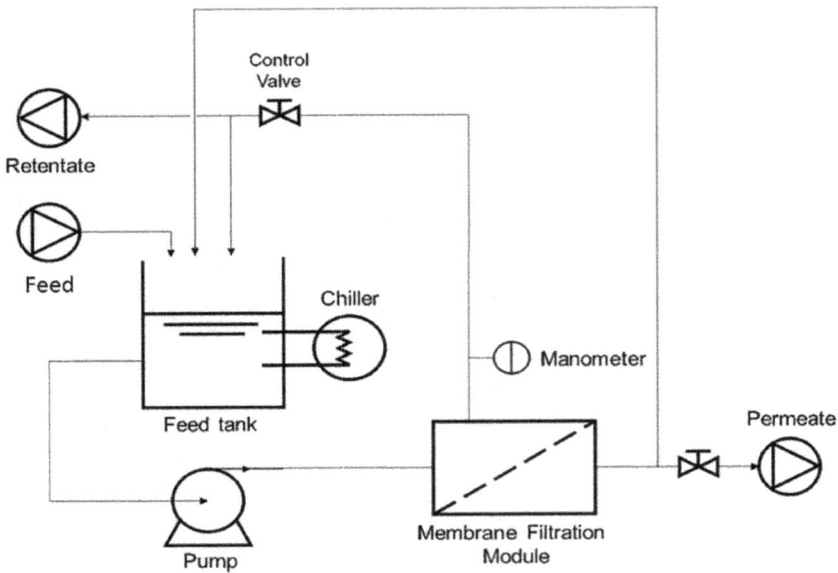

FIGURE 15.6   Schematic of a standalone NF system [146].

15.6.2 FEED PRETREATMENT

An industrial effluent usually consists mixture of suspended solids, organic and inorganic solutes, colloidal particles, hardness and TDSs. These solutes are concentrated near the membrane surface, deposit and form a fouling layer. A proper pretreatment before NF is necessary to prolong the service lifespan of the membrane.

The selection of pretreatment methods depends on the characteristic of the solutes (sizes, biological degradability) in the feed. The primary pretreatment involves the removal of large particles in the feed. A sieving-type filter such as coarse filter could retain particles with sizes larger than 1 μm [147]. Meanwhile, the pH of the feed shall be adjusted before NF process [148,149]. A long period of exposure to an acidic medium causes hydrolysis to the active layer of the polymeric membrane [142]. Meanwhile, colloidal particles such as silica, iron corrosion products, bacteria and clay in the feed can be removed using coagulants such as alum, ferric chloride or cationic polyelectrolytes [150]. The colloids will agglomerate, forming larger complex sediment which is easily screened off. Worth highlighting also that the reagents used in coagulation or flocculation potentially become a source of fouling if not removed. For instance, excessive dosage of antiscalant could agglomerate the calcium ions and solutes in the feed, followed by deposition on membrane surface [151]. A solution containing arsenate was pretreated using calcium precipitation leading to a thick cake layer of about 23.1 μm on Desal-HL membrane [152]. Wang et al. [153] compared the efficiency of adsorption using powdered activated carbon and coagulation using aluminium sulphate as pretreatment of secondary effluent of wastewater treatment. Both processes reduced the amount of organic solutes in the feed; however, coagulation severely fouled the NF membrane due to Al^{3+} inorganic scaling. Dasgupta et al. [150] evidenced the roles of coagulation before NF in treating tannery effluent. Fouling is mitigated as natural organic matter such as humic substances was removed, and 98% of chromium(VI) rejection (98%) was obtained.

Worth noting also that the water in the feed is reclaimed and hence the concentration of solutes in retentate gradually built up during the continuous filtration. Given a water recovery of 80%, the concentration of heavy metals such as chromium in the retentate could be quadrupled [154]. Scaling would occur when sparingly soluble salts such as calcium sulphate, carbonate and fluoride; barium sulphate; and strontium sulphate are concentrated beyond their solubility limit [155]. Therefore, antiscalant such as polyacrylic acid, polyacrylamide, polymaleic anhydride, and polyphosphates was added to halt the nucleation and growth of crystal on the membrane surface [156]. Andrade et al. [157] demonstrated the use of commercial antiscalant Acumer 4300 (Dow FilmTec) at a dosage of 10 ppm as one pretreatment before NF on gold mining effluent. The subsequent NF process exhibited reduced flux decay up to 36%.

Before the NF process, the feed could be pretreated with other classes of pressure-driven membrane with larger pores such as microfiltration and UF to remove dissolved organic solutes [158] and submicron solutes which are not able to be removed during the first-stage screening [159]. For instance, acid mine drainage was microfiltered using Kerasep ceramic membrane of MWCO 0.45 µm in a Rhodia Orelis unit to remove particles and biological contamination [160]. UF of MWCO 100 kDa before the NF process could remove up to 98% of oil and lubricants in waste emulsions generated in the cable factory resulting in 99% of copper removal in the subsequent NF [161].

Several attempts to integrate pretreatment processes were reported. Chang et al. [162] integrated coagulation, UF and NF to treat flowback and produced water. An optimal dosage of iron coagulant (900 mg/L) to remove particulates and organic matters could reduce the extent of fouling of UF by 60%. A combination of both coagulation-UF elevated the permeation flux of NF by 23%. The integrated design rendered a clean permeate with high removal on turbidity (99.9%), COD (94.2%), Ca^{2+} (72.8%), Mg^{2+} (86.3%), Ba^{2+} (82.8%), Sr^{2+} (80.1%) and SO_4^{2-} (91.7%). Zolfaghari and Kargar [163] integrated carbon filter, sand filter, MF and NF in a pilot-scale treatment process to remove the chromium(VI) and sulphate. A rejection rate of 99.6% on both chromium and sulphate was achieved at pH 10 and 1 bar. The rejection of chromium was decreased with the increasing sulphate concentration. An integrated treatment of reverse osmosis concentrate to achieve near-zero liquid discharge in a textile plant was studied based on coagulation using iron chloride ($FeCl_3$) and chemical precipitation, followed by filtration using ceramic microfiltration and NF [164]. The proposed method successfully removed calcium (Ca^{2+}), magnesium (Mg^{2+}), silica (SiO_2) and TOC up to 97%, 83%, 92% and 87%, respectively. The authors also claimed that the design would be able to reduce the overall water consumption by around 15%.

15.6.3 OPERATING VARIABLES

The performance of the NF membrane is usually evaluated in terms of permeation flux and rejection, governed by operating variables such as feed pH, transmembrane pressure, crossflow velocity, temperature and composition of the feedwater (concentration of heavy metals and presence of competing ions or solutes). Tremendous works have been reported to optimize the separation efficiency by manipulating these variables, at the same time aiming to minimize the costs while maximizing permeate quality and recovery. Table 15.6 provides a comprehensive summary of the effect of operating variables on the flux and rejection of a few selected commercial NF membranes.

15.6.3.1 pH

The pH of the solution significantly affects both the solution chemistry and membrane surface charge. First, the pH would determine the speciation of the ionic species in the solution, which would further influence the interaction between the ions with the membrane surface. Figure 15.7 shows the ions speciation of heavy metals such as Cd(II), Cu(II) and Cr(III) in pH ranging from 3 to 8. Most heavy metals are divalent or trivalent cationic in acidic solution, turn into hydroxide complexes and eventually form insoluble metal hydroxide precipitates in alkaline conditions. The filtration in the alkaline range should be avoided since the metal hydroxide precipitates would form a thick scale over the membrane surface that cause severe flux decline. Most of the NF membranes are claimed to be amphoteric, specifically on polyamide NF which usually contains negatively charged carboxylic acid group and positively charged amine functional group. Depending on its

TABLE 15.6

The Effect of Operating Variables on the NF Separation Performances

Types of Membrane/ Manufacturer	Heavy Metals/ Sources of Wastewater	Operating Variables	Membrane Separation Performance	Reference
AFC40/PCI membrane system	Co (II)/ synthetic wastewater	Feed concentration: 15–1,000 mg/L Transmembrane pressure: 5–45 bar pH: 3–6	Higher the feed concentration, lower the permeation flux and higher the rejection rate. The permeation flux is proportionally increased with the pressure, while the rejection is increased until an optimal pressure (15 bar), then it is slightly declined with further increase of pressure. Permeate flux was increased from 188 to 213 LMH. The highest observed metal ion rejection (≈97%) at pH 3 and the rejection was reduced to 75% at pH 6.	[181]
NF99HF/Alfa Laval AB	Ge, Mo, Co, Cu, Zn	Transmembrane pressure: 10–20 bar Flow velocity: 0.5–1.1 m/s	No effect on the rejection Retention for Ge and Mo was increased, and insignificant on Co, Cu and Zn.	[173]
NF99/Alfa Laval	Cu(II), Fe (III), Mn(II)/ synthetic acid mine drainage	Feed flow rate: 400–900 L/h Temperature: 20°C–60°C Feed pressure: 5–20 bar	The permeate flux is increased, while the rejection on all metals remained at around 94%. Slight reduction in metal rejection and significant increment of permeation flux. Pure water flux is linearly increased with pressure, while the rejection of metals is increased slightly.	[170]
DK/GE Osmonics	Cu(II), Fe (III), Mn(II)/ synthetic acid mine drainage	Feed flow rate: 400–900 L/h Temperature: 20°C–60°C Feed pressure: 5–20 bar	The permeate flux is increased, while the rejection on all metals remained Slight reduction in metal rejection and significant increment of permeation flux. Pure water flux is linearly increased with pressure. The pressure has no effect on the rejection of metals.	[170]
NF90-2540 (Spiral wound)/Dow FilmTec	As	Metal concentration: 100–1,000 ppb pH: 3.5–10 Transmembrane pressure: 2–12 bar Feed temperature: 15°C–40°C	No effect on permeate flux, rejection was slightly reduced from 80% to 75% Flux was increased from 18 to 36 LMH, but reduced when pH further increases to 10. Rejection was increased from 74% to 88% Permeate flux was increased from 20 to 80 LMH Permeate flux was increased from 35 to 75 LMH, rejection was reduced from 90% to 70%	[166]
NF270–2540 (Spiral wound)/Dow FilmTec	Cu, Al, Zn	Transmembrane pressure: 4–30 bar Feed flow rate: 700–1,400 L/h	Permeate flux was increased. However, permeate flux density ratio began to decrease after 15–20 bar. Rejection was marginally increased from 85% to 95%. No effect on the permeate flux. Rejection on metal was increased by about 20%.	[160]
NF270 (Flat sheet)/Dow FilmTec	Al (III), Ni (II), Cr/Aluminium anodic oxidation wastewaters	Transmembrane pressure: 10–20 bar	Rejection of all the metals were increased from 91%, 97%, and 66% at 10 bar to 99%, 99% and 94% for aluminium, nickel, and chromium, respectively	[158]

(Continued)

TABLE 15.6 (*Continued*)
The Effect of Operating Variables on the NF Separation Performances

Types of Membrane/ Manufacturer	Heavy Metals/ Sources of Wastewater	Operating Variables	Membrane Separation Performance	Reference
NF270 (Flat sheet)/Dow FilmTec	Ni (II), Cr(VI)/ Metal plating effluent	Transmembrane pressure: 10–30 bar pH: 3.5–10	Rejection of Ni (II) increased from 88% to 94% and 90% to 92% for Cr(VI) when pressure was increased from 10 to 30 bar. Rejection of Ni (II) increased from 93.0% to 98.7% and 94.0% to 95.4% for Cr(VI) when pH was increased from 3.5 to 10.	[14]
N30F-2440 (Spiral wound)/ Microdyn Nadir	As	Metal concentration: 100–1,000 ppb pH: 3.5–10 Transmembrane pressure: 2–12 bar Feed temperature: 15°C–40°C	No effect on permeate flux and rejection No effect on permeate flux, rejection was increased from 95% to 97% Permeate flux was increased from 20 to 50 LMH, but no effect on pressure Permeate flux was increased from 27.5 to 52.5 LMH, rejection was slightly reduced from 97% to 95%	[166]
Desal-HL/ Osmonics	Ni(II)/synthetic wastewater	Nickel concentration: 10–100 ppm pH: 3–5.5	Rejection of nickel was increased from 64% to 80%. No significant changes on the rejection. The permeation flux was slightly decreased from 50 to 48 LMH	[141]
NF200/Dow FilmTec	Ni(II)/synthetic wastewater	Nickel concentration: 10–100 ppm pH: 3–5.5	Rejection of nickel was declined from 86% to 80%. The rejection of nickel is declined when pH increases until reaching minimal point at pH 4.5 (near IEP of membrane). The rejection was increased beyond pH 4.5.	[141]
Nanomax 50/ Millipore USA	Zn(II)/synthetic wastewater	Transmembrane pressure: 2–14 bar Zinc concentration: 10–100 ppm Recirculation flow: 100–350 L/h	Rejection of zinc increases with pressure, reaches a maximum then decreases. Permeation flux is linearly increased with pressure. Zn^{2+} retention increases from 70% to 90% when the concentration increase from 10 to 100 ppm at pH 6, 14 bar. Permeate flux is proportionally increased. The rejection of zinc is increased from 20% to 90% when flow rate is increased from 100 to 350 L/h	[172]

IEP, the membrane turns positive when soaked in acidic solution and turns negatively charged in alkaline conditions.

The rejection of metal in NF depends greatly on the repulsive forces between the membrane surface and the metal cations. Therefore, the highest metal rejection could be obtained when the metals are presented in a cationic state, mostly existing in acidic conditions. The rejection of metals shows a reverse relationship with pH until reaching the minimal when the membrane is in an electroneutral state. The membrane possesses the loosest steric hindrance and the weakest electrostatic repulsion at the IEP [152]. On contrary, the rejection of the metal cations increases again when the pH of the feed solution is above the IEP. Despite a negative charge surface promoting adsorption of

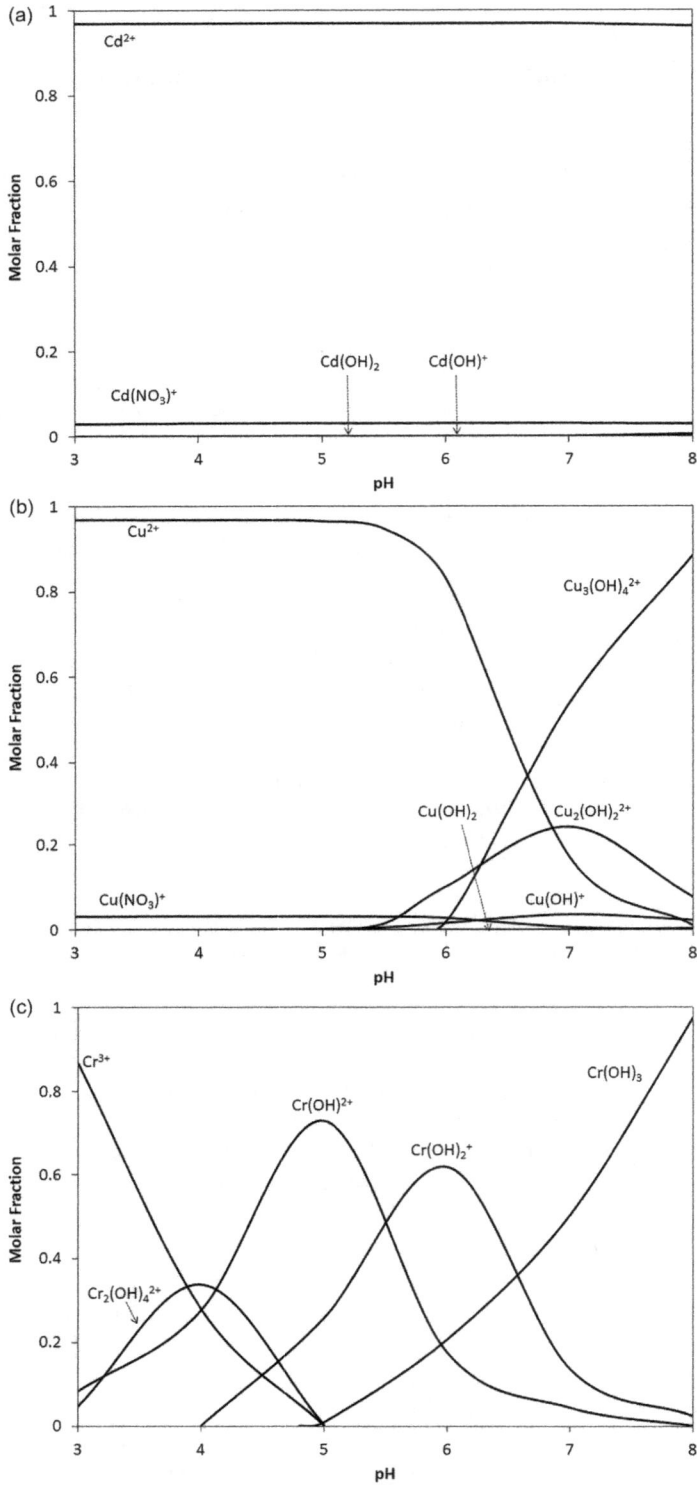

FIGURE 15.7 Ionic speciation of (a) cadmium(II), (b) copper(II) and (c) chromium(III) in pH ranging from 3 to 8 [144].

counterionic metal cations on the membrane surface, the membrane may experience a reduction in pore sizes which improves the rejection through sieving effect [165]. Indifferent from cationic heavy metals, NF rejection on oxyanions arsenic shows a proportional relationship with the pH [166]. By knowing monovalent $H_2AsO_4^-$ is dominant in the range of pH 4–6, while the divalent ion $HAsO_4^{2-}$ is dominant above pH 7; therefore, a stronger repulsion on the divalent than the monovalent anionic arsenic by the negatively charged N30F membrane can be observed at elevated pH. Meanwhile, divalent ions have larger hydrated radii compared to monovalent ions which are easily sieved by the membrane.

The pH of the feed solution could be controlled to selectively recover minerals from a mixture. Schütte et al. [167] demonstrated the recovery of neutral and negatively charged phosphorus compounds from multivalent metal cations in pretreated sewage sludge. The authors found that the selectivity between heavy metals and phosphorous compounds increased by reducing the feed pH 94% of heavy metals and 30% of phosphorous were rejected under optimal conditions. Acid recovery could be important in the metallurgical industry. Under high acidity conditions (pH < 1), H^+ and HSO_4^- were found to permeate across the NF270 while the heavy metals were rejected [142]. MPF-34 (Koch Membrane Systems) even exhibited 100% permeability to the sulphuric acid and about 90% of metal (Co^{2+} and Ni^{2+}) rejection when being used to treat gold ore processing effluent at pH around 0 and 1 [168]. The selectivity was reduced with increasing pH due to the higher rejection of proton (H^+). Chen et al. [169] demonstrated a cascade NF process operated at different pH to selectively recover chromium(VI) from a spent plating solution. The first-stage NF was operated at pH 3; therefore, the membrane could reject the multivalent Ni^{2+} while allowing the monovalent chromate ($HCrO_4^-$) to permeate across. $HCrO_4^-$ was converted to CrO_4^{2-} in the second-stage NF by elevating the pH to 8, and retained by the NF. The process rendered 90% separation among the chromium and nickel.

15.6.3.2 Transmembrane Pressure

Transmembrane pressure measures the pressure differences across the membrane on the feed and permeates side. It significantly affects the permeation flux and mildly affects the rejection of heavy metals. In a typical NF membrane, a transmembrane pressure of 3–10 bars is used. Water permeability shows a linear relationship to the transmembrane pressure. Such linear relationship is due to the osmotic pressure difference between the concentrate and permeate. For instance, NF90 and NF270 demonstrated a linear flux increment from 10 to 60 LMH and 20 to 100 LMH, respectively, when the pressure was increased from 4 to 10 bar [160]. Low pressure is unfavourable as solutes could deposit on the membrane surface and cause flux decline [152]. Nonetheless, excessive high pressure may cause over compression to the membrane, causing the membrane structure to be collapsed. Pino et al. [160] reported the decay of permeation flux density during the filtration on acid mine drainage beyond 15–20 bar. The permeation flux showed a linear relationship with the pressure in the low-pressure range. However, the gradient of flux increment was reduced at high pressure. Under the high-pressure range, CP, counter-diffusive flux and fouling could be significant which affect the concentration of solutes at the surface and hence the resistance of flow across the membrane. There are a few reports on the effect of pressure on the rejection rate of the metal cations. High pressure leads to higher permeation flux, while the permeability of the cations remains. Therefore, there would be a dilution to the permeate streams and a lower concentration of the metal cations could be obtained.

15.6.3.3 Crossflow Velocity

Crossflow velocity, sometimes expressed in volumetric flow rate, measures the circulation rate of the feed stream over the crossflow filtration system. Higher crossflow velocity minimizes CP and deposition of foulant on membrane surface [157,170]. Therefore, permeate flux increases with increasing crossflow velocity due to a lesser extent of fouling and higher overall osmotic pressure [171]. Some may think that the crossflow velocity has minimal impact on the permeation flux since the flux is governed by the transmembrane pressure [172]. On the contrary, the rejection of the heavy metal could be increased with an increase in feed velocity. For example, filtration using Nanomax 50 (Millipore USA) at a pressure of 14 bar, the rejection on Zn^{2+} could be greatly improved from 20% to nearly 90% when the circulation rate is increased from 100 to 350 L/h [172]. Such observation could be explained based on the reduced CP effect at elevated crossflow velocities. As the polarization layer is gradually built on the membrane surface, the tangential flow could impose shear stress over the membrane surface, decreasing the electric double layer and the CP effect [173]. Subsequently, the osmotic pressure is reduced, which improves the water permeability. Meanwhile, a lower concentration gradient exists between retentate and permeate; hence the solute is less favourable to diffuse across the membrane. This explains why the rejection is increased with high crossflow velocity. The optimal range of crossflow velocity is determined by considering the permeation flux, membrane retention and specific energy consumption [171]. Despite having increment in both permeate flux and rejection, high crossflow velocity also increases the energy consumption due to higher pump recirculation rate is needed.

15.6.3.4 Temperature

Most NF is performed at ambient temperature. However, the feed temperature could fluctuate and varies at different point source in the treatment plant. For instance, discharged effluent from bleaching and dyeing baths in the textile industry could reach 90°C, while in the pulp and paper industry, the water temperature is often above 60°C.

The temperature could affect the water permeation flux, solute flux and retention. First, the water flux across the NF increases with an increase in temperature. The water flux could be 1.5 times higher when the temperature is increased from 10°C to 30°C [174]. The changes in solute flux depend on the nature of the solute. A neutral solute and monovalent ions (chloride, sodium and potassium) will have higher passage across the membrane which gives lower rejection. On the other hand, the rejection of divalent ions such as heavy metals, hardness and sulphate is not significantly affected by the temperature [175]. Although the metal cations tend to adsorb onto the membrane surface in warmer feed and diffuse at a faster rate across the membrane, the metals in the permeate stream will be diluted by the increased water permeate. A similar observation was reported using DL, DK and NTR-7450 in treating electroplating rinse wastewater, where the rejection of Cr and Cu did not significantly change when the temperature is raised from 20°C to 45°C [176].

The variation of filtration performance at different temperatures is explained by the changes in membrane physicochemical properties and the kinematic viscosity of the solution. Amar et al. [177] reported a 1.72% increment in pore radius when the temperature was increased from 22°C to 40°C, while the membrane thickness decreased by 53% and 4% for solute and water transport, respectively. The effective pore radius of Desal 5 DK NF membrane was increased from 0.58 to 0.67 nm when the temperature increased from 22°C to 50°C [178]. The temperature of feed alters the surface charge density, by changing the dissociation of the surface functional group and adsorption of ionic solutes onto the membrane surface [179]. A warmer feed usually has a lower viscosity and higher ion diffusivity also thus contributed to the increasing permeation flux [180].

15.6.3.5 Metal Concentration in Feed

The amount of metal present in the effluent would affect the water permeability and the rejection rate of the metal cations. First, a higher metal concentration would cause higher osmotic pressure over the retentate side of the membrane, which cancel the effect of hydraulic pressure. Therefore, lower water permeability is observed in a system that contains high metal concentrations. In the meantime, high metal concentration induces CP over the membrane surface, which drifts the diffusion of metal across the membrane. Most of the commercial membranes showed decline in metal rejection when the concentration of metal is increased [141]. The excessive metal cations were claimed to shield the membrane surface charge, causing weaker electrostatic repulsion towards the multivalent anions in the solution [141]. The shielding effect is insignificant when the concentration of metals is below 100 ppm. A filtration was performed at pH 6 and 15 bar using feed containing zinc chloride [172]. The rejection of zinc was increased from 70% to 90% when the concentration of zinc chloride was increased from 10 to 100 ppm.

In a continuous NF process, the feed stream will be concentrated, causing a gradual increment of solute concentration. The increasing concentration in the feed might induce CP over the membrane surface which eventually affects the permeate flux and rejection. A relationship between the volume reduction factor (VRF) and its effect on the permeate flux and rejection was reported using NF270 in treating copper acid mine drainage [160]. 45% of permeation flux was reduced when VRF reaches 4, yet there is a 7% increment in observed rejection of the metals.

15.6.4 POST-TREATMENT

After a long filtration process, the NF membrane might experience reduction in permeability due to fouling and increasing concentration of solutes in the retentate. Occasionally, the pressure is increased to compensate the flux lost up to an extent where the pressure increase does not proportionally improve the flux. Hence, regeneration of membrane is needed to restore its performance. Reversible fouling could be easily mitigated by means of water flushing and chemical cleaning, whereas fouling by pore blockage might not be reversible. On the other hand, a continuous recycling of retentate across the NF system is causing a gradual increasing of solute concentration. The concentrated feed is rich in the minerals that are worth to be recovered. Therefore, this section provides summary on the NF membrane restoration and options available to handle the concentrated retentate.

15.6.4.1 Membrane Cleaning

Membrane cleaning is necessary to remove the foulant on the membrane surface after filtration. Membrane could suffer from irreversible fouling induced by pore blocking and deposition of metal ions and its hydroxides on the membrane surface that make it difficult to be restored. Restoration of membrane could be done through physical or chemical washing using acid or alkaline. In situ water washing was claimed to be able to restore the permeation flux of the spent PSF-based NF membrane (after treating clarifier tank effluent from steel plant) by 97% [182]. Alternatively, the filtrate could be reversely pumped back through the membrane to give a periodic backwash, or backpulsing, a more frequent periodic backwashing. Backwash could be done osmotically, by circulating a high salinity solution and pure water over the active layer and support, respectively [183,184]. Osmotic backwash using 1.0 M NaCl draw solution was proven to be effective in relieving organic bridging fouling caused by Ca^{2+} ions.

Forward flush or backwashing is meant to remove the deposited foulants by means of tangential shearing force or reverse flux. Nonetheless, the physical washing might not be effective when salts begin to precipitate, deposit and adsorb on the membrane or clog the internal pores. Several works demonstrated membrane restoration via acid cleaning approach. Acid reduces the medium pH, which leads to solubilization and cleavage of carbonate, hydroxide and oxide deposited on membrane surface [185]. Amaral el al. [171] examine the efficiency of using 0.2% hydrochloric acid, drochloric acid ency of using 0.2% oxalic acid, 2% citric acid and EDTA in restoration of NF. All acids gave almost 100% recovery, while EDTA has cleaning efficiency above 100%. Ates and Uzal [158] developed a protocol in restoring the membrane after treating aluminium anodic oxidation wastewaters. The spent NF270 membrane was flushed with nitric acid solution at pH 3, distilled water, sodium hydroxide solution at pH of 9, and again distilled water at 2–3 bar, 1 hour for each flushing cycle.

Most works have shown that acid cleaning did not damage the membrane, while caustic soda cleaning on the NF membrane can cause swelling of the polypiperazineamide active layer and pore enlargement by 23% after 18 hours cleaning [186]. Chemical cleaning aided with sonication could further improve the recovery efficiency. For instance, 1.0 M NaOH treatment on NF membrane with

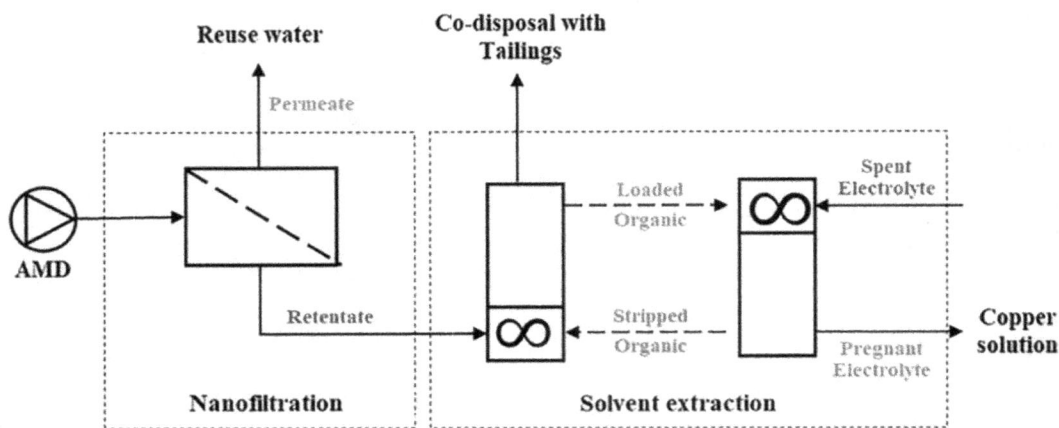

FIGURE 15.8 Schematic of solvent extraction of the retentate of NF on acid mine drainage [146].

4 minutes of sonication frequency of 24 kHz and power dissipation of 135 W renders a 90% water recovery after the membrane was fouled with organic foulant [187]. The spent NF3A (Zhejiang Mey Technology) after treatment of arsenic-rich brackish water could reach water recovery of 99.99% when regenerated via ultrasonic-assisted acid cleaning using citric acid at pH 3 [188].

15.6.4.2 Disposal of Retentate

The retentate is the concentrated stream produced during NF process. It contains concentrated rejected solutes, such as multivalent metal cations, large organic compounds and some impermeable monovalent ions. Considering water recovery of 80%, the concentration of metal cations could be quadrupled [146].

The NF retentate could be recycled to the primary physicochemical treatment followed by ozonation [189] or concentrated using reverse osmosis to further reduce the volume [190], or else sent to the metal recovery stage. The retentate could be subjected to chemical precipitation [159] or solvent extraction [146] to recover the metal for further reuse. Referring to Figure 15.8, Pino et al. [146] demonstrated solvent extraction of the NF retentate of acid mine drainage using hydroxyoxime extractants. About 97% of copper could be retrieved from the retentate using 10% (v/v) LIX 84-IC. In other NF retentate-containing organic compounds, advanced oxidation process such as ozonation [191] and electrochemical oxidation [192] could be applied.

15.6.5 INTEGRATED DESIGN OF NF SYSTEM

The standalone NF system might have its own limitations. First, the feed water is usually rich in other solutes/nutrients which can potentially foul the membrane. A pretreatment to remove the foulant is suggested prior to the NF process. Typically, the industrial effluent contains high initial concentration of heavy metals. If this effluent is directly treated using NF, the membrane could suffer from severe CP which limits the removal efficiency. The used membrane needs frequent regeneration to restore its performance. Worth mentioning also, feed such as electroplating effluent is acidic, which most polymeric membrane could not withstand. Therefore, a pH adjustment shall be done. Here in this section, a summary based on integrated design of NF system coupled with some other conventional processes in removal of heavy metals from industrial effluent is provided.

To further polish the quality of permeate from a feed with moderate high concentration of metals, a NF cascade was proposed. Simulated work showing a feed containing 120 ppm lead could be

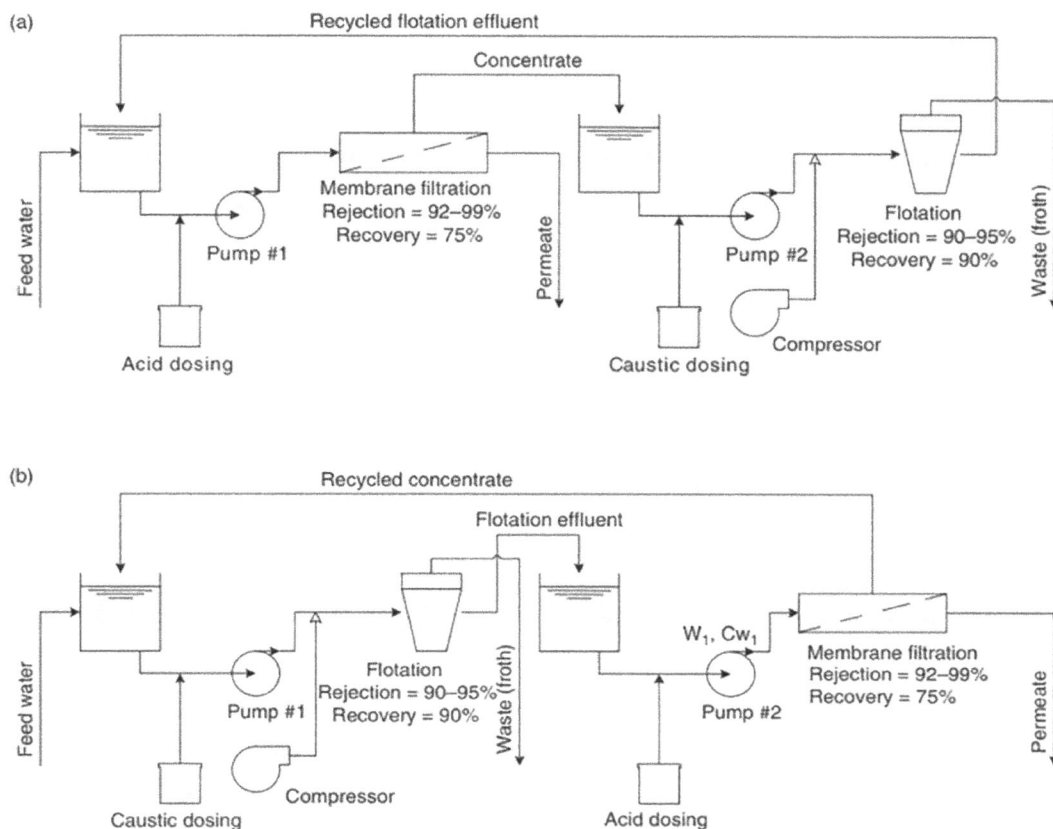

FIGURE 15.9 Integrated design of electroflotation and NF/RO membrane filtration for the removal of copper with (a) membrane filtration prior to flotation and (b) flotation prior to membrane filtration [195].

reduced to 10 ppb in less than four stages NF process using AFC80 [193]. Several works reported on the integrated design of NF-RO in treating effluent-containing heavy metals [148]. Zhu et al. [194] probed the feasibility of hybrid RO-NF process in treating electroplating rinse wastewater. Both setup has almost complete rejection on Ni^{2+}, yet the RO-NF setup has slight edge over the NF-RO in terms of water recovery, lower energy consumption and unit product cost.

Sudilovskiy et al. [195] proposed membrane-electroflotation integrated design in treating of effluent-containing copper. In their work, polyacrylamide was used as flocculant, while conventional spiral wound RO (TFM-75, GE Osmonics) and NF spiral wound (ERN-B-45–300, Vladipor) were used. They found that the sequence of the process could be significant to the copper rejection and the recovery of the water. Based on their design illustrated in Figure 15.9, membrane filtration prior to electroflotation would give slightly lower rejection (95%), but higher recovery (96.8%) and lower energy consumption (up to 1.1 kW h/m³) and reagent dosage. On the contrary, electroflotation prior to membrane filtration gives a higher rejection (99%), but lower recovery (87%), higher energy consumption (up to 1.3 kW h/m³) and reagent dosage.

Peyravi et al. [196] reported the integration of submerged aerobic powdered activated carbon-membrane bioreactor (MBR) with lab-fabricated polyamide NF in treating old landfill leachate. His work demonstrated that the NF pushed further the rejection of COD and heavy metals (zinc and nickel) to 94% and 99%, respectively, compared to 74% and around 70% removal using MBR alone.

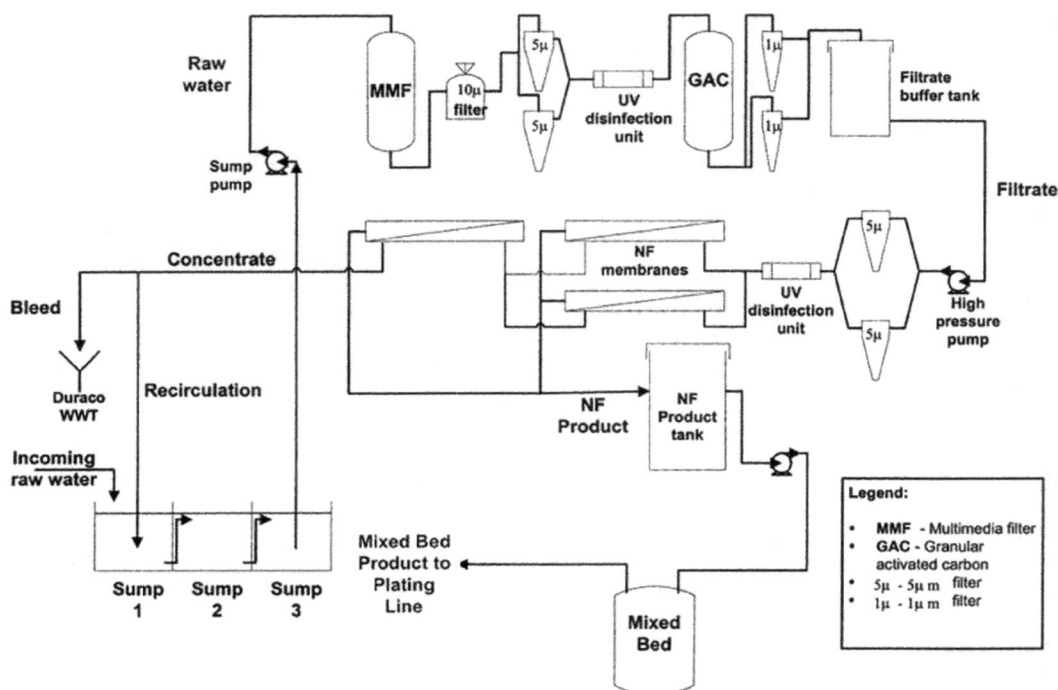

FIGURE 15.10   Schematic of integrated design for treating electroless plating effluent [198].

Wang et al. [197] developed a municipal landfill leachate treatment system based on combination of MBR-NF-RO process. No heavy metals (Pb, Cd and Cr) were detected in the NF effluent.

Wong et al. [198] developed a pilot plant for treating electroless plating as shown in Figure 15.10. The design consists of segregation of wastewater, pretreatment includes multimedia filter, UV disinfection, granular activated carbon adsorption and microfiltration to remove particulates, microorganisms, free chlorine, and to reduce TOC. The pretreatment effluent was subjected to NF and final polishing through ion exchange. The system rendered water recovery of 90% and the discharge was free of heavy metals and low conductivity (<5 µS/cm).

In addition to a conventional wastewater treatment consisting physical separation, chemical treatment and biological treatment, NF was suggested as the advanced treatment unit to further remove the remaining heavy metals in both textile and tannery effluent [199]. The conventional wastewater treatment facilities are sufficient to remove most suspended solids, organic solute and nutrients (such as nitrogen and phosphorus), but there is still high amount of heavy metals such as Cu, Pb, Sb, V, Si in textile manufacturing effluent, and Cr, Cu, Sb, Si for leather effluent. The NF was proven to be more effective than any other advanced treatment process such as adsorption and ozonation, especially when treating high concentration of heavy metals.

A hybrid membrane system involving pressure-retarded process (forward osmosis) and NF at lab-scale in the removal of heavy metals was reported. Choi et al. [200] examined the feasibility of FO-NF hybrid process in treating acid mine drainage, heavy metals such as Mn, As, Cd and Pb were completely rejected but Fe, Cu, Zn had rejection efficiencies 80%–85%. Hussain et al. [201] reported treatment of acidic molybdenum mining industry wastewater using hybrid selective electrodialysis and NF. The process gives coupling NF with SED could reach an acid recovery rate of over 94.2%. Hanif et al. [202] demonstrated NF as a final polishing process in hybrid biosorbent-NF processes for lead removal. The hybrid process rendered 98% of lead removal.

15.7 CONCLUSIONS

NF has been applied in treating industrial effluent-containing heavy metals, discharged from sources like electroplating, metallurgical industries, battery production, mine drainage and textile. NF membrane, either polymeric or inorganic, is available in several forms such as flat sheet, tubular and hollow fibre. These membranes perform separation through size exclusion, Donnan exclusion and selective adsorption mechanism. The process produces clean water permeate, while the valuable solutes like metals and acids could be recovered for further reuse. It can be easily scale up and its selectivity can be fine-tuned by varying the operating conditions.

NF membrane faces several critical issues such as fouling, scaling, CP and poor selectivity. They might cause irreversible deterioration to the permeation flux, which may not be fully restored either through physical or chemical cleaning approach. Nonetheless, early prevention of fouling could be done via pretreatment of feed by screening off suspended particulates using sand filter, removal of potential foulant through coagulation, adsorption or other classes of membrane like microfiltration and UF. Modification to the current NF membrane is needed, to make it more hydrophilic and antimicrobial, which can help to mitigate the adhesion of foulant and bacteria onto membrane surface. Process variables such as transmembrane pressure, crossflow velocity and feed pH could be manipulated to minimize the deposition of foulant on the membrane surface. The membrane should be modified with additional functionalities to prevent the deposition of foulant on membrane surface. Lastly, a cleaning process at the end of filtration could cleave the foulant from the membrane surface and restore the membrane performance.

Most commercial membranes are negatively charged. Therefore, the cationic metal cations could be easily adsorbed onto membrane surface and a concentration gradient could be existing across the feed and permeate side, which drift the diffusion of metals to the permeate. Current research focuses fabrication of positively charged NF membrane using amine-rich additives and monomers specifically for the removal of multivalent heavy metal cations. The positively charged NF membrane has a stronger repulsion towards metal cations; therefore, a higher removal efficiency could be obtained.

REFERENCES

1. A. Otero-Fernández, J.A. Otero, A. Maroto, J. Carmona, L. Palacio, P. Prádanos, A. Hernández, Concentration-polarization in nanofiltration of low concentration Cr(VI) aqueous solutions. Effect of operative conditions on retention, *J. Clean. Prod.* 150 (2017) 243–252. https://doi.org/10.1016/j.jclepro.2017.03.014.
2. A.S. Mohammed, A. Kapri, R. Goel, *Heavy Metal Pollution: Source, Impact, and Remedies BT – Biomanagement of Metal-Contaminated Soils*, in: M.S. Khan, A. Zaidi, R. Goel, J. Musarrat (Eds.), Springer, Dordrecht, 2011: pp. 1–28. https://doi.org/10.1007/978-94-007-1914-9_1.
3. V. Masindi, *Environmental Contamination by Heavy Metals*, in: K.L.M.E.-H.E.-D.M.S.E.-R.F. Aglan (Ed.), IntechOpen, Rijeka, 2018: Ch. 7. https://doi.org/10.5772/intechopen.76082.
4. Environmental Quality Act 1974, Environmental quality (industrial effluent) regulations 2009, n.d. https://www.doe.gov.my/portalv1/wp-content/uploads/2015/01/Environmental_Quality_Industrial_Effluent_Regulations_2009_-_P.U.A_434-2009.pdf (accessed January 17, 2022).
5. United States Environmental Protection Agency, Effluent Guidelines Database, n.d. https://www.epa.gov/eg/effluent-guidelines-database (accessed January 17, 2022).
6. J.A. Pandya, Nanofiltration for recovery of heavy metal from waste water, 2015.
7. P.B. Tchounwou, C.G. Yedjou, A.K. Patlolla, D.J. Sutton, Heavy metal toxicity and the environment, *Mol. Clin. Environ. Toxicol.* (2012) 133–164.
8. D.N. Wilson, Association cadmium. Cadmium-market trends and influences, in: *Cadmium 87 Proc. 6th Int. Cadmium Conf.*, London, 1988: pp. 9–16.
9. P.N. Gabby, Lead: In mineral commodity summaries, Reston, VA US, *Geol. Surv.* (2006).
10. J. Musilova, J. Arvay, A. Vollmannova, T. Toth, J. Tomas, Environmental contamination by heavy metals in region with previous mining activity, *Bull. Environ. Contam. Toxicol.* 97 (2016) 569–575.
11. L. Dresner, Some remarks on the integration of the extended Nernst-Planck equations in the hyperfiltration of multicomponent solutions, *Desalination.* 10 (1972) 27–46. https://doi.org/10.1016/S0011–9164(00)80245–3.

12. E.R. Nightingale, Phenomenological theory of ion solvation. Effective radii of hydrated ions, *J. Phys. Chem.* 63 (1959) 1381–1387. https://doi.org/10.1021/j150579a011.
13. R. Wang, S. Lin, Pore model for nanofiltration: History, theoretical framework, key predictions, limitations, and prospects, *J. Memb. Sci.* 620 (2021) 118809. https://doi.org/10.1016/j.memsci.2020.118809.
14. G. Basaran, D. Kavak, N. Dizge, Y. Asci, M. Solener, B. Ozbey, Comparative study of the removal of nickel(II) and chromium(VI) heavy metals from metal plating wastewater by two nanofiltration membranes, *Desalin. Water Treat.* 57 (2016) 21870–21880. https://doi.org/10.1080/19443994.2015.1127778.
15. G. Moradi, S. Zinadini, L. Rajabi, Development of the tetrathioterephthalate filler incorporated PES nanofiltration membrane with efficient heavy metal ions rejection and superior antifouling properties, *J. Environ. Chem. Eng.* 8 (2020) 104431. https://doi.org/10.1016/j.jece.2020.104431.
16. H. Zhang, S. Zhu, J. Yang, A. Ma, W. Chen, Enhanced removal efficiency of heavy metal ions by assembling phytic acid on polyamide nanofiltration membrane, *J. Memb. Sci.* 636 (2021) 119591. https://doi.org/10.1016/j.memsci.2021.119591.
17. B. Tansel, J. Sager, T. Rector, J. Garland, R.F. Strayer, L. Levine, M. Roberts, M. Hummerick, J. Bauer, Significance of hydrated radius and hydration shells on ionic permeability during nanofiltration in dead end and cross flow modes, *Sep. Purif. Technol.* 51 (2006) 40–47. https://doi.org/10.1016/j.seppur.2005.12.020.
18. T. Balanyà, J. Labanda, J. Llorens, J. Sabaté, Separation of metal ions and chelating agents by nanofiltration, *J. Memb. Sci.* 345 (2009) 31–35. https://doi.org/10.1016/j.memsci.2009.08.009.
19. B. Thabo, B.J. Okoli, S.J. Modise, S. Nelana, Rejection capacity of nanofiltration membranes for nickel, copper, silver and palladium at various oxidation states, *Membr.* 11 (2021). https://doi.org/10.3390/membranes11090653.
20. Y.H. Teow, J.Y. Sum, K.C. Ho, A.W. Mohammad, Principles of nanofiltration membrane processes, in: *Osmosis Eng.*, Elsevier, 2021: pp. 53–95.
21. M. Dalwani, N.E. Benes, G. Bargeman, D. Stamatialis, M. Wessling, Effect of pH on the performance of polyamide/polyacrylonitrile based thin film composite membranes, *J. Memb. Sci.* 372 (2011) 228–238. https://doi.org/10.1016/j.memsci.2011.02.012.
22. R.R. Sharma, R. Agrawal, S. Chellam, Temperature effects on sieving characteristics of thin-film composite nanofiltration membranes: Pore size distributions and transport parameters, *J. Memb. Sci.* 223 (2003) 69–87. https://doi.org/10.1016/S0376-7388(03)00310-7.
23. B.A.M. Al-Rashdi, D.J. Johnson, N. Hilal, Removal of heavy metal ions by nanofiltration, *Desalination.* 315 (2013) 2–17. https://doi.org/10.1016/j.desal.2012.05.022.
24. S. Lari, S.A.M. Parsa, S. Akbari, D. Emadzadeh, W.J. Lau, Fabrication and evaluation of nanofiltration membrane coated with amino-functionalized graphene oxide for highly efficient heavy metal removal, *Int. J. Environ. Sci. Technol.* 19 (2022) 4615–4626. https://doi.org/10.1007/s13762-021-03464-2.
25. X. Ren, C. Zhao, S. Du, T. Wang, Z. Luan, J. Wang, D. Hou, Fabrication of asymmetric poly (m-phenylene isophthalamide) nanofiltration membrane for chromium (VI) removal, *J. Environ. Sci.* 22 (2010) 1335–1341.
26. A.K. Shukla, J. Alam, M. Alhoshan, L. Arockiasamy Dass, F.A.A. Ali, M. M. R, U. Mishra, M.A. Ansari, Removal of heavy metal ions using a carboxylated graphene oxide-incorporated polyphenylsulfone nanofiltration membrane, *Environ. Sci. Water Res. Technol.* 4 (2018) 438–448. https://doi.org/10.1039/C7EW00506G.
27. M. Mulder, J. Mulder, *Basic Principles of Membrane Technology*, Springer Science & Business Media, Berlin, Ger. X. 79234247, 1996.
28. Y. Qi, L. Zhu, X. Shen, A. Sotto, C. Gao, J. Shen, Polyethyleneimine-modified original positive charged nanofiltration membrane: Removal of heavy metal ions and dyes, *Sep. Purif. Technol.* 222 (2019) 117–124. https://doi.org/10.1016/j.seppur.2019.03.083.
29. B. Al-Rashdi, C. Somerfield, N. Hilal, Heavy metals removal using adsorption and nanofiltration techniques, *Sep. Purif. Rev.* 40 (2011) 209–259. https://doi.org/10.1080/15422119.2011.558165.
30. R.G. Pearson, Absolute electronegativity and hardness: Application to inorganic chemistry, *Inorg. Chem.* 27 (1988) 734–740. https://doi.org/10.1021/ic00277a030.
31. T.L. Ho, Hard soft acids bases (HSAB) principle and organic chemistry, *Chem. Rev.* 75 (1975) 1–20. https://doi.org/10.1021/cr60293a001.
32. G. Abdi, A. Alizadeh, S. Zinadini, G. Moradi, Removal of dye and heavy metal ion using a novel synthetic polyethersulfone nanofiltration membrane modified by magnetic graphene oxide/metformin hybrid, *J. Memb. Sci.* 552 (2018) 326–335. https://doi.org/10.1016/j.memsci.2018.02.018.
33. G. Zeng, Y. He, Y. Zhan, L. Zhang, Y. Pan, C. Zhang, Z. Yu, Novel polyvinylidene fluoride nanofiltration membrane blended with functionalized halloysite nanotubes for dye and heavy metal ions removal, *J. Hazard. Mater.* 317 (2016) 60–72.

34. S.E. Demirel, J. Li, M.M.F. Hasan, Membrane separation process design and intensification, *Ind. Eng. Chem. Res.* 60 (2021) 7197–7217. https://doi.org/10.1021/acs.iecr.0c05072.

35. M. Jain, D. Attarde, S.K. Gupta, Removal of thiophenes from FCC gasoline by using a hollow fiber pervaporation module: Modeling, validation, and influence of module dimensions and flow directions, *Chem. Eng. J.* 308 (2017) 632–648. https://doi.org/10.1016/j.cej.2016.09.043.

36. S.A. Younssi, *Alumina Membranes for Desalination and Water Treatment*, in: M. Breida (Ed.), IntechOpen, Rijeka, 2018: Ch. 11. https://doi.org/10.5772/intechopen.76782.

37. S. Sridhar, S. Bee, S. Bhargava, Membrane-based gas separation: Principle, applications and future potential, *Chem. Eng. Dig.* 1 (2014) 1–25.

38. M. Scholz, M. Wessling, J. Balster, Chapter 5: Design of membrane modules for gas separations, in: *Membr. Eng. Treat. Gases Vol. 1 Gas-Separation Probl. with Membr.*, The Royal Society of Chemistry, 2011: pp. 125–149. https://doi.org/10.1039/9781849733472-00125.

39. Z. Berk, Membrane processes, in: *Food Process Eng. Technol.*, Academic Press, 2009: pp. 233–257.

40. J. Balster, *Plate and Frame Membrane Module BT – Encyclopedia of Membranes*, in: E. Drioli, L. Giorno (Eds.), Springer, Berlin, Heidelberg, 2015: pp. 1–3. https://doi.org/10.1007/978-3-642-40872-4_1584-1.

41. A.G. (Tony) Fane, R. Wang, Y. Jia, *Membrane Technology: Past, Present and Future BT – Membrane and Desalination Technologies*, in: L.K. Wang, J.P. Chen, Y.-T. Hung, N.K. Shammas (Eds.), Humana Press, Totowa, NJ, 2011: pp. 1–45. https://doi.org/10.1007/978-1-59745-278-6_1.

42. J. López, M. Reig, X. Vecino, J.L. Cortina, Arsenic impact on the valorisation schemes of acidic mine waters of the Iberian Pyrite Belt: Integration of selective precipitation and spiral-wound nanofiltration processes, *J. Hazard. Mater.* 403 (2021) 123886. https://doi.org/10.1016/j.jhazmat.2020.123886.

43. E. Obotey Ezugbe, S. Rathilal, Membrane technologies in wastewater treatment: A review, *Membranes (Basel).* 10 (2020) 89.

44. K.K. Chen, W. Salim, Y. Han, D. Wu, W.S.W. Ho, Fabrication and scale-up of multi-leaf spiral-wound membrane modules for CO2 capture from flue gas, *J. Memb. Sci.* 595 (2020) 117504. https://doi.org/10.1016/j.memsci.2019.117504.

45. A.J. Karabelas, C.P. Koutsou, M. Kostoglou, The effect of spiral wound membrane element design characteristics on its performance in steady state desalination – A parametric study, *Desalination.* 332 (2014) 76–90. https://doi.org/10.1016/j.desal.2013.10.027.

46. J. Balster, Spiral Wound Membrane Module BT – Encyclopedia of Membranes, in: E. Drioli, L. Giorno (Eds.), Springer, Berlin, Heidelberg, 2016: pp. 1812–1814. https://doi.org/10.1007/978-3-662-44324-8_1586.

47. W.S. Tan, C.K. Chua, T.H. Chong, A.G. Fane, A. Jia, 3D printing by selective laser sintering of polypropylene feed channel spacers for spiral wound membrane modules for the water industry, *Virtual Phys. Prototyp.* 11 (2016) 151–158. https://doi.org/10.1080/17452759.2016.1211925.

48. C.P. Koutsou, A.J. Karabelas, A novel retentate spacer geometry for improved spiral wound membrane (SWM) module performance, *J. Memb. Sci.* 488 (2015) 129–142. https://doi.org/10.1016/j.memsci.2015.03.064.

49. W.S. Tan, S.R. Suwarno, J. An, C.K. Chua, A.G. Fane, T.H. Chong, Comparison of solid, liquid and powder forms of 3D printing techniques in membrane spacer fabrication, *J. Memb. Sci.* 537 (2017) 283–296. https://doi.org/10.1016/j.memsci.2017.05.037.

50. G.S.S. Goon, O. Labban, Z.H. Foo, X. Zhao, J.H. Lienhard, Deformation-induced cleaning of organically fouled membranes: Fundamentals and techno-economic assessment for spiral-wound membranes, *J. Memb. Sci.* 626 (2021) 119169. https://doi.org/10.1016/j.memsci.2021.119169.

51. C. Dong, R. He, S. Xu, H. He, H. Chen, Y.-B. Zhang, T. He, Layer-by-layer (LBL) hollow fiber nanofiltration membranes for seawater treatment: Ion rejection, *Desalination.* 534 (2022) 115793. https://doi.org/10.1016/j.desal.2022.115793.

52. N. Fazil, S. Saqib, A. Mukhtar, M. Younas, M. Rezakazemi, Module design and membrane materials, *Membr. Contactor Technol.* (2022) 99–142. https://doi.org/10.1002/9783527831036.ch4.

53. P.M. Visakh, O. Nazarenko, *Nanostructured Polymer Membranes, Volume 2: Applications*, John Wiley & Sons, 2016.

54. Z.F. Cui, Y. Jiang, R.W. Field, *Fundamentals of Pressure-Driven Membrane Separation Processes*, First Edition, Elsevier Ltd, 2010. https://doi.org/10.1016/B978-1-85617-632-3.00001-X.

55. M.M. Martín, *Industrial Chemical Process Analysis and Design*, Elsevier, 2016.

56. J. Balster, *Tubular Membrane Module BT – Encyclopedia of Membranes*, in: E. Drioli, L. Giorno (Eds.), Springer, Berlin, Heidelberg, 2015: p. 1. https://doi.org/10.1007/978-3-642-40872-4_1588-1.

57. R.W. Baker, *Membrane Technology and Applications*, John Wiley & Sons, 2012.

58. R.P. Singh, G.J. Dahe, K.W. Dudeck, C.F. Welch, K.A. Berchtold, High temperature polybenzimidazole hollow fiber membranes for hydrogen separation and carbon dioxide capture from synthesis gas, *Energy Procedia.* 63 (2014) 153–159. https://doi.org/10.1016/j.egypro.2014.11.015.

59. N. Zhang, X. Chen, Y. Su, H. Zheng, O. Ramadan, X. Zhang, H. Chen, S. Riffat, Numerical investiga-
    tions and performance comparisons of a novel cross-flow hollow fiber integrated liquid desiccant dehu-
    midification system, *Energy*. 182 (2019) 1115–1131. https://doi.org/10.1016/j.energy.2019.06.036.
60. J. Balster, *Tubular Membrane Module BT – Encyclopedia of Membranes*, in: E. Drioli, L. Giorno (Eds.),
    Springer, Berlin, Heidelberg, 2016: pp. 1939–1940. https://doi.org/10.1007/978-3-662-44324-8_1588.
61. J. Balster, *Hollow Fiber Membrane Module BT – Encyclopedia of Membranes*, in: E. Drioli, L. Giorno
    (Eds.), Springer, Berlin, Heidelberg, 2016: pp. 955–957. https://doi.org/10.1007/978-3-662-44324-8_1583.
62. M.A. Alaei Shahmirzadi, S.S. Hosseini, G. Ruan, N.R. Tan, Tailoring PES nanofiltration membranes
    through systematic investigations of prominent design, fabrication and operational parameters, *RSC
    Adv*. 5 (2015) 49080–49097. https://doi.org/10.1039/C5RA05985B.
63. Y. Liu, J. Zhu, J. Zheng, X. Gao, J. Wang, X. Wang, Y.F. Xie, X. Huang, B. Van der Bruggen, A facile and
    scalable fabrication procedure for thin-film composite membranes: Integration of phase inversion and inter-
    facial polymerization, *Environ. Sci. Technol*. 54 (2020) 1946–1954. https://doi.org/10.1021/acs.est.9b06426.
64. Y. Li, J. Zhu, S. Li, Z. Guo, B. Van der Bruggen, Flexible aliphatic–aromatic polyamide thin film com-
    posite membrane for highly efficient organic solvent nanofiltration, *ACS Appl. Mater. Interfaces*. 12
    (2020) 31962–31974. https://doi.org/10.1021/acsami.0c07341.
65. Z. Zhou, D. Lu, X. Li, L.M. Rehman, A. Roy, Z. Lai, Fabrication of highly permeable polyamide mem-
    branes with large "leaf-like" surface nanostructures on inorganic supports for organic solvent nanofil-
    tration, *J. Memb. Sci*. 601 (2020) 117932. https://doi.org/10.1016/j.memsci.2020.117932.
66. D.M. Warsinger, S. Chakraborty, E.W. Tow, M.H. Plumlee, C. Bellona, S. Loutatidou, L. Karimi, A.M.
    Mikelonis, A. Achilli, A. Ghassemi, L.P. Padhye, S.A. Snyder, S. Curcio, C.D. Vecitis, H.A. Arafat, J.H.
    Lienhard, A review of polymeric membranes and processes for potable water reuse, *Prog. Polym. Sci*.
    81 (2018) 209–237. https://doi.org/10.1016/j.progpolymsci.2018.01.004.
67. Y. Shi, C. Li, D. He, L. Shen, N. Bao, Preparation of graphene oxide–cellulose acetate nanocomposite
    membrane for high-flux desalination, *J. Mater. Sci*. 52 (2017) 13296–13306. https://doi.org/10.1007/
    s10853-017-1403-0.
68. E.N. Durmaz, P. Zeynep Çulfaz-Emecen, Cellulose-based membranes via phase inversion using
    [EMIM]OAc-DMSO mixtures as solvent, *Chem. Eng. Sci*. 178 (2018) 93–103. https://doi.org/10.1016/j.
    ces.2017.12.020.
69. A. Anand, B. Unnikrishnan, J.Y. Mao, H.J. Lin, C.C. Huang, Graphene-based nanofiltration membranes
    for improving salt rejection, water flux and antifouling–A review, *Desalination*. 429 (2018) 119–133.
    https://doi.org/10.1016/j.desal.2017.12.012.
70. S. Bandehali, F. Parvizian, H. Ruan, A. Moghadassi, J. Shen, A. Figoli, A.S. Adeleye, N. Hilal, T.
    Matsuura, E. Drioli, S.M. Hosseini, A planned review on designing of high-performance nanocompos-
    ite nanofiltration membranes for pollutants removal from water, *J. Ind. Eng. Chem*. 101 (2021) 78–125.
    https://doi.org/10.1016/j.jiec.2021.06.022.
71. M.C. Hacıfazlıoğlu, H.R. Tomasini, L. Bertin, T.Ö. Pek, N. Kabay, Concentrate reduction in NF and
    RO desalination systems by membrane-in-series configurations-evaluation of product water for reuse in
    irrigation, *Desalination*. 466 (2019) 89–96. https://doi.org/10.1016/j.desal.2019.05.011.
72. L. Gao, H. Wang, Y. Zhang, M. Wang, Nanofiltration membrane characterization and application:
    Extracting lithium in lepidolite leaching solution, *Membranes (Basel)*. 10 (2020) 178.
73. P. Religa, A. Kowalik, P. Gierycz, Effect of membrane properties on chromium(III) recirculation
    from concentrate salt mixture solution by nanofiltration, *Desalination*. 274 (2011) 164–170. https://doi.
    org/10.1016/j.desal.2011.02.006.
74. I. Kowalska, A. Klimonda, Application of nanofiltration membranes for removal of surfactants from
    water solutions, in: *E3S Web Conf*., EDP Sciences, 2017: p. 44.
75. P. Pasqualin, P. Davies, Experimental and numerical investigations of nanofiltration membranes for
    liquid desiccant regeneration in air conditioning applications, Available SSRN 4075669. (n.d.).
76. J. Zdarta, A. Thygesen, M.S. Holm, A.S. Meyer, M. Pinelo, Direct separation of acetate and furfural
    from xylose by nanofiltration of birch pretreated liquor: Effect of process conditions and separation
    mechanism, *Sep. Purif. Technol*. 239 (2020) 116546. https://doi.org/10.1016/j.seppur.2020.116546.
77. S.S. Wadekar, Separation potential characterization and its role in selecting nanofiltration membranes
    for the removal of inorganic ions, (2018).
78. D. Peng, Disinfection by-products and the application potential of nanofiltration in swimming pool
    water treatment, (2016).
79. N.C. Magalhães, A.F.R. Silva, P.V.M. Cunha, J.E. Drewes, M.C.S. Amaral, Role of nanofiltration or
    reverse osmosis integrated to ultrafiltration-anaerobic membrane bioreactor treating vinasse for the con-
    servation of water and nutrients in the ethanol industry, *J. Water Process Eng*. 36 (2020) 101338. https://
    doi.org/10.1016/j.jwpe.2020.101338.

80. J. Ding, H. Liang, X. Zhu, D. Xu, X. Luo, Z. Wang, L. Bai, Surface modification of nanofiltration membranes with zwitterions to enhance antifouling properties during brackish water treatment: A new concept of a "buffer layer", *J. Memb. Sci.* 637 (2021) 119651. https://doi.org/10.1016/j.memsci.2021.119651.
81. D.J. Miller, D.R. Dreyer, C.W. Bielawski, D.R. Paul, B.D. Freeman, Surface modification of water purification membranes, *Angew. Chemie Int. Ed.* 56 (2017) 4662–4711. https://doi.org/10.1002/anie.201601509.
82. C. Wang, M.J. Park, D.H. Seo, H.K. Shon, Inkjet printing of graphene oxide and dopamine on nanofiltration membranes for improved anti-fouling properties and chlorine resistance, *Sep. Purif. Technol.* 254 (2021) 117604. https://doi.org/10.1016/j.seppur.2020.117604.
83. A. Li, Q. Rao, F. Liang, L. Song, X. Zhan, F. Chen, J. Chen, Q. Zhang, Polyhydroxy group functionalized zwitterion for a polyamide nanofiltration membrane with high water permeation and antifouling performance, *ACS Appl. Polym. Mater.* 2 (2020) 3850–3858. https://doi.org/10.1021/acsapm.0c00551.
84. F. Oulad, S. Zinadini, A.A. Zinatizadeh, A.A. Derakhshan, Fabrication and characterization of a novel tannic acid coated boehmite/PES high performance antifouling NF membrane and application for licorice dye removal, *Chem. Eng. J.* 397 (2020) 125105. https://doi.org/10.1016/j.cej.2020.125105.
85. Y.S. Khoo, W.J. Lau, Y.Y. Liang, M. Karaman, M. Gürsoy, A.F. Ismail, A green approach to modify surface properties of polyamide thin film composite membrane for improved antifouling resistance, *Sep. Purif. Technol.* 250 (2020) 116976. https://doi.org/10.1016/j.seppur.2020.116976.
86. X. Lai, C. Wang, L. Wang, C. Xiao, A novel PPTA/PPy composite organic solvent nanofiltration (OSN) membrane prepared by chemical vapor deposition for organic dye wastewater treatment, *J. Water Process Eng.* 45 (2022) 102533. https://doi.org/10.1016/j.jwpe.2021.102533.
87. T. Hong Anh Ngo, K. Dinh Do, D. Thi Tran, Surface modification of polyamide TFC membranes via redox-initiated graft polymerization of acrylic acid, *J. Appl. Polym. Sci.* 134 (2017) 45110. https://doi.org/10.1002/app.45110.
88. T. Hong Anh Ngo, D.T. Tran, C. Hung Dinh, Surface photochemical graft polymerization of acrylic acid onto polyamide thin film composite membranes, *J. Appl. Polym. Sci.* 134 (2017). https://doi.org/10.1002/app.44418.
89. V. Vatanpour, M.E. Pasaoglu, H. Barzegar, O.O. Teber, R. Kaya, M. Bastug, A. Khataee, I. Koyuncu, Cellulose acetate in fabrication of polymeric membranes: A review, *Chemosphere.* 295 (2022) 133914. https://doi.org/10.1016/j.chemosphere.2022.133914.
90. F. Galiano, K. Briceño, T. Marino, A. Molino, K.V. Christensen, A. Figoli, Advances in biopolymer-based membrane preparation and applications, *J. Memb. Sci.* 564 (2018) 562–586. https://doi.org/10.1016/j.memsci.2018.07.059.
91. L.H.T. Lyly, B.S. Ooi, W.J. Lim, Y.S. Chang, C.J.C. Derek, J.K. Lim, Desalinating microalgal-rich water via thermoresponsive membrane distillation, *J. Environ. Chem. Eng.* 9 (2021) 105897. https://doi.org/10.1016/j.jece.2021.105897.
92. L.H.T. Lyly, Y.S. Chang, W.M. Ng, J.K. Lim, C.J.C. Derek, B.S. Ooi, Development of membrane distillation by dosing SiO2-PNIPAM with thermal cleaning properties via surface energy actuation, *J. Memb. Sci.* 636 (2021) 119193. https://doi.org/10.1016/j.memsci.2021.119193.
93. L.H.T. Lyly, B.S. Ooi, J.K. Lim, C.J.C. Derek, S.C. Low, Correlating the membrane surface energy to the organic fouling and wetting of membrane distillation at elevated temperature, *J. Environ. Chem. Eng.* 9 (2021) 104627. https://doi.org/10.1016/j.jece.2020.104627.
94. S. Gao, J. Sun, P. Liu, F. Zhang, W. Zhang, S. Yuan, J. Li, J. Jin, A robust polyionized hydrogel with an unprecedented underwater anti-crude-oil-adhesion property, *Adv. Mater.* 28 (2016) 5307–5314. https://doi.org/10.1002/adma.201600417.
95. C. Liu, W. Bi, D. Chen, S. Zhang, H. Mao, Positively charged nanofiltration membrane fabricated by poly(acid–base) complexing effect induced phase inversion method for heavy metal removal, *Chinese J. Chem. Eng.* 25 (2017) 1685–1694. https://doi.org/10.1016/j.cjche.2017.06.001.
96. J. Zhu, J. Zheng, C. Liu, S. Zhang, Ionic complexing induced fabrication of highly permeable and selective polyacrylic acid complexed poly (arylene ether sulfone) nanofiltration membranes for water purification, *J. Memb. Sci.* 520 (2016) 130–138. https://doi.org/10.1016/j.memsci.2016.07.059.
97. S. Emonds, J. Kamp, J. Borowec, H. Roth, M. Wessling, Polyelectrolyte complex tubular membranes via a salt dilution induced phase inversion process, *Adv. Eng. Mater.* 23 (2021) 2001401. https://doi.org/10.1002/adem.202001401.
98. A.K. Rana, V.K. Gupta, A.K. Saini, S.I. Voicu, M.H. Abdellattifaand, V.K. Thakur, Water desalination using nanocelluloses/cellulose derivatives based membranes for sustainable future, *Desalination.* 520 (2021) 115359. https://doi.org/10.1016/j.desal.2021.115359.
99. S. Akram, V. Naddeo, Z. Rehan, M. Zahid, A. Rashid, W. Razzaq, A comprehensive review on polymeric nano-composite membranes for water treatment, *J. Membr. Sci. Technol.* 8 (2018). https://doi.org/10.4172/2155–9589.1000179.

100. N. Angel, S. Li, F. Yan, L. Kong, Recent advances in electrospinning of nanofibers from bio-based carbohydrate polymers and their applications, *Trends Food Sci. Technol.* 120 (2022) 308–324. https://doi.org/10.1016/j.tifs.2022.01.003.

101. M. Cha, C. Boo, C. Park, Simultaneous retention of organic and inorganic contaminants by a ceramic nanofiltration membrane for the treatment of semiconductor wastewater, *Process Saf. Environ. Prot.* 159 (2022) 525–533. https://doi.org/10.1016/j.psep.2022.01.032.

102. F. Lu, D. Astruc, Nanocatalysts and other nanomaterials for water remediation from organic pollutants, *Coord. Chem. Rev.* 408 (2020) 213180. https://doi.org/10.1016/j.ccr.2020.213180.

103. S.M. Samaei, S. Gato-Trinidad, A. Altaee, The application of pressure-driven ceramic membrane technology for the treatment of industrial wastewaters – A review, *Sep. Purif. Technol.* 200 (2018) 198–220. https://doi.org/10.1016/j.seppur.2018.02.041.

104. E.S. Mirdamadi, M.H. Nazarpak, M. Solati-Hashjin, Metal oxide-based ceramics 10, *Struct. Biomater. Prop. Charact. Sel.* (2021) 301.

105. Z. Yang, Y. Zhou, Z. Feng, X. Rui, T. Zhang, Z. Zhang, A review on reverse osmosis and nanofiltration membranes for water purification, *Polymers (Basel).* 11 (2019) 1–22. https://doi.org/10.3390/polym11081252.

106. P. Pandey, R.S. Chauhan, Membranes for gas separation, *Prog. Polym. Sci.* 26 (2001) 853–893.

107. K.K.O.S. Silva, C.A. Paskocimas, F.R. Oliveira, J.H.O. Nascimento, A. Zille, Development of porous alumina membranes for treatment of textile effluent, *Desalin. Water Treat.* 57 (2016) 2640–2648. https://doi.org/10.1080/19443994.2015.1018333.

108. Z. Wang, Y.-M. Wei, Z.-L. Xu, Y. Cao, Z.-Q. Dong, X.-L. Shi, Preparation, characterization and solvent resistance of γ-Al2O3/α-Al2O3 inorganic hollow fiber nanofiltration membrane, *J. Memb. Sci.* 503 (2016) 69–80. https://doi.org/10.1016/j.memsci.2015.12.039.

109. H. Guo, S. Zhao, X. Wu, H. Qi, Fabrication and characterization of TiO2/ZrO2 ceramic membranes for nanofiltration, *Microporous Mesoporous Mater.* 260 (2018) 125–131. https://doi.org/10.1016/j.micromeso.2016.03.011.

110. C. Li, W. Sun, Z. Lu, X. Ao, C. Yang, S. Li, Systematic evaluation of TiO2-GO-modified ceramic membranes for water treatment: Retention properties and fouling mechanisms, *Chem. Eng. J.* 378 (2019) 122138. https://doi.org/10.1016/j.cej.2019.122138.

111. N. Rao, R. Singh, L. Bashambu, Carbon-based nanomaterials: Synthesis and prospective applications, *Mater. Today Proc.* 44 (2021) 608–614. https://doi.org/10.1016/j.matpr.2020.10.593.

112. P.S. Goh, A.F. Ismail, A review on inorganic membranes for desalination and wastewater treatment, *Desalination.* 434 (2018) 60–80. https://doi.org/10.1016/j.desal.2017.07.023.

113. L. Bazli, M. Siavashi, A. Shiravi, A review of carbon nanotube/TiO2 composite prepared via sol-gel method 1 (2019) 1–12. https://doi.org/10.29252/jcc.1.1.1.

114. N. Lavoine, L. Bergström, Nanocellulose-based foams and aerogels: Processing, properties, and applications, *J. Mater. Chem. A.* 5 (2017) 16105–16117. https://doi.org/10.1039/C7TA02807E.

115. R. Gusain, N. Kumar, S.S. Ray, Recent advances in carbon nanomaterial-based adsorbents for water purification, *Coord. Chem. Rev.* 405 (2020) 213111. https://doi.org/10.1016/j.ccr.2019.213111.

116. M. Faustini, L. Nicole, E. Ruiz-Hitzky, C. Sanchez, History of organic–inorganic hybrid materials: Prehistory, art, science, and advanced applications, *Adv. Funct. Mater.* 28 (2018) 1704158. https://doi.org/10.1002/adfm.201704158.

117. C. Li, W. Sun, Z. Lu, X. Ao, S. Li, Ceramic nanocomposite membranes and membrane fouling: A review, *Water Res.* 175 (2020) 115674. https://doi.org/10.1016/j.watres.2020.115674.

118. S.T. Muntha, A. Kausar, M. Siddiq, A review featuring fabrication, properties, and application of polymeric mixed matrix membrane reinforced with different fillers, *Polym. Plast. Technol. Eng.* 56 (2017) 2043–2064. https://doi.org/10.1080/03602559.2017.1298801.

119. M.S. Sri Abirami Saraswathi, A. Nagendran, D. Rana, Tailored polymer nanocomposite membranes based on carbon, metal oxide and silicon nanomaterials: A review, *J. Mater. Chem. A.* 7 (2019) 8723–8745. https://doi.org/10.1039/C8TA11460A.

120. P. Sherugar, N.S. Naik, M. Padaki, V. Nayak, A. Gangadharan, A.R. Nadig, S. Déon, Fabrication of zinc doped aluminium oxide/polysulfone mixed matrix membranes for enhanced antifouling property and heavy metal removal, *Chemosphere.* 275 (2021) 130024. https://doi.org/10.1016/j.chemosphere.2021.130024.

121. G. Moradi, S. Zinadini, L. Rajabi, A. Ashraf Derakhshan, Removal of heavy metal ions using a new high performance nanofiltration membrane modified with curcumin boehmite nanoparticles, *Chem. Eng. J.* 390 (2020) 124546. https://doi.org/10.1016/j.cej.2020.124546.

122. Y.X. Li, Y. Cao, M. Wang, Z.L. Xu, H.Z. Zhang, X.W. Liu, Z. Li, Novel high-flux polyamide/TiO2 composite nanofiltration membranes on ceramic hollow fibre substrates, *J. Memb. Sci.* 565 (2018) 322–330. https://doi.org/10.1016/j.memsci.2018.08.014.

123. S. Roy, S.A. Ntim, S. Mitra, K.K. Sirkar, Facile fabrication of superior nanofiltration membranes from interfacially polymerized CNT-polymer composites, *J. Memb. Sci.* 375 (2011) 81–87. https://doi.org/10.1016/j.memsci.2011.03.012.

124. V. Grosso, D. Vuono, M.A. Bahattab, G. Di Profio, E. Curcio, S.A. Al-Jilil, F. Alsubaie, M. Alfife, J. B.Nagy, E. Drioli, E. Fontananova, Polymeric and mixed matrix polyimide membranes, *Sep. Purif. Technol.* 132 (2014) 684–696. https://doi.org/10.1016/j.seppur.2014.06.023.

125. M.H. Davood Abadi Farahani, D. Hua, T.-S. Chung, Cross-linked mixed matrix membranes consisting of carboxyl-functionalized multi-walled carbon nanotubes and P84 polyimide for organic solvent nanofiltration (OSN), *Sep. Purif. Technol.* 186 (2017) 243–254. https://doi.org/10.1016/j.seppur.2017.06.021.

126. M.H. Davood Abadi Farahani, T.-S. Chung, Solvent resistant hollow fiber membranes comprising P84 polyimide and amine-functionalized carbon nanotubes with potential applications in pharmaceutical, food, and petrochemical industries, *Chem. Eng. J.* 345 (2018) 174–185. https://doi.org/10.1016/j.cej.2018.03.153.

127. A.W. Mohammad, Y.H. Teow, W.L. Ang, Y.T. Chung, D.L. Oatley-Radcliffe, N. Hilal, Nanofiltration membranes review: Recent advances and future prospects, *Desalination.* 356 (2015) 226–254. https://doi.org/10.1016/j.desal.2014.10.043.

128. S. Shirazi, C.J. Lin, D. Chen, Inorganic fouling of pressure-driven membrane processes - A critical review, *Desalination.* 250 (2010) 236–248. https://doi.org/10.1016/j.desal.2009.02.056.

129. J.M. Ochando-Pulido, A. Martínez-Ferez, Fouling modelling on a reverse osmosis membrane in the purification of pretreated olive mill wastewater by adapted crossflow blocking mechanisms, *J. Memb. Sci.* 544 (2017) 108–118. https://doi.org/10.1016/j.memsci.2017.09.018.

130. R.R. Choudhury, J.M. Gohil, S. Mohanty, S.K. Nayak, Antifouling, fouling release and antimicrobial materials for surface modification of reverse osmosis and nanofiltration membranes, *J. Mater. Chem. A.* 6 (2018) 313–333. https://doi.org/10.1039/c7ta08627j.

131. M. El Batouti, N.F. Alharby, M.M. Elewa, Review of new approaches for fouling mitigation in membrane separation processes in water treatment applications, *Separations.* 9 (2022). https://doi.org/10.3390/separations9010001.

132. N. Alsawaftah, W. Abuwatfa, N. Darwish, G. Husseini, A comprehensive review on membrane fouling : Mathematical, *Water.* 13 (2021) 1–37.

133. P.S. Goh, W.J. Lau, M.H.D. Othman, A.F. Ismail, Membrane fouling in desalination and its mitigation strategies, *Desalination.* 425 (2018) 130–155. https://doi.org/10.1016/j.desal.2017.10.018.

134. M.A. Abdel-Fatah, Nanofiltration systems and applications in wastewater treatment: Review article, *Ain Shams Eng. J.* 9 (2018) 3077–3092. https://doi.org/10.1016/j.asej.2018.08.001.

135. W. Zhang, J. Luo, L. Ding, M.Y. Jaffrin, A review on flux decline control strategies in pressure-driven membrane processes, *Ind. Eng. Chem. Res.* 54 (2015) 2843–2861. https://doi.org/10.1021/ie504848m.

136. A. Matin, T. Laoui, W. Falath, M. Farooque, Fouling control in reverse osmosis for water desalination & reuse: Current practices & emerging environment-friendly technologies, *Sci. Total Environ.* 765 (2021) 142721. https://doi.org/10.1016/j.scitotenv.2020.142721.

137. A. Ali, M.C. Nymann, M.L. Christensen, C.A. Quist-Jensen, Industrial wastewater treatment by nanofiltration—a case study on the anodizing industry, *Membranes (Basel).* 10 (2020). https://doi.org/10.3390/membranes10050085.

138. M. Mullett, R. Fornarelli, D. Ralph, Nanofiltration of mine water: Impact of feed pH and membrane charge on resource recovery and water discharge, *Membranes (Basel).* 4 (2014) 163–180. https://doi.org/10.3390/membranes4020163.

139. B.S. Ooi, Inorganic acids recovery by nanofiltration BT – Encyclopedia of membranes, in: E. Drioli, L. Giorno (Eds.), Springer, Berlin, Heidelberg, 2016: pp. 1040–1041. https://doi.org/10.1007/978-3-662-44324-8_1780.

140. J.E. Efome, D. Rana, T. Matsuura, C.Q. Lan, Effects of operating parameters and coexisting ions on the efficiency of heavy metal ions removal by nano-fibrous metal-organic framework membrane filtration process, *Sci. Total Environ.* 674 (2019) 355–362. https://doi.org/10.1016/j.scitotenv.2019.04.187.

141. G.T. Ballet, A. Hafiane, M. Dhahbi, Influence of operating conditions on the retention of nickel in water by nanofiltration, *Desalin. Water Treat.* 9 (2009) 28–35. https://doi.org/10.5004/dwt.2009.749.

142. J. López, M. Reig, O. Gibert, J.L. Cortina, Increasing sustainability on the metallurgical industry by integration of membrane nanofiltration processes: Acid recovery, *Sep. Purif. Technol.* 226 (2019) 267–277. https://doi.org/10.1016/j.seppur.2019.05.100.

143. A. Maher, M. Sadeghi, A. Moheb, Heavy metal elimination from drinking water using nanofiltration membrane technology and process optimization using response surface methodology, *Desalination.* 352 (2014) 166–173. https://doi.org/10.1016/j.desal.2014.08.023.

144. J.Y. Sum, A.L. Ahmad, B.S. Ooi, Selective separation of heavy metal ions using amine-rich polyamide TFC membrane, *J. Ind. Eng. Chem.* 76 (2019). https://doi.org/10.1016/j.jiec.2019.03.052.

145. D. Qadir, H.B. Mukhtar, L.K. Keong, Rejection of divalent ions in commercial tubular membranes: Effect of feed concentration and anion type, *Sustain. Environ. Res.* 27 (2017) 103–106. https://doi.org/10.1016/j.serj.2016.12.002.

146. L. Pino, E. Beltran, A. Schwarz, M.C. Ruiz, R. Borquez, Optimization of nanofiltration for treatment of acid mine drainage and copper recovery by solvent extraction, *Hydrometallurgy.* 195 (2020) 105361. https://doi.org/10.1016/j.hydromet.2020.105361.

147. S.I. Lazarev, I. V Khorokhorina, K. V Shestakov, D.S. Lazarev, Recovery of zinc, copper, nickel, and cobalt from electroplating wastewater by electro-nanofiltration, *Russ. J. Appl.* Chem. 94 (2021) 1105–1110. https://doi.org/10.1134/S1070427221080127.

148. L. Wang, G.L. Zhang, H.B. Jiang, X.Z. Wei, B.S. Lv, Water recycling from electroplating effluent using membranes, *Adv. Mater. Res.* 233–235 (2011) 435–438. https://doi.org/10.4028/www.scientific.net/AMR.233-235.435.

149. A. Ali, M.C. Nymann, M.L. Christensen, C.A. Quist-Jensen, Industrial wastewater treatment by nanofiltration – A case study on the anodizing industry, *Membr.* 10 (2020). https://doi.org/10.3390/membranes10050085.

150. J. Dasgupta, D. Mondal, S. Chakraborty, J. Sikder, S. Curcio, H.A. Arafat, Nanofiltration based water reclamation from tannery effluent following coagulation pretreatment, *Ecotoxicol. Environ. Saf.* 121 (2015) 22–30. https://doi.org/10.1016/j.ecoenv.2015.07.006.

151. W.L. Ang, A.W. Mohammad, A. Benamor, N. Hilal, C.P. Leo, Hybrid coagulation–NF membrane process for brackish water treatment: Effect of antiscalant on water characteristics and membrane fouling, *Desalination.* 393 (2016) 144–150. https://doi.org/10.1016/j.desal.2016.01.010.

152. F.F. Chang, W.J. Liu, Arsenate removal using a combination treatment of precipitation and nanofiltration, *Water Sci. Technol.* 65 (2012) 296–302.

153. J. Wang, X. Tang, Y. Xu, X. Cheng, G. Li, H. Liang, Hybrid UF/NF process treating secondary effluent of wastewater treatment plants for potable water reuse: Adsorption vs. coagulation for removal improvements and membrane fouling alleviation, *Environ. Res.* 188 (2020) 109833. https://doi.org/10.1016/j.envres.2020.109833.

154. M. Giagnorio, S. Steffenino, L. Meucci, M.C. Zanetti, A. Tiraferri, Design and performance of a nanofiltration plant for the removal of chromium aimed at the production of safe potable water, *J. Environ. Chem. Eng.* 6 (2018) 4467–4475. https://doi.org/10.1016/j.jece.2018.06.055.

155. A. Antony, J.H. Low, S. Gray, A.E. Childress, P. Le-Clech, G. Leslie, Scale formation and control in high pressure membrane water treatment systems: A review, *J. Memb. Sci.* 383 (2011) 1–16. https://doi.org/10.1016/j.memsci.2011.08.054.

156. Y.M. Al-Roomi, K.F. Hussain, Potential kinetic model for scaling and scale inhibition mechanism, *Desalination.* 393 (2016) 186–195. https://doi.org/10.1016/j.desal.2015.07.025.

157. L.H. Andrade, A.O. Aguiar, W.L. Pires, L.B. Grossi, M.C.S. Amaral, Comprehensive bench- and pilot-scale investigation of NF for gold mining effluent treatment: Membrane performance and fouling control strategies, *Sep. Purif. Technol.* 174 (2017) 44–56. https://doi.org/10.1016/j.seppur.2016.09.048.

158. N. Ates, N. Uzal, Removal of heavy metals from aluminum anodic oxidation wastewaters by membrane filtration, *Environ. Sci. Pollut. Res.* 25 (2018) 22259–22272. https://doi.org/10.1007/s11356-018-2345-z.

159. E. Piedra, J.R. Álvarez, S. Luque, Hexavalent chromium removal from chromium plating rinsing water with membrane technology, *Desalin. Water Treat.* 53 (2015) 1431–1439. https://doi.org/10.1080/19443994.2014.943058.

160. L. Pino, C. Vargas, A. Schwarz, R. Borquez, Influence of operating conditions on the removal of metals and sulfate from copper acid mine drainage by nanofiltration, *Chem. Eng. J.* 345 (2018) 114–125. https://doi.org/10.1016/j.cej.2018.03.070.

161. K. Karakulski, A.W. Morawski, Recovery of process water from spent emulsions generated in copper cable factory, *J. Hazard. Mater.* 186 (2011) 1667–1671. https://doi.org/10.1016/j.jhazmat.2010.12.041.

162. H. Chang, B. Liu, B. Yang, X. Yang, C. Guo, Q. He, S. Liang, S. Chen, P. Yang, An integrated coagulation-ultrafiltration-nanofiltration process for internal reuse of shale gas flowback and produced water, *Sep. Purif. Technol.* 211 (2019) 310–321. https://doi.org/10.1016/j.seppur.2018.09.081.

163. G. Zolfaghari, M. Kargar, Nanofiltration and microfiltration for the removal of chromium, total dissolved solids, and sulfate from water, *MethodsX.* 6 (2019) 549–557. https://doi.org/10.1016/j.mex.2019.03.012.

164. M.D. Çelebi, M. Dilaver, M. Kobya, A study of inline chemical coagulation/precipitation-ceramic microfiltration and nanofiltration for reverse osmosis concentrate minimization and reuse in the textile industry, *Water Sci. Technol.* 84 (2021) 2457–2471. https://doi.org/10.2166/wst.2021.439.

165. M.R. Teixeira, M.J. Rosa, M. Nyström, The role of membrane charge on nanofiltration performance, *J. Memb. Sci.* 265 (2005) 160–166. https://doi.org/10.1016/j.memsci.2005.04.046.

166. A. Figoli, A. Cassano, A. Criscuoli, M.S.I. Mozumder, M.T. Uddin, M.A. Islam, E. Drioli, Influence of operating parameters on the arsenic removal by nanofiltration, *Water Res.* 44 (2010) 97–104. https://doi.org/10.1016/j.watres.2009.09.007.

167. T. Schütte, C. Niewersch, T. Wintgens, S. Yüce, Phosphorus recovery from sewage sludge by nanofiltration in diafiltration mode, *J. Memb. Sci.* 480 (2015) 74–82. https://doi.org/10.1016/j.memsci.2015.01.013.

168. B.C. Ricci, C.D. Ferreira, L.S. Marques, S.S. Martins, M.C.S. Amaral, Assessment of nanofiltration and reverse osmosis potentialities to recover metals, sulfuric acid, and recycled water from acid gold mining effluent, *Water Sci. Technol.* 74 (2016) 367–374. https://doi.org/10.2166/wst.2016.206.

169. S.S. Chen, B.C. Hsu, C.H. Ko, P.C. Chuang, Recovery of chromate from spent plating solutions by two-stage nanofiltration processes, *Desalination.* 229 (2008) 147–155. https://doi.org/10.1016/j.desal.2007.08.015.

170. H. Al-Zoubi, A. Rieger, P. Steinberger, W. Pelz, R. Haseneder, G. Härtel, Optimization study for treatment of acid mine drainage using membrane technology, *Sep. Sci. Technol.* 45 (2010) 2004–2016. https://doi.org/10.1080/01496395.2010.480963.

171. M.C.S. Amaral, L.B. Grossi, R.L. Ramos, B.C. Ricci, L.H. Andrade, Integrated UF–NF–RO route for gold mining effluent treatment: From bench-scale to pilot-scale, *Desalination.* 440 (2018) 111–121. https://doi.org/10.1016/j.desal.2018.02.030.

172. N. Ben Frarès, S. Taha, G. Dorange, Influence of the operating conditions on the elimination of zinc ions by nanofiltration, *Desalination.* 185 (2005) 245–253. https://doi.org/10.1016/j.desal.2005.02.079.

173. K. Meschke, N. Hansen, R. Hofmann, R. Haseneder, J.-U. Repke, Influence of process parameters on separation performance of strategic elements by polymeric nanofiltration membranes, *Sep. Purif. Technol.* 235 (2020) 116186. https://doi.org/10.1016/j.seppur.2019.116186.

174. J. Schaep, B. Van der Bruggen, S. Uytterhoeven, R. Croux, C. Vandecasteele, D. Wilms, E. Van Houtte, F. Vanlerberghe, Removal of hardness from groundwater by nanofiltration, *Desalination.* 119 (1998) 295–301. https://doi.org/10.1016/S0011-9164(98)00172-6.

175. X. Wei, X. Kong, S. Wang, H. Xiang, J. Wang, J. Chen, Removal of heavy metals from electroplating wastewater by thin-film composite nanofiltration hollow-fiber membranes, *Ind. Eng. Chem. Res.* 52 (2013) 17583–17590. https://doi.org/10.1021/ie402387u.

176. Z. Wang, G. Liu, Z. Fan, X. Yang, J. Wang, S. Wang, Experimental study on treatment of electroplating wastewater by nanofiltration, *J. Memb. Sci.* 305 (2007) 185–195. https://doi.org/10.1016/j.memsci.2007.08.011.

177. N. Ben Amar, H. Saidani, A. Deratani, J. Palmeri, Effect of temperature on the transport of water and neutral solutes across nanofiltration membranes, *Langmuir.* 23 (2007) 2937–2952. https://doi.org/10.1021/la060268p.

178. N. Ben Amar, H. Saidani, J. Palmeri, A. Deratani, Effect of temperature on the rejection of neutral and charged solutes by Desal 5 DK nanofiltration membrane, *Desalination.* 246 (2009) 294–303. https://doi.org/10.1016/j.desal.2008.03.056.

179. R.R. Sharma, S. Chellam, Temperature and concentration effects on electrolyte transport across porous thin-film composite nanofiltration membranes: Pore transport mechanisms and energetics of permeation, *J. Colloid Interface Sci.* 298 (2006) 327–340. https://doi.org/10.1016/j.jcis.2005.12.033.

180. Y. Roy, D.M. Warsinger, J.H. Lienhard, Effect of temperature on ion transport in nanofiltration membranes: Diffusion, convection and electromigration, *Desalination.* 420 (2017) 241–257. https://doi.org/10.1016/j.desal.2017.07.020.

181. C.V. Gherasim, K. Hancková, J. Palarčík, P. Mikulášek, Investigation of cobalt(II) retention from aqueous solutions by a polyamide nanofiltration membrane, *J. Memb. Sci.* 490 (2015) 46–56. https://doi.org/10.1016/j.memsci.2015.04.051.

182. R. Mukherjee, R. Mondal, A. Sinha, S. Sarkar, S. De, Application of nanofiltration membrane for treatment of chloride rich steel plant effluent, *J. Environ. Chem. Eng.* 4 (2016) 1–9. https://doi.org/10.1016/j.jece.2015.10.038.

183. H. Lee, S.J. Im, H. Lee, C.M. Kim, A. Jang, Comparative analysis of salt cleaning and osmotic backwash on calcium-bridged organic fouling in nanofiltration process, *Desalination.* 507 (2021) 115022. https://doi.org/10.1016/j.desal.2021.115022.

184. W. Jiang, Y. Wei, X. Gao, C. Gao, Y. Wang, An innovative backwash cleaning technique for NF membrane in groundwater desalination: Fouling reversibility and cleaning without chemical detergent, *Desalination.* 359 (2015) 26–36. https://doi.org/10.1016/j.desal.2014.12.025.

185. Y.S. Li, L.C. Shi, X.F. Gao, J.G. Huang, Cleaning effects of oxalic acid under ultrasound to the used reverse osmosis membranes with an online cleaning and monitoring system, *Desalination*. 390 (2016) 62–71. https://doi.org/10.1016/j.desal.2016.04.008.

186. S.S. Wadekar, Y. Wang, O.R. Lokare, R.D. Vidic, Influence of chemical cleaning on physicochemical characteristics and ion rejection by thin film composite nanofiltration membranes, *Environ. Sci. Technol.* 53 (2019) 10166–10176. https://doi.org/10.1021/acs.est.9b02738.

187. N. V Thombre, A.P. Gadhekar, A. V Patwardhan, P.R. Gogate, Ultrasound induced cleaning of polymeric nanofiltration membranes, *Ultrason. Sonochem.* 62 (2020) 104891. https://doi.org/10.1016/j.ultsonch.2019.104891.

188. J. Wang, X. Gao, Y. Xu, Q. Wang, Y. Zhang, X. Wang, C. Gao, Ultrasonic-assisted acid cleaning of nanofiltration membranes fouled by inorganic scales in arsenic-rich brackish water, *Desalination*. 377 (2016) 172–177. https://doi.org/10.1016/j.desal.2015.09.021.

189. F. Ferella, I. De Michelis, C. Zerbini, F. Vegliò, Advanced treatment of industrial wastewater by membrane filtration and ozonization, *Desalination*. 313 (2013) 1–11. https://doi.org/10.1016/j.desal.2012.11.039.

190. C. García-Figueruelo, A. Bes-Piá, J.A. Mendoza-Roca, J. Lora-García, B. Cuartas-Uribe, Reverse osmosis of the retentate from the nanofiltration of secondary effluents, *Desalination*. 240 (2009) 274–279. https://doi.org/10.1016/j.desal.2008.01.052.

191. S. Van Geluwe, J. Degrève, C. Vinckier, L. Braeken, C. Creemers, B. Van der Bruggen, Kinetic study and scaleup of the oxidation of nanofiltration retentates by O_3, *Ind. Eng. Chem. Res.* 51 (2012) 7056–7066. https://doi.org/10.1021/ie202065x.

192. I. Salmerón, G. Rivas, I. Oller, A. Martínez-Piernas, A. Agüera, S. Malato, Nanofiltration retentate treatment from urban wastewater secondary effluent by solar electrochemical oxidation processes, *Sep. Purif. Technol.* 254 (2021) 117614. https://doi.org/10.1016/j.seppur.2020.117614.

193. A. Otero-Fernández, J.A. Otero, A. Maroto-Valiente, J.I. Calvo, P. Palacio, P. Prádanos, A. Hernández, Reduction of Pb(II) in water to safe levels by a small tubular membrane nanofiltration plant, *Clean Technol. Environ. Policy*. 20 (2018) 329–343. https://doi.org/10.1007/s10098-017-1474-2.

194. J. Zhu, Q. Jin, L. Dongming, Investigation on two integrated membrane systems for the reuse of electroplating wastewater, *Water Environ. J.* 32 (2018) 267–275. https://doi.org/10.1111/wej.12324.

195. P.S. Sudilovskiy, G.G. Kagramanov, V.A. Kolesnikov, Use of RO and NF for treatment of copper containing wastewaters in combination with flotation, *Desalination*. 221 (2008) 192–201. https://doi.org/10.1016/j.desal.2007.01.076.

196. M. Peyravi, M. Jahanshahi, M. Alimoradi, E. Ganjian, Old landfill leachate treatment through multistage process: Membrane adsorption bioreactor and nanofitration, *Bioprocess Biosyst. Eng.* 39 (2016) 1803–1816. https://doi.org/10.1007/s00449-016-1655-0.

197. G. Wang, Z. Fan, D. Wu, L. Qin, G. Zhang, C. Gao, Q. Meng, Anoxic/aerobic granular active carbon assisted MBR integrated with nanofiltration and reverse osmosis for advanced treatment of municipal landfill leachate, *Desalination*. 349 (2014) 136–144. https://doi.org/10.1016/j.desal.2014.06.030.

198. F.S. Wong, J.J. Qin, M.N. Wai, A.L. Lim, M. Adiga, A pilot study on a membrane process for the treatment and recycling of spent final rinse water from electroless plating, *Sep. Purif. Technol.* 29 (2002) 41–51. https://doi.org/10.1016/S1383–5866(02)00002–3.

199. B.H. Gursoy-Haksevenler, E. Atasoy-Aytis, M. Dilaver, Y. Karaaslan, Treatability of hazardous substances in industrial wastewater: Case studies for textile manufacturing and leather production sectors, *Environ. Monit. Assess.* 194 (2022) 383. https://doi.org/10.1007/s10661-022-09982-x.

200. J. Choi, S.J. Im, A. Jang, Application of volume retarded osmosis – Low pressure membrane hybrid process for recovery of heavy metals in acid mine drainage, *Chemosphere*. 232 (2019) 264–272. https://doi.org/10.1016/j.chemosphere.2019.05.209.

201. A. Hussain, H. Yan, N. Ul Afsar, H. Wang, J. Yan, C. Jiang, Y. Wang, T. Xu, Acid recovery from molybdenum metallurgical wastewater via selective electrodialysis and nanofiltration, *Sep. Purif. Technol.* 295 (2022) 121318. https://doi.org/10.1016/j.seppur.2022.121318.

202. A. Hanif, S. Ali, M.A. Hanif, U. Rashid, H.N. Bhatti, M. Asghar, A. Alsalme, D.A. Giannakoudakis, A novel combined treatment process of hybrid biosorbent – Nanofiltration for effective Pb(II) removal from wastewater, *Water*. 13 (2021). https://doi.org/10.3390/w13233316.

16 Hybrid Nanofiltration Membrane Technology for Water Reclamation

Leow Hui Ting Lyly
Universiti Kebangsaan Malaysia

Zhen Hong Chang
UCSI University

Yeit Haan Teow
Universiti Kebangsaan Malaysia

Abdul Wahab Mohammad
Universiti Kebangsaan Malaysia
University of Sharjah

CONTENTS

DOI: 10.1201/9781003261827-16

16.1 INTRODUCTION

Water stress is the pressure exerted on freshwater resources by human activities [1]. Attributed to the poor management of water resources, climate change, depletion and pollution, water stress has been increasing around the globe [2]. Generally, water stress in industrialized countries is mainly due to water pollution, while in developing countries, water depletion is the main cause of water stress [2,3]. According to the Food and Agriculture Organization of United Nations (FAO) and United Nations Water (UN-Water) [1], approximately 10% of the global population is living in countries experiencing high and critical water stress. With consistent growth in the global population, the demand for clean water continues unabated. Not to mention, the outbreak of COVID-19 has further intensified the demand for clean water to ensure a safe and healthy life [4]. For that reason, FAO and UN-Water had claimed various water sources, including storm run-off, desalination and wastewater, as an alternative to relieve the water stress. Since most of the alternative resources cannot be consumed and/or discharged directly, water and wastewater treatment systems are urgently needed to remove the unwanted contaminants from the water sources.

Over the years, various water and wastewater treatment methods have been studied and well implemented at an industrial scale. The treatment methods could be divided into physical, chemical and biological methods. Examples of physical methods are flotation [5], flocculation [6], adsorption [7] and membrane filtration [8]. The physical methods are based on the mechanism of mass transfer [9,10], with regard to the difference in size and density of the contaminants or the formation of Van der Waals interaction between the adsorbent and contaminants without manipulating the biochemical properties of the contaminants. Chemical methods include coagulation [6,11], adsorption [7,12] and electrochemical treatment [13]. The chemical methods involve changes in surface charges and the formation of chemical bonding via a series of chemical reactions. Biological methods include composting [14], biological oxidation [15] and anaerobic digestion [16]. The biological methods incorporate microorganisms to break down the organic contaminants using normal cellular processes [10]. All these methods mentioned have been proven to remove pollutants from various water sources. Among them, membrane separation has advantages in terms of high removal efficiency on a wide range of contaminants (e.g., suspended solids, macromolecules, microorganisms, organic and inorganic compounds and metal ions), easily integrated with various treatment processes (as pretreatment or post-treatment), simple operation with simple instrumentation, continuous separation under mild conditions, easy to up-scale and flexibility in membrane fabrication and modification.

The membrane separation process involves the permeation of water molecules across a semipermeable membrane, with or without the presence of other co-existing substances. The water permeation could be driven by the difference in hydraulic pressure, osmotic pressure and thermal. Pressure-driven membrane separation process required the externally applied pressure at the feed side to force the diffusion of water across the membrane. Example of pressure-driven membrane separation process includes microfiltration (MF), ultrafiltration (UF), nanofiltration (NF) and reverse osmosis (RO)). Each of them can be differentiated by the membrane pore size and the operating pressure [8]. Both MF and UF processes are commonly used to separate macromolecules, colloids and suspended solids from the feed stream. NF membrane could reject solutes with a molecular weight not less than 200 g/mol, while RO membrane can withstand the greatest operating pressure (15–75 bar). Osmotically driven membrane processes incorporate draw solution to induce an osmotic gradient across the membrane to drive the diffusion of water. In this case, a draw solution with high osmotic pressure is in favour. Example of osmotically driven membrane processes includes forward osmosis (FO) and pressure-assisted osmosis (PAO). Generally, both FO and PAO processes use membranes such as UF, NF and RO to recover water from the feed stream [17–20]. Thermally driven membrane separation processes, such as membrane distillation (MD) and pervaporation (PV), could utilize low-grade heat to drive the separation process [21]. These membrane processes involve three main stages: evaporation, diffusion and condensation. The MD process is more commonly used in desalination, while PV process is used for the separation of water and various organics.

In general, the membrane separation process could be conducted as a standalone process or as an integrated membrane process [22]. The standalone process is a single membrane process, incorporating membranes such as MF, UF and NF in treating the water, while the integrated membrane processes involve pre-treatment process(es) before or post-treatment process(es) after the membrane filtration unit. In academic research, the standalone process is more commonly conducted intending to determine the performance of membrane that varied in synthesis and modification methods under controlled operating conditions [23], while the integrated membrane process is more commonly implemented to study the feasibility of a separation system in terms of energy consumption and techno-economic analysis when recovering valuable materials from effluent collected from industrial sites [19]. In an integrated membrane separation system, the pre-treatment membrane separation process is essential to protect the following membrane unit (e.g., NF and RO) suffering from severe membrane fouling. It acts as a barrier to prevent those large-size solids and compounds from clogging the tight membranes, hence preserving a longer serving time. On the other hand, the post-treatment membrane separation process is used to further enhance the quality of the permeate by rejecting most of the unwanted compounds, producing clean and hygienic water. A study conducted by Ramos et al. [24] designed an integrated UF-NF-RO process to reclaim valuable components (i.e., metals, acid and water) from the gold mining effluent. Each of the membrane processes plays an important role during the separation process: UF removed suspended solids; NF recovered heavy metals; RO recovered acid and produced clean water. Without the UF and NF processes, the high content of suspended solids and heavy metals in the effluent would induce huge stress on the RO process, diminishing the membrane performance. On the other hand, by having solely UF or NF process, the permeate is not reusable due to the high acid content in the permeate.

The chapter studies hybrid membrane separation technology for water reclamation, which focuses on the integration of NF membrane with other membrane processes. The chapter starts with the review of standalone NF applied in desalination and wastewater treatment in various industries which include oil refinery, textile, semiconductor, mining, paper and pulp and food processing. This is followed by the discussion on the limitation of a standalone NF system during water reclamation. Next, the working principle of the hybrid membrane process of NF with other membrane processes such as MF, UF, RO, FO and MD for water reclamation is discussed together with their application. Lastly, an outlook and conclusion based on the membrane hybrid technology for future research in the water reclamation field are provided. It is presumed that the systematic explanation in this chapter will brief readers on the understanding of NF membrane hybrid technology in the application of water reclamation.

16.2 NANOFILTRATION FOR WATER RECLAMATION

16.2.1 Nanofiltration for Desalination

Seawater accounts for the largest proportion of the earth's water resources. To meet the high demand for clean water, seawater is being processed via membrane desalination and thermal desalination. However, thermal desalination has limitations in terms of high energy consumption as well as high investment and maintenance cost [25]. For that reason, membrane desalination (i.e., RO process) has dominated the seawater desalination process over the years. Generally, seawater is known for its high salt content (i.e., Na^+, Mg^{2+}, Cl^- and SO_4^{2-}) [26]. Despite the high ionic rejection efficiency of the RO process, it suffers severe membrane fouling from a wide range of metal ions, which prohibits the long-term applicability of the RO membrane.

For that reason, researchers incorporated the NF process as a pre-treatment before the RO process to reduce the fouling tendency induced by the divalent ions (e.g., Mg^{2+}, Ca^{2+}, SO_4^{2-}). For example, Kumar et al. [27] developed a thin film composite (TFC) NF membrane modified with a layer of trimesic acid in the interlayer to desalinate the seawater. The seawater was collected from the Doha east desalination power station had a pH of 7.4, a total dissolved solids (TDS) of 43.697 g/L and contained various ions, such as K^+ (0.458 g/L), Na^+ (15.037 g/L), Mg^{2+} (1.55 g/L), Ca^{2+} (0.5449 g/L), B^{3+} (3.8 g/L), SO_4^{2-} (4.352 g/L) and Cl^- (26.084 g/L). After the NF treatment, the membrane showed an overall ionic rejection of 41.3%, 36.5%, 52.4%, 56.7%, 40.0%, 80.2% and 44.9%, respectively. Another study was conducted by Wang et al. [28], separating metal ions from the seawater using a NF membrane modified with molybdenum disulphide. The seawater collected from the East China Sea had a conductivity of 60,000 mS/m and contained 7.31 g/L of Na^+, 1.19 g/L of Mg^{2+}, 29.91 g/L of Cl^- and 3.93 g/L of SO_4^{2-}. The fabricated membrane demonstrated an ionic rejection of 26.0%, 65.8%, 21.6% and 92.8%, respectively, resulting in a reduction (22.4%) in overall conductivity. Hence, by just having a standalone NF system in desalinating sweater, it is not sufficient to produce high-quality clean water. However, due to its lower pressure consumption, as well as lower operation cost, it is a good candidate used to pre-treat seawater.

16.2.2 Nanofiltration for Wastewater Treatment

Generally, the wastewater discharged from the industries is a combination of effluents from various processing units, including but not limited to solvent processing, chemical reaction, cooling, transporting of products, washing or cleaning. The initial high concentrated effluent from a particular process will then be diluted and/or neutralized after they are collected in the effluent pool, generating an enormous amount of wastewater that are containing mostly water with multiple components. For instance, in the study reported by Núñez et al. [29], the wastewater generated from the process of dyed linen was characterized by a chemical oxygen demand (COD) value of 800 mg/L, a colour intensity of 10,000 Pt-Co mg/L, a conductivity of 512,000 μS/cm, a pH of 11.6 and turbidity of 105 NTU. After it was mixed with wastewater from other processing lines in the residual effluent pool, the resulted wastewater demonstrated a lower value for each parameter: a COD value of 478.7 mg/L, a colour intensity of 401 Pt-Co mg/L, a conductivity of 668.8 μS/cm, a pH of 7.1 and turbidity of 12.5 NTU. The high water content in the wastewater associated with varied contaminants makes the wastewater treatment process a crucial step to treat the effluent before discharging to the environment or recycling back to the plant. Table 16.1 summarizes the application of NF membrane in treating various real wastewater collected from industrial sites for water reclamation.

16.2.2.1 Oil Refinery

In the oil refining process, sodium hydroxide is added to eliminate the production of hydrogen sulphide. As a result, the produced effluent, also known as spent caustic, is characterized as a highly toxic solution, possessing a relatively high value in alkalinity (pH > 12), salinity (5–12 wt%), sulphide content (1–4 wt%) and COD [30]. Therefore, proper treatment is required prior to discharging spent caustic to the water bodies. Rita et al. [31] applied polymeric NF membrane with a molecular weight cut-off (MWCO) of 200 Da in treating spent caustic collected from a petroleum refinery plant. The wastewater had a conductivity of 60.7 mS/cm, pH 13.9, polar and oil grease content of 18,970 mg/L and COD of 102,190 mg O_2/L. In the first week of NF process, the applied membrane showed a high rejection of COD (99.6%), phenolic compounds (98.1%), oil and grease (99.5%), followed by a continuous decrease to 90.0%, 78.9% and 88.3%, respectively, with the ageing time for 12 weeks. The reduction in rejection is believed, attributing to the deterioration of the membrane active layer over time. Despite that the spent caustic showed a high conductivity, the performance of the NF200 membrane in reducing the conductivity is not evidenced in the study.

Apart from spent caustic, the desalter effluent from the oil refining process also produces hazardous wastewater containing oils, petroleum hydrocarbon, phenols, sulphides, ammonia and metal ions [32]. Dadari et al. [33] prepared a NF membrane embedding adipate ferroxane nanoparticles to treat the desalter wastewater with a hardness of 95,000 mg/L and containing multiple metal ions (i.e., Ca^{2+}, Mg^{2+}, Fe^{2+}, Na^+ and K^+). After the filtration, the hardness of effluent was reduced to 665 mg/L (approximately 93% of reduction). The fabricated membrane also demonstrated a better rejection on multivalent ions (>92%), as compared to monovalent ions (~40%).

TABLE 16.1

Application of NF Membrane in Treating Real Industrial Wastewater

NF Membrane	Operating Condition	Feed	Performance	Reference
Self-assembly NF TFC membrane modified with trimesic acid interlayer	P: 10.34 bar T: 25°C–26°C Mode: crossflow	Seawater from desalination plant	Overall rejection: Na^+: 36.5% K^+: 41.3% Ca^{2+}: 56.7% Mg^{2+}: 52.4% B^{3+}: 40.8% Cl^-: 44.9% SO_4^{2-}: 80.2%	[27]
Self-assembly NF TFC membrane modified with molybdenum disulphide	P: 4 bar T: N/A Mode: crossflow	Seawater from East China Sea	Overall rejection: Conductivity: 22.4 % Na^+: 26% Mg^{2+}: 65.8% Cl^-: 21.6% SO_4^{2-}: 92.8%	[28]
NF200	P: 10 bar T: N/A Mode: crossflow	Spent caustic solution from petroleum refinery plant	Overall rejection: COD: 99.6% Phenolic compound: 98.1% Sulphide: 100% O&G: 99.5%	[31]
NF modified with adipate ferroxane nanoparticles	P: 4.5–6 bar T: N/A Mode: dead end	Desalter effluent from oil field desalting unit	Overall rejection: Ca^{2+}: 95% Mg^{2+}: 92.5% Fe^{2+}: 93.8% Na^+: 41% K^+: 38.5% Cl^-: 40% Hardness: 93% COD: >90%	[33]
Self-assembly NF membrane modified with gelatin biopolymer	P: 2 bar T: N/A Mode: dead end	Wastewater from textile factory	Overall rejection: COD: 67.6% TDS: 60.7% TSS: 86.6% Oil and grease: 55% Conductivity: 45.5% Colour: 85%	[37]
Self-assembly NF membrane modified with cerium oxide nanoparticles	P: 5 bar T: N/A Mode: dead end	Effluent from textile plant	Overall rejection: COD: 79% Turbidity: 99% TDS: 6% TSS: 93% Conductivity: 3%	[38]
Ceramic NF450	P: 15 bar T: 25°C Mode: crossflow	Wastewater from semiconductor wastewater treatment plant	Overall rejection: TOC: 100% TN: 32% NH_4^+: 36% Ca^{2+}: 44% SO_4^{2-}: 56%	[40]

(Continued)

TABLE 16.1 (*Continued*)
Application of NF Membrane in Treating Real Industrial Wastewater

NF Membrane	Operating Condition	Feed	Performance	Reference
Self-assembled NF membrane modified with composited multi-walled carbon nanotubes-poly(amidoamine)	P: 4 bar T: 25°C Mode: crossflow	Wastewater from mine site	Overall Rejection: Cu^{2+}: ~99% Zn^{2+}: ~87% Ni^{2+}: ~90% Pb^{2+}: ~94% Na^{+}: ~62%	[43]
Self-assembled NF membrane modified with dopamine-coated titanium oxide nanoparticles	P: N/A T: N/A Mode: N/A	Wastewater from paper mill factory	Overall rejection: COD: 62% BOD: 50% TDS: 15% TSS: 80% Total hardness: 64% Conductivity: 61% Cl^{-}: 61% SO_4^{2-}: 63%	[46]
NF270	P: 24 bar T: N/A Mode: N/A	Industrial lupin beans cooking wastewater	Overall rejection: Lupanine content: 99.5% COD: 94% Conductivity: 52%	[51]

P, pressure; *T*, temperature; N/A, not available.

16.2.2.2 Textile

Textile wastewater is generated from various wet processing operations (e.g., bleaching, dyeing, washing) in the textile industries [34]. During the wet process, a large quantity of water is required to wash and clean the dyed textile products, subsequently producing wastewater with high loadings of dyes, salts, heavy metals and surfactants [35,36]. Nasr and Ali [37] synthesized a NF membrane modified with gelatin biopolymer to treat the textile wastewater. The wastewater was first decanted to remove macro contaminants prior to the NF process. The removal efficiencies on TDS, total suspended solids (TSS), COD, conductivity, colour, oil and grease at 2 bar were reported to be 60.7%, 86.6%, 67.6%, 45.5%, 85% and 55%, accordingly. Despite the advantage in terms of low applied pressure, the efficiency of fabricated NF membrane is not high enough to produce high-quality clean water, as there are a relatively high content of dissolved solids, heavy metals and oil and grease present in the permeate, and thus post-treatment is required to reclaim the clean water.

Another study conducted by Tavangar et al. [38] treated the textile wastewater using cerium oxide nanoparticles-modified NF membrane at 5 bar. The wastewater showed a COD of 3,460 mg/L, turbidity of 1,340 NTU, a TDS of 6,300 mg/L, a TSS of 250 mg/L and a conductivity of 12.75 µS/cm. After the NF process, the removal efficiency was found to be 79%, 99%, 6%, 93% and 3%, respectively. The extremely low rejection efficiency in TDS and conductivity indicated that the fabricated membrane was large enough to allow the diffusion of heavy metals and salt across the membrane. Similar to Nasr and Ali [37], here also post-treatment is applied to remove the salt and heavy metals from the textile effluent before discharging into the environment.

16.2.2.3 Semiconductor

In the manufacturing of semiconductors, water is used abundantly to remove the organic and inorganic compounds emitted from processes, such as chemical mechanical polishing, wafer planarization and photoresist processing [39]. Consequently, the resulted wastewater contains a wide range of contaminants, including acids, bases, organic solvents, fluoride, ammonia, salts and heavy metals [40]. Thus, it is required to treat the wastewater before recycling back into the system and/or discharging it into the environment. Cha et al. [40] implemented a ceramic NF membrane (MWCO of 450 Da) to retain the organic and inorganic compounds from the semiconductor wastewater. The collected wastewater had been processed by multi-stage chemical coagulation and sedimentation prior to the study. At 10 bar, the ceramic NF membrane was found to achieve approximately 100% rejection on total organic carbon (TOC), while it had an average rejection efficiency on total nitrogen (TN) (~32%), NH^{4+} (~36%), Ca^{2+} (~44%) and SO_4^{2-} (~56%). Besides, the ceramic NF membrane experienced a reduction in normalized flux by 40% over 55 hours filtration due to the membrane fouling. After the chemical cleaning, the membrane showed an irreversible fouling resistance of 40.9%. As no further characterization of the fouled layer was conducted, the composition of fouled layer is unclear. Perhaps, integrating the NF with other pre-treatment membrane processes could reduce the irreversible fouling on the NF membrane and enhance the overall performance of the NF membrane.

16.2.2.4 Mining

The constant exploitation of minerals from the soil generates an enormous amount of liquid effluents, also known as acid mine drainage. During the mining process, the sulphide minerals (e.g., such as pyrrhotite, pyrite, marcasite) are oxidized by oxygen and water, producing sulphuric acids [41,42]. Subsequently, the acid solution dissolves the surrounded soil minerals, leading to the release of metals (e.g., iron, copper, zinc) and rare earth elements. For that reason, researchers find ways to recover valuable by-products while producing clean water to eliminate the negative impact on the environment. For instance, Zhang et al. [43] developed an NF membrane modified with composite multi-walled carbon nanotubes-poly(amidoamine) to treat the mine wastewater collected from a mine site. The wastewater had been processed on-site and filtered with a UF membrane prior to the NF process. The wastewater was found to contain 78.83 mg/L of Cu^{2+}, 16.56 mg/L of Ni^{2+}, 18.28 mg/L of Zn^{2+}, 9.96 mg/L of Pb^{2+}, 372.76 mg/L of Na^+, 10.56 mg/L of SO_4^{2-} and 706.31 mg/L of Cl^-. The NF process was conducted at 4 bar in an acidic condition (pH 4.77), and the fabricated membrane showed an overall high rejection (>80%) on divalent ions, while only possessing about 62% of rejection on Na^+ ions. The low rejection on monovalent ions was in agreement with other studies when NF membrane was being used to treat the wastewater.

Another study conducted by López et al. [41] implemented the NF process in treating the synthetic acid mine drainage using the NF270 membrane. The synthetic wastewater had a pH of 1.0 and consisted of varying amounts of metal-chloride, nitrate, oxide and sulphate salts as well as sulphuric acid. They reported relatively higher (>98%) metal ion rejections as compared to the rejection (35%) on sulphur at a crossflow velocity of 0.7 m/s. They further reported that the membrane became less stable in highly acidic solution (1 M sulphuric acid) due to the occurrence of membrane ageing. The membrane active was partially hydrolysed, enlarging the free pore volumes. Subsequently, the membrane rejection of metal ions was reduced to 70%. Thus, the stability of polymeric NF membrane in treating highly acidic mine effluent is an important factor for a long-term wastewater treatment process. Not to mention, the raw mine effluent contains a large number of total solids (~2,700 mg/L) and TSS (56 mg/L) that could severely damage the NF membrane [44]. A pre-treatment process is required to remove most of the solids before the NF process.

16.2.2.5 Paper and Pulp

The manufacturing process of pulp and paper involves multiple washing and dewatering processes [45]. Water is required to clean the products as well as to transport chemicals and fibres [46].

Consequently, a huge amount of liquid effluent is produced, containing various contaminants (e.g., cellulose, lignin, resins, phenols, tannins) [45]. To recover the useful compounds and water, Khalid et al. [46] developed a poly(vinyl alcohol) NF membrane modified with titanium dioxide to treat the wastewater collected from the local paper milling industry. The wastewater had a COD of 326 mg/L, BOD of 128 mg/L, a conductivity of 1,350 µS/cm and a TDS of 1,100 mg/L. After the NF treatment, the membrane demonstrated an overall rejection on COD (62%), BOD (50%), conductivity (61%) and TDS (15%), indicating that the MWCO of the NF membrane was too large to reject most of the ions and dissolved solids.

Valderrama et al. [47], on the other hand, implemented a pH-stable NF membrane in treating the synthetic pulping wastewater. The NF membrane had an active layer of hydrophilized polysulfone with a MWCO of 600–800 Da. Unlike real industrial wastewater, synthetic wastewater only consisted of lignin (8.65 g/L), xylan (1.17 g/L), NaOH (12 g/L), Na_2CO_3 (3.5 g/L) and $MgSO_4$ (13.24 g/L). The lignin and xylan content contributed to a TOC of 4,226.67 mg/L, while the high NaOH content contributed to a highly alkaline (pH 12) solution. After the NF treatment, the membrane showed an overall rejection on TOC (92.5%), Na^+ (73.2%), Mg^{2+} (99.9%), CO_3^{2-} (84.1%) and SO_4^{2-} (88.7%). However, in reality, the real pulping wastewater would contain a higher amount of lignin and other solid constituents (e.g., cellulose, ash, microorganism) [48], which could easily clog the NF membrane without a pre-treatment process.

16.2.2.6   Food

Unlike wastewater generated from other manufacturing and mining industries, the liquid effluent from the food processing industry is generally excluded from hazardous and toxic substances. Instead, most of the food processing wastewater contains a high concentration of suspended and dissolved organic compounds (e.g., fats, proteins, carbohydrates, fibres, food-graded dyes) as well as minerals, contributing to the high level of COD and BOD [49,50]. NF process had been extensively implemented in treating the food processing effluent, not only to recover the valuable ingredients but also to produce clean water for recycling and/or reusing. For example, Esteves et al. [51] treated the industrial lupanine bean wastewater that was generated during the hydration, cooking and cooling processes of the beans using NF90, NF200 and NF270 membranes, respectively. The wastewater showed a pH value of 4 and had a COD of 27.85 gO_2/L and a lupanine concentration of 3.26 g/L. Before the NF process, the wastewater was centrifuged to remove the suspended solids. At 24 bar, NF270 demonstrated the greatest rejection (>95%) on lupanine and COD. However, due to the intrinsic large MWCO, NF270 had a lower rejection (52%) on conductivity as compared to NF90 (92%).

16.2.3   Limitation of Standalone Nanofiltration System

Depending on the nature of the activity, the wastewater produced consists of a wide range of contaminants; some might be highly toxic and hazardous to humans and the environment. Among the studies, both synthetic and real wastewater was used to study the applicability of the NF process to produce clean water while recovering valuable compounds. As compared to synthetic wastewater, real wastewater is commonly consisting of a high level of suspended solids and/or macro-sized molecules that could severely clog and damage the NF membrane if it is applied directly to the NF process. Due to its intrinsically small pore size, the NF membrane could remove most of the contaminants without the aid of the pre-treatment process. However, a pre-treatment process prior to the NF process is still well recommended to separate the large-sized particles to ensure a long-term operation of the NF membrane. On the other hand, despite the low MWCO of the NF membrane, it is unable to reject most of the monovalent ions (i.e., Na^+ and Cl^-) as well as some organic solvents. This limits the usability of NF permeates in recycling back to the system as raw materials or for washing purposes. A post-treatment process is required followed by the NF process to increase the quality of the clean water. In other words, the NF process could be used as a pre-treatment for some high energy consumption processes (e.g., RO, MD and ED) to eliminate most of the multivalent ions, subsequently reducing the fouling tendency of the following process and thus leading to a better long-term operability.

16.3 HYBRID NANOFILTRATION MEMBRANE TECHNOLOGY FOR WATER RECLAMATION

Hybrid membrane system involves minimum two different membrane processes for better system performance and also final product quality. Various types of hybrid membrane systems have been established based on the types of wastewater feed and also problems faced in conventional treatment plant. In general, there are some advantages of employing hybrid membrane systems. Firstly, water quality is improved, whereby it cannot be accomplished by a standalone process. Besides, proper design of hybrid system can aid in energy-saving owing to lower energy consumption. Additionally, reduction of waste disposal can be easily achieved with hybrid membrane systems, which then provide environmental friendliness. Furthermore, hybrid membrane systems have higher efficiency in treatment plants which help in reduction of operating costs. Briefly, types of hybrid membrane systems in water treatment plant are flexible. However, there are combinations of hybrid membrane systems that still have not been fully discovered [22]. The focus of this chapter is the hybrid system of NF with other membrane processes such as pressure-driven process such as MF, UF and RO, osmotic pressure-driven process such as FO and thermal-driven process such as MD. During the discussion of hybrid system in the following section, examples are given in the field of desalination and wastewater treatment.

16.3.1 HYBRID NANOFILTRATION WITH OTHER PRESSURE-DRIVEN MEMBRANE PROCESSES

Membrane filtration is a pressure-driven process. Membrane will act as a selective barrier that is resistance towards contaminants such as microorganism, organic and inorganic matter, metal ions, nutrient and turbidity. Pressure-driven membrane process has been classified based on its pore sizes, namely MF, UF, NF and RO [52]. The retain diameters of MF membrane ranging between 1 and 10 μm with operating pressure ranging from 1 to 3 bar. UF membrane can retain contaminants ranging between 0.01 and 1 μm with operating pressure ranging from 2 to 5 bar. NF membrane can retain 0.01–0.1 μm of contaminants with lower operating pressure between 5 and 15 bar. In a pressure-driven membrane process, RO has the highest operating pressure which is from 15 to 75 bar and lowest retain diameters ranging from 0.001 to 0.01 μm [53].

MF and UF are very efficacious to remove colloids, microorganisms, fine particles and suspended solid. In a hybrid system, MF/UF is normally used as pre-treatment to harvest solid-free permeate, followed by other treatment processes. Although MF/UF is cost-effective but the pore sizes of MF/UF is not enough to retain dissolved organics. On the other hand, the smaller pore sizes of NF and RO membrane compared to MF and UF are able to retain dissolved organics or ions. Although both NF and RO can produce high quality of water but they faced challenges such as required large energy, fouling and scaling problems, few trace organics cannot be removed using NF/RO systems and also management of RO concentrate [54]. In most of the hybrid membrane system applications, MF, UF and NF were used as pre-treatment steps of RO [55–57]. This sequence serves as a multi-barrier in removing wastewater contaminants; at the same time, it can reduce fouling on RO membrane and maintain the flux [58]. Figure 16.1 shows the schematic diagram of hybrid system with NF-MBR and RO system for high water recovery and low fouling propensity for water reclamation [59].

MF-NF hybrid system in wastewater treatment is often used to produce clean water for recycling purpose. For instance, in the work of Shanmuganathan et al. [54] treating biological-treated sewage effluent, the hybrid MF-NF can produce high-quality water for reuse purpose. In the first MF

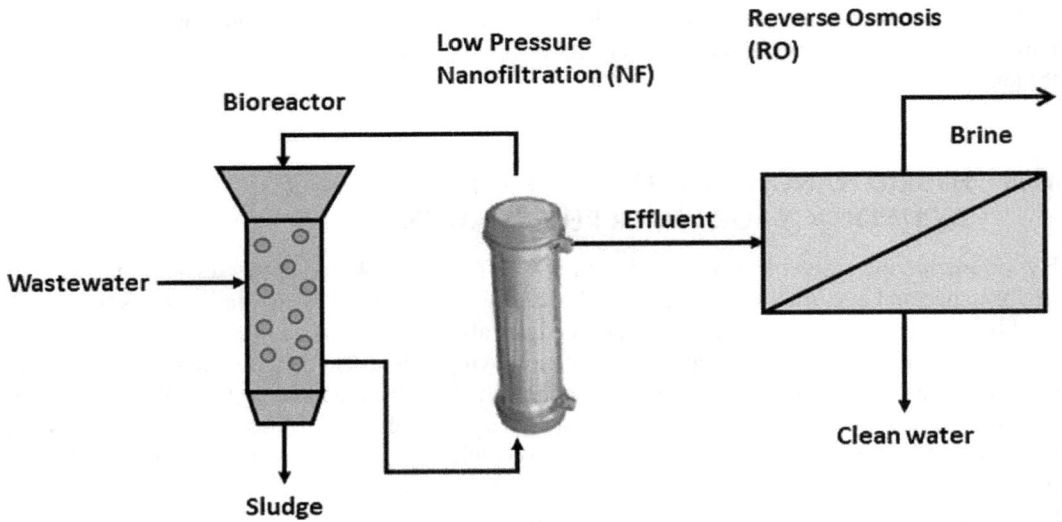

FIGURE 16.1 Hybrid system with NF-MBR and RO system for high water recovery with low fouling propensity [59].

membrane system, >80% of organic matters were removed but only little of inorganic matter can be removed. However, in the second NF membrane system, >95% of dissolved organic carbon, 99% of sulphates, 70% of calcium and 60% of magnesium were rejected. In another MF-NF hybrid system adopted by Amaral et al. [60] in treating landfill leachate, the application has allowed production of colourless permeates. By adopting the hybrid system, removal of organic and inorganic matters were achieved up to 94% and 91%, respectively. Besides, MF as a pre-treatment can reduce fouling of humic substance on NF membrane.

Despite NF sharing many characteristics with RO, NF demonstrates high rejection on multivalent ions at a lower hydraulic pressure as compared to RO, while having a lower rejection on monovalent ions (i.e., Na^+, Cl^-, etc.). Bortoluzzi et al. [61] has conducted an research to compare the performance of MF-NF and MF-RO hybrid systems. In this work, dairy wastewater was used as feed to compare the performance between the hybrid MF-NF and hybrid MF-RO. It was found that hybrid MF-RO has better performance with complete removal of colour and turbidity, while hybrid MF-NF has completely removed turbidity but with only 96% of colour removal. In another comparison study of UF-NF and UF-RO hybrid systems by Sun et al. [62] using paper mill wastewater, it was found that in the first membrane treatment using UF, 56.4% of COD and 9.6% of TDS were removed. In the subsequent treatment with UF permeate with NF, there was 94.1% of COD and 92.3% of phenol being rejected, while treatment with RO has 95.5% and 94.9% rejection, respectively.

16.3.2 HYBRID NANOFILTRATION-FORWARD OSMOSIS (NF-FO)

Currently, RO process is considered as the most efficient technology in desalination. However, RO required high energy consumption and is expensive. Among several desalination technologies, FO has become a promising technology for water reclamation application, especially water that contains high concentration of organic compounds due to its low fouling propensity. Besides, researchers have combined FO with other membrane filtration systems such as NF and RO for draw solution recovery to improve overall system performance. In FO system, draw solution is important to create osmotic pressure gradient, while, in NF system, high recovery rates could be achieved because NF has high rejection of multivalent ions with low pressure. The hybrid FO-NF system has better advantages when compared with standalone RO process in desalination performance [63].

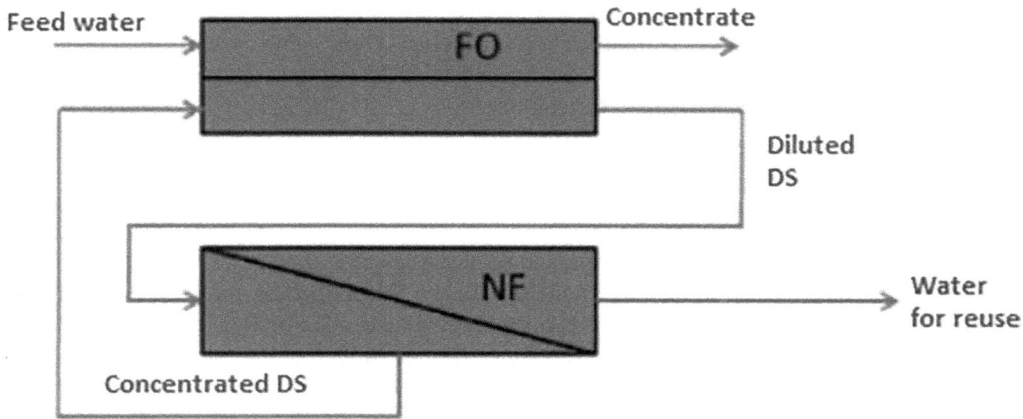

FIGURE 16.2   Schematic diagram of forward osmosis-nanofiltration (FO-NF) hybrid system. (Adapted from Ref. [67].)

In the water treatment hybrid process for NF and FO system, the feed would first pass through the FO system, while the purpose of NF is to recover the draw solution of the FO system as shown in Figure 16.2. Permeate from NF could recycle for other usage. Thus, employing FO-NF system not only reduces the energy consumption during operation but also minimizes the waste disposal to the environment. However, in the case of fabrication of novel membrane for FO system, prior of FO process, the performance of membrane was first evaluated using NF system [64]. This is due to the membrane properties required for NF and FO processes are mostly similar. Thus, there are researchers who use NF membrane for FO wastewater treatment [65,66].

In the comparison study conducted by Kim et al. [68] between FO-NF, MF-RO and UF-RO hybrid system treating coal mine impaired water, it was found that FO-NF has the lowest energy consumption (1.08 kWh/m^3) and lowest environmental impact. The optimum feed recovery of NF in FO-NF hybrid system process was 84% with lowest operating cost which was $0.41/m^3. Besides, a full-scale FO-NF system has been studied by Giagnorio et al. [69] to extract high-quality water from industrial wastewater and contaminated groundwater. It was found that the water recovery achieved using FO system with countercurrent configuration can be 85% with minimum or reversible fouling by using MgCl$_2$ or Na$_2$SO$_4$ as draw solution, followed by the recovery of draw solution via NF process to maintain the osmotic pressure in FO stage. In a long-term hybrid FO-NF continuous operation pilot-scale study conducted by Corzo et al. [67] to reuse water for irrigation from MBR effluent from urban wastewater treatment plant, they use three different draw solutions which were magnesium sulphate, sodium polyacrylate and magnesium chloride to examine the quality of water. Magnesium sulphate leads to irreversible fouling on NF membrane; therefore, it cannot be used for irrigation. Although sodium polyacrylate has reversible fouling on both FO and NF membrane, the permeate produced is also not suitable for agriculture purpose. Magnesium chloride as draw solution has the best performance in the FO-NF hybrid system where it has low propensity of fouling and good permeability for long-hours operation (480 days).

In the pilot-scale study of integrated FO-NF hybrid system for water reclamation from tannery wastewater, permeate flux of 52–55 L/m^2.h and high rejection of COD, chloride and sulphate (<97%) were achieved. Besides, long-hours operation was achieved due to crossflow action, which reduces fouling, where no membrane cleaning or replacement needed. In this hybrid system, the purpose of FO membrane module is to remove pollutants from tannery wastewater feed and produce clean water using draw solution. The NF membrane module is to recovery the draw solution to generate water for reuse or recycle purpose [70]. Dutta et al. [71] have proven the effectiveness of FO-NF system in dewatering of brackish water with low cost. In the first FO system, an ultralow-pressure RO membrane was utilized to concentrate the brackish water. Then, in NF system, inorganic fertilizer draw solution was partially recovered. The integrated FO-NF system was employed to produce quality water for the application of direct fertigation.

16.3.3 HYBRID NANOFILTRATION MEMBRANE DISTILLATION (NF-MD)

MD is a thermally driven process where temperature difference is generated across a hydrophobic membrane during operation. In MD, volatile vapour are allowed to penetrate through the hydrophobic MD membrane, while non-volatile components are maintained on the feed side. The low operating temperature in MD system could harness the available solar energy, geothermal energy, nuclear energy or low-grade heat from industrial stream. In MD process, high-quality water is able to produce from high-salinity feed. In general, MD process can be categorized into four basic configurations based on the separation efficiency, namely direct contact membrane distillation (DCMD), sweep gas membrane distillation, air gap membrane distillation and vacuum membrane distillation (VMD).

Application of NF in purification, separation and concentration of products has been applied in various fields, for example, sugar fractionation, fermentation product separation and sugar concentration. It is hard to prepare concentrated solution with only NF system. However, in MD process, solution can be concentrated to the state of saturation, which is useful in the application of volatile compound separation and non-volatile solute concentration. Additionally, by employing hybrid NF-MD separation process, NF is not only used to pre-concentrate the solution before applied in MD system but could also recycle the process water, which is an economic advantage. Figure 16.3 shows the schematic diagram of hybrid NF-MD system where the solution was first pre-concentrated with NF, followed by further concentration with MD system.

In a recent study of lithium recovery from brine via NF-DCMD hybrid system by Park et al. [73], NF was employed to remove divalent ions (Ca^{2+} and Mg^{2+}) from brine solution followed by the DCMD process to concentrate the brine solution which contains lithium. Lithium was successfully concentrated from 100 to 1,200 ppm with high water flux of 22.5 $L/m_2.h$. In a similar work of lithium (Li) recovery from salt lake water brine by Pramanik et al. [74], integrated NF-DCMD

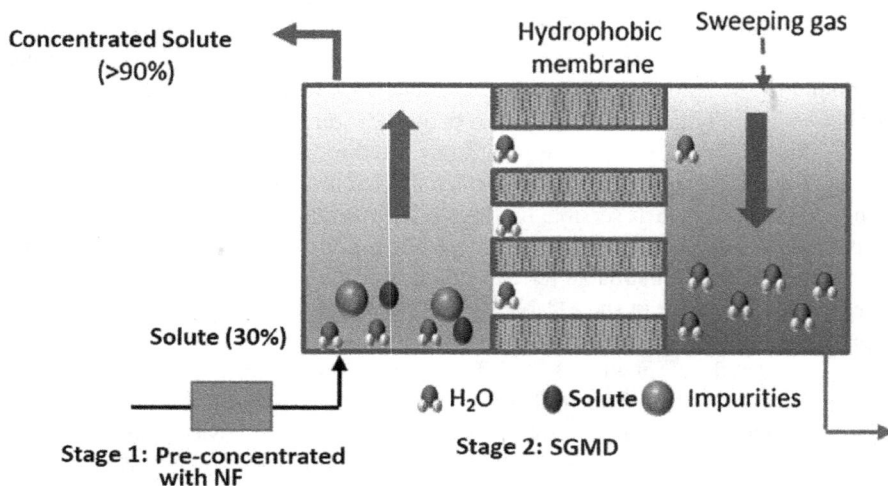

FIGURE 16.3 Schematic of hybrid filtration of NF-MD system in concentrating solution. (Adapted from Ref. [72].)

was employed. In the hybrid system, Li was first separated from salt lake brine via NF system and then enriched via DCMD. Two commercial NF membranes, namely NF90 and NF270, were used in NF process. Results showed that NF90 could reject 77% of Li, while NF270 could reject 56% of Li; after that, Li was further concentrated with DCMD system to 80%. Besides, NF-MD hybrid system has been applied in the treatment of flue gas desulfurization wastewater. The wastewater was pre-treated with NF to reduce membrane scaling followed by VMD process. In VMD system, high purity of brine was achieved, and the permeate was reused by thermal power plant; therefore, there is no wastewater discharged during the operation. The combination of NF and MD in the wastewater treatment could achieve 99% of salt rejection and 92% of water recovery ratio.

In the purification and recycling treatment of radioactive wastewater, Chen et al. [75] proposed using the hybrid NF-VMD system with ceramic membranes. The NF ceramic membrane was first use to purify boric acid, followed by concentrating the boric acid with hydrophobic ceramic membrane via VMD process. In NF system, two-stage NF process was implemented with high removal of Co^{2+} and Ag^+; on the other hand, in MD system, the boric acid was concentrated from 1 to 107 g/L with the permeate flux of 20 $L/m_2.h$ and 99% of rejection during VMD operation. Bastrzyk et al. [76] employed MF-NF-DCMD system with broths from fermentation process. The broths were first treated with MF to remove biomass and feed clarification, and then the broths were further separated via NF process to reject salt, citric acid and multivalent ions. 100% of rejection was achieved in NF process. The solution from NF was further concentrated via MD system. The quality permeate of MD was close to clean water; therefore, it could be reused as process water.

16.4 CONCLUSION AND OUTLOOK

In this chapter, examples of hybrid membrane processes have been discussed, which focus on NF membrane process associated with other membrane processes such as MF, UR, RO, FO and MD. Most of the membrane hybrid process studies are pilot scale, because obtained results are hard to correlate with large-scale applications. Therefore, the focus could be on the large-scale and long-term operation to provide understanding on the feasibility of hybrid membrane process at large scale. In the application of hybrid process of pressure-driven processes such as MF-NF system and UF-NF system, MF and UF act as a pre-treatment before NF process to remove colloids, micro-organisms, fine particles and suspended solid from raw water before NF process to improve the NF membrane performance and also its lifespan. Nevertheless, the implementation of MF-NF or UF-NF process required fouling control, especially in the beginning of hybrid process to reduce the cost of membrane replacement due to fouling. On the other hand, in NF-RO system, NF acts as pre-treatment of RO system, especially in the application of desalination and wastewater treatment to reduce water hardness and multivalent ions before RO process. The retention of monovalent species in NF process could reduce the osmotic pressure of RO feed and retentate which then allow high water recovery. Besides, the combination of FO with NF membrane process results in high-quality permeate due to its low fouling propensity in FO system, and the permeate could also be reused for other applications in long term. However, optimization of FO-NF process is required in terms of energy consumption to transform the hybrid technology into a competitive technology in water reclamation. The integration of NF-MD system is often reported for the purpose of purification and concentration of wastewater with excellent solute rejection. In short, membrane hybrid processes not only improve the process design but also lead to more economical process for water reclamation to safeguard the environment.

ACKNOWLEDGEMENT

The authors are thankful to the financial support provided by Geran Translasional UKM (TR-UKM, UKM-TR-009).

REFERENCES

1. Fao, U.N. Water, Progress on the level of water stress. Global status and acceleration needs for SDG Indicator 6.4.2, 2021, Progress on the level of water stress (2021).
2. M. Peydayesh, J. Vogt, X. Chen, J. Zhou, F. Donat, M. Bagnani, C.R. Müller, R. Mezzenga, Amyloid-based carbon aerogels for water purification, *Chem. Eng. J.*, 449 (2022) 137703–137703.
3. M.A. Shannon, P.W. Bohn, M. Elimelech, J.G. Georgiadis, B.J. Marīas, A.M. Mayes, Science and technology for water purification in the coming decades, *Nature*, 452 (2008) 301–310.
4. A. Boretti, L. Rosa, Reassessing the projections of the world water development report, *npj Clean Water*, 2 (2019) 1–6.
5. C. Wang, Y. Lü, C. Song, D. Zhang, F. Rong, L. He, Separation of emulsified crude oil from produced water by gas flotation: A review, *Sci. Total Environ.*, 845 (2022) 157304–157304.
6. M.H. Mohamed Noor, S. Wong, N. Ngadi, I. Mohammed Inuwa, L.A. Opotu, Assessing the effectiveness of magnetic nanoparticles coagulation/flocculation in water treatment: A systematic literature review, *Int. J. Environ. Sci. Technol.*, 19 (2022) 6935–6956.
7. R. Sinha, R. Kumar, P. Sharma, N. Kant, J. Shang, T.M. Aminabhavi, Removal of hexavalent chromium via biochar-based adsorbents: State-of-the-art, challenges, and future perspectives, *J. Environ. Manag.*, 317 (2022) 115356–115356.
8. E. Obotey Ezugbe, S. Rathilal, Membrane technologies in wastewater treatment: A review, *Membranes*, 10 (2020) 89.
9. V. Katheresan, J. Kansedo, S.Y. Lau, Efficiency of various recent wastewater dye removal methods: A review, *J. Environ. Chem. Eng.*, 6 (2018) 4676–4697.
10. S.F. Ahmed, M. Mofijur, S. Nuzhat, A.T. Chowdhury, N. Rafa, M.A. Uddin, A. Inayat, T.M.I. Mahlia, H.C. Ong, W.Y. Chia, P.L. Show, Recent developments in physical, biological, chemical, and hybrid treatment techniques for removing emerging contaminants from wastewater, *J. Hazard. Mater.*, 416 (2021) 125912.
11. D. Nayeri, S. Alireza, A comprehensive review on the coagulant recovery and reuse from drinking water treatment sludge, *J. Environ. Manag.*, 319 (2022) 115649–115649.
12. M. Marcińczyk, Y.S. Ok, P. Oleszczuk, From waste to fertilizer: Nutrient recovery from wastewater by pristine and engineered biochars, *Chemosphere*, 306 (2022) 135310–135310.
13. O. Garcia-Rodriguez, E. Mousset, H. Olvera-Vargas, O. Lefebvre, Electrochemical treatment of highly concentrated wastewater: A review of experimental and modeling approaches from lab- to full-scale, *Crit. Rev Environ. Sci. Technol.*, 52 (2022) 240–309.
14. A. Hosseinzadeh, M. Baziar, H. Alidadi, J.L. Zhou, A. Altaee, A.A. Najafpoor, S. Jafarpour, Application of artificial neural network and multiple linear regression in modeling nutrient recovery in vermicompost under different conditions, *Bioresour. Technol.*, 303 (2020) 122926–122926.
15. S. Amin, H. Naja, S. Zolgharnian, S. Shari, N. Asasian-kolur, Biological oxidation methods for the removal of organic and inorganic contaminants from wastewater: A comprehensive review, *Sci. Total Environ.*, 843 (2022).
16. Q. Wu, D. Zou, X. Zheng, F. Liu, L. Li, Z. Xiao, Science of the total environment effects of antibiotics on anaerobic digestion of sewage sludge : Performance of anaerobic digestion and structure of the microbial community, *Sci. Total Environ.*, 845 (2022) 157384–157384.
17. J. Li, Q. Wang, L. Deng, X. Kou, Q. Tang, Y. Hu, Fabrication and characterization of carbon nanotubes-based porous composite forward osmosis membrane: Flux performance, separation mechanism, and potential application, *J. Membr. Sci.*, 604 (2020) 118050.
18. Z. Yao, L.E. Peng, H. Guo, W. Qing, Y. Mei, C.Y. Tang, Seawater pretreatment with an NF-like forward osmotic membrane: Membrane preparation, characterization and performance comparison with RO-like membranes, *Desalination*, 470 (2019) 114115.
19. H. Wang, Y. Gao, B. Gao, K. Guo, H.K. Shon, Q. Yue, Z. Wang, Comprehensive analysis of a hybrid FO-NF-RO process for seawater desalination: With an NF-like FO membrane, *Desalination*, 515 (2021) 115203.
20. G. Blandin, D.T. Myat, A.R. Verliefde, P. Le-Clech, Pressure assisted osmosis using nanofiltration membranes (PAO-NF): Towards higher efficiency osmotic processes, *J. Membr. Sci.*, 533 (2017) 250–260.
21. Z. Yuan, Y. Yu, X. Sui, Y. Yao, Y. Chen, Carbon composite membranes for thermal-driven membrane processes, *Carbon*, 179 (2021) 600–626.
22. W.L. Ang, A.W. Mohammad, N. Hilal, C.P. Leo, A review on the applicability of integrated/hybrid membrane processes in water treatment and desalination plants, *Desalination*, 363 (2015) 2–18.

23. D.J. Johnson, N. Hilal, Nanocomposite nanofiltration membranes: State of play and recent advances, *Desalination*, 524 (2022) 115480–115480.
24. R.L. Ramos, L.B. Grossi, B.C. Ricci, M.C.S. Amaral, Membrane selection for the Gold mining pressure-oxidation process (POX) effluent reclamation using integrated UF-NF-RO processes, *J. Environ. Chem. Eng.*, 8 (2020) 104056–104056.
25. S.F. Anis, R. Hashaikeh, N. Hilal, Reverse osmosis pretreatment technologies and future trends: A comprehensive review, *Desalination*, 452 (2019) 159–195.
26. A. Shokri, M.S. Fard, Corrosion in seawater desalination industry: A critical analysis of impacts and mitigation strategies, *Chemosphere* (2022) 135640.
27. R. Kumar, M. Ahmed, S. Ok, B. Garudachari, J.P. Thomas, Boron selective thin film composite nanofiltration membrane fabricated: Via a self-assembled trimesic acid layer at a liquid-liquid interface on an ultrafiltration support, *New J. Chem.*, 43 (2019) 3874–3883.
28. X. Wang, Q. Xiao, C. Wu, P. Li, S. Xia, Fabrication of nanofiltration membrane on MoS2 modified PVDF substrate for excellent permeability, salt rejection, and structural stability, *Chem. Eng. J.*, 416 (2021) 129154–129154.
29. J. Núñez, M. Yeber, N. Cisternas, R. Thibaut, P. Medina, C. Carrasco, Application of electrocoagulation for the efficient pollutants removal to reuse the treated wastewater in the dyeing process of the textile industry, *J. Hazard. Mater.*, 371 (2019) 705–711.
30. M. Delnavaz, H. Khoshvaght, A. Sadeghi, K. Ghasemipanah, M.H. Aliabadi, Experimental, statistical and financial analysis of the treatment of organic contaminants in naphthenic spent caustic soda using electrocoagulation process modified by carbon nanotubes, *J. Clean. Prod.*, 327 (2021) 129515–129515.
31. A.I. Rita, A.R. Nabais, L.A. Neves, R. Huertas, M. Santos, L.M. Madeira, S. Sanches, Assessment of the potential of using nanofiltration polymeric and ceramic membranes to treat refinery spent caustic effluents, *Membranes*, 12 (2022).
32. H. Ye, B. Liu, Q. Wang, Z.T. How, Y. Zhan, P. Chelme-Ayala, S. Guo, M. Gamal El-Din, C. Chen, Comprehensive chemical analysis and characterization of heavy oil electric desalting wastewaters in petroleum refineries, *Sci. Total Environ.*, 724 (2020) 138117–138117.
33. S. Dadari, M. Rahimi, S. Zinadini, Crude oil desalter effluent treatment using high flux synthetic nanocomposite NF membrane-optimization by response surface methodology, *Desalination*, 377 (2016) 34–46.
34. H. Özgün, H. Sakar, M. Ağtaş, I. Koyuncu, Investigation of pre-treatment techniques to improve membrane performance in real textile wastewater treatment, *Int. J. Environ. Sci. Technol.* (2022).
35. E. Hassanzadeh, M. Farhadian, A. Razmjou, N. Askari, An efficient wastewater treatment approach for a real woolen textile industry using a chemical assisted NF membrane process, *Environ. Nanotechnol. Monit. Manage.*, 8 (2017) 92–96.
36. D.A. Yaseen, M. Scholz, *Textile Dye Wastewater Characteristics and Constituents of Synthetic Effluents: A Critical Review*, Springer: Berlin Heidelberg, 2019.
37. R.A. Nasr, E.A.B. Ali, Polyethersulfone/gelatin nano-membranes for the Rhodamine B dye removal and textile industry effluents treatment under cost effective condition, *J. Environ. Chem. Eng.*, 10 (2022) 107250–107250.
38. T. Tavangar, M. Karimi, M. Rezakazemi, K.R. Reddy, T.M. Aminabhavi, Textile waste, dyes/inorganic salts separation of cerium oxide-loaded loose nanofiltration polyethersulfone membranes, *Chem. Eng. J.*, 385 (2020) 123787–123787.
39. S.A. An, J. Lee, J. Sim, C.G. Park, J.S. Lee, H. Rho, K.D. Park, H.S. Kim, Y.C. Woo, Evaluation of the advanced oxidation process integrated with microfiltration for reverse osmosis to treat semiconductor wastewater, *Process Saf. Environ. Prot.*, 162 (2022) 1057–1066.
40. M. Cha, C. Boo, C. Park, Simultaneous retention of organic and inorganic contaminants by a ceramic nanofiltration membrane for the treatment of semiconductor wastewater, *Process Saf. Environ. Prot.*, 159 (2022) 525–533.
41. J. López, M. Reig, O. Gibert, E. Torres, C. Ayora, J.L. Cortina, Application of nanofiltration for acidic waters containing rare earth elements: Influence of transition elements, acidity and membrane stability, *Desalination*, 430 (2018) 33–44.
42. V.R. Moreira, Y.A.R. Lebron, A.F.S. Foureaux, L.V.D.S. Santos, M.C.S. Amaral, Acid and metal reclamation from mining effluents: Current practices and future perspectives towards sustainability, *J. Environ. Chem. Eng.*, 9 (2021).
43. H.Z. Zhang, Z.L. Xu, J.Y. Sun, Three-channel capillary NF membrane with PAMAM-MWCNT-embedded inner polyamide skin layer for heavy metals removal, *RSC Adv.*, 8 (2018) 29455–29463.

44. A. Aguiar, L. Andrade, L. Grossi, W. Pires, M. Amaral, Acid mine drainage treatment by nanofiltration: A study of membrane fouling, chemical cleaning, and membrane ageing, *Sep. Purif. Technol.*, 192 (2018) 185–195.

45. A.K. Singh, R. Chandra, Pollutants released from the pulp paper industry: Aquatic toxicity and their health hazards, *Aquat. Toxicol.*, 211 (2019) 202–216.

46. F. Khalid, M. Tabish, K.A.I. Bora, Novel poly(vinyl alcohol) nanofiltration membrane modified with dopamine coated anatase TiO_2 core shell nanoparticles, *J. Water Process Eng.*, 37 (2020) 101486–101486.

47. O.J. Valderrama, K.L. Zedda, S. Velizarov, Membrane filtration opportunities for the treatment of black liquor in the paper and pulp industry, *Water (Switzerland)*, 13 (2021).

48. A. Kumar, N.K. Srivastava, P. Gera, Removal of color from pulp and paper mill wastewater- methods and techniques – A review, *J. Environ. Manag.*, 298 (2021) 113527–113527.

49. K. Nath, H.K. Dave, T.M. Patel, Revisiting the recent applications of nanofiltration in food processing industries: Progress and prognosis, *Trends Food Sci. Technol.*, 73 (2018) 12–24.

50. Y. Wang, L. Serventi, Sustainability of dairy and soy processing: A review on wastewater recycling, *J. Clean. Prod.*, 237 (2019) 117821–117821.

51. T. Esteves, A.T. Mota, C. Barbeitos, K. Andrade, C.A.M. Afonso, F.C. Ferreira, A study on lupin beans process wastewater nanofiltration treatment and lupanine recovery, *J. Clean. Prod.*, 277 (2020).

52. H.K. Shon, S. Phuntsho, D.S. Chaudhary, S. Vigneswaran, J. Cho, Nanofiltration for water and wastewater treatment - A mini review, *Drink. Water Eng. Sci.*, 6 (2013) 47–53.

53. E. Obotey Ezugbe, S. Rathilal, Membrane Technologies in Wastewater Treatment: A Review, *Membranes*, 10 (2020) 89.

54. S. Shanmuganathan, M.A.H. Johir, T.V. Nguyen, J. Kandasamy, S. Vigneswaran, Experimental evaluation of microfiltration–granular activated carbon (MF–GAC)/nano filter hybrid system in high quality water reuse, *J. Membr. Sci.*, 476 (2015) 1–9.

55. S. Šostar-Turk, M. Simonič, I. Petrinić, Wastewater treatment after reactive printing, *Dyes Pigments*, 64 (2005) 147–152.

56. S.K. Nataraj, K.M. Hosamani, T.M. Aminabhavi, Distillery wastewater treatment by the membrane-based nanofiltration and reverse osmosis processes, *Water Res.*, 40 (2006) 2349–2356.

57. A. Salahi, R. Badrnezhad, M. Abbasi, T. Mohammadi, F. Rekabdar, Oily wastewater treatment using a hybrid UF/RO system, *Desal. Water Treat.*, 28 (2011) 75–82.

58. C.R. Bartels, M. Wilf, K. Andes, J. Iong, Design considerations for wastewater treatment by reverse osmosis, *Water Sci. Technol.*, 51 (2005) 473–482.

59. High performance nanofiltration-membrane bioreactor and reverse osmosis system for high recovery water reclamation, in, Tplus Technology Exchange and Investment (2019).

60. M.C.S. Amaral, H.V. Pereira, E. Nani, L.C. Lange, Treatment of landfill leachate by hybrid precipitation/microfiltration/nanofiltration process, *Water Sci. Technol.*, 72 (2015) 269–276.

61. A.C. Bortoluzzi, J.A. Faitão, M. Di Luccio, R.M. Dallago, J. Steffens, G.L. Zabot, M.V. Tres, Dairy wastewater treatment using integrated membrane systems, *J. Environ. Chem. Eng.*, 5 (2017) 4819–4827.

62. X. Sun, C. Wang, Y. Li, W. Wang, J. Wei, Treatment of phenolic wastewater by combined UF and NF/RO processes, *Desalination*, 355 (2015) 68–74.

63. S. Zhao, L. Zou, D. Mulcahy, Brackish water desalination by a hybrid forward osmosis–nanofiltration system using divalent draw solute, *Desalination*, 284 (2012) 175–181.

64. M.F. Hamid, N. Abdullah, N. Yusof, N.M. Ismail, A.F. Ismail, W.N.W. Salleh, J. Jaafar, F. Aziz, W.J. Lau, Effects of surface charge of thin-film composite membrane on copper (II) ion removal by using nanofiltration and forward osmosis process, *J. Water Process Eng.*, 33 (2020) 101032.

65. W.N.A.S. Abdullah, W.-J. Lau, F. Aziz, D. Emadzadeh, A.F. Ismail, Performance of nanofiltration-like forward-osmosis membranes for aerobically treated palm oil mill effluent, *Chem. Eng. Technol.*, 41 (2018) 303–312.

66. Y.N. Hu, Q.M. Li, Y.F. Guo, L.J. Zhu, Z.X. Zeng, Z. Xiong, Nanofiltration-like forward osmosis membranes on in-situ mussel-modified polyvinylidene fluoride porous substrate for efficient salt/dye separation, *J. Polym. Sci.*, 59 (2021) 1065–1071.

67. B. Corzo, T. de la Torre, C. Sans, R. Escorihuela, S. Navea, J.J. Malfeito, Long-term evaluation of a forward osmosis-nanofiltration demonstration plant for wastewater reuse in agriculture, *Chem. Eng. J.*, 338 (2018) 383–391.

68. J.E. Kim, S. Phuntsho, L. Chekli, S. Hong, N. Ghaffour, T. Leiknes, J.Y. Choi, H.K. Shon, Environmental and economic impacts of fertilizer drawn forward osmosis and nanofiltration hybrid system, *Desalination*, 416 (2017) 76–85.

69. M. Giagnorio, F. Ricceri, M. Tagliabue, L. Zaninetta, A. Tiraferri, Hybrid forward osmosis–nanofiltration for wastewater reuse: System design, *Membranes*, 9 (2019) 61.

70. P. Pal, S. Chakrabortty, J. Nayak, S. Senapati, A flux-enhancing forward osmosis–nanofiltration integrated treatment system for the tannery wastewater reclamation, *Env. Sci. Pollut. Res.*, 24 (2017) 15768–15780.

71. S. Dutta, K. Nath, Dewatering of brackish water and wastewater by an integrated forward osmosis and nanofiltration system for direct fertigation, *Arab. J. Sci. Eng.*, 44 (2019) 9977–9986.

72. P.M. Duyen, P. Jacob, R. Rattanaoudom, C. Visvanathan, Feasibility of sweeping gas membrane distillation on concentrating triethylene glycol from waste streams, *Chem. Eng. Process.: Process Intensif.*, 110 (2016) 225–234.

73. S.H. Park, J.H. Kim, S.J. Moon, J.T. Jung, H.H. Wang, A. Ali, C.A. Quist-Jensen, F. Macedonio, E. Drioli, Y.M. Lee, Lithium recovery from artificial brine using energy-efficient membrane distillation and nanofiltration, *J. Membr. Sci.*, 598 (2020) 117683.

74. B.K. Pramanik, M.B. Asif, S. Kentish, L.D. Nghiem, F.I. Hai, Lithium enrichment from a simulated salt lake brine using an integrated nanofiltration-membrane distillation process, *J. Env. Chem. Eng.*, 7 (2019) 103395.

75. X. Chen, T. Chen, J. Li, M. Qiu, K. Fu, Z. Cui, Y. Fan, E. Drioli, Ceramic nanofiltration and membrane distillation hybrid membrane processes for the purification and recycling of boric acid from simulative radioactive waste water, *J. Membr. Sci.*, 579 (2019) 294–301.

76. J. Bastrzyk, M. Gryta, Separation of post-fermentation glycerol solution by nanofiltration membrane distillation system, *Desalin. Water Treat.*, 53 (2015) 319–329.

17 The Use of Nanofiltration in Food and Biotech Industry

Nazlee Faisal Ghazali and Ki Min Lim
Universiti Teknologi Malaysia

CONTENTS

17.1 INTRODUCTION

In this era of global industrialization, various food, nutritional and pharmaceutical products have been developed and produced to provide for human consumption and treatment of diseases. Various products have been produced globally such as edible oil, dairy products, beverages, sugar, nutritional dietary products and pharmaceutical products. The raw materials undergo a series of refining processes before the final product is obtained. In the manufacturing of these food or pharmaceutical products, waste or by-products were generated as a result of the refining or production processes. These waste or by-products were mostly low in value, as they were normally not entirely pure to be used for other purposes. Besides that, the effluents from the manufacturing processes need to be treated before it can be released to the environment. Moreover, as most of the current refining processes employ distillation method, a huge amount of energy is required to obtain the desired refined products.

Therefore, researchers have investigated ways to obtain refined products with low energy expenditure, and reduced waste or by-products from processes. In recent years, researchers have proposed membrane technology as an alternative to the conventional refining methods.

17.2 MEMBRANE TECHNOLOGY

The use of membrane technology is certainly not uncommon, especially when a process requires the separation of two or more compounds from a mixture of solutes and solvents. Initial application of membrane technology has been used mostly in the filtering of large compounds from the mixtures, usually as strainer. These membranes are usually microporous membranes which allow high solvent recovery and retention of the bulk compounds. In recent years, membranes of smaller pore diameter such as ultrafiltration membrane and nanofiltration (NF) membrane have been studied. Ultrafiltration membranes are capable of filtering compounds with sizes in between 1,000 Da and 10 kDa. On the other hand, NF membranes are capable of filtering compounds with size from 200 to 1,000 Da, and finally, reverse osmosis membranes can filter compounds with size smaller than 200 Da [1]. NF particularly is useful, due to lower energy expenditure and higher flux when

DOI: 10.1201/9781003261827-17

compared to reverse osmosis, in separating small solutes, in which fractionation or concentration of desired products can be achieved.

There are two main types of membranes, which are ceramic membrane and polymeric membrane. Ceramic membranes are made of inorganic materials such as titanium, alumina, zirconium oxides, silicon carbides and other inorganic-based materials. Polymeric membranes, on the other hand, as the name suggests, are made of polymers, which includes polyimide, polyacrylonitrile, polydimethylsiloxane, polybenzimidazole, polysulfone, cellulose acetate and other organic polymers. Ceramic membrane generally has a better permeability and stability, especially when involves the separation of compounds in harsh solvents. However, ceramic membranes are costly and sensitive to physical impact, making it difficult to be handled and developed. In contrast, the flexibility of polymeric membrane to be modified at different pore sizes, and the ease of manufacture and handling, make it a popular choice to be studied [2].

In recent years, the developments and investigations on NF membrane for different industries have bloomed, due to its ability to separate small molecules without the use of complicated experimental set-up [3–5]. Besides that, it is also more cost-effective as it only requires pressure difference to drive the separation between the different molecules. The set-up for the NF process can be constructed by having the feed solution to be pressurized and flow parallel to the surface of the membrane (or known as crossflow configuration) or pressurized and flow perpendicular to the surface of the membrane (or known as dead-end configuration). This would allow the separation of molecules to occur through diffusion or convection [1].

17.3 APPLICATION OF NANOFILTRATION

NF has been actively applied on different fields of study, due to its ability of segregating different molecules with sizes in between 200 and 1,000 Da. The segregations by NF are typically affected by steric hindrances, molecular sizes, interactions between solutes, solvent and membranes, hydrophilicity of membrane and pore size of membrane [1]. As there are numerous possible factors that may affect the separation between molecules, studies have been performed by researchers in different fields of study, to identify suitable membranes for each of the separation processes.

17.3.1 FOOD INDUSTRY

In the food industry, membrane technology has become a useful method in separating undesired compounds and also in recovering solvents. In the dairy industry, membrane technology is employed to obtain and separate different components and compounds in dairy products. In the cheese production, for example, whey, which is the by-product from the process, is separated and removed [6]. Whey contains components such as protein and lactose which have nutritional values in human diet. However, whey needs to be refined, as it also consists of unwanted salts and acids, which dissociate into organic or inorganic ions in water. The removal of inorganic and organic ions in sweet whey and acid whey can be performed by NF (Figure 17.1).

In the NF of sweet whey, the removal of single-charged ions and multi-charged ions are 78% and 9%, respectively. The low removal of multi-charged ions is due to the large radius of hydrated ions and high hydration energy. On the other hand, the mineral contents in acid whey, namely calcium and phosphate ions, are 2.9 and 2.3 times higher than those in sweet whey, respectively. There is also a high content of organic acid such as lactic acid and citric acid in acid whey, which contributes to a lower pH of acid whey. As the membrane used was negatively charged at a given pH range, the lower pH of acid whey resulted in a less-dissociated citric acid, hence allowing its removal [7]. Raw acid whey processed by using ultrafiltration (UF) membranes, generate a protein-rich stream retentate and lactose stream permeate. The lactose stream permeate can be further filtered by using a thin-film composite NF membrane (150–300 Da). About 71% of lactic acid and 48% of total cations were removed in the NF process [8].

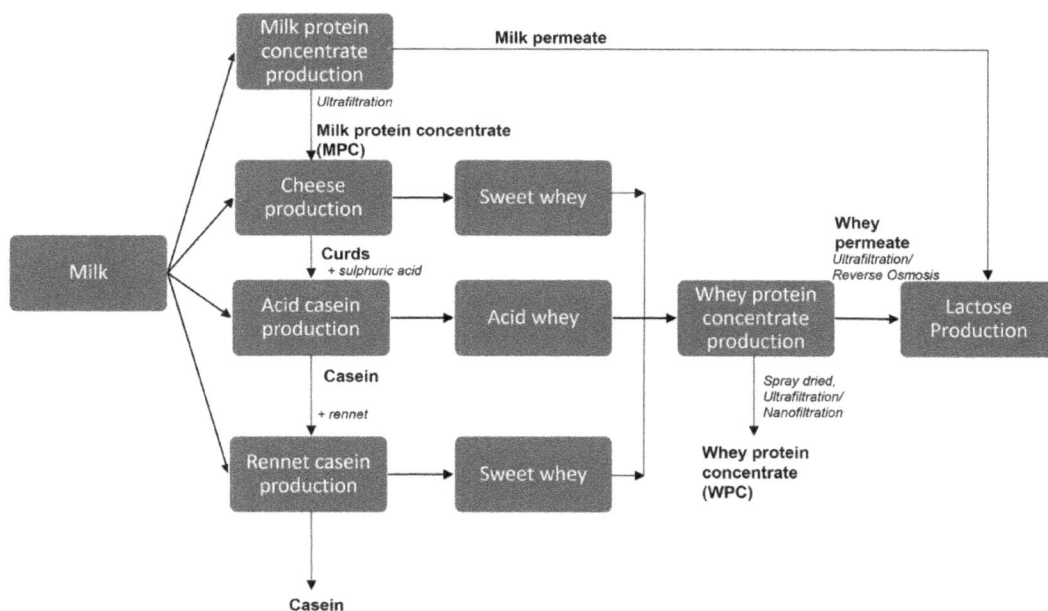

FIGURE 17.1 Production of lactose from milk.

In the process of concentrating fruit juices, conventionally, the fruit juices were obtained through series of multistage distillation process. In recent years, membrane technology has been investigated for its uses in obtaining concentrated fruit juices and nutrients [9]. In the recent investigation, procyanidin from grape juice was concentrated by using NF. Polyamide NF membrane (800 Da) in crossflow configuration was used to filter the grape juice. It was found that the membrane was able to reject >85% of the procyanidin in grape juice when it is operated at pH of 8.0 [10]. This suggests that the membrane is capable of recovering essential compounds from grapes. Besides that, NF was also employed in the concentration of phenolic compounds after extraction process. After the extraction of apple pomace by using ethanol, a hydrophilicnthin film composite (TFC) type of membrane (150–300 Da) was used to filter the phenolic compounds from ethanol. The NF membrane was able to obtain an average retention of 98%–99% of phenolic compounds [11]. The concentration of bioactive compounds can also be performed by using NF membrane. NF of elderberry was performed by Tundis et al. by using polymeric membranes with the membrane molecular-weight cutoff (MWCO) of 400 and 1,000 Da. It was found that all the investigated membranes have a high rejection rate towards anthocyanins, astragalin and other antioxidant compounds [12]. Figure 17.2 shows illustration of the fruit concentration process through membrane technology.

In the wine-making industry, acetic acid, which is part of the volatile acids content in wine, is by-product of alcoholic and lactic fermentation. The excessive amount of acetic acid would lead to undesirable aroma and wine spoilage [13]. In the investigation of the application of NF in red wine Cabernet Sauvignon, Ivić et al. [14] had found that the NF membrane was able to remove acetic acid from the feed retentate. Acetic acid was not able to be detected in the feed retentate after NF, which denotes that the acetic acid was mostly removed.

NF can also be used for separating caffeine from spent coffee grounds (SCG). The caffeine can be used to produce cosmetics and energy foods. In an investigation of a caffeine separation by NF, caffeine and chlorogenic acid were initially extracted from SCG with water. It was then subjected to NF by using NF membrane with MWCO of 400 Da by Microdyn-Nadir GmbH. The NF was able to achieve a high rejection of chlorogenic acid (75.7%) and low rejection of caffeine (2%) from the water solutions at 30 bar pressure [15].

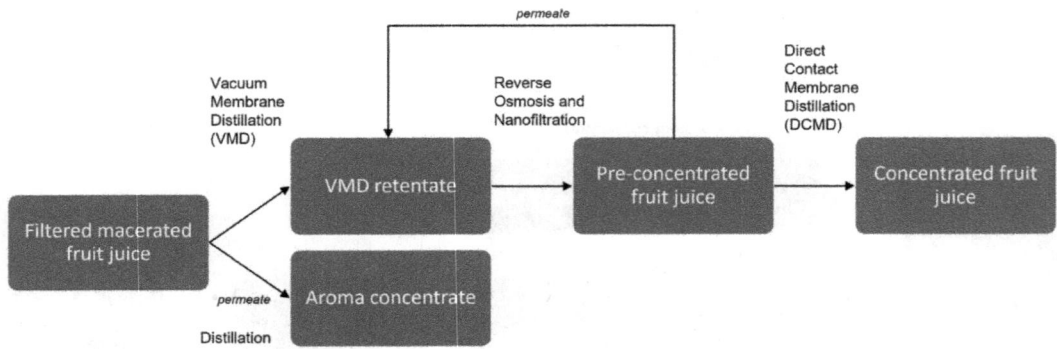

FIGURE 17.2 Illustration of fruit concentration process.

In the vegetable processing industry, NF membranes were studied for their uses in deacidification of free fatty acids and nutrient recovery. In the conventional deacidification process, degummed vegetable oil is deacidified through chemical treatment and steam distillation process. In the chemical treatment process, alkali is added to the degummed vegetable oil to neutralize the free fatty acid in the vegetable oil. The neutralization process resulted in the production of soap stocks and loss of oil in the process [16]. The chemical treatment process is usually applicable to vegetable oil containing small amount of free fatty acids. In the case of vegetable oil with high free fatty acids content, steam distillation is used. Steam distillation is efficient in removing free fatty acids in the vegetable oil; however, essential nutrients were removed in the process as well [17]. Moreover, high energy requirements of the steam distillation process inspired researchers to search for an alternative to the process. Figure 17.3 shows the illustration of vegetable oil refined through conventional process and membrane technology.

As crude vegetable oil was extracted by using solvent prior to the distillation process, Shi and Chung [18] had developed a novel Teflon AF2400/polyethylene NF membrane for the separation of oil from commonly used solvent such as acetone and hexane. It was found that the membrane was able to reject 99.95% of triglyceride at 0.36 LMH/bar and 99.84% of triglyceride at 1.45 LMH/bar. In another study, Werth et al. [19] had performed investigations of commercial solvent-resistant membrane in deacidification of rapeseed oil. It was found that PuraMem 280 membrane was able to reject 97% of triglyceride and 34.7% of oleic acid in rapeseed oil at 40 bar and 4.54 kg/m^{2}h when ethanol was used as the solvent. Shi et al. [20] also had studied the use of different commercial membranes in deacidification of soybean oil. It was found that DuraMem 300 and DuraMem 150 rejected both glyceryl trilinoleate and linoleic acid at a high rejection rate of 100% and 83.7%–100%, respectively. On the other hand, PuraMem 600 and DuraMem 500 were able to separate linoleic acid and glyceryl trilinoleate at rejection rate of 13.4%–35.1% and 85.8%–88.5%, respectively, with permeance of 1.02–1.23 LMH/bar.

17.3.2 PHARMACEUTICAL INDUSTRY

In the pharmaceutical manufacturing industry, the membrane technology can be applied for the purification of active pharmaceutical ingredients (APIs). Currently, the API purification relies on distillation, solvent extraction, fractionation and chromatography processes. However, these processes are expensive and require a large quantity of chemical solvents for processing. With membrane technology, the solvent used can be reduced and recycled, which is not only cost-effective, but also reduces the chemical waste produced during the purification process. In investigating the feasibility of membranes in purifying API, various membranes were developed [21]. The membranes are usually chemically inert membrane, in which membranes are made of solvent-resistant polymers

Conventional process

Membrane Technology

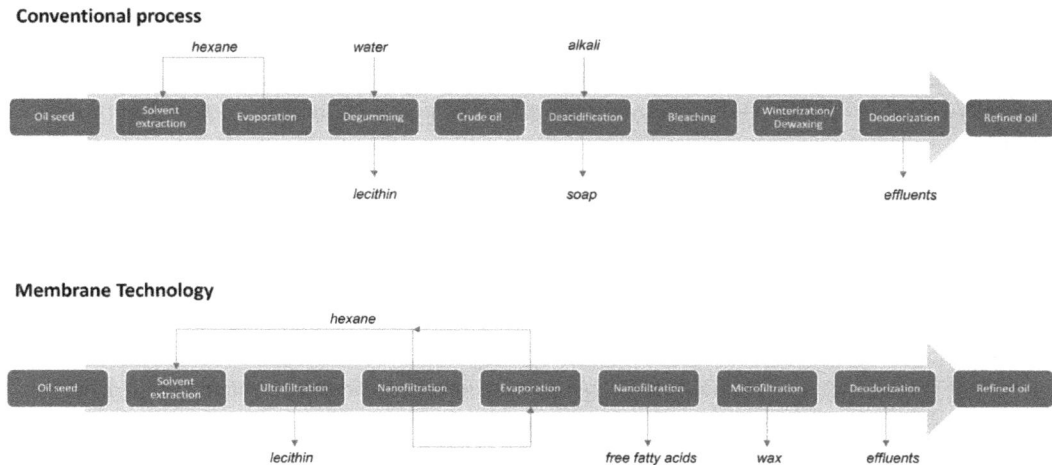

FIGURE 17.3 Illustration of vegetable oil refining process through conventional method and membrane technology.

such as polyimide, polybenzimidazole and polydimethylsiloxane, to name a few. In a recent study, NF membranes which were developed from hydrophilic perfluorinated polymer was found to be able to reject 95% of the API compound and withstand a high temperature (75°C) and 4,000 kPa [22]. The higher temperature of 75°C yielded at least 2–3 times higher solvent fluxes than at 25°C. This proves that membrane technology is capable of recovering API compounds. In a different study by Foureaux et al. [23], a study was performed to investigate the rejection of pharmaceutical compounds from the surface of water by NF. It was found that NF was able to reject pharmaceutical compounds at a high rejection rate, which was comparable to that of the performance of a reverse osmosis membrane. Besides that, the NF was also able to achieve a higher permeate flux as compared to reverse osmosis, which will bring about cost-effectiveness during operation. In the study by Goh et al. [24], they had developed a 100-piece hollow-fibre thin-film composite membrane. The membrane was organic solvent NF type of membrane, in which it can withstand the harsh property of organic solvent during filtration process. In their study, the membrane was tested by using acetone and methyl red, to determine the molecular weight cut-off of the membrane. Subsequently, the novel membrane was used to concentrate levofloxacin (361 Da) in acetonitrile, whereby an excellent rejection rate of 98.2% was achieved.

17.3.3 BIOTECHNOLOGY

In the field of biotechnology, NF can be used for the separation and purification of biocompounds from fermentation broth. Cabrera-González et al. [25] had performed the investigation of lactic acid purification by using NF membranes. In their investigation, it was found that NF270 membrane had the rejection rate of lactic acid of 71% at 25°C and pH 2.8. On the contrary, MPF-36 membrane had the best performance among all the tested membranes, whereby yield of 93% (permeate) was obtained at pH 2.8 and 25°C. In order to improve the efficiency of the membrane separation process, membrane separation cascades were often used. In a study by McDowall et al. [26], they developed a separation cascade for the separation of medium-chain fatty acids (MCFAs). MCFAs are fatty acids with 6–12 carbon atoms which are usually found in coconut, palm kernel oil and dairy products [27]. From the results of study by McDowall et al. [26], it was found that the membrane cascade was able to retain more than 90% of the MCFAs. However, it was also found that membrane fouling caused flux decline, which limits the separation process. Diafiltration was also performed; however, it was not able to improve the condition caused by membrane fouling. In a different study by Alves et al. [28], they had performed a study on the xylitol purification by using different NF membranes. It was found that the "NF" membrane (polypiperazine amide-based membrane) performed most optimally and was further optimized by using surface response methodology. After optimization, the membrane was able to remove 86% of proteins and peptides from the fermentation broth. In the study by Alves et al. [28], they also performed microfiltration and ultrafiltration prior to the NF process to minimize the effect of membrane fouling.

17.4 CHALLENGES AND FUTURE OUTLOOK

As discussed in the chapter, it can be said that NF technology is a potential alternative for the purification of dairy products, beverages, vegetable oil, nutritional and pharmaceutical products and in the field of biotechnology. The main challenges of implementing NF are the membrane fouling mitigations, and the use predictive modelling for a certain purification process. As of the present time, there are limited literatures on predictive modelling applied to the purification of food and biotech products. It is crucial as the performance prediction of a purification process relies on the model that describes it. With the availability of process-specific models for an NF process and databases of membrane filtration for related process, simultaneous optimization of the process and fabrication can be achieved. From the optimization, it would greatly improve the compatibility of a membrane towards a certain process. Besides, a greener alternative for the membrane fabrication process is desirable for future considerations, in order to achieve a more sustainable membrane fabrication and scale-up process. To date, there are various developments of modified membranes for the purification process. However, there are only a handful research studies which were performed in the area of food and biotechnology products. Therefore, the development of process-specific and a green membrane would bring about a more sustainable implementation of NF technology.

REFERENCES

1. Mulder, J. (2012). *Basic Principles of Membrane Technology*, Springer Science & Business Media, The Netherlands.
2. Cui, Z. and Muralidhara, H. (2010). *Membrane Technology: A Practical Guide to Membrane Technology and Applications in Food and Bioprocessing*, Elsevier.
3. Nguyen Thi, H.Y., Nguyen, B.T.D. and Kim, J.F. "Sustainable fabrication of organic solvent nanofiltration membranes." *Membranes* 11(2021): 19.
4. Gunawan, F.M., Mangindaan, D., Khoiruddin, K. and Wenten, I.G. "Nanofiltration membrane crosslinked by m-phenylenediamine for dye removal from textile wastewater." *Polymers for Advanced Technologies* 30(2019): 360–367.
5. Werth, K., Kaupenjohann, P. and Skiborowski, M. "The potential of organic solvent nanofiltration processes for oleochemical industry." *Separation and Purification Technology* 182(2017): 185–196.
6. Mohammad, A., Teow, Y., Ho, K. and Rosnan, N. (2019). "Recent developments in nanofiltration for food applications." *Nanomaterials for Food Applications*, Elsevier, pp. 101–120.
7. Merkel, A., Voropaeva, D. and Ondrušek, M. "The impact of integrated nanofiltration and electrodialytic processes on the chemical composition of sweet and acid whey streams." *Journal of Food Engineering* 298(2021): 110500.
8. Talebi, S., Suarez, F., Chen, G.Q., Chen, X., Bathurst, K. and Kentish, S.E. "Pilot study on the removal of lactic acid and minerals from acid whey using membrane technology." *ACS Sustainable Chemistry & Engineering* 8(2020): 2742–2752.
9. Belleville, M.P., Sanchez-Marcano, J., Bargeman, G. and Timmer, M. "Nanofiltration in the food industry." *Nanofiltration: Principles, Applications, and New Materials* 2(2021): 499–542.
10. Li, C., Ma, Y., Li, H. and Peng, G. "Exploring the nanofiltration mass transfer characteristic and concentrate process of procyanidins from grape juice." *Food Science & Nutrition* 7(2019): 1884–1890.

11. Uyttebroek, M., Vandezande, P., Van Dael, M., Vloemans, S., Noten, B., Bongers, B., Porto-Carrero, W., Muñiz Unamunzaga, M., Bulut, M. and Lemmens, B. "Concentration of phenolic compounds from apple pomace extracts by nanofiltration at lab and pilot scale with a techno-economic assessment." *Journal of Food Process Engineering* 41(2018): e12629.
12. Tundis, R., Loizzo, M.R., Bonesi, M., Sicari, V., Ursino, C., Manfredi, I., Conidi, C., Figoli, A. and Cassano, A. "Concentration of bioactive compounds from elderberry (*Sambucus nigra* L.) juice by nanofiltration membranes." *Plant Foods for Human Nutrition* 73(2018): 336–343.
13. Conidi, C., Castro-Muñoz, R. and Cassano, A. (2020). "Nanofiltration in beverage industry." *Nanotechnology in the Beverage Industry*, Elsevier, pp. 525–548.
14. Ivić, I., Kopjar, M., Obhođaš, J., Vinković, A., Pichler, D., Mesić, J. and Pichler, A. "Concentration with nanofiltration of red wine cabernet sauvignon produced from conventionally and ecologically grown grapes: Effect on volatile compounds and chemical composition." *Membranes* 11(2021): 320.
15. Peshev, D., Mitev, D., Peeva, L. and Peev, G. "Valorization of spent coffee grounds – A new approach." *Separation and Purification Technology* 192(2018): 271–277.
16. Hamm, W., Hamilton, R.J. and Calliauw, G. (2013). *Edible Oil Processing*, John Wiley & Sons,
17. Ghazali, N.F., Md Hanim, K., Pahlawi, Q.A. and Lim, K.M. "Enrichment of carotene from palm oil by organic solvent nanofiltration." *Journal of the American Oil Chemists' Society* 99(2022): 189–202.
18. Shi, G.M. and Chung, T.-S. "Teflon AF2400/polyethylene membranes for organic solvent nanofiltration (OSN)." *Journal of Membrane Science* 602(2020): 117972.
19. Werth, K., Kaupenjohann, P., Knierbein, M. and Skiborowski, M. "Solvent recovery and deacidification by organic solvent nanofiltration: Experimental investigation and mass transfer modeling." *Journal of Membrane Science* 528(2017): 369–380.
20. Shi, G.M., Farahani, M.H.D.A., Liu, J.Y. and Chung, T.-S. "Separation of vegetable oil compounds and solvent recovery using commercial organic solvent nanofiltration membranes." *Journal of Membrane Science* 588(2019): 117202.
21. Buonomenna, M. and Bae, J. "Organic solvent nanofiltration in pharmaceutical industry." *Separation & Purification Reviews* 44(2015): 157–182.
22. Chau, J., Sirkar, K.K., Pennisi, K.J., Vaseghi, G., Derdour, L. and Cohen, B. "Novel perfluorinated nanofiltration membranes for isolation of pharmaceutical compounds." *Separation and Purification Technology* 258(2021): 117944.
23. Foureaux, A.F.S., Reis, E., Lebron, Y., Moreira, V., Santos, L., Amaral, M. and Lange, L. "Rejection of pharmaceutical compounds from surface water by nanofiltration and reverse osmosis." *Separation and Purification Technology* 212(2019): 171–179.
24. Goh, K.S., Chen, Y., Chong, J.Y., Bae, T.H. and Wang, R. "Thin film composite hollow fibre membrane for pharmaceutical concentration and solvent recovery." *Journal of Membrane Science* 621(2021): 119008.
25. Cabrera-González, M., Ahmed, A., Maamo, K., Salem, M., Jordan, C. and Harasek, M. "Evaluation of nanofiltration membranes for pure lactic acid permeability." *Membranes* 12(2022): 302.
26. McDowall, S.C., Braune, M. and Nitzsche, R. "Recovery of bio-based medium-chain fatty acids with membrane filtration." *Separation and Purification Technology* 286(2022): 120430.
27. Roopashree, P., Shetty, S.S. and Kumari, N.S. "Effect of medium chain fatty acid in human health and disease." *Journal of Functional Foods* 87(2021): 104724.
28. Alves, Y.P.C., Antunes, F.A.F., da Silva, S.S. and Forte, M.B.S. "From by-to bioproducts: Selection of a nanofiltration membrane for biotechnological xylitol purification and process optimization." *Food and Bioproducts Processing* 125(2021): 79–90.

Index

Note: **Bold** page numbers refer to tables and *italic* page numbers refer to figures.

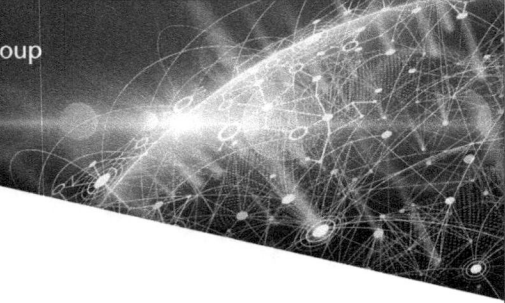

For Product Safety Concerns and Information please contact our EU
representative GPSR@taylorandfrancis.com
Taylor & Francis Verlag GmbH, Kaufingerstraße 24, 80331 München, Germany